Communications
in Computer and Information Science 370

Editorial Board

Simone Diniz Junqueira Barbosa
  Pontifical Catholic University of Rio de Janeiro (PUC-Rio),
  Rio de Janeiro, Brazil
Phoebe Chen
  La Trobe University, Melbourne, Australia
Alfredo Cuzzocrea
  ICAR-CNR and University of Calabria, Italy
Xiaoyong Du
  Renmin University of China, Beijing, China
Joaquim Filipe
  Polytechnic Institute of Setúbal, Portugal
Orhun Kara
  TÜBİTAK BİLGEM and Middle East Technical University, Turkey
Igor Kotenko
  St. Petersburg Institute for Informatics and Automation
  of the Russian Academy of Sciences, Russia
Krishna M. Sivalingam
  Indian Institute of Technology Madras, India
Dominik Ślęzak
  University of Warsaw and Infobright, Poland
Takashi Washio
  Osaka University, Japan
Xiaokang Yang
  Shanghai Jiao Tong University, China

Andrzej Kwiecień   Piotr Gaj
Piotr Stera (Eds.)

Computer Networks

20th International Conference, CN 2013
Lwówek Śląski, Poland, June 17-21, 2013
Proceedings

 Springer

Volume Editors

Andrzej Kwiecień
Piotr Gaj
Piotr Stera

Silesian University of Technology
Institute of Informatics
ul. Akademicka 16
44-100 Gliwice, Poland

E-mail: {andrzej.kwiecien, piotr.gaj, piotr.stera}@polsl.pl

ISSN 1865-0929   e-ISSN 1865-0937
ISBN 978-3-642-38864-4   e-ISBN 978-3-642-38865-1
DOI 10.1007/978-3-642-38865-1
Springer Heidelberg Dordrecht London New York

Library of Congress Control Number: 2013939777

CR Subject Classification (1998): C.2, C.4, C.3, C.5.3, G.3, H.4, D.2, H.3, K.4.4, I.6

© Springer-Verlag Berlin Heidelberg 2013
This work is subject to copyright. All rights are reserved by the Publisher, whether the whole or part of
the material is concerned, specifically the rights of translation, reprinting, reuse of illustrations, recitation,
broadcasting, reproduction on microfilms or in any other physical way, and transmission or information
storage and retrieval, electronic adaptation, computer software, or by similar or dissimilar methodology
now known or hereafter developed. Exempted from this legal reservation are brief excerpts in connection
with reviews or scholarly analysis or material supplied specifically for the purpose of being entered and
executed on a computer system, for exclusive use by the purchaser of the work. Duplication of this publication
or parts thereof is permitted only under the provisions of the Copyright Law of the Publisher's location,
in ist current version, and permission for use must always be obtained from Springer. Permissions for use
may be obtained through RightsLink at the Copyright Clearance Center. Violations are liable to prosecution
under the respective Copyright Law.
The use of general descriptive names, registered names, trademarks, service marks, etc. in this publication
does not imply, even in the absence of a specific statement, that such names are exempt from the relevant
protective laws and regulations and therefore free for general use.
While the advice and information in this book are believed to be true and accurate at the date of publication,
neither the authors nor the editors nor the publisher can accept any legal responsibility for any errors or
omissions that may be made. The publisher makes no warranty, express or implied, with respect to the
material contained herein.

Typesetting: Camera-ready by author, data conversion by Scientific Publishing Services, Chennai, India

Printed on acid-free paper

Springer is part of Springer Science+Business Media (www.springer.com)

Preface

Over 20 years have passed since the very first edition of the 'Computer Networks' conference was organized in Poland. The idea of the conference came from the Institute of Informatics belonging to the Department of Automatic Control, Electronics and Computer Science of the Silesian University of Technology with Professor Andrzej Grzywak as head. In the beginning, it was a national conference aiming at academic and industry integration. Currently, for six years, it has been an international event with the purpose of delivering a science platform for exchanging information, experiences, and knowledge. The event is intended for academics, scientists, researchers, developers, industrialists, engineers, management staff, and many others interested in computer networks, teleinformatics, and communications. In 2013 there was the 20th, jubilee edition. Many interesting papers were submitted for this edition. The 58 carefully selected ones are included in these proceedings and were discussed during the conference.

20th International Science Conference
Computer Networks

Computer networks are one of the most important parts of the contemporary IT systems. In industry, distributed architecture is very popular and all data exchange is mainly based on network solutions. At home and in entertainment, office, management, and business solutions, the important and growing share of services is based on local, wide, and public networks with significant usage of mobile devices and wireless systems. Improvements to existing solutions as well as the creation of brand new ones are possible owing to the new methods and tools for designing, modeling, testing, etc. The progress in this domain enables networking technologies to be applied in every kind of human activity. Engineers, researchers, developers, and inventors, whose eminent representatives are coauthors of this volume, ensure the continuous growth of computer networking and communications domains. The contents of the proceedings cover a wide spectrum of technical issues that are connected with the following general matters:

- Computer Networks
- Teleinformatics and Communications
- New Technologies
- Queues Theory
- Innovative Applications
- Networking in e-Business

Within the range of the domains mentioned above, there are many topics currently considered as valid and up to date, such as: network architectural issues, Internet and wireless solutions, security aspects of hardware and software, industrial systems, quantum and bio-informatics, queues theory and queueing networks, cloud networking and services, and others.

Generally, the content of the proceedings is focused on the mentioned subjects in the order presented above. However, we have decided not to create separate parts, because the described themes are not strictly separated from one another but overlap partially and are mentioned in various contexts.

The papers related to the computer networks issues are presented at the beginning of the book and comprise among others: load scheduling, usage of discrete-event simulators, distributed processing, APRS traffic nature, simulation of periodic processes with perturbation, spatial regression models, and Web performance analyses and forecasts. The great effort in the research of the new generation networks is mostly put into wireless solutions. Some of the following papers present topics on networking without a cable, e.g., solutions for geographic routing scenarios, sensor networks in greenhouse data-collecting applications, analysis of the LEACH protocol, performance analysis of the 802.11 EDCA and DCF, and discussions about routing protocols. There are some topics regarding important security issues, such as: quantitative risk analysis for storage, Web traffic anonymization, usage of fuzzy logic with botnets, and method of rootkit operation detection. Moreover, some issues referring to networking in industrial systems are also presented (e.g., topics on failure detection in dual bus, multi-protocol nodes designing, bandwidth optimization method, tests for RT communication performance, and SOA utilization in networked control systems).

The next group of papers concerns communication issues. Among others, there are papers where authors discuss problems such as the simulation of an LDPC-coded OFDM system, dynamical access to the radio spectrum by devices, modeling of information propagation within the opportunistic network, modification of the DPPM based on MPPM coding, replacing the TLS handshake by the EAP authentication, automatic PCI assignment in LTE self-organized networks, securing minimum-cost multicast with network coding, and reliable data transmission between PC and FPGA device over Ethernet.

The third group is related to new technologies in the networking domain. Among others, topics about bio-informatics and quantum technology are presented, as well as valid issues related to stream data processing systems and cloud computing.

Next, in view of the importance of the theory of queues and queueing networks, a few chapters related to this area are included in the volume. Among others, there are topics about the TCP connection with CHOKe or gCHOKe AQM policy, modeling of the data stream intensity in fault-tolerant distributed processing systems, modeling operation of Web service, multiserver queueing systems with random arrivals, tandem queueing system with multiserver stations and without buffers, performance evaluation and the analysis of QoS, modeling of a multiservice queueing, and a scheduler for creating virtual links on physical link.

The volume also contains chapters of the character of innovative application. The reader can find topics related to network security, wireless systems, DSP, and cloud networking.

At the end of the book there are papers connected with the e-Business area.

We would like to take this opportunity to express our appreciation to all the authors for sharing their research results and for their assistance in creating the proceedings. In our belief, this book is a valuable reference on computer networks and communications. We would also like to thank the members of the international Program Committee for their participation in multiple reviews of each paper.

April 2013   Piotr Gaj
                                                        Andrzej Kwiecień

Organization

CN 2013 was organized by the Institute of Informatics from the Faculty of Automatic Control, Electronics and Computer Science, Silesian University of Technology (SUT), and supported by the Committee of Informatics of the Polish Academy of Sciences (PAN), Section of Computer Network and Distributed Systems in technical cooperation with the IEEE and consulting support of iNEER organization.

Executive Committee

All members of the Executing Committee are from the Silesian University of Technology, Poland.

Honorary Member	Halina Węgrzyn
Organizing Chair	Piotr Gaj
Technical Volume Editor	Piotr Stera
Technical Support	Aleksander Cisek
Technical Support	Arkadiusz Jestratjew
Technical Support	Jacek Stój
Office	Małgorzata Gładysz
Web Support	Piotr Kuźniacki
PAN Coordinator	Tadeusz Czachórski
IEEE PS Coordinator	Jacek Izydorczyk
iNEER Coordinator	Win Aung

Program Committee

Program Chair

Andrzej Kwiecień	Silesian University of Technology, Poland

Honorary Members

Klaus Bender	TU München, Germany
Adam Czornik	Silesian University of Technology, Poland
Andrzej Karbownik	Silesian University of Technology, Poland
Bogdan M. Wilamowski	Auburn University, USA

Technical Program Committee

Anoosh Abdy	Realm Information Technologies, USA
Iosif Androulidakis	University of Ioannina, Greece
Tülin Atmaca	Institut National de Télécommunication, France
Win Aung	iNEER, USA
Zbigniew Banaszak	Warsaw University of Technology, Poland
Leszek Borzemski	Wrocław University of Technology, Poland
Markus Bregulla	University of Applied Sciences Ingolstadt, Germany
Tadeusz Czachórski	Silesian University of Technology, Poland
Andrzej Duda	INP Grenoble, France
Alexander N. Dudin	Belarusian State University, Belarus
Max Felser	Bern University of Applied Sciences, Switzerland
Jean-Michel Fourneau	Versailles University, France
Rosario G. Garroppo	University of Pisa, Italy
Natalia Gaviria	Universidad de Antioquia, Colombia
Erol Gelenbe	Imperial College, UK
Roman Gielerak	University of Zielona Góra, Poland
Adam Grzech	Wrocław University of Technology, Poland
Edward Hrynkiewicz	Silesian University of Technology, Poland
Zbigniew Huzar	Wrocław University of Technology, Poland
Jacek Izydorczyk	Silesian University of Technology, Poland
Jürgen Jasperneite	Ostwestfalen-Lippe University of Applied Sciences, Germany
Jerzy Klamka	IITiS Polish Academy of Sciences, Gliwice, Poland
Demetres D. Kouvatsos	University of Bradford, UK
Stanisław Kozielski	Silesian University of Technology, Poland
Henryk Krawczyk	Gdańsk University of Technology, Poland
Wolfgang Mahnke	ABB, Germany
Francesco Malandrino	Politecnico di Torino, Italy
Kevin M. McNeil	BAE Systems, USA
Michele Pagano	University of Pisa, Italy
Nihal Pekergin	Université de Paris, France
Piotr Pikiewicz	College of Business in Dąbrowa Górnicza, Poland

Jacek Piskorowski	West Pomeranian University of Technology, Poland
Bolesław Pochopień	Silesian University of Technology, Poland
Oksana Pomorova	Khmelnitsky National University, Ukraine
Silvana Rodrigues	Integrated Device Technology, Canada
Akash Singh	IBM Corp, USA
Mirosław Skrzewski	Silesian University of Technology, Poland
Maciej Stasiak	Poznań University of Technology
Kerry-Lynn Thomson	Nelson Mandela Metropolitan University, South Africa
Oleg Tikhonenko	IITiS Polish Academy of Sciences, Gliwice, Poland
Arnaud Tisserand	IRISA, France
Bane Vasic	University of Arizona, USA
Miroslaw Voznak	VSB-Technical University of Ostrava, Czech Republic
Sylwester Warecki	Peregrine Semiconductor Inc., USA
Tadeusz Wieczorek	Silesian University of Technology, Poland
Józef Woźniak	Gdańsk University of Technology, Poland
Hao Yu	Auburn University, USA
Grzegorz Zaręba	University of Arizona, USA

Referees

Iosif Androulidakis	Edward Hrynkiewicz	Oksana Pomorova
Tülin Atmaca	Zbigniew Huzar	Akash Singh
Zbigniew Banaszak	Jacek Izydorczyk	Mirosław Skrzewski
Leszek Borzemski	Jürgen Jasperneite	Maciej Stasiak
Tadeusz Czachórski	Jerzy Klamka	Kerry-Lynn Thomson
Andrzej Duda	Demetres D. Kouvatsos	Oleg Tikhonenko
Alexander N. Dudin	Stanisław Kozielski	Arnaud Tisserand
Max Felser	Henryk Krawczyk	Bane Vasic
Jean Michel Fourneau	Andrzej Kwiecień	Miroslaw Voznak
Rosario G. Garroppo	Wolfgang Mahnke	Sylwester Warecki
Natalia Gaviria	Michele Pagano	Tadeusz Wieczorek
Erol Gelenbe	Nihal Pekergin	Józef Woźniak
Roman Gielerak	Piotr Pikiewicz	Hao Yu
Adam Grzech	Jacek Piskorowski	Grzegorz Zaręba

Sponsoring Institutions

Organizer: Institute of Informatics, Faculty of Automatic Control, Electronics and Computer Science, Silesian University of Technology
Coorganizer: Committee of Informatics of the Polish Academy of Sciences, Section of Computer Network and Distributed Systems.
Technical cosponsor: IEEE Poland Section.

Technical Partner

Conference partner: iNEER

Table of Contents

Synchronization Algorithm for Timed Colored Petri Nets and Ns-2
Simulators 1
   Wojciech Rząsa

Experimental Results of Dynamic Load Scheduling in the CMS Data
Acquisition System 11
   Michal Simon, Hannes Sakulin, and Stanislaw Kozielski

Client-Side Processing Environment Based on Component Platforms
and Web Browsers 21
   Adam Piórkowski and Przemysław Szemla

Study of the Character of APRS Traffic in AX.25 Network 31
   Remigiusz Olejnik

Modelling the Pertubation of Traffic Based on Ateb-functions 38
   *Mykola Medykovsky, Ivanna Droniuk, Maria Nazarkevich, and
Olga Fedevych*

Spatial Econometrics Models in Web Server's Performance 45
   Leszek Borzemski and Anna Kamińska-Chuchmała

Empirical Web Performance Evaluation with MWING System and
Turning Bands Method 55
   Leszek Borzemski, Michał Danielak, and Anna Kamińska-Chuchmała

A Note on the Local Minimum Problem in Wireless Sensor Networks . . . 64
   Adam Czubak

Lessons Learned from the Deployment of Wireless Sensor Networks 76
   *Tomasz Surmacz, Mariusz Słabicki, Bartosz Wojciechowski, and
Maciej Nikodem*

Analysis and Optimization of LEACH Protocol for Wireless Sensor
Networks 86
   Błażej Adamczyk

Performance Analysis of IEEE 802.11 EDCA for a Different Number
of Access Categories and Comparison with DCF 95
   Olga Leontyeva and Kvitoslava Obelovska

Simulation Comparison of LEACH-Based Routing Protocols
for Wireless Sensor Networks 105
   Agnieszka Brachman

Routing Protocols for Border Surveillance Using ZigBee-Based Wireless
Sensor Networks 114
  Hoda Sharei-Amarghan, Alireza Keshavarz-Haddad, and
  Gaëtan Garraux

Quantitative Risk Analysis for Data Storage Systems 124
  Tomasz Bilski

Onion Routing Efficiency for Web Anonymization in Various
Configurations 136
  Tomas Sochor

Multi-agent Based Approach for Botnet Detection in a Corporate Area
Network Using Fuzzy Logic 146
  Oksana Pomorova, Oleg Savenko, Sergii Lysenko, and
  Andrii Kryshchuk

Monitoring System's Network Activity for Rootkit Malware
Detection 157
  Mirosław Skrzewski

Research of Failure Detection Algorithms of Transmission Line and
Equipment in a Communication System with a Dual Bus 166
  Błażej Kwiecień and Marcin Sidzina

The Concept of Using Multi-protocol Nodes in Real-Time Distributed
Systems for Increasing Communication Reliability 177
  Andrzej Kwiecień, Marcin Sidzina, and Michał Maćkowski

Bandwidth Optimization Method for Non-critical Data Transmission in
Real-Time Communication Systems 189
  Rafał Cupek, Kamil Folkert, and Mateusz Starzyk

Communication Performance Tests in Distributed Control Systems 200
  Marcin Jamro, Dariusz Rzońca, and Bartosz Trybus

Some Problems of Integrating Industrial Network Control Systems
Using Service Oriented Architecture 210
  Marcin Fojcik and Joar Sande

OFDM Transmission with Non-binary LDPC Coding in Wireless
Networks 222
  Grzegorz Dziwoki, Marcin Kucharczyk, and Wojciech Sulek

Spectrum Access Game for Cognitive Radio Networks with Incomplete
Information 232
  Jerzy Martyna

Performance Modeling of Opportunistic Networks 240
  Jerzy Martyna

Differential Two-Pulses Position Modulation for Synchronized Wireless
Optical Communications   252
   Mehdi Rouissat and Riad Ahmed Borsali

Extending the TLS Protocol by EAP Handshake to Build a Security
Architecture for Heterogenous Wireless Network   258
   Krzysztof Grochla and Piotr Stolarz

Conservative Graph Coloring: A Robust Method for Automatic PCI
Assignment in LTE   268
   Lukasz Chrost and Krzysztof Grochla

Network Coding-Based QoS and Security for Dynamic Interference-
Limited Networks   277
   Amin Mohajer, Mojtaba Mazoochi, Freshteh Atri Niasar,
   Ali Azami Ghadikolayi, and Mohammad Nabipour

Simple Communication with FPGA Device over Ethernet Interface   290
   Marcin Kucharczyk and Grzegorz Dziwoki

Evaluation and Development Perspectives of Stream Data Processing
Systems   300
   Marcin Gorawski, Anna Gorawska, and Krzysztof Pasterak

The Use of a Cloud Computing and the CUDA Architecture in
Zero-Latency Data Warehouses   312
   Marcin Gorawski, Damian Lis, and Michal Gorawski

MViewer: Visualization of Protein Molecular Structures Stored in the
PDB, mmCIF and PDBML Data Formats   323
   Dawid Stanek, Dariusz Mrozek, and Bożena Małysiak-Mrozek

CASSERT: A Two-Phase Alignment Algorithm for Matching 3D
Structures of Proteins   334
   Dariusz Mrozek and Bożena Małysiak-Mrozek

Transfers of Entangled Qudit States in Quantum Networks   344
   Marek Sawerwain and Joanna Wiśniewska

An Analysis of the Ping-Pong Protocol Operation in a Noisy Quantum
Channel   354
   Piotr Zawadzki

Comparison of CHOKe and gCHOKe Active Queues Management
Algorithms with the Use of Fluid Flow Approximation   363
   Adam Domański, Joanna Domańska, and Tadeusz Czachórski

Modeling Data Stream Intensity in Distributed Stream Processing
System   372
   Marcin Gorawski, Pawel Marks, and Michal Gorawski

Modeling Operation of Web Service 384
  Krzysztof Zatwarnicki and Anna Zatwarnicka

Total Volume Distribution for Multiserver Queueing Systems with
Random Capacity Demands 394
  Oleg Tikhonenko and Magdalena Kawecka

Queueing System $MAP|PH|N|R$ with Session Arrivals Operating in
Random Environment 406
  Chesoong Kim, Alexander Dudin, Sergey Dudin, and Olga Dudina

Tandem Queueing System with Correlated Input and Cross-Traffic 416
  Valentina Klimenok, Alexander Dudin, and Vladimir Vishnevsky

Analytical and Numerical Means to Model Transient States in
Computer Networks 426
  Tadeusz Czachórski, Monika Nycz, Tomasz Nycz, and
  Ferhan Pekergin

The Queueing Model of a Multiservice System with Dynamic Resource
Sharing for Each Class of Calls 436
  Sławomir Hanczewski, Maciej Stasiak, and Joanna Weissenberg

Scheduler for Virtualization of Links with Partial Performance
Isolation 446
  Tomasz Fortuna and Andrzej Chydzinski

Biometric Voice Identification Based on Fuzzy Kernel Classifier 456
  Adam Dustor and Piotr Kłosowski

Automatic Speech Segmentation for Automatic Speech Translation 466
  Piotr Kłosowski and Adam Dustor

WSN Power Conservation Using Mobile Sink for Road Traffic
Monitoring 476
  Marcin Bernaś

Optimizing Data Collection for Object Tracking in Wireless Sensor
Networks 485
  Bartłomiej Płaczek and Marcin Bernaś

Data Security in Microprocessor Units 495
  Andrzej Kwiecień, Michał Maćkowski, and Marcin Sidzina

The Concept of Software-Based Techniques of Increasing Immunity
of Microprocessor Unit to Electromagnetic Disturbances 507
  Andrzej Kwiecień, Michał Maćkowski, and Krzysztof Skoroniak

Hardware Aspects of Data Transmission in Coal Mines with Explosion
Hazard 517
  Marek Kryca

Planning-Based Method for Communication Protocol Negotiation in
a Composition of Data Stream Processing Services 531
    *Paweł Stelmach, Paweł Świątek, Łukasz Falas, Patryk Schauer,
    Adam Kokot, and Maciej Demkiewicz*

Users in IT Product Development Process 541
    Malgorzata Pankowska

Automatic Customer Segmentation for Social CRM Systems 552
    Adam Czyszczoń and Aleksander Zgrzywa

Practical Aspects of Log File Analysis for E-Commerce 562
    Grażyna Suchacka and Grzegorz Chodak

Multi-criteria Index Selection for Grouped SQL Queries 573
    Radosław Boroński and Grzegorz Bocewicz

Applying the Bidding Mechanism in Web Services with Quality of
Service 582
    Jolanta Wrzuszczak-Noga and Leszek Borzemski

Author Index 593

Synchronization Algorithm for Timed Colored Petri Nets and Ns-2 Simulators

Wojciech Rząsa

Rzeszow University of Technology,
al. Powstańców Warszawy 12, 35-959 Rzeszów, Poland
wrzasa@kia.prz.edu.pl
http://www.prz.edu.pl

Abstract. This paper describes the method of orchestrating behavior of two discrete-event simulators: Timed Colored Petri Net (TCPN) simulator and popular Ns-2. The simulators are connected to provide reliable model of distributed systems based on TCPN formalism supported with precise TCP model from the Ns-2, based on real implementation. Consequently, precise results of network transmission time estimation can be combined with reliable model of distributed application in order to obtain credible estimations of its efficiency on the basis of simulation. Data passing interface together with clock synchronization algorithm is described and discussed. Results of experiments are compared with simulations.

Keywords: Petri nets, simulation, distributed systems, efficiency, TCP.

1 Introduction

Petri nets (PN) are graphical and mathematical tool designed for modeling and analysis of concurrent activities in different scientific domains. They were first proposed by C. Petri in his PhD thesis [1] and got their name from the inventor. The PN are designed to model concurrent activities and thus are frequently used in informatics and automatics to enhance development of critical software and devices. There is large number of different extensions of the original concept with Timed Colored Petri Nets (TCPN) proposed by K. Jensen [2] being one of them. The TCPN are convenient formalism that allow to create compact and clear models of large and complex systems and to take time into consideration. Petri net models can be analyzed with formal methods as well as using simulation. In this work the formalism is used to reliably model and conveniently simulate distributed computer systems.

In analysis of distributed systems network transmission time should necessarily be considered. From the whole stack of the Internet-related protocols the *Transmission Control Protocol* (TCP) is the one that has an important impact on efficiency of data transmissions, due to its flow control mechanisms and responsibility for detecting and avoiding congestion of the network links.

A. Kwiecień, P. Gaj, and P. Stera (Eds.): CN 2013, CCIS 370, pp. 1–10, 2013.
© Springer-Verlag Berlin Heidelberg 2013

TCP is a complex protocol that has been successively developed over the years and is described in a number of different documents. Consequently, different implementations join the specifications in their peculiar way. There is significant number of available options (e.g. congestion control algorithms) that modify behavior of the protocol. Reliable analysis of efficiency of TCP data transmissions is thus a substantial challenge.

Petri nets are convenient and reliable formalism for modeling and simulation of distributed systems, but taking TCP details into consideration can be a complex issue. Especially if the work is not meant to analyze behavior of a theoretical TCP (as e.g. [3]), but to provide simulation results consistent with a real implementation. Thus it was decided to join advantages of TCPN for modeling distributed applications and concurrency with the Ns-2 simulator that includes real TCP implementations from Linux. The goal was to combine advantages of formalism-based modeling with accuracy of simulation based real protocol implementation to provide reliable simulator for the concepts presented in [4].

TCPN model of distributed application used e.g. in [5] was connected with two different external modules implementing network transmission protocols. *Ns2TCP* is TCP module created from the Linux-TCP [6] implementation from Ns-2 simulator[1], *Java Native Interface* (JNI) [7] was used to couple Java and C/C++ code. *SimpleTransport* is implementation of the simplest possible transport module which is only passing data between *application elements* and the network.

The first module is a real TCP implementation designed for the well-known Ns-2 network simulator. It consists of a Linux-like implementation of the protocol itself and real Linux code of congestion control modules, that are crucial for the TCP efficiency. Thus as closest to real as possible implementation of the transport protocol is included in the simulation method. The concept of interconnection between TCPN and external network transport model was briefly described in [8].

2   Related Work

Similar problem concerning synchronization of discrete-event simulators was solved by Quiang and MacGregor in [9] where a formally verified TCP model is connected to the Ns-2 infrastructure. Authors use Design/CPN as TCPN simulator, communication is implemented using Comms/CPN – a socket based communication library for Design/CPN. Clock synchronization method between the simulators is tersely described in the cited work. It consists in *exchanging synchronization messages* between the simulators by *synchronization agents* placed in every simulator whenever *a related event occurs in either of the simulation models*. The agents communicate over the Comms/CPN interface. Unfortunately description of the solution in the paper presents concepts rather then details and is neither sufficient to apply it to the problem presented in this work, nor to compare its details to the above described one.

---

[1] http://www.isi.edu/nsnam/ns/

3   Data Flow between the Simulators

The architecture of data passing between TCPN and Ns-2 is presented in the Fig. 1 When an *element* of simulated application sends a data package to another element over a network the external module is exploited. The information about data package is passed from the TCPN model to the *data send* part of the network protocol module. The module performs the sender-side processing.

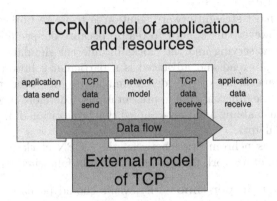

Fig. 1. Interconnection between TCPN based model of application and external model of network protocol on the example of TCP

Thereafter the *data segments* produced by the external module are delivered to the TCPN based network model. The external module decides when and how many *data segments* are delivered for network transmission. Network transmission is modeled by the use of Petri nets since it involves resource usage and the PNs are the formalism enabling reliable simulation of this aspect. The limited bandwidth of the network, delay introduced by subsequent *network segments* and possibly data loss are modeled by the use of the TCPN. After the transmission is finished the *data segments* are delivered to the external module that is supposed to perform the transport protocol receiver-side data processing. Thereafter the received part of the external module delivers the data to the TCPN based model of another *element* of simulated application. Thus the model of application and resources (including the networks) exploits Timed Colored Petri Nets, while the transport protocol itself is modeled by an external module.

The architecture of the simulator does not force any kind of data processing by the external module. The module is supposed to: (1) receive data from the TCPN application model, (2) transmit segments of data over the TCPN based model of the network according to its needs, (3) ensure delivery of the whole data volume required by modeled application to the receiving part of the TCPN based application model. Therefore the transport protocol module can be as complicated as real implementation of the TCP, but also as simple as a trivial model that immediately passes unchanged data from the *application element* to the model of the network and after network transmission from the network model to the receiving *application element.*

4   Synchronization of Simulation Clocks

The aim of the simulation clock interface is to synchronize time in the coupled discrete-event simulators: the simulator of Timed Colored Petri Net and the simulator of an external module.

Certainly the most straightforward solution is to make both simulators use a single clock. However this is hard to realize because of specific interfaces between events, event serving and the clocks in different simulators. The way in which the clocks are read and advanced is a particular solution strongly connected with the specificity of considered simulators. Moreover, frequently different simulators use different time representations. Therefore it was decided to develop a solution enabling a loosely-coupled synchronization of the independent clocks of the simulators.

The method of synchronization between the TCPN clock and the clock of an external model of network protocol should retain following conditions:

passing events at proper time – the events should be passed between the simulators as if both of them were governed by a single clock,

monotonicity of time flow (at least for external module) – some discrete-event simulators do no tolerate putting simulation clock back, even if it does not embrace already served events; TCPN simulator used in this work handles putting simulation clock back properly if it does not affect already fired transitions,

transparency for the simulators – the solution may not disturb normal behavior of the simulators, that should be able to properly handle their usual events.

4.1   The Algorithm

The clock time ticks in discrete-events simulators are not even, since they depend on the time of the next available event in an event queue. The simulation is performed in two steps for each clock tick. The first step is done by the TCPN and in the second step the external module catches up with the TCPN model. After the second step the time ticks of the two clocks are adjusted. Thus when a single iteration is finished both clocks are synchronized and all events that should be passed between the simulators are on the proper side of the border. The synchronization algorithm is schematically presented on Fig. 2.

The first step is finished when the TCPN simulator has fired all transitions enabled at specified time and thus decides to advance its clock. The pace of the TCPN simulator is stopped after the TCPN clock is advanced, but before any transition enabled at the new time is fired. At this moment it is known what is the new value of the TCPN simulation time, the only change in state of the

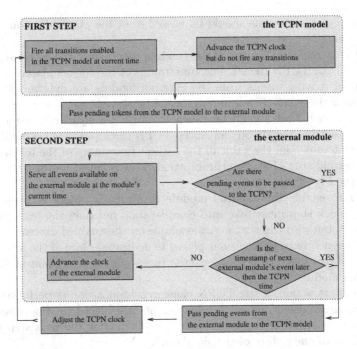

Fig. 2. Algorithm of synchronization of clocks and events between TCPN based model of application and external model of network protocol

TCPN model is the change in the value of the clock. When the first step is finished the clock of the external module is still set to the value representing the previous moment in simulation time, responding the previous value of the TCPN clock, used while the first step.

Before the second step can be performed all available tokens (events) that concern the external module must be passed from the TCPN to the event queue of the external module. The time at which each passed event should occur in the external module is determined by timestamps of the tokens in the TCPN.

The second step begins when all required tokens are passed from the TCPN to the external module. Then the external module is enabled to serve its events available at the time determined by the current state of its clock. The clock of the external module is not advanced first. After all events available at current time are served by the external module it is verified if the events that already occurred result in a need to create new events (tokens) in the TCPN model, if it is true then the second step is finished. Otherwise the second condition is checked in order to verify if the timestamp of next event available in the external module's queue can be served without the need to advance the module's simulation clock to the value later then the actual TCPN time. If it is possible the clock of external module is advanced to serve next available event. The second step can be continued until the external module's clock reaches the time corresponding to the new value to which the TCPN clock was advanced at the beginning of the first step or some events should be passed from the external module to the TCPN model.

When the second step is finished the pending events from the external module are passed to the TCPN model. The timestamps of the tokens placed in TCPN as a result of moving events from the external module are determined by the time at which the events being moved should be served.

The last action that should be taken to finish the synchronization of the simulators is setting the already advanced TCPN clock to the value of minimum of the current value and the value of the earliest timestamp of the tokens added while the second step of the synchronization. This is called "Adjust the TCPN clock" on the Fig. 2.

In order to enable the external module to serve its events at proper time the TCPN clock step must take into consideration not only the next available TCPN event, but also the next event available on the enabled external module. Therefore a synchronization token is placed in dedicated place of the model. The timestamp of this token corresponds to the timestamp of next event available in the external module.

After the last action of the TCPN clock synchronization is performed both clocks are synchronized and all pending events are placed on the proper side of the border separating the two simulators. Thereafter the first step can be performed for the new state of the simulators.

4.2   Discussion of Correctness of the Algorithm

Before the first step of the algorithm both clocks (TCPN simulator clock and the clock of the external module) are set to the values that correspond to the same time, however their values may not be equal if the representation of time in both simulators is different. Before the first iteration of the simulation both clocks are set to the starting point in time.

The first step is finished when the TCPN model has served all events available at the moment indicated by both clocks. This first step could produce events that should be passed to the external module. The tokens from the TCPN are passed in each time tick (between the steps). The new events that appear while the first step can have timestamps not before the current time determined by both clocks (the starting time of the step) since the TCPN does not produce tokens with timestamp value less then current simulation time. Therefore the events passed to the external module does not have their timestamps before the current time and the module is capable of handling them correctly. The values of timestamps are converted while the TCPN tokens are translated to the external module's events.

In the second step the external module is able to handle all events pending in current moment of time (the moment already considered by the TCPN model). If the TCPN produced any events that should be handled by the external module with no delay, at the same moment in time, this is possible in the second step. The external module continues serving its events in subsequent moments of time until a change in the TCPN model is required. The demanded change may be twofold.

First the external module might produce events that should be handled by the TCPN, second the external module's clock may exceed the value corresponding to next event available for the TCPN. If any of the two happens the second step is finished. Thus the external module is allowed to proceed only until its activity is independent from the TCPN. Thereafter the pending results of the external module's activity are passed to the TCPN to be handled.

When the external module finishes its step its clock may be set to the value from before the step (if it was not advanced) or might have been advanced by the time that does not require TCPN activity. In both cases all events available till the moment of time indicated by the external modules clock have been handled in the proper moments of time.

The events served by the external module can produce another events that should be handled in the future (in relation to the time at which the causing event is served). Thus the serving of the external module's events results in placing in the TCPN model tokens with timestamps that may be not less then the previous value of the TCPN clock (since the earliest events are served by the external module at the time to which both clocks were set before the first step), but may be less then its current value (which depends only on the timestamps of tokens that has been in the TCPN before the second step, but obviously neither on the events from the external module nor their results).

Therefore the last phase of the iteration may require putting the TCPN clock back in time, but as described before the new value will not be less then the previous one (from before the first step). Since no transition is fired after the TCPN clock is advanced it can be regarded that in fact the TCPN clock was advanced to the finally determined value, while the value set after the first step of the synchronization was only used to determine the upper limit of the value of external module's clock for the second step. The final value of the TCPN clock corresponds to the earliest event that is available in both simulators, thus in next iteration each simulator is able to serve its events at required time.

Application of this synchronization solution may however be considered not valid in general case. If there is significant number of events that should be passed between the simulators in both directions, without advancing the clocks then from the global point of view, considering equally events from both simulators, the synchronization method forces to some extent sequence in which the events are served. This behavior could be considered invalid for cases where the simulators are tightly coupled and maybe, to some extent their areas of responsibility overlap, since not every sequence of events carrying the same timestamp is equally probable. In such a case a finer grained strategy of event exchange should be applied. However in the problem considered in this work the scopes of responsibility of both simulators are well separated and in fact present independent areas. Additionally time pace is important in the simulation and thus the forced exchange of events is relatively frequent in our application of the algorithm. We should also notice that frequently in discrete-event simulators, with Ns-2 among them, sequence in which events scheduled for the same time occur is arbitrarily chosen in a manner determined by design of the simulator (or even by some implementation details) and their fairness is not considered as carefully as it is in Petri nets. Therefore described issue can probably be omitted in majority of possible applications of the algorithm.

5   State of the TCPN and an External Module

It is required from the Petri net models to store all information about state of the model in the Petri net. It is natural for low level, black-white Petri nets. However it is an important requirement also in the models using e.g. Timed Colored Petri Nets, where models include functions implemented by the use of general purpose programing languages that may enable storing the state information on the other levels of the model. In consequence two equal calls from the Petri net to a guard or inscription function could produce different results. This is not a valid behavior of the Petri net based model.

In this work we aimed at being in consent to this requirement and all functions used by the TCPN model are stateless. However the connection between the TCPN model and an external module of network protocol is a substantially different situation. For the sake of complexity of both models and simulators this case cannot be considered as a call from the Petri net to a stateless function. In fact it is a connection between two distinct simulators, with their details orchestrated to enable them to work together.

The connection between the TCPN and an external module is in fact similar to a connection between different pages of Hierarchical Colored Petri Net model [2]. The *substitution transitions* and *socket places* are used to pass tokens between different parts or levels of abstraction of a PN model. The difference in our case is reduced to the fact that a part of the model is not implemented as a Petri net. This obviously has its consequences, but does not pose a formal problem in case of simulation.

In the Hierarchical Petri nets pages of PN are allowed to store information that is not available on the other ones until it is computed, collected, aggregated and may be passed to subsequent parts of the model. This strategy is applied in the connection between the TCPN model and an external module of network protocol in this research. An external module connected to the TCPN behaves as separate pages of TCPN. As such an external module is allowed to store part of the state information of the model being simulated, until the information is passed to the TCPN model in required level of details.

6   Experiments

The experiments were performed by the use of test application running on laboratory infrastructure consisting on three nodes and connected with network. The packages were sent from node01 to node02 that passed it to node03, then packages were returned through node02 to the node01. Round trip time was measured as a result. The nodes were configured to pass the data as fast as possible consequently the network was the only significant cause of transmission delay. Parameters of the network were set using NetEm [10], that was used in a number of other network-related works [11,12,13]. The bandwidth of the network segments was controlled using *Token Bucket Filter* (TBF)[2].

In all experiments 1 MB packages were transmitted over the network links with different parameters. Two important tests were carried out. In the first one no significant delay was introduced in the network and thus the links had no noticeable capacity, the measurements were carried on for subsequent bandwidths: 1 Mbps, 2 Mbps, 5 Mbps and 10 Mbps. In the second experiment each network segment introduced 100 ms delay resulting in capacity from 12.8 kB for 1 Mbps to 128 kB for 10 Mbps. Simulations were performed for the same configurations using the two network transport modules. For Ns2TCP we performed also simulation that took into consideration specifics of the TBF. Comparison of the experiments and the simulations is presented in the Fig. 3 and Fig. 4.

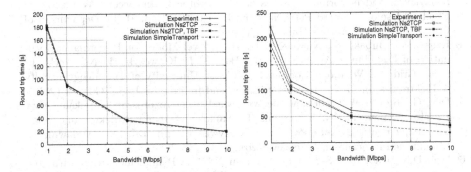

Fig. 3. Results for low-delay network   **Fig. 4.** Results for high capacity network

The experiments show that for the basic example with insignificant network capacity, the simple network transport model is sufficient. However, for the high-capacity links exploiting the Ns-2 TCP module improves accuracy of the results. The accuracy depends on precision of the TCP model in the Ns-2 simulator and certainly could be improved.

7 Summary

The goal of the work was to connect Timed Colored Petri Net model with the Ns-2 simulator. Algorithm of synchronization between TCPN and Ns-2 events was described and correctness of this algorithm was discussed. Results of the experiments showed that the orchestrated TCPN and Ns-2 simulators produce correct results. They are also capable of providing improved precision for simulations of Petri net models of distributed applications, using specialized network

---

[2] http://lartc.org/

models. Certainly precision of the Ns-2 TCP model could be improved, it is however, outside the scope of this work.

The presented algorithm is sufficiently general, to enable connecting the other discrete-event simulators to Petri nets, enabling more precise and thus more valuable simulations using this formalism.

References

1. Petri, C.A.: Kommunikation mit Automaten. Bonn: Institut für Instrumentelle Mathematik. Schriften des IIM Nr. 3 (1962) Also, English translation, Communication with Automata. New York: Griffiss Air Force Base. Tech. Rep. RADC-TR-65-377, vol. 1, (suppl. 1) (1966)
2. Jensen, K.: Coloured Petri Nets. Basic Concepts, Analysis Methods and Practical Use. Basic Concepts. EATCS Monographs on Theoretical Computer Science, vol. 1. Springer (1994)
3. De Figueiredo, J.C.A., Kristensen, L.M.: Using Coloured Petri nets to investigate behavioural and performance issues of TCP protocols. In: Second Workshop on Practical Use of Coloured Petri Nets and Design/CPN, pp. 21–40. Department of Computer Science, Aarhus University (1999)
4. Rząsa, W., Bubak, M., Nawarecki, E.: High-Level Model for Performance Evaluation of Distributed Applications. In: Balicki, J., Krawczyk, H., Nawarecki, E. (eds.) Grid and Volunteer Computing, Gdansk University of Technology Faculty of Elektronics, pp. 7–23. Telecomunication and Informatics Press, Gdańsk (2012)
5. Dec, G., Rząsa, W.: Modelowanie wielowarstwowej rozproszonej aplikacji www z zastosowaniem TCPN. In: Trybus, L., Samolej, S. (eds.) Projektowanie, analiza i implementacja systemów czasu rzeczywistego, pp. 137–148. Wydawnictwa Komunikacji i Łączności, Warszawa (2011)
6. Wei, D.X., Cao, P.: NS-2 TCP-Linux: an NS-2 TCP implementation with congestion control algorithms from Linux. In: WNS2 2006: Proceeding from the 2006 Workshop on ns-2: the IP Network Simulator. ACM Press, Pisa (2006)
7. Liang, S.: The Java^{TM} Native Interface Programmer's Guide and Specification. Prentice Hall (June 1999)
8. Rząsa, W.: Combining Timed Colored Petri Nets and Real TCP Implementation to Reliably Simulate Distributed Applications. In: Kwiecień, A., Gaj, P., Stera, P. (eds.) CN 2009. CCIS, vol. 39, pp. 79–86. Springer, Heidelberg (2009)
9. Ye, Q., MacGregor, M.H.: Combining Petri Nets and ns-2: A Hybrid Method for Analysis and Simulation. In: Proc. of the 4th Annual Communication Networks and Services Research Conference, pp. 139–148. IEEE Computer Society Press, Washington, DC (2006)
10. Hemminger, S.: Network Emulation with NetEm. In: Network Emulation with NetEm. In: Proc. of the 2005 Linux Conference Australia, LCA-2005 (April 2005)
11. Burger, M., Kielmann, T.: MOB: zero-configuration high-throughput multicasting for grid applications. In: Proc. of the 16th International Symposium on High Performance Distributed Computing, Monterey, California, USA (June 2007)
12. Sangtae, H., Injong, R., Lisong, X.: CUBIC: a new TCP-friendly high-speed TCP variant. ACM SIGOPS Operating Systems Review 42(5), 64–74 (2008)
13. Choe, Y.R., Schuff, D.L., Dyaberi, J.M., Pai, V.S.: Improving VoD server efficiency with bittorrent. In: Proc. of the 15th International Conference on Multimedia, Augsburg, Germany (September 2007)

Experimental Results of Dynamic Load Scheduling in the CMS Data Acquisition System

Michal Simon[1,2], Hannes Sakulin[1], and Stanislaw Kozielski[2]

[1] CERN, Geneva, Switzerland
{michal.simon,hannes.sakulin}@cern.ch
[2] Silesian University of Technology, Gliwice, Poland
stanislaw.kozielski@polsl.pl

Abstract. The online Data Acquisition system of the Compact Muon Solenoid (CMS) experiment at CERN's Large Hadron Collider is designed to collect data corresponding to a single collision of particles, referred to as an event, from about 500 detector Front-Ends. Each of those Front-Ends delivers event-fragments of an average size of 2 KB at a rate of 100 kHz. The event-fragments are statically distributed (usually in round robin fashion) between 8 identical computing farms, which construct the whole events. In this paper we present experimental results of employing a distributed, asynchronous load scheduling algorithm in place of the static event allocation mechanism. The research focuses in particular on balancing the event flow in case of degradations in computing power or network throughput. The discussed studies prove that the proposed method meets the requirements of CMS experiment and has a positive impact on the resource utilization and overall fault tolerance.

Keywords: load scheduling, data acquisition, distributed computing.

1 Introduction

The Compact Muon Solenoid (CMS) is one of two big general-purpose experiments at CERN's Large Hadron Collider (LHC) for studying proton-proton and heavy ion collisions at TeV scale [1]. Since the probability that such a collision results in a phenomena of interest is very low, a collision rate in MHz range is needed. However, not all registered data can be sent to the persistent storage due to the huge amount of space that would be required. Therefore a drastic filtration has to be achieved. Data that correspond to a single collision of particles referred to as an event are acquired from millions of readout channels. Preliminary filtration is performed by dedicated hardware based only on the input from the most important channels. The readout channels are then merged into several hundreds of detector front-ends, which act as the data sources for the online Data Acquisition (DAQ) system [2]. Each of those front-ends delivers event-fragments of an average size of 2 KB. The DAQ system reads out all the front-ends at a reduced rate of 100 kHz. The event-fragments are statically allocated (usually in round robin fashion) between eight identical filtering farms

A. Kwiecień, P. Gaj, and P. Stera (Eds.): CN 2013, CCIS 370, pp. 11–20, 2013.
© Springer-Verlag Berlin Heidelberg 2013

(called DAQ Slices) and then transferred over a non-blocking network [3] (implemented in Myrinet technology [4]). Each filtering farm is organized around a Terascale Force10 switch, where parallelization is achieved through SPMD (Single Process, Multiple Data [5]) technique. Such computing farm receives the event-fragments through distributed readout consisting of computing nodes called Readout Units (RUs). In a first event construction stage, event fragments are being assembled into so called super fragments inside RUs. Subsequently, in the second stage, a supervisor node called Event Manager (EVM) dynamically generates schedules for the event fragments based on the requests received from Builder Units (BUs). All the super-fragments corresponding to a single event are allocated to a BU that builds the whole event. The complete event is then being passed further to a Filter Unit (FU), which is carrying out the task of selecting the events of interest for persistent storage (BU and FU processes are hosted on the same computing node). The accepted events are then transferred to Storage Manager (SM) nodes that are connected to a Storage Area Network.

In this paper we aim to explore the possibility of improving the CMS online Data Acquisition system by replacing the static event allocation mechanism in the first stage of event building with a dynamic workload scheduling algorithm. The currently employed scheduling policy relies on an assumption that the capacities of all the computing farms participating in a data taking run are known in advance. In most cases, the events are allocated in round robin fashion because the filtering farms are identical. If a filtering farm has a reduced capacity, for example due to a malfunction, a smaller fraction of the workload can be assigned to it, however this has to be predefined at the start of the data acquisition process. Whenever a computing farm becomes less efficient during a data taking run it slows down other farms. As a result, the aggregated capacity may drop under the level that is necessary to handle the incoming data, even though sufficient resources are available. Our goal is to introduce a scheduling mechanism that would allow for dynamic adjustment to capacity fluctuations, in particular those that are caused by software and hardware failures.

2   Related Work

Although the available literature discussing workload scheduling and fault tolerance in general is very rich, it is very modest when it comes to addressing requirements comparable to those of the CMS DAQ system. Nevertheless, there is a category of algorithms that is especially interesting for our studies, namely, the multisource scheduling that employs the Divisible Load Theory (DLT) [6].

Jia et al proposed in [7] a dynamic load scheduling algorithm for multisource loads that follows the network partitioning concept. The algorithm is designed for a system of m load sources and n processing nodes. Each of the load sources has an independent workload inflow, and also participates in load processing. Processing nodes are assigned to the source for which the communication time is minimal. This way, m regions are created, each of them being the shortest path spanning tree. Subsequently, the source with the smallest workload processing time t_{min} is determined. The source nodes distribute the workload fraction inside their regions, in such a way that the expected processing time in each region is t_{min}. From that point on, all incoming workloads are queued in source's buffers. After the distributed portion of workload has been processed new partitioning is carried out, and again the source with lowest load processing time is being determined. This process is repeated until the entire workload is consumed. Note that after a load processing cycle is finished the source that has been previously recognized as the one with lowest processing time t_{min}, not necessarily needs to own any further workload. In this case, the resources assigned to this source, along with this source node itself, will be reallocated to other regions. On the other hand, if an idle source found itself in possession of new workload a new region will be created. Thus the number of sources and regions may fluctuate.

Yu and Robertazzi, in turn, proposed in [8] a dynamic load scheduling strategy for multisource loads that follows the idea of superposition. Like in electric circuit theory, they conduct the workload distribution analysis for each load source separately, as if other sources would not exist. Subsequently, the single-source workload is being assigned to available computing nodes proportionally to their computing power. Since the DLT is linear [6], and there is a pre-assumption made that the load is arbitrary divisible, and that the inflowing workload in different sources is not related, the solutions obtained for each source can be superimposed algebraically. Thus, the multi-source workloads have been allocated to computing nodes proportionally to their computing power. What is more, the amount of workload that has to be transferred between nodes is minimized.

Although both of the algorithms discussed above assume that the loads originating from different sources are not related (which is not the case when it comes to the CMS DAQ system), they both feature a very desirable advantage. That is to say, they generate schedules for multisource workloads assuring that the expected processing time for all the computing nodes will be the same. This feature, in turn, is crucial for avoiding idleness periods in more capacious resources. Beside the previously mentioned methods, our scope of interest also includes the dedicated solutions that have been applied in DAQ systems of other High Energy Physics (HEP) experiments.

ATLAS [9] is the second general purpose experiments at CERN's LHC. Like CMS, it registers particle collisions at a rate of 40 MHz and therefore has to fulfil similar design requirements. After the registered data are pre-filtered by dedicated hardware (the rate is reduced to 100 kHz), but before the final event reconstruction happens, an additional filtering step has been introduced. The selection decision is taken based on partial event data, so called Region of Interest (RoI). The RoI is allocated in round robin fashion (for redundancy and performance reasons) to one of a number of supervising nodes, which, in turn, assigns it dynamically to one of a number of filtering nodes. The additional filtering step causes significant reduction of the rate at which the full detector is read out down to 3 kHz. For each accepted event a central load scheduling node allocates an event-building node according to pull-requests obtained from those nodes. This way, a demand driven load scheduling has been obtained. The event-building node requests all event fragments, assembles the whole event (expected event size is 1.5 MB) and then transfers it to the Event Filter that performs the final selection step. It should be noted that at the initial inflowing rate the events are not scheduled dynamically but are distributed in round robin way. Dynamic load scheduling is performed only in the final reconstruction step, at a rate of 3 kHz. Our goal, on the other hand, is to introduce a dynamic event allocation method that could handle the initial rate of 100 kHz.

Smaller, but still having similar requirements, is the LHCb [10] detector at CERN. In this case, the online DAQ system accommodates the event fragments (of average size of 120 B) that are delivered by about 320 sources at a rate of 1 MHz. The event reconstruction process is carried out by transferring all the event data from the data sources to the same computing node in the filtering farm. In order to improve network utilization a packing of event fragments into Multi-Event Packets (MEP) is performed (typical packing factor is ten). The process of scheduling event fragments is being controlled by a supervisor node (implemented in FPGAs), which uses a dedicated optical network to broadcast commands to the data sources. The filtering node is chosen according to a credit scheme. Initially, at the beginning of a data taking run, as well as after processing of each MEP, each filtering node asks for new events throughout sending a MEP request to the supervisor. The credits corresponding to a filtering node are incremented based on the MEP requests and decremented whenever a respective node is used as the destination for the next MEP. The initial event input rate for the online DAQ system of the LHCb experiment is an order of magnitude greater than those that are being handled by ATLAS and CMS experiments. This has been possible not only due to the considerably smaller event size, but also because a packaging factor has been applied, meaning that scheduling is performed for blocks of events rather than on an event after event basis. Moreover, dedicated hardware is employed to perform the load scheduling.

3   Proposed Method

In this subsection we aim to give an overview of the proposed workload scheduling method (previously described in more details in [11]). A schematic view of all important components (computing nodes, networks, etc.) along with the general workflow of the scheduling algorithm is shown in Fig. 1. Firstly, the capacities of the computing farms that are participating in the data acquisition process are being estimated. All the Builder Units in a given farm are passing the information about their local event building efficiency to the Event Manager, which is merging those data. Afterwards, the EVMs are exchanging their estimates (so called load-data) in order to achieve redundancy (Fig. 1, step 2). Subsequently, the redundant load-data are transferred to the data sources, also known as FRLs (Fig. 1, step 3, part 1 and 2). Based on the load data received from all EVMs the FRLs are making the allocation decision for blocks of events (step 4). It can be noticed that there are two more components (shown in Fig. 1), which were not discussed previously, namely the Filter Units (FUs) and the Storage

Managers (SMs). Although they are not directly involved in load scheduling activities, they are key components of the CMS online DAQ system, and likewise have a crucial impact on data taking rate (in the production system usually the computing power of the BU-FU nodes is the limiting factor). Due to the event building protocol that has been employed in the dedicated software running on a filtering farm (described in [12]), the capacities of those components are reflected in the load-data that are based on the information obtained from BUs.

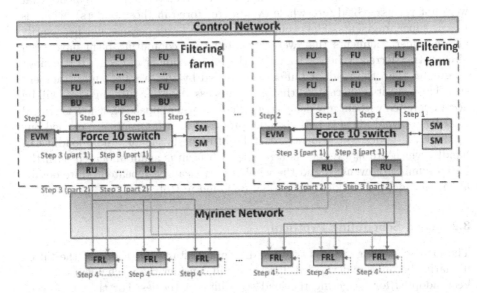

Fig. 1. Schematic view of DAQ components and the scheduling algorithm's workflow

3.1   Workload Index

The data sources deliver data at a rate of 100 kHz, which means that approximately every 10 μs there are about 500 new event fragments that need to be allocated to a computing farm. Sending a single message between computing nodes in the discussed system, depending on the network type, takes about 10 to 100 μs[1]. As a result, the workload has to be calculated and exchanged for blocks of events, rather than for single events. In order to avoid congestions, it has to be assured that the filtering farm can accommodate the assigned group of events. Therefore, the workload index has to carry the information about the occupancy of the readout buffers in the readout nodes. Furthermore, the available capacity and workload on each computing farm have to be estimated so the farm can be allocated with a respective fraction of the incoming load.

---

[1] The latest networking technologies are an order of magnitude faster and are considered for a future upgrade of CMS online DAQ System [13].

We aim to measure the workload on a filtering farm by the size of the data it owns (as proposed by Fonlupt et al. in [14]). The capacity of a farm, in turn, can be estimated as the number of events built in a given period of time. Those measurements can be easily combined by assigning an initial number of events n to each farm, and then when the first one becomes underloaded, by checking the numbers of events n_c that have been already constructed. This way, both the workload ($n - n_c$ events) and the capacity (n_c events per measurement time) have been estimated. The proposed measurement gives us also the information about worst case readout buffer occupancy (the received super fragments that were not yet assembled into whole events are stored in those buffers), which is $n - n_c$ super fragments. The occupancy might be lower because not necessarily all the super fragments that were allocated to a computing farm have been already transferred. When it comes to the initial number of events n it cannot be greater than the readout buffer size divided by the average super fragment size. This way, it is guaranteed that all events assigned to a filtering farm will be accommodated without any delay (regardless of farm's capacity), which is crucial in order to avoid idleness periods. The proposed measurements are carried out by the EVM nodes. Subsequently, each EVM requests the maximum load it can handle, which are n_c events. As a result, each farm is again allocated with the initial number of events n and the whole process can be repeated. More details about the adopted workload index can be found in [15].

3.2   Load Scheduling Protocol

The load scheduling process is event driven and starts when one of the filtering farms becomes underloaded. First, we will consider the mechanism that has been adopted for triggering the load scheduling activities. For the purpose of this discussion a filtering farm will be represented by its supervisor node (EVM). Onedirectional ring topology has been chosen to implement the communication between EVMs. When a farm becomes underloaded the EVM notifies its successor by transferring the load-data. The recipient, in turn, notifies its successor, and so on, until the ring is closed. However, if the recipient already sent the notification (because its farm also became underloaded) it will not be re-sent for a second time. In addition, each EVM maintains a bitmask of all the filtering farms. If a fatal error is detected, a supervisor node masks out the respective farm in its bitmask. The bitmask is then propagated together with the load-data and is used to exclude corrupted computing farms from a data taking run. The second action that is being triggered when a notification is received is sending the load-data to the data sources. The transfer only begins after receiving the notification and includes the load-data of the EVM itself and of its predecessor, thereby ensuring redundancy in order to cope with a potential EVM failure. The measurements are sent in two steps. First, the EVM performs a multicast to the readout nodes and then the RUs transmit the load-data further to the data-sources over the non-blocking network. Even if a readout node fails in one of the filtering farms all the load-data (from all the farms) will be delivered due to the redundancy that has been achieved. More details about the employed workload communication pattern can be found in [15].

3.3   Event Fragment Allocation Algorithm

Each data source makes the event allocation decision asynchronously (distributed, asynchronous load balancer [16]) based on a full load-data set (the measurements from all computing farms performed in a given load scheduling round). Since all the data sources deliver event fragments in the same sequence and receive the same loaddata, it is guaranteed that they will take the same allocation decision regarding event fragments corresponding to the same event. As the algorithm will be executed for each event fragment separately (on average once per $10\,\mu s$), it is crucial to keep its complexity as low as possible. For the purpose of the experiments that will be presented in the next section the following allocation mechanism has been employed. There is a counter corresponding to each filtering farm and it is initially assigned with the number of events that was request by the farm. The events are allocated in round robin fashion to all the farms whose counters are greater than zero (those whose counters are equal to zero are omitted). Every time an event fragment is allocated to a farm its counter is decremented. At the point when all the counters reach zero, a new set of counters is created using the requests from the subsequent load scheduling round.

4   Results and Conclusions

A prototype has been developed and then studied in the CMS DAQ production system during a technical stop in August 2011. Each of the computing farms participating in those tests was having 1 EVM, 63 RUs, 82 BUs and 2 SMs. The goal of our experiments was to verify if the prototype meets the requirements of the CMS experiment, if the fault tolerance of the system has been enhanced and if the resource utilization was more efficient.

The network throughput overhead of the load scheduling protocol has been measured for a setup of 8 filtering farms. The throughput was measured for expected event fragment sizes in the range from 128 B to 10 KB. The measurements were carried out for variable event fragment sizes from log-normal distribution (stdev = 0.5). The initial number of events n was set to 3000. The obtained results (shown in Fig. 2) confirm that the algorithm meets the design specifications of the CMS DAQ System (for 2 KB event fragments the throughput is greater than 200 MB/s). Although, the overhead in the range from 2.5 KB to 5 KB is substantial, it is acceptable because during data taking the limiting factor lies in the available computing power and not in the network throughput. The overhead is caused by the event allocation procedure (that is executed for each event fragment), which can be further optimized. Subsequently, the throughput has been measured in case of a network-link failure (one of the two) in one of the

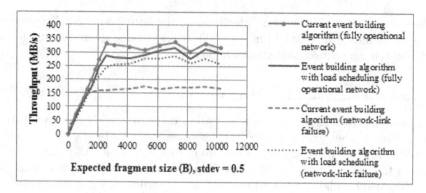

Fig. 2. The throughput in case of a fully operational network and in case of a network-link failure in one of the readout nodes for the static and dynamic scheduling mechanism

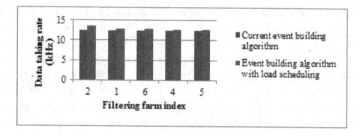

Fig. 3. The capacities of the individual computing farms (in a setup of 5 farms) respectively for the static and dynamic scheduling mechanism

readout nodes. As shown in Fig. 2, in the system employing the static scheduling mechanism (dashed curve), the discussed malfunction caused a drastic capacity loss below the compulsory threshold, meaning that the data acquisition rate of 100 kHz was no longer sustainable and had to be throttled. In order to recover from this type of failure a restart of the data taking process and reconfiguration of the DAQ system are required. On the other hand, the capacity of the system running with dynamic load scheduling (dotted curve) remained above the required level and as a result the data taking run has been continued. It is important to notice that since the data acquisition process was not interrupted it has been possible to avoid a downtime.

The next experiment was conducted on a setup of 5 computing farms. The standard system was tuned so the throughput limitation came from the event filter farms and the maximum possible data taking rate per farm was 12.5 kHz (the nominal speed). Afterwards the same settings have been applied to a system employing dynamic scheduling method. As shown in Fig. 3, the system running with dynamic scheduling slightly outperformed the standard system, which proves that the resource utilization is more efficient.

The response of the system to a fault occurrence in a particular filtering farm has been studied as shown in the Figs. 4 and 5. Again a setup of 8 computing farms has been used. Both for the static and dynamic scheduling, the system was tuned so that the throughput limitation would come from event filter farms and the maximum possible data taking rate would be 100 kHz (12.5 kHz per farm). The initial data acquisition rate was set to maximum (100 kHz). In the first experiment, after a certain period of time SM nodes were powered off one after another. It can be noticed that the standard system lost 50 % of its original capacity after switching off the first SM. Then, after second one was turned off the whole data acquisition process was stopped. On the other hand, the system running with dynamic workload scheduling algorithm lost, as expected, only 6.25 % of its capacity per SM node. In the second experiment, a RU node (which is a single point of failure for a computing farm) was powered off. In case of the standard system, the experiment resulted in an immediate termination of data taking. The system employing dynamic scheduling however lost only 12.5 % of its capacity (which corresponds to one computing farm).

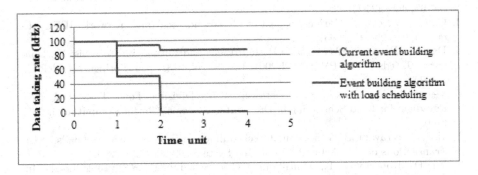

Fig. 4. System response to failing SM nodes in one filtering farm

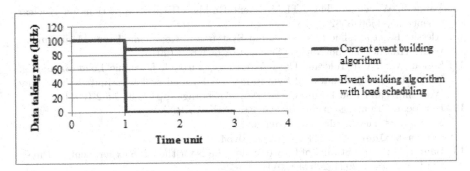

Fig. 5. System response to failing RU nodes in one computing farm

The above discussed experiments were repeated with faults introduced into more than one filtering farm, and also gave positive outcomes. Analogical analysis was carried out for the BU nodes (as described previously in [11] and [15]) and likewise confirmed the robustness of the dynamic load scheduling. During the course of our studies, it has been proven that the performance of the system has been improved in case of degradation of one (or more) of the computing farms participating in a data taking run. It has been also possible to decouple the farms from each other and as a result to limit the effects of some fault occurrences just to the concerned farm. This feature, in turn, is especially important when it comes to single points of failure of an individual farm, which otherwise would became critical for the whole DAQ system.

References

1. The CMS collaboration et al: The CMS experiment at the CERN LHC. JINST 3, 1–5, 261–282 (2008)
2. Bauer, G., et al.: The Terabit/s Super-Fragment Builder and Trigger Throttling System for the Compact Muon Solenoid Experiment at CERN. IEEE Transactions on Nuclear Science 55, 190–197 (2008)
3. Clos, C.: A study of non-blocking switching networks. Bell System Technical Journal 32, 406–424 (1952)
4. Cohen, D., et al.: Myrinet: a gigabit-per-second local area network. IEEE Micro 15(10), 29–36 (1995)
5. Darema, F.: *TheSPMDModel*: Past, Present and Future. In: Cotronis, Y., Dongarra, J. (eds.) Euro PVM/MPI 2001. LNCS, vol. 2131, p. 1. Springer, Heidelberg (2001)
6. Veeravalli, B., Ghose, D., Robertazzi, T.G.: Divisible Load Theory: A New Paradigm for Load Scheduling in Distributed Systems. Cluster Computing 6, 7–17 (2003)
7. Jia, J., Veeravalli, B., Weissman, J.: Scheduling Multisource Divisible Loads. IEEE Transactions on Parallel and Distributed Systems 99, 520–531 (2009)
8. Yu, D., Robertazzi, T.: Multi-Source Grid Scheduling for Divisible Loads. In: 40th IEEE Annual Conference on Information Sciences and Systems, Princeton, pp. 188–191 (2006)
9. Vandelli, W., et al.: The ATLAS Event Builder. IEEE Transactions on Nuclear Science 55, 3556–3562 (2008)
10. Alessio, F., et al.: The LHCb Readout System and Real-Time Event Management. IEEE Transactions on Nuclear Science 57, 663–668 (2010)
11. Simon, M.: Fault Tolerant Data Acquisition through Dynamic Load Balancing. In: 2011 IEEE International Symposium on Parallel and Distributed Processing Workshops and Phd Forum (IPDPSW), Anchorage, pp. 2049–2052 (2011)
12. European Organizasion for Nuclear Research (CERN), http://cms-ru-builder.web.cern.ch/cms-ru-builder/RUBUILDER_G_V1_12_0.pdf
13. Baurer, G., et al.: Studies of future readout links for the CMS experiment. J. Phys.: Conf. Ser. 331, 22004–22009 (2011)
14. Fonlupt, C., Marquet, P., Dekeyser, J.-l.: Data-Parallel Load Balancing Strategies. Parallel Computing 24, 1665–1684 (1996)
15. Simon, M., Sakulin, H., Kozielski, S.: Studies on load metric and communication for a load balancing algorithm in a distributed data acquisition system. J. Phys.: Conf. Ser. 331, 22040–22046 (2011)
16. Osman, A., Ammar, H.: Dynamic Load Balancing Strategies for Parallel Computers. Sci. Ann. Comput. Sci. 11, 110–120 (2002)

Client-Side Processing Environment Based on Component Platforms and Web Browsers

Adam Piórkowski and Przemysław Szemla

AGH University of Science and Technology,
Faculty of Geology, Geophysics and Environment Protection,
Department of Geoinformatics and Applied Computer Science,
al. A. Mickiewicza 30, 30-059 Kraków, Poland
pioro@agh.edu.pl, pszemla@geol.agh.edu.pl
http://www.geoinf.agh.edu.pl

Abstract. Distributed processing is an important issue of numerical calculations, in particular concerning the problems of time-consuming calculations. Solving this problem requires appropriate software, which is more complicated than the implementation in parallel environments. This article presents a proposal of distributed processing solution based on web browsers. This method, unlike the commonly used, does not require installing any software on the compute nodes. This is achieved through the distribution and execution of the code in the container, which is a web browser.

Keywords: parallel computing, distributed computing, numerical computing, client-side processing, web browsers, component technologies, clusters, domain decomposition.

1 Introduction

Distributed processing is an important method of resolving time-consuming calculations. In the past decades, methods of implementation of distributed processing have changed many times. The basic method of implementation of distributed computing is the design of remote processes, communicating through network mechanisms, for example network protocols such as TCP/IP are being commonly used for that purpose. Unfortunately, this method has significant drawbacks:

- for every application you need to implement a network connection layer,
- there is a full set of distributed settings required – often the static allocation of resources requires creating a new solution for every problem.

PVM (Parallel Virtual Machine) and MPI (Message Passing Interface) are the next methods of communication in parallel and distributed systems. A characteristic feature is the strong binding of a program that uses these technologies [1,2].

A. Kwiecień, P. Gaj, and P. Stera (Eds.): CN 2013, CCIS 370, pp. 21–30, 2013.
© Springer-Verlag Berlin Heidelberg 2013

The next approach is to create components that allow for distributed processing. Examples of such environments are CORBA (Common Object Request Broker Architecture) and DCOM (Distributed Component Object Model). Both the environments can help to implement a network communication, but are associated with a particular execution platform (system and hardware). Also in this case, the calculation for a specific problem requires a dedicated system [3,4,5].

Another approach allows for a hardware and a system independence. Component environments (Java ME, .NET Framework) allow for providing of mechanisms for calling methods of objects (Java RMI, .NET Remoting). This solution has many advantages, however, it still needs to install software on the computing nodes [6,7].

There are special distributed computing environments [8]. Examples of solutions are the environments: openSSI, XtreemOS and Apache Hadoop [9,10,11].

Distributed processing, as well as parallel processing, requires the ability to decompose the problem. The basic decomposition methods for parallel processing are the functional and the domain decomposition. Functional decomposition requires the ability to implement various functions in the nodes. Domain decomposition usually performs the same code on different nodes, sharing data between these nodes. This approach, although more difficult to implement, allows for much greater scalability than functional decomposition.

The research on distributed computing in component environments and on the calculations in the network [6,7], both on server- and client-side [12,13], has resulted in the concept of system that processes data in browsers (as containers). Therefore, this solution does not require any software installation on the compute nodes. Due to using component environments a code can be run on any computer. Data transmission is performed by using the serialization mechanisms specific to those environments. The idea of applet client-side computing is also a topic of the research [14], but the way of decomposition of numerical problem is different than described in this article.

2   The Concept of the Solution

Flexible environment for numerical computations should meet the demands:

- Portability of code,
- Dynamic adjoining nodes to the system,
- A minimum of complicated configuration of nodes.

Code portability feature can be achieved using technologies such as Java Component VM, NET Framework and Mono. There are environments for different hardware platforms for these technologies.

Dynamic adjoining nodes to the system is an important feature. Static assignment of nodes can be a cause of problems if one node fails. The system is waiting for a result that nodes had to provide. If the system can dynamically add computational units, overcoming this problem is available (Fig. 1). Minimum complicated configuration is another challenge. The best solution involves that

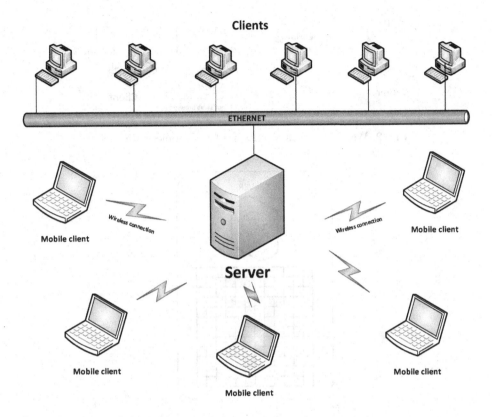

Fig. 1. Schema of the network for the proposed system

a client should only report to the system access to participate in the calculation. Other activities, such as the transfer of input and output data and executable code should not occupy his attention.

The proposed solution that satisfies these assumptions is the client-side application, in particular an applet executed inside the container, which is a Web browser (Fig. 2). The use of Java-applet or MS Silverlight (Moonlight for Linux, Mac) allows to run precompiled and managed code in browser [13]. Attaching nodes to the system can be made at any time, by user actions or service (e.g. screensaver). Once connected, the system should maintain connections to objects. By using browser system should not require any configuration at the client side.

The scheme should have an object interface, that allows for easy consolidation of its classes with classes of a numerical problem solution. An analogy to the Parallel. For loop (.NET 4.0) can be used – Distributed.For. Such a loop would allow for the automatic division of the data domains to computational units in the system. This applies only to numerical problems that allow for the proper domain decomposition (Fig. 3).

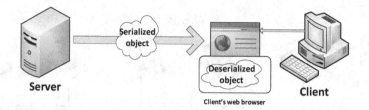

Fig. 2. Web browser as a container for an applet with code

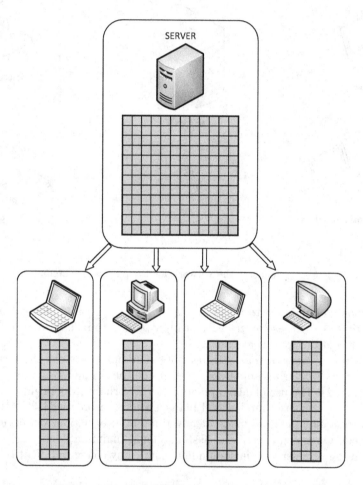

Fig. 3. Schema of the domain decomposition

3 Principles of Implementation

The loading of the numerical code at the client side takes place at the applet downloading. Next the data should be passed. The way of passing data is to use serialization feature for component objects (Fig. 2). After deserialization the application can start processing the data at the range pointed by server for the current node. The output values are returned to the server. The action's scenario is shown on the Fig. 4. The first version of presented solution was implemented for Java VM environment.

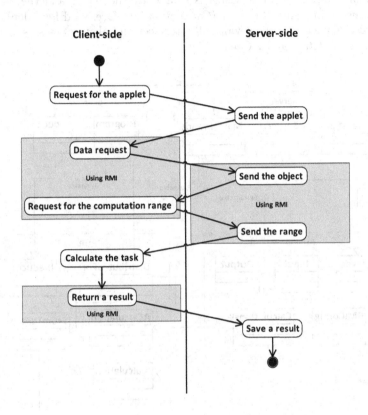

Fig. 4. Action schema

3.1 Implementation of a Server

The server of an application is a three-threaded [15]:

- starting thread – it initiates a connection with RMI register,
- programmer's thread – the code written by a programmer that uses a solution for an implementation of own numerical problem,
- main thread – it controls the programmer's thread.

The Figure 5 shows a class diagram a server of solution [15]. There are four parts of server:

- *server* package – it implements two of the threads mentioned above,
- *program* package – it contains a code of numerical calculations, it uses the *DistributedFor* function, that allows to divide a problem into domains based on parts of input data, the *reduce* class merges the parts of solution delivered by nodes,
- *common* package – this package is shared by the server and the client, it contains an implementation of *ICalculation* interface, used for calculations,
- *distributed* package – it provides all interfaces and abstract classes, e.g. *IReduce*, *ICalculation* and *ACalculation*.

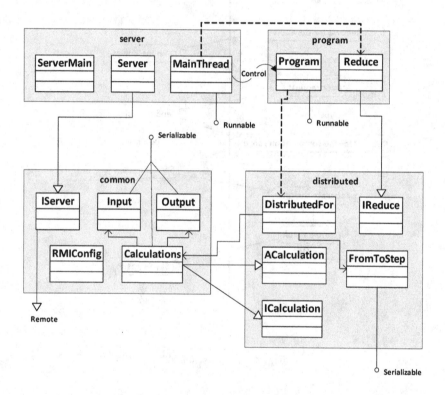

Fig. 5. Class diagram of a server

3.2   Implementation of a Client

The main code of calculations is realized as a self-signed applet (Swing technology). It contains the *client* package, also shares *common* and *distributed* packages, excluding *DistributedFor* class and *IReduce* interface.

At the start of connection the applet takes the calculation class, takes the current state and data of this class by deserialization of server data and gets the range of calculations (From, To, Step). Next the processing is performed and the results are uploaded back to the server. This procedure is repeated for the next range for calculations.

4   Tests

To assess a performance of proposed solution the tests have been carried out [15]. 8 PC (Intel i7 Core 2600, 4 GB DDR3, 1333 MHz) were used as client nodes and the ninth PC as a web server. The Gigabit Ethernet connection with switch was used. As a test algorithm a median filter with mask of 101×101 for gray scale images was implemented (as an example of time-consuming calculations). The tests were performed for two images:

- Picture A of 1920×1200,
- Picture B of 1200×900.

The results of tests are presented in the table (Table 1) and on the plot (Fig. 6).

Table 1. Execution time of distributed image processing [s] (and speedup)

	Serial	1 node	2 nodes	4 nodes	8 nodes
Pic A, 1920×1200	381	391 (102 %)	197 (51 %)	103 (27 %)	65 (17 %)
Pic B, 1200×900	150	151 (101 %)	77 (51 %)	42 (28 %)	22 (14 %)

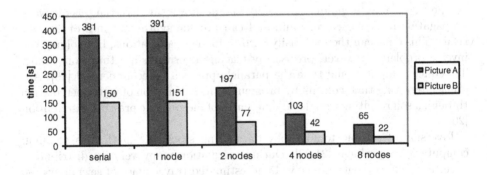

Fig. 6. The results of tests

The results of tests proved that proposed solution is an effective way of parallelization. There was a very small network overhead in case of Picture B, estimated as a comparison of serial mode and calculations on a single node, although the transmitted data after decompression was not negligible. An important factor was the speed of the network. The proposed solution can process in different environments, so the network overhead should be taken into account. The problem of network speed in cluster computations was considered in [16].

5   Conclusions and the Future Work

The proposed solution meets the conditions of flexible environment for client-side distributed numerical calculations. This way of processing enables to set a distributed environment that contains a dynamic number of nodes, that are based on different hardware and software platforms. Numerous problems and applications can be processed at the client side. It is also possible to provide computational services by individual clients. The future work involves the implementation for MS .NET framework and MS Silverlight. Although Java based applets will most likely be a very common solution due to mobile market penetration, MS Silverlight allows to easier implementation using provided interfaces, especially a compression of transmitted data stream. It reduces a network overhead [17,18]. Further improvements (like compression of input data) should allow to reach better efficiency because in case of Picture B processing, simple extrapolation of the measured data for current implementation shows less improvement in execution time in case of 16 processes and significant drop in efficiency for larger numbers. For 8 nodes the speedup is 17 % instead 12.5 %. Another problem is memory allocation – we tested max. 2 GB of input data – such amount of data needs changes in VM configuration. More detailed and extensive testing is needed to accurately assess the solution.

An important issue is the use of the proposed method in real numerical problems. The research on algorithms for distributing task for distributed systems involves time-consuming algorithms used in Geophysics.

The ray tracing technique can easily and accurate simulate the seismic wave propagation in geological medium and can provide synthetic times of first arrivals. This data can then be easily used in further calculations, like solving the inverse problem. However, precise ray tracing algorithms are time consuming. To reduce computational time the parallel approach is recommended [19]. Another interesting time-consuming problem is the calculation of water percolation through a soil resulting from local differences of piezometric pressure distribution [20].

Inversion of seismic tomography data using stochastic method is common computational problem [21,22]. During calculation many velocity distribution were tested by comparing received and estimated travel times of seismic waves. Additional computational problem is a long time for estimation travel times.

The proposed solution is not limited to use in Geophysics. It can be also used for solving time-consuming numerical problems in other domains, like simulation of energy production and distribution for electricity market [23].

Acknowledgments. This work was co-financed by the AGH – University of Science and Technology, Faculty of Geology, Geophysics and Environmental Protection, Department of Geoinformatics and Applied Computer Science as a part of statutory project.

References

1. Sunderam, V.S.: PVM: A Framework for Parallel Distributed Computing. Concurrency: Practice and Experience 2(4), 315–339 (1990)
2. Gropp, W., Lusk, E., Skjellum, A.: Using MPI: Portable Parallel Programming with the Message-Passing Interface. MIT Press (1994)
3. Object Management Group. The Common Object Request Broker: Architecture and Specification. OMG Document, Ver. 2.0 (1995)
4. Onderka, Z.: The Efficiency Analysis of the Object Oriented Realization of the Client-Server Systems Based on the CORBA Standard. Schedae Informaticae 20, 181–194 (2011)
5. Onderka, Z.: DCOM and CORBA Efficiency in the Wireless Network. In: Kwiecień, A., Gaj, P., Stera, P. (eds.) CN 2012. CCIS, vol. 291, pp. 448–458. Springer, Heidelberg (2012)
6. Kowal, A., Piorkowski, A., Danek, T., Pieta, A.: Analysis of selected component technologies efficiency for parallel and distributed seismic wave field modeling. In: Proceedings of the 2008 International Conference on Systems, Computing Sciences and Software Engineering (SCSS), part of the International Joint Conferences on Computer, Information, and Systems Sciences, and Engineering, CISSE 2008, Bridgeport, Connecticut, USA. Innovations and Advances in Computer Sciences and Engineering, pp. 359–362. Springer (2010)
7. Piorkowski, A., Pieta, A., Kowal, A., Danek, T.: The Performance of Geothermal Field Modeling in Distributed Component Environment. In: Sobh, T., et al. (eds.) Innovations in Computing Sciences and Software Engineering. Proceedings of the 2009 International Conference on Systems, Computing Sciences and Software Engineering (SCSS), part of the International Joint Conferences on Computer, Information, and Systems Sciences, and Engineering (CISSE 2009), Bridgeport, Connecticut, pp. 279–283. Springer (2010)
8. Czerwinski, D.: Numerical performance in the grid network relies on a Grid-Appliance. In: Kwiecień, A., Gaj, P., Stera, P. (eds.) CN 2011. CCIS, vol. 160, pp. 214–223. Springer, Heidelberg (2011)
9. http://hadoop.apache.org/
10. Krauzowicz, Ł., Szostek, K., Dwornik, M., Oleksik, P., Piórkowski, A.: Numerical Calculations for Geophysics Inversion Problem Using Apache Hadoop Technology. In: Kwiecień, A., Gaj, P., Stera, P. (eds.) CN 2012. CCIS, vol. 291, pp. 440–447. Springer, Heidelberg (2012)
11. Kim, H., Kim, W., Lee, K., Kim, Y.: A Data Processing Framework for Cloud Environment Based on Hadoop and Grid Middleware. In: Kim, T.-H., Adeli, H., Cho, H.-S., Gervasi, O., Yau, S.S., Kang, B.-H., Villalba, J.G. (eds.) GDC 2011. CCIS, vol. 261, pp. 515–524. Springer, Heidelberg (2011)
12. Piorkowski, A., Plodzien, D.: Efficiency Analysis of the Server-Side Numerical Computations. In: Kwiecień, A., Gaj, P., Stera, P. (eds.) CN 2009. CCIS, vol. 39, pp. 225–232. Springer, Heidelberg (2009)
13. Szatan, P., Piorkowski, A., Danek, T., Pieta, A.: Client side web-based simulations of geophysical phenomena. Mineralia Slovaca 43, 187 (2011)
14. Jin, H., Sullivan, G.F., Masson, G.M.: Distributed Applet-Based Certifiable Processing in Client/Server Environments. In: Proceedings of the the the 7th Symposium on the Frontiers of Massively Parallel Computation (FRONTIERS 1999), p. 44. IEEE Computer Society, Washington, DC (1999)

15. Szemla, P.: Web based environment for distributed calculations. Engineering Thesis, WGGiOS, AGH (2013)
16. Wrzuszczak-Noga, J., Borzemski, L.: Comparison of MPI Benchmarks for Different Ethernet Connection Bandwidths in a Computer Cluster. In: Kwiecień, A., Gaj, P., Stera, P. (eds.) CN 2010. CCIS, vol. 79, pp. 342–348. Springer, Heidelberg (2010)
17. Piorkowski, A.: Methods of creating database applications in.NET environment. In: Kwiecien, A. (ed.) Computer Networks 2007, Computer Networks – Applications and Uses, vol. 2, pp. 195–202. WKiL, Warsaw (2007)
18. Flak, J., Gaj, P., Tokarz, K., Wideł, S., Ziębiński, A.: Remote Monitoring of Geological Activity of Inclined Regions – The Concept. In: Kwiecień, A., Gaj, P., Stera, P. (eds.) CN 2009. CCIS, vol. 39, pp. 292–301. Springer, Heidelberg (2009)
19. Szostek, K., Leśniak, A.: Parallelization of the seismic ray trace algorithm. In: Wyrzykowski, R., Dongarra, J., Karczewski, K., Waśniewski, J. (eds.) PPAM 2011, Part II. LNCS, vol. 7204, pp. 411–418. Springer, Heidelberg (2012)
20. Onderka, Z., Schaefer, R.: Markov chain based management of large scale distributed computations of earthen dam leakages. In: Palma, J.M.L.M., Dongarra, J. (eds.) VECPAR 1996. LNCS, vol. 1215, pp. 49–64. Springer, Heidelberg (1997)
21. Dwornik, M., Pięta, A.: Parallel Implementation of Stochastic Inversion of Seismic Tomography Data. In: Wyrzykowski, R., Dongarra, J., Karczewski, K., Waśniewski, J. (eds.) PPAM 2011, Part II. LNCS, vol. 7204, pp. 353–360. Springer, Heidelberg (2012)
22. Danek, T., Slawinski, M.A.: Bayesian inversion of VSP traveltimes for linear inhomogeneity and elliptical anisotropy. Geophysics 77(6), R239–R243 (2012)
23. Pałka, P.: Multilateral negotiations in distributed, multi-agent environment. In: Jędrzejowicz, P., Nguyen, N.T., Hoang, K. (eds.) ICCCI 2011, Part II. LNCS, vol. 6923, pp. 80–89. Springer, Heidelberg (2011)

Study of the Character of APRS Traffic in AX.25 Network

Remigiusz Olejnik

Faculty of Computer Science and Information Technology
West Pomeranian University of Technology, Szczecin
ul. Żołnierska 49, 71-210 Szczecin, Poland
r.olejnik@ieee.org

Abstract. The paper presents the study of the traffic nature collected from APRS port in AX.25 network. Theoretical foundations of stochastic self-similar processes are presented. Four samples of collected traffic are analysed for the presence of self-similarity; the research used two common methods: R/S method and variance analysis method. The article ends with conclusions based on correlations found in the study.

Keywords: self-similarity, computer networks.

1 Introduction

Many years of research have shown that traffic in computer networks is characterized by long-term dependencies, ie. in a large set of time scales the traffic generated by the sources has the characteristics of self-similarity. These dependences were observed at several layers, as well as for a variety of network topologies and technologies [1–6].

Character of the traffic has a significant impact on the performance of computer networks such as the length of queues in the switches, which results in a higher probability of delay or packet loss due to buffers overflow. Knowledge of the traffic nature allows to design algorithms that could provide adequate quality of service.

The vast majority of previous work on network traffic character concerned wired networks. This paper presents study of nature of the traffic found in amateur radio AX.25 network, specifically at the APRS port.

The further part of this paper is divided into following parts: second section shows the theoretical foundations of stochastic self-similar processes, whereas third section presents the results of AX.25 network traffic analysis. The fourth section recapitulates the study.

2 Self-similarity Evaluation Methods

The definition of self-similar stochastic process is assumed as follows [1, 7]:

A. Kwiecień, P. Gaj, and P. Stera (Eds.): CN 2013, CCIS 370, pp. 31–37, 2013.
© Springer-Verlag Berlin Heidelberg 2013

Definition 1. *A stochastic process X_t is self-similar with the self-similarity parameter H, if for any positive g rescaled process $g^{-H}X_{gt}$ has the same distribution as the original process X_t.*

The H parameter is called a *Hurst exponent*. A stochastic process for which $0.5 < H < 1$ is a process in which there are Long Range Dependence.

There are many well known methods for verification and evaluation of self-similarity; the most popular are based on experiments and analysis of Hurst, originally on the fluctuations in the water level of the Nile River [8]. Three most popular network traffic evaluation methods are presented below.

2.1   R/S Method

Normalized dimensionless measure characterizing the variability is rescaled range R/S [9]. For a set of observations $X = \{X_m, m \in Z^+\}$, with a mean $\overline{X}(m)$, a variance $S^2(m)$ and a range $R(m)$, the R/S is defined as [8]:

$$\frac{R(m)}{S(m)} = \frac{\max(0, \Delta_1, \Delta_2, \ldots, \Delta_m) - \min(0, \Delta_1, \Delta_2, \ldots, \Delta_m)}{S(m)} , \qquad (1)$$

where $\Delta_k = \sum_{i=1}^{k} X_i - k\overline{X}$ for $k = 1, 2, \ldots, m$.

Hurst also observed that for many natural phenomena, for $m \to \infty$ and for constant c independent from m following formula is valid:

$$E\left[\frac{R(n)}{S(n)}\right] \sim cm^H . \qquad (2)$$

After logarithmisation, for $m \to \infty$ we obtain:

$$\log_{10}\left\{E\left[\frac{R(m)}{S(m)}\right]\right\} \sim H\log_{10}(m) + \log_{10}(c) . \qquad (3)$$

Hurst exponent value can be estimated using that formula – to do so, we must take into account the characteristics of $\log_{10}\left\{E\left[\frac{R(m)}{S(m)}\right]\right\}$ as a function of $\log_{10}(m)$, then approximate it with a straight line, which slope coefficient gives an estimated value of H. Value of H in range $0.5 \le H \le 1$ is a premise of the presence of self-similarity.

2.2   Aggregated Process Time Variance Analysis Method (aggvar)

The variance of aggregated process $X_k^{(m)}$ is defined as [1]:

$$Var(X_k^{(m)}) = \sigma^2 m^{-\beta} , \qquad (4)$$

where $0 < \beta < 1$, and σ^2 is variance of $X_k^{(m)}$ process for $m = 1$. Value of β is related with Hurst exponent H value:

$$H = 1 - \frac{\beta}{2} , \tag{5}$$

where value of β corresponds to the slope of the graph $\log_{10}\frac{Var(X_k^{(m)})}{\sigma^2}$ to $\log_{10}(m)$.

2.3 Graphical Method

In graphical method graph's temporal variability is evaluated (here: the number of packets in a time window of length m) when changing the time scale a correlation between the properties of the graph for different time scales is sought. However it is not an estimation method because it gives only a visual evidence about self-similarity. This method is not used in further part of this work.

3 Study of APRS Traffic Network Nature

As mentioned before, studies on the nature of computer networks traffic are not new. Summary of previous research carried out on the wired networks can be found in [1].

3.1 APRS Service

APRS (Automatic Packet Reporting System) [10] is a real time system used by radio amateurs to transmit data containing messages, alerts, announcements and bulletins. Nowadays most of the messages convey also GPS coordinates allowing quick localization of transmitting station. At network level APRS utilizes AX.25 unnumbered frames which are sent in unconnected broadcast manner. In physical layer 1200 baud AFSK (modified Bell 202) modulation is used when operating in VHF (2 meters – 145 MHz) band, which is most popular. APRS frames received by digipeaters are forwarded according to specific "routing" schemes, in almost all cases they are also entered into APRS Internet backbone where many useful online tools help tracing messages and their senders.

3.2 Research Methodology

The traffic was collected from APRS port of AX.25 network, ie. all received network packets. The configuration of collecting node was:

- AX.25 network connection: SR1BSZ (http://www.sr1bsz.ampr.org) node IP-tunnel,
- APRS radio port: 144.800 MHz.

Collected traffic (ASCII data) was saved to the file, then time of arrival for each packet was exported to another file. Finally CSV file has been created that contained number of collected bytes for each minute. That file was the source of analysis in MATLAB R2012a environment. Output of MATLAB script were plots and values of H for two Hurst exponent estimation methods (presented in 2.1 and 2.2).

3.3 Experimental Results

Table 1 contains results of measurements: number of received bytes, collecting time (in minutes) and values of Hurst exponent H: H_1 and H_2 obtained with methods: R/S and *aggvar* along with H value variance. The Figures 1–4 present plots generated by both methods for sample 2 and sample 4 (which is a sum of samples 1–3) respectively.

Table 1. Measurements results

sample number	number of bytes	time [mins]	H_1 R/S	H_2 *aggvar*	H variance
1	17436	120	**0.6644**	**0.5001**	0.0135
2	51028	240	**0.6222**	**0.6717**	0.0012
3	167054	480	**0.8714**	**0.7528**	0.0070
4	235518	840	**0.7275**	**0.7994**	0.0026

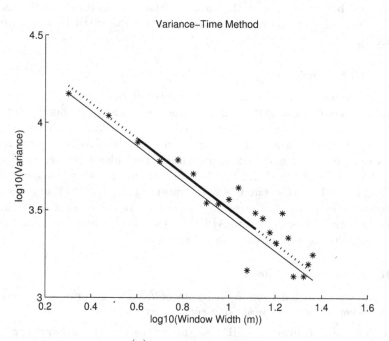

Fig. 1. $\log_{10} \dfrac{Var(X_k^{(m)})}{\sigma^2}$ as a function of $\log_{10}(m)$ for sample 2

Table 1 clearly shows that the Hurst exponent H values given by various methods are different, nevertheless H values for all of the samples are in the range $0.5 < H < 1$. For increasing numbers of packets in the sample we observe increasing H_2 value. Larger samples have not such a clear influence with R/S method and H_1 values.

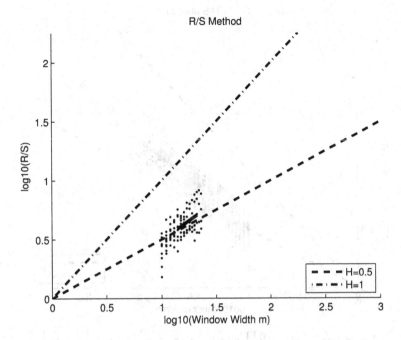

Fig. 2. $\log_{10}\left\{E\left[\frac{R(m)}{S(m)}\right]\right\}$ as a function of $\log_{10}(m)$ for sample 2

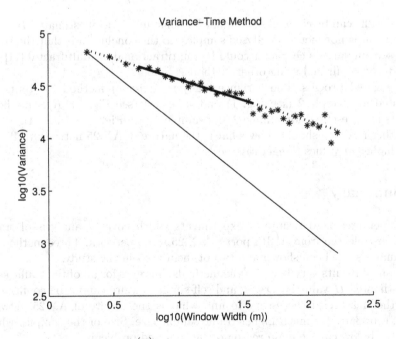

Fig. 3. $\log_{10}\frac{Var(X_k^{(m)})}{\sigma^2}$ as a function of $\log_{10}(m)$ for sample 4

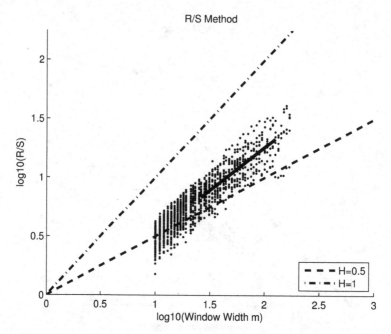

Fig. 4. $\log_{10}\left\{E\left[\frac{R(m)}{S(m)}\right]\right\}$ as a function of $\log_{10}(m)$ for sample 4

The traffic can be claimed self-similar for constant value of estimated H. That requirement is not met for analysed samples, so the conclusion is that the traffic is not self-similar. Its character could be multifractional or multifractal [11] but is has to be confirmed using other methods.

Aggregated process time variance analysis ($aggvar$) method's outputs are presented for sample 2 (see Fig. 1) and sample 4 (see Fig. 3). R/S method's outputs are presented for sample 2 and sample 4 (see Fig. 2 and Fig. 4).

Further research should show what is the nature of AX.25 network traffic for much higher m values (longer samples).

4   Summary

The paper presented results of experiments which covered analysis of traffic character collected from APRS port in AX.25 radio network. Three methods of such analysis has been shown and two of them used in the study.

Obtained results vary between the methods, however for all of the traffic samples estimated H value is varying and self-similar nature cannot be confirmed.

Further research is necessary to find what is the nature of AX.25 network traffic, especially for much higher traffic sample size. Size of the samples which is relatively low can be main reason of encountered problems.

Another possible problem is potential incompleteness of collected data due to a "hidden node problem" that is present in every wireless network. That issue can be prevented using more nodes for data collecting, but such approach needs also additional effort to compare and select unique packets. Moreover other known methods of Hurst exponent evaluation (eg. the wavelet estimator [12]) can be used and compared along with R/S and *aggvar* methods which were used in this study.

References

1. Czachórski, T., Domańska, J., Sochan, A.: Samopodobny charakter natężenia ruchu w sieciach komputerowych. Studia Informatica 22(1) (43), 93–108 (2001) (in Polish)
2. Zatwarnicki, K., Zatwarnicka, A.: Estimation of Web Page Download Time. In: Kwiecień, A., Gaj, P., Stera, P. (eds.) CN 2012. CCIS, vol. 291, pp. 144–152. Springer, Heidelberg (2012)
3. Leland, W.E., Taqqu, M.S., Willinger, W., Wilson, D.V.: On the self-similar nature of Ethernet traffic (extended version). IEEE/ACM Transactions on Networking 2(1), 1–15 (1994)
4. Crovella, M.E., Bestavros, A.: Self-Similarity in World Wide Web traffic: evidence and possible causes. IEEE/ACM Transactions on Networking 5(6), 835–846 (1997)
5. Feldmann, A., Gilbert, A.C., Willinger, W., Kurtz, T.G.: The Changing Nature of Network Traffic: Scaling Phenomena. ACM SIGCOMM Computer Communication Review 28(2), 5–29 (1998)
6. Paxson, V., Floyd, S.: Wide Area Traffic: The failure of Poisson modeling. IEEE/ACM Transactions on Networking 3(3), 226–244 (1995)
7. Beran, J.: Statistics for Long-Memory Processes. Chapman and Hall (1994)
8. Bassingthwaighte, J., Raymond, G.: Evaluating rescaled range analysis for time series. Annals of Biomedical Engineering 22(4), 432–444 (1994)
9. Grabowski, F.: Procesy $1/f$ w systemach rozproszonych. Studia Informatica 23(2A) (48), 143–153 (2002) (in Polish)
10. Automatic Packet Reporting System, http://www.aprs.org
11. Abry, P., Baraniuk, R., Flandrin, P., Riedi, R., Veitch, D.: Multiscale nature of network traffic. IEEE Signal Processing Magazine 19(3), 28–46 (2002)
12. Abry, P., Veitch, D.: Wavelet Analysis of Long-Range-Dependent Traffic. IEEE Transactions on Information Theory 44(1), 2–15 (1998)

Modelling the Pertubation of Traffic Based on Ateb-functions

Mykola Medykovsky, Ivanna Droniuk, Maria Nazarkevich, and Olga Fedevych

Lviv Polytechnik National University

Abstract. Traffic's periodic processes in computer networks based on Ateb-functions without perturbation were simulated. The types of small perturbations that influence the traffic's vibration were considered. Appropriate software that outputs the results in graphical and tabular forms was developed.

Keywords: Ateb-function, traffic's modelling, small perturbation.

1 Introduction

Two areas of telecommunications have recently become widespread: the Internet and cellular mobile radio communication. The number of hosts on the Internet increases every day, as well as the number of Internet users. Existing modern communication networks do not use their functional capacities to the fullest because of their internal structure. One of the reasons is the complexity of the behavior of network traffic.

At the same time, the number of subscribers of cellular wireless communication systems is increasing. Currently these two areas are merged into the technology of packet data submission called GPRS (General Packet Radio Service) as a superstructure in cellular radio networks [1]. The main goal of GPRS is to provide uninterrupted access of the subscribers of cellular mobile radio to the Internet. The efficiency of access to packet networks is determined by two key indicators – high transmission capacity and small delay. To fully use the potential of mobile network access and guarantee its successful operation it is necessary to ensure an effective relationship between QoS, the network resources used, and the characteristics of traffic in the network.

Classical Markov models with Poisson distribution, designed for circuit-switched channels, were previously used owing to their simplicity and high convergence for predicting the behavior of network traffic with packet commutation. But, as the result, the requirements for the communication channel and buffers sizes in package system were underestimated. The theory of nonlinear oscillation systems with one degree of freedom was offered to model the traffic in the network. Introduction of small perturbations into the system is one of the investigation methods. Simulation of such systems is analyzed on the basis of partial solutions of differential equations without perturbation. These solutions can be

A. Kwiecień, P. Gaj, and P. Stera (Eds.): CN 2013, CCIS 370, pp. 38–44, 2013.
© Springer-Verlag Berlin Heidelberg 2013

built through periodic Ateb-functions [2]. Asymptotic methods of nonlinear mechanics are the most effective methods of investigating oscillating processes of nonlinear systems with small perturbation.

2 Problem Definition

The problem of modeling periodic processes in nonlinear systems and the problem of constructing and predicting the behavior of such systems is of great practical importance. Therefore, development of methods of modeling such systems is topical. A model of a nonlinear oscillatory system with one degree of freedom, whose motion is described by a system of ordinary differential equations of the first order, was developed, and the problem of modeling based on the theory of Ateb-functions was solved. Results of the authors work upon the simulation of behavior of oscillatory system without perturbation and the one with the small perturbations are presented in this article.

3 Analysis of Recent Research and Publications

Nonlinear oscillating system with one degree of freedom with a small perturbation, whose motion is described by a system of ordinary differential equations of the first order with a small parameter ε,

$$\dot{y} - \alpha_1 x^n = \varepsilon f(x, y, \chi t) \; ; \quad \dot{x} - \alpha_2 y^m = \varepsilon f(x, y, \chi t) \; , \tag{1}$$

where x, y are phase coordinates of oscilation; α_1, α_2 constants, with $\alpha_1 > 0$, $\alpha_2 > 0$; $f(x, y, \chi t)$, $g(x, y, \chi t)$ – some analytic functions; n, m – numbers of the following kind:

$$m = \frac{2\mu_1^* + 1}{2\mu_2^* + 1} \; ; \quad n = \frac{2\mu_1^{**} + 1}{2\mu_2^{**} + 1} \tag{2}$$

where μ_1^*, μ_2^*, μ_1^{**}, μ_2^{**} are natural numbers or zero. It is proved that the analytical solution of system (1) is represented as Ateb-function.

4 Purposes of the Article

The aim of this work is to develop a method of modeling periodic processes in oscillating systems (as traffic in computer network, which has day,week, month oscillation) without disturbances and modeling of small perturbations. It is shown that the motion of such system is described by a system of ordinary differential equations of the first order with a small parameter and that analytical solutions of this system are based on Ateb-functions theory. Although the solution is recorded analytically, rendering of the system behavior needs special methods of results computation, which is related to the implicit representation of the solution in an analytical form. It is of practical importance to model the behavior of the solution with different input parameters. These tasks are solved in this paper, and the results of the calculations are presented in the form of corresponding diagrams.

5   Summary of the Main Content

Solution of unperturbed Equations (1) will be sought in amplitude – phase variables by replacing variables of the following kind:

$$y = u(\alpha, \varphi) \ , \quad x = v(\alpha, u) \ , \tag{3}$$

where $u(\alpha, \varphi), v(\alpha, u) - 2\Pi(m, n)$ periodic on φ which satisfy the conditions:

$$v(\alpha, u)_{\varphi = k\Pi(m.n)} = (-1)^k a \ , \quad v(\alpha, u)_{u=\pm a} = 0 \tag{4}$$

where $2\Pi(m, n)$ – the period of Ateb-function. From the first correlation substitute (2) and conditions (3) imposed on the function $u(\alpha, \varphi)$, it follows that $a = max y$, therefore the parameter a will further be called an amplitude of the phase coordinate y (amplitude of oscillation), while φ – will be referred to as a phase of oscillation. Inserting the solution (2) in the original unperturbed system of Equations (1), we obtain a system of differential equations to determine the unknown functions in the following form:

$$\frac{\partial u}{\partial \varphi} \omega(\alpha) - \alpha_1 v^n(\alpha, u) = 0 \ ; \quad \frac{\partial v}{\partial u} \frac{\partial u}{\partial \varphi} \omega(\alpha) - \alpha_2 u^m(\alpha, \varphi) = 0 \tag{5}$$

Integrating the system of differential Equations (4) under the conditions (3), we obtain:

$$v(\alpha, u) = (-1)^k \left(\frac{\alpha_2}{\alpha_1} \frac{n+1}{m+1} (\alpha^{m+1} - u^{m+1}) \right)^{\frac{1}{n+1}} \ ; \tag{6}$$

$$\phi = 2(r-1)\pi - \frac{\omega(a)}{m} \int_a^U \left(\frac{\alpha_2}{\alpha_1} \frac{n+1}{m+2} (\alpha^{m+1} - u^{-m+1}) \right)^{\frac{n}{n+1}} d\bar{u} \tag{7}$$

attached to

$$-a \le u \le a \ ; \quad (2r-1)\Pi \le \varphi \le 2r\Pi \tag{8}$$

where $k = 1$ for $a \le u \le a$ and $k = 2$ for $a \ge u \ge -a$. Relations (5), (6) are implicit expressions to find the function $u(\alpha, \varphi)$. This function can be expressed through Ateb-function as [3]:

$$u = Ca(m, n, l\phi) \ . \tag{9}$$

In the last dependence the constant l is determined by the condition $2\Pi(m, n)$ – periodicity in φ of obtained function. Similarly, we obtain an expression for $v(a, u)$ through Ateb-sine function:

$$v(\alpha, u) = \alpha^{\frac{m+1}{n+1}} \left(\frac{\alpha_2}{\alpha_1} \frac{n+1}{m+1} \right)^{\frac{1}{n+1}} Sa(n, m, l\varphi) \tag{10}$$

from $2\Pi(m,n)$ – periodicity through φ solution unperturbed system gives:

$$w(a) = \frac{\alpha_1(m+1)}{2}\left(\frac{\alpha_2}{\alpha_1}\frac{n+1}{m+2}\right)^{\frac{n}{n+1}} a^{\frac{nm-1}{1+n}} .$$ (11)

Thus it is shown that the dynamic processes of the researched nonlinear systems are described by Ateb-functions. Computation of the solutions of unperturbed system (1) in the form (9), (10), which was implemented in this paper, as well as calculation of the effect of small perturbations on the solution of the system, is of a significant scientific and practical interest.

Ratios (9) and (10) were selected as a basis for the software implementation of mathematical models of periodic processes in nonlinear systems such as traffic in computer network. The algorithm for searching values of Ateb-functions like cosine and sine, $Sa(n,m,l\varphi)$, $Ca(n,m,l\varphi)$ which is presented in [4] was used for calculations. On this basis an appropriate software was developed. The input data used in the calculations includes the values of parameters m, n, α_1, α_2 and the value of amplitude a. The result of the calculation is the phase coordinates of traffic's oscillation x, y and the function $w(a)$ is presented in Table 1. The implemented algorithm is described below. Initially, parameters m, n, α_1, α_2 and a are specified while the periodicity condition is checked. Next the Ateb-functions dependent on the values of defined parameters are tabulated. The developed program calculates the values of phase coordinates and according to Formulas (9) and (10), and populates the table of results.

Table 1. Calculated values of the phase coordinates both x and y and function $w(a)$ for values $n = 0.01$; $m = 0.1$

No	a	cos	sin	x	y	$w(a)$
1	0	1	0	1	0	0.5498140978
2	0.1	0.84765625	0.182273942	0.84765625	0.167502123	0.5498140978
3	0.2	0.67578125	0.361836980	0.67578125	0.332513040	0.5498140978
4	0.3	0.48828125	0.547223680	0.48828125	0.502875658	0.5498140978
5	0.4	0.30078125	0.725708336	0.30078125	0.666895586	0.5498140978
6	0.5	0.10546875	0.905795063	0.10546875	0.832387751	0.5498140978
7	0.555167030	0	1	0	0.918958144	0.5498140978

The initial schedule is built only for a quarter of a period, but as the given function is periodic with period value $2\Pi(m,n)$, the schedule is extended on the entire period. The results of the calculation were used to construct three-dimensional diagrams of dependence of phase coordinates traffic oscilation values on the parameters α_1 (Fig. 1).

After the systems behavior during unperturbed oscillations was modeled, the processes in nonlinear oscillatory systems with small perturbation are simulated. The following types of perturbation presented in Fig. 2 will be considered. Formulas, according to which the disturbances were conducted, are presented in (12–17).

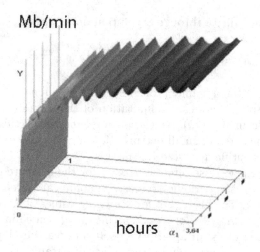

Fig. 1. Schedule showing traffic changes depending on the parameter α_1

$$f(x,y) = A_\varepsilon \qquad\qquad g(x,y) = B_\varepsilon \ , \qquad\qquad (12)$$

$$f(x,y) = \begin{cases} A_\varepsilon & \text{if} \quad x < \text{const} \\ 0 & \text{if} \quad x \geq \text{const} \end{cases} \qquad g(x,y) = \begin{cases} B_\varepsilon & \text{if} \quad y < \text{const} \\ 0 & \text{if} \quad y \geq \text{const} \end{cases} ; \qquad (13)$$

$$f(x,y) = \begin{cases} x & \text{if} \quad x < \text{const} \\ 0 & \text{if} \quad x \geq \text{const} \end{cases} \qquad g(x,y) = \begin{cases} y & \text{if} \quad y < \text{const} \\ 0 & \text{if} \quad y \geq \text{const} \end{cases} ; \qquad (14)$$

$$f(x,y) = A_\varepsilon sin(x) \qquad\qquad g(x,y) = B_\varepsilon sin(y) \ , \qquad (15)$$

$$f(x,y) = (x+y)^{A_\varepsilon} \qquad\qquad g(x,y) = (x+y)^{B_\varepsilon} \ , \qquad (16)$$

$$f(x,y) = x\exp(y) \qquad\qquad g(x,y) = y\exp(x) \ . \qquad (17)$$

Formula (12) shows a constant perturbation, (13) – an abrupt perturbation, (14) a linear perturbation with a particular interval, (15) – a periodic perturbation, (16) – a power perturbation, (17) – an exponential perturbation. These perturbations are introduced into the system (1), and the system's behavior is modeled. There were simulated daily cycle traffic oscillation in telecommunication company. By choosing parametres m, n big traffic oscillations were modeled for 2 different traffic's data Fig. 3 but local oscillations were simulated using small perturbation. Figure 1 shows the traffic's simulation using periodic perturbation (14). Comparison of modeling results is presented in Table 2. There were taking into account model without perturbation ($M1$), model with periodic perturbation (14) ($M2$) and model with periodic perturbation as Ateb-function which can has another m, n parameters ($M3$). Correlation ρ and coefficient k standard deviation concerning to maximum deviation were calculated in a half period of Ateb-function interval.

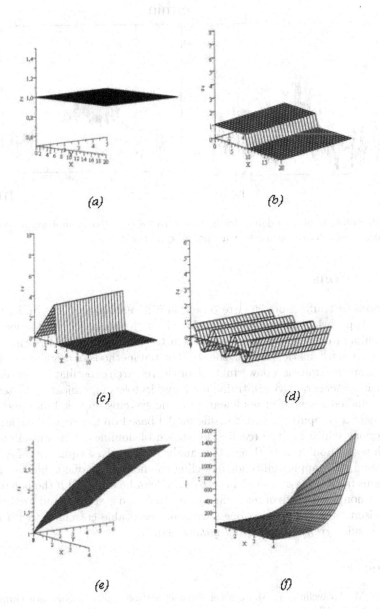

(a)

(b)

(c)

(d)

(e)

(f)

Fig. 2. Graphical representation of the perturbation

Table 2. The calculation of correlation ρ and coefficient k standard deviation concerning to maximum deviation

$m = 1/3,\ n = 1/7$	M1	M2	M3$(m = 1, n = 7)$
k	40 %	16 %	12 %
ρ	0.52	0.78	0.82

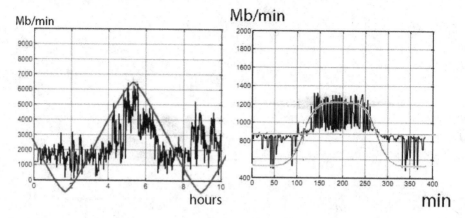

Fig. 3. 2 diffent examples modeling traffic in heterogenous telecommunication system and the results of calculation model with periodic perturbation

6   Conclusions

The methods of traffic modeling are based on differential equations and Ateb-functions. Appropriate software was developed for the numerical simulation.

The results of calculations are presented in tabular and graphical forms. So, the developed mathematical model, the constructed methods of calculations and the appropriate software is a powerful tool for the research of oscillatory processes in nonlinear systems of network traffic modeling in telecommunication systems.

The conducted research of nonlinear dynamic systems provided an opportunity to create a computer network traffic model based on the representation of Ateb-functions which are the results of solution of nonlinear differential equations with power nonlinearity. The article analyzed the solved equations of Ateb-functions and their applicability for modeling the network traffic behavior. The biggest gains from using this model (up to 12 %) can be obtained if the behavior of the most nonlinear parts of telecommunication traffic is studied and predicted in comparison with the existing ones, with an acceptable correlation with the real traffic and current models being maintained.

References

1. Ilchenko, M., Kravchuk, S.: Modern telecommunication systems. Naukova Dumka, Kyiv (2008) (in Ukrainian)
2. Senyk, P.: Appeal incomplete Beta-function. Eng. Math. Chem. 21(3), 325–333 (1969) (in Ukrainian)
3. Sokol, B.: Asymptotic approximations solutions for a non-autonomous nonlinear equation. Ukr. Math. Magazine 49(11), 1580–1583 (1997) (in Ukrainian)
4. Hrytsyk, V., Nazarkevych, M.: Mathematical models of algorithms and implementation Ateb-functions. Reports of the National Academy of Sciences of Ukraine 12, 37–43 (2007) (in Ukrainian)

Spatial Econometrics Models in Web Server's Performance

Leszek Borzemski and Anna Kamińska-Chuchmała

Institute of Informatics, Wrocław University of Technology,
Wrocław, Poland
{leszek.borzemski,anna.kaminska-chuchmala}@pwr.wroc.pl

Abstract. In recent years we saw, how desirable is possibility of mobile communications in modern society. The consequence of this is to have Internet more reliable and predictable in context of Web access. Thus, there is a need to analyzing Web server's performance and trying to predict future demand on given server's. This kind of research requires spatial methods of analysis of such data. Therefore we decided using spatial econometrics methods to explore Web server's performance.

This paper contains description the spatial regression models: Classic Regression Model (CRM), Spatial Lag Model (SLM) and Spatial Error Model (SEM). We use these models to predict total download time of data from Web servers. The real-life dataset was obtained in active experiments performed by the Multiagent Internet Measurement System (MWING), which monitored web transactions issued by MWING's agent located in Gdańsk, Poland and targeting Web servers in Europe. Data analyzed in this paper contains the measurements, which were taken every day at the same time: at 6:00 a.m., 12:00 a.m. and 6:00 p.m. We presented our analysis of measurement data and created spatial econometric models. Next, influences on prediction errors in regression models were described. After that we compared econometric with geostatistical methods. At the end, conclusions and future research directions to Web performance predictions were given.

Keywords: web server's performance prediction, spatio-temporal data mining, spatial econometrics, regression models.

1 Introduction

Increased request for access to the Internet by mobile devices have influence on the working of the whole network. The main investors and operators should take into account the necessity of good Web performance, especially during near-term planning of Web of Things. In order to properly analyze Web performance, appropriate methods should be chosen to perform such research. In this paper we presented our approach to research on Web performance with using spatial econometrics methods at the first time. We decided to use these methods because they give us possibility to see heterogeneity of spatial data and their autocorrelation. This methods with using neighborhood matrix could give a very

A. Kwiecień, P. Gaj, and P. Stera (Eds.): CN 2013, CCIS 370, pp. 45–54, 2013.
© Springer-Verlag Berlin Heidelberg 2013

good estimation of considered regression models. Currently, to the best of our knowledge the spatial econometric approach to Web performance prediction as presented in this paper is unique, leaving no similar problem statement in the literature.

In the next section we described econometrics methods used in this research. After that, we presented our results of research and conclusions.

2   Econometrics Methods and Models

When in 1970 Waldo Tobler formulated first law of geography, that is: "Everything is related to everything else, but near things are more related than distant things" [1] it was the beginning of spatial econometrics. Four years later Jean Paelinck introduced the notion of spatial econometrics. Professor Luc Anselin wrote the monograph about spatial econometrics [2], which is one of the most cited positions in literature. The author takes into account the spatial effects – the main characteristics of spatial econometrics, which can be divided into two groups: spatial dependence (or autocorrelation) and spatial heterogeneity. Also Anselin founded and up to now directs the GeoDa Center for Geospatial Analysis and Computation [3], and developed the software OpenGeoDa, which was used in our research presented in this paper.

Spatial econometrics due to [4] have a five characteristic features it deals with: the role of spatial interdependence in spatial models, the asymmetry in spatial relations, the importance of explanatory factors located in other spaces, differentiation between ex post and ex ante interaction and explicit modeling of space.

Below three models of spatial regression are described that we use in the next sections.

2.1   Classic Regression Model

The simple model of linear regression can be expressed in the following form:

$$y = X\beta + \epsilon \ , \tag{1}$$

where y is dependent variable, $X = [X_1 X_2 \ldots X_k]$ are independent variables, β is model's coefficient and ϵ is a random component. In practice such model is not always sufficient, hence it uses hybrid models and models with weights.

Spatial weight matrices W are formed on the basis of the distance matrix or neighborhood matrix and treated as independent variables in econometric model. Interaction strength between two points i and j depends on the distance between them. For larger distances between two points, the interaction strength should be smaller. This can be interpreted as the presence and strength of a link between nodes (the observations) in a network representation that matches the spatial weights structure. Therefore, in weight matrix W, computed on the basis of the distances d_{ij}, particular elements are in general inverse functions or exponential-inverse functions of distances:

$$w_{ij} = d_{ij}^{-\alpha}, w_{ij} = e^{-\alpha d_{ij}} \; , \tag{2}$$

where α is predetermined parameter.

2.2 Spatial Lag Model

Spatial Lag Model (SLM), also known as Spatial Autoregressive Model (SAR), in matrix notation, can be written as follows:

$$y = \rho W y + X\beta + \epsilon \; , \tag{3}$$

where ρ is the autoregressive parameter and $\epsilon \sim N(0, \sigma^2 I)$.

SLM is usually considered as the formal specification for the stability outcome of a spatial interaction process, in which the value of the dependent variable for one node is jointly determined with that of the neighboring nodes [5]. Generally SLM is used to obtain empirical estimates for the parameters of a spatial reaction function [6] or social multiplier [7]. The modeling of complex neighborhood and network effects for example in [8] requires considerable attention to identification issues. Therefore, the choice of the weights in a spatial lag model is very important.

2.3 Spatial Error Model

As opposed to the SLM, a Spatial Error Model (SEM) does not require a theoretical model for spatial interaction, but is a special case of a non-spherical error covariance matrix. SEM is a model with spatial linear autocorrelation of random component:

$$y = X\beta + \xi \; , \tag{4}$$

$$\xi = \lambda W \xi + \epsilon \; , \tag{5}$$

where ξ is a random component, λ is a parameter of model and $\epsilon \sim N(0, \sigma^2 I)$.

3 Data Analysis

The database using to predict Web performance, was collected during active measurements made by MWING system. This is the Internet measurement infrastructure developed in our Institute [9,10]. The data was collected by MWING agent located in Gdansk, whose main task was to target, by means of HTTP transactions, European Web servers. The database contained the information about web performance (Z), which was the total downloading time of rfc1945.txt file, a server's geographical location, which the Gdansk agent targeted and the time stamp of taking a measurement. The measurements were done between

7th and 28th of February 2009 and they were taken every day at the same time: at 6:00 a.m., 12:00 a.m., and 6:00 p.m.

The statistics of download times from considered Web servers in Table 1 were presented. The high value of variability coefficient, especially for 6:00 a.m., indicates the large dispersion of the data. The maximum values in comparison to average values are clearly high, for example the largest span of data values are presented for 6:00 a.m., where difference between average and maximum values is equal to 28.44 seconds. In addition, high value of kurtosis and skewness coefficient prove the changeability of the examined process. Moreover this skewness indicates big right side asymmetry of performance distribution for each moment of measurement.

Table 1. Elementary statistics of download times from web servers between 7–28.02.2009 [11]

Statistical parameters	6:00 a.m	12:00 a.m.	6:00 p.m.
Minimum value Z_{min} [s]	0.11	0.12	0.12
Maximum value Z_{max} [s]	29.06	12.15	7.93
Average value Z [s]	0.60	0.62	0.60
Standard deviation S [s]	1.59	1.07	0.77
Variability coefficient V [%]	265.00	172.58	128.33
Skewness coefficient G	15.35	7.27	4.99
Kurtosis coefficient K	265.65	64.48	34.61

As one can see in Fig. 1, there are two Web servers in Europe, which are outliers on the cartogram in 28th of February 2009.

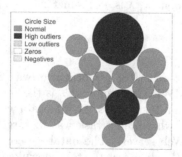

Fig. 1. Cartogram of total download time for 28th of February 2009

We made multivariate Moran's *I* statistics presented in Fig. 2 to verify auto-correlation between exemplary days: for input data 14th and 15th February 2009 – Fig. 2a and between real data 2nd and 3rd March 2009 – Fig. 2b, which will be used to verify data from prediction. As we can see, positive autocorrelation in input data is very small – close to zero. For data from days which will be predicted there is negative autocorrelation and also nearly zero. These results from Moran's statistics indicates on large heterogeneity among Web servers for particularly days. There is no similar behavior of Web server during file down-loading.

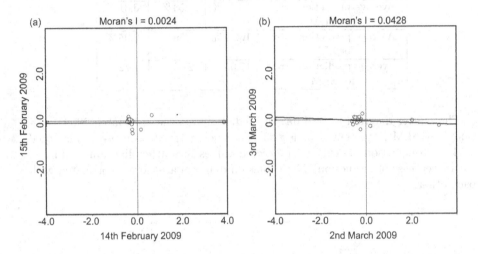

Fig. 2. Multivariate Moran's *I* statistics for exemplary days

4   Prediction of Web Server's Performance Calculated with the Spatial Econometrics Methods

There were three type of models used to predict the total time of resource down-load from the Internet: CRM, SLM and SEM. Prediction was calculated with a four day advance, i.e. it encompassed the period between 1st and 4th March 2009 for 6:00 a.m., 12:00 a.m. and 6:00 p.m., respectively. In our models it was assumed, that dependent variable was a day, which we want to predict and in-dependent variables were the days of history in database (February 2009). For three of best obtained models, the data comprising 16 days history were used. To create matrix weights Euclidean distance was used. Threshold distance was equal 13.41° and variables for x and y coordinates was assumed as centroids. All predictions were performed with using OpenGeoDa software [3].

The quality of econometrics methods is presented Table 2. On the basis of the accuracy of prediction we can conclude that SLM method is the best. Error of prediction in SLMs are smaller than about 1 % to 2 % than SEMs and even about 4 % than CRMs. Skewness for 6:00 a.m. was the highest in input data, therefore error of prediction is larger than for other hours computing all of given methods. CRM is the worse method, because in all cases it gives the smallest accuracy of prediction.

Table 2. Average prediction error *ext post* for all Web servers in a four-day forecast for 16 days history, compared methods

Spatial econometric method	6:00 a.m.	12:00 a.m.	6:00 p.m.
Average prediction error for CRM	18.82 %	3.54 %	6.31 %
Average prediction error for SLM	16.59 %	1.56 %	2.05 %
Average prediction error for SEM	17.55 %	2.83 %	4.72 %

The scatter diagram for 3rd day of prediction, i.e. the 3rd March 2009, computed by SLM is presented in Fig. 3. The average prediction error was equal 1.54 %. Comparison of real and predicted values in scatter diagram confirmed the correctness of prediction. The values on diagram are closely correlated with each other.

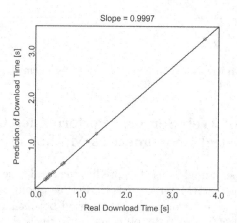

Fig. 3. Scatter diagram for 3rd day of prediction made by SLM

5   The Impact of Various Factors on the Accuracy of Prediction

Very interesting situation can be observed during studying the impact of the number of days used as independent variable on the prediction accuracy. The prediction error has increased with the systematic decreasing the number of days in the history of the database. This situation has been appeared for all studied spatial econometrics methods. Unfortunately, the maximum number of days in history could be 16 days due to the OpenGeoDa software running error. The average prediction error vs. number of history days for CRM is shown in Fig. 4. It should be mentioned that CRM was the least accurate method to predict a given Web server's performance. Nevertheless there is interesting situation and question is, what was the reason such behavior. In our opinion, SLM utilizes the autoregression, whereas SEM – autocorrelation of random component, thus the former gives better prediction. Moreover, the weak autocorrelation of Moran's I negatively influences the prediction with spatial econometric methods.

Other factors, such as appropriately selected matrix weight has influence on quality of prediction too. For example, properly chosen threshold distance decrease prediction error by about 4 %. Also variables for x and y coordinates should be correctly defined.

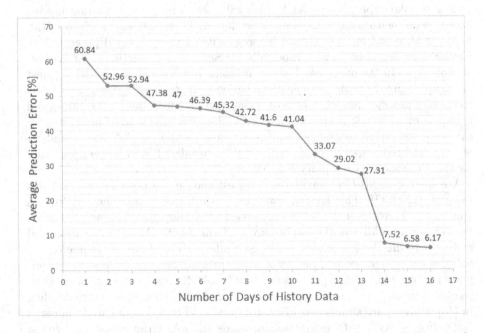

Fig. 4. Prediction error relative to numbers of days of history data for 12:00 a.m. of 1st March 2009 with using CRM

6   Spatial Econometrics Methods vs. Geostatistical Methods

Usually, the majority of the research in spatial data analysis can be divided into two branches: the model-driven approach and the data-driven approach. The

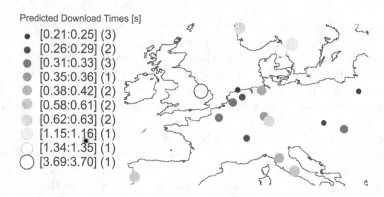

Predicted Download Times [s]
- [0.21:0.25] (3)
- [0.26:0.29] (2)
- [0.31:0.33] (3)
- [0.35:0.36] (1)
- [0.38:0.42] (2)
- [0.58:0.61] (2)
- [0.62:0.63] (2)
- [1.15:1.16] (1)
- [1.34:1.35] (1)
- [3.69:3.70] (1)

Fig. 5. Unique map of download time values from the Internet on 03.03.2009 at 6:00 p.m. determined with SLM

spatial econometric methods use the model-driven approach where the crucial issue is to take appropriate model which will takes into account relationship between servers with using spatial autocorrelation. These methods such as CRM, SLM, SEM, could analyze spatial heterogeneity of the examined process. Prediction accuracy is very good, especially for SLM, and prediction error is equal at most 3 %. On the other hand, there are some restrictions – in model we must know approximate value of dependent variable. If we want to only verify correctness of assumed model, that is ok, otherwise we must "guess" the values and trying to match using Moran's I statistics. Moreover this kind of prediction we can made iteratively, only one step (in our case one day) ahead. As a result we obtain admittedly spatial prediction for this considered Web servers, but only for them (please see exemplary Fig. 5).

We used geostatistical simulation and estimation methods in our previous research [12,13,11]. This spatial statistics – geostatistics uses the data-driven approach, where the search ellipsoid is used to consider neighborhood of given Web server. In addition, the studies are performed with the 3D simulation grid and as a result these methods could give information about performance not only for considered servers, but for a whole investigated area (please see exemplary Fig. 6). This spatial prediction based on 3D directional variogram gives us results as much ahead as we want without any hints about the predicted value. In [14] we compare simulation (Turning Bands – TB and Sequential Gaussian Simulation – SGS) with estimation methods (Simple Kriging) by performing spatial prediction on the same database as we use in this work. The results give us information about advantage simulation over estimation methods, but the least prediction error was equal 16 %, much larger than for econometrics methods. On the other hand it is difficult compare econometrics and geostatistical methods in this context, because these give us different quality information about such prediction: econometric about autocorrelations between spatial data (Web servers) and simulation about Web server performance for a whole considered area. However, there were the attempts to compare this methods in other domain – to analyzing the crime rate in Columbus (Ohio) [15]. Author showed

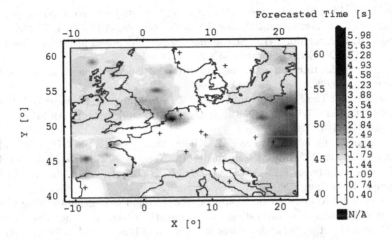

Fig. 6. Sample raster map of download time values from the Internet on 01.03.2009 at 6:00 p.m. determined with SGS [11]

advantage of geostatistical estimation methods over spatial econometric methods. Mean absolute error was higher at least 2 %.

7 Summary

In this paper, we proposed an approach for predicting Web performance with using spatial econometrics methods: CRM, SLM and SEM. We presented our results and discussed them. Furthermore, we refer to our previous research with using geostatistical simulation and estimation methods.

We claim that presented econometrics methods to prediction Web server performance could be helpful in spatial analysis of Internet and Web performance. The future research includes an active measurement experiment performed in a new manner to gain new measurement attributes. Also other methods from the domain of spatial statistics to predict Web performance will be studied.

References

1. Tobler, W.: A Computer Model Simulating Urban Growth in the Detroit Region. Economic Geography 46(2), 236 (1970)
2. Anselin, L.: Spatial Econometrics: Methods and Models. Kluwer Academic Publishers, Dordrecht (1988)
3. https://geodacenter.asu.edu
4. Paelinck, J.H.P., Klaassen, L.H.: Spatial Econometrics. Farnborough, Saxon House (1979)
5. Anselin, L., Le Gallo, J., Jayet, H.: Spatial Panel Econometrics. In: Mátyás, L., Sevestre, P. (eds.) The Econometrics of Panel Data, pp. 625–660. Springer, Heidelberg (2008)

6. Brueckner, J.K.: Strategic Interaction Among Governments: An Overview of Empirical Studies. International Regional Science Review 26(2), 175–188 (2003)
7. Glaeser, E.L., Sacerdote, B.I., Scheinkman, J.A.: The Social Multiplier. Technical Report 9153, NBER, Cambridge, MA 02138 (2002)
8. Topa, G.: Social Interactions, Local Spillover and Unemployment. Review of Economic Studies 68(2), 261–295 (2001)
9. Borzemski, L.: The Experimental Design for Data Mining to Discover Web Performance Issues in a Wide Area Network. Cybernetics and Systems: An International Journal 41, 31–45 (2010)
10. Borzemski, L., Cichocki, Ł., Fraś, M., Kliber, M., Nowak, Z.: MWING: A Multiagent System for Web Site Measurements. In: Nguyen, N.T., Grzech, A., Howlett, R.J., Jain, L.C. (eds.) KES-AMSTA 2007. LNCS (LNAI), vol. 4496, pp. 278–287. Springer, Heidelberg (2007)
11. Borzemski, L., Kamińska-Chuchmała, A.: Knowledge Engineering Relating to Spatial Web Performance Forecasting with Sequential Gaussian Simulation Method. In: Graña, M., et al. (eds.) Advances in Knowledge-Based and Intelligent Information and Engineering Systems. Frontiers in Artificial Intelligence and Applications, vol. 243, pp. 1439–1448. IOS Press, Amsterdam (2012)
12. Borzemski, L., Kamińska-Chuchmała, A.: Distributed Web Systems Performance Forecasting Using Turning Bands Method. IEEE Transactions on Industrial Informatics 9(1), 254–261 (2013)
13. Borzemski, L., Kamińska-Chuchmała, A.: Client-Perceived Web Performance Knowledge Discovery through Turning Bands Method. Cybernetics and Systems: An International Journal 43(4), 354–368 (2012)
14. Borzemski, L., Kamińska-Chuchmała, A.: Web Performance Forecasting with Kriging Method. In: Ali, M., Bosse, T., Hindriks, K.V., Hoogendoorn, M., Jonker, C.M., Treur, J. (eds.) Contemporary Challenges & Solutions in Applied AI. SCI, vol. 489, pp. 149–154. Springer, Heidelberg (2013)
15. Fernández-Avilés Calderón, G.: Spatial Regression Analysis vs. Kriging Methods for Spatial Estimation. International Advances in Economic Research 15(1), 44–58 (2009)

Empirical Web Performance Evaluation with MWING System and Turning Bands Method

Leszek Borzemski, Michał Danielak, and Anna Kamińska-Chuchmała

Institute of Informatics, Wrocław University of Technology,
ul. Wybrzeze Wyspianskiego 27, 50-370 Wrocław, Poland
{leszek.borzemski,michal.danielak,anna.kaminska-chuchmala}@pwr.wroc.pl
http://www.ii.pwr.wroc.pl/index.php/en/institute

Abstract. This paper presents the Turning Bands method (TB) as a geostatistical approach for making spatio-temporal Web performance analyses and forecasts. The first part of this paper briefly introduces and characterizes the Multiagent Internet Measuring System (MWING) whose main task involves collecting data necessary for making the afore-mentioned analyses and forecasts. MWING comprises of many agents located in different parts of the globe; however, for the purpose of this paper the data that had been collected only by the agent in Wrocław were used. The measurements were taken every Monday in two periods: the first, three times a day (at 6:00 am, 12:00 pm and 6:00 pm), between 1 February 2009 and 28 February 2009; the second, at the same hours, but between 1 December 2012 and 24 December 2012. The subsequent part of this paper briefly elucidates TB algorithm; the final part commences with data analysis which is followed by presenting sample analyses and forecasts of web servers performance made using TB.

Keywords: web performance analyses, Turning Bands method, spatio-temporal forecasts.

1 Introduction

The Internet performance evaluation has always been an important issue addressed and discussed in scientific literature in many various ways. Mirza et al., for instance, use machine learning in TCP throughput prediction [1]. Other approaches try to answer the question whether prediction models can be applied at the moment of measurement to predict bandwidth available in the future [2], or tackle the problem of Internet performance by measuring either download times of Web resources or loading responses of given Web pages [3,4]. These approaches, however, usually deal with this problem only by taking into account either its temporal or spatial aspect. The spatial aspect usually consists in performance estimation for unknown network nodes using data collected for their neighbours, while the temporal aspect means making performance forecasts for different time horizons by analysing historical data of nodes. Nevertheless, the evaluation of Internet performance is certainly a challenging task, especially in

A. Kwiecień, P. Gaj, and P. Stera (Eds.): CN 2013, CCIS 370, pp. 55–63, 2013.
© Springer-Verlag Berlin Heidelberg 2013

its backbone where not only is it difficult to measure and accurately estimate values of backbone performance, but it is also an uphill struggle to model traffic of such magnitude.

The method presented in this paper combines the two aforementioned aspects. Namely, by analysing series of collected samples it allows to carry out spatio-temporal forecast. The usage of TB for the purpose of such research can be justified, because this method has already proven itself not only in areas such as mining, geology, geography or hydrology, but also in load forecasting in electrical networks [5]. However, to the best of the authors' knowledge, the use of this method in forecasting of Web performance is an innovative solution, leaving no similar statement in the literature. More information about this approach may be found in [6,7]. The manner in which the necessary data have been collected is presented in the subsequent section herein.

2   Multiagent Internet Measuring System

Web users have always been expecting high efficiency, especially when the same web services are available on many different web servers. Therefore, a subjective quality indicator, which is seen from the perspective of a given agent[1] may be introduced. This subjective quality indicator is represented by download times needed for the agents to obtain the same resource from a group of evaluated web servers. One can distinguish four such MWING agents: Las Vegas, Gdańsk, Gliwice and Wrocław. The latter is the main agent where all collected measurements are combined and processed for further analyses.

In the next step, these measurements with pieces of information such as servers' locations (i.e. their latitudes and longitudes) and timestamps for each measurement are added to a database; all this information is necessary to make spatio-temporal analysis and forecasts of Web performance. For the purpose of this research, only data collected by the agent in Wrocław were used. The measurements were taken every Monday in two periods: the first, three times a day (at 6:00 am, 12:00 pm and 6:00 pm, between 1 February 2009 and 28 February 2009; the second, at the same hours but between 1 December 2012 and 24 December 2012.

MWING is also an active experiment. This means that new data have to be generated in network in order to obtain measurements; this results in the problem of selection a proper resource; namely its size cannot be neither too small (download times must be greater than committed measurement error) nor too large (transmission cannot excessively burden evaluated web servers). Moreover, a selected resource must be legal and readily available on the Internet. To satisfy all these conditions, the text document rfc1945.txt (size: 137 582 bytes) was selected. More information about MWING may be found in [8,9], and [10].

---

[1] Computers with Web performance measurement software.

3   Turning Bands Method

Some covariance models may be simulated directly in R^n. However, every now and then it is much more convenient to use the Turning Bands method[2] which reduces given multi-dimensional simulation R^n with covariance $C(h)$ (n is the number of dimensions) to N one-dimensional realizations of that simulation with unidimensional covariances C_{θ_i}, $i \in [1, N]$. This is done by adding up a large number of independent simulations defined on lines spanning the n-dimensional space; namely the simulated value at a point x of the space is a sum of the values obtained at the projections of x on different one-dimensional lines, called bands. The first usage of the Turning Bands method for the purpose of simulation was introduced by Matheron [12,13].

TB requires the minimum amount of input data to perform spatio-temporal forecasts; in this research, it needs only such information as resource download times, a timestamp of each measurement taken, and servers' locations (i.e. their latitudes and longitudes). Another advantage of using TB is the fact that it deals with both the temporal and the spatial aspect of the empirical Web performance evaluation problem; namely, forecasts made by means of TB are carried out not only for the considered web servers, but for the whole examined area. For all Web performance forecasts and analyses made in this paper, it was assumed that $n = 3$ (one can distinguish three dimensions: latitude, longitude and time), $N = 100$ (the number of independent one-dimensional simulations) and Europe was the considered area of forecasts.

Let us assume that the field that is to be simulated is second-order stationary and isotropic. It is said that a second order stationary stochastic process is a process that satisfies the following conditions:

1. The mean does not depend on the position of each point in R^n:

$$E[Z(x)] = m(x) = m, \forall x \in R \tag{1}$$

   where $E[]$ is the expectation operator and $Z(x)$ is a random variable corresponding to a point x.
2. The covariance function $C(x_i, x_j)$ does not depend on each particular vector x_i, x_j, yet it depends only on the vector difference:

$$C(x_i, x_j) = C(x_i - x_j) = C(h) \tag{2}$$

   where $h = x_i - x_j$. Moreover, a second-order stationary process is called isotropic when covariance function $C(x_i, x_j)$ depends only on the length of the vector $|h|$ and is not affected by its direction.

The algorithm of the Turning Bands method may be presented in the following way [14]:

---

[2] TB and conditional simulations are discussed in more detail in [11].

1. Transform input data using Gaussian anamorphosis.
2. Select directions $\theta_1, \ldots, \theta_n$ so that $\frac{1}{n}\sum_{k=1}^{n} \delta_{\theta_k} \approx \varpi$.
3. Generate standard, independent stochastic processes $X_1, \ldots, X_n$ with co-variance functions $C_{\theta_1}, \ldots, C_{\theta_n}$.
4. Calculate $\frac{1}{\sqrt{n}}\sum_{k=1}^{n} X_k(< x, \theta_k >)$ for every $x \in D$.
5. Make kriged estimate $y^*(x) = \sum_c \lambda_c(x)y(c)$ for each $x \in D$.
6. Simulate a Gaussian random function with mean 0, covariance C in domain D on condition points. Let $(z(c), c \in C)$ and $(z(x), x \in D)$ be the obtained results.
7. Make kriged estimate $z^*(x) = \sum_c \lambda_c(x)z(c)$ for each $x \in D$.
8. Obtain the random function $W(x) = (y^*(x) + z(x) - z^*(x), x \in D)$ as the result of conditional simulation.
9. Perform a Gaussian back transformation to return to the original data.

The problem of the determination of unidimensional covariances C_{θ_i}, $i \in [1, N]$ (Step 3) was solved by Matheron who found the relationship between unidimensional and multi-dimensional covariances:

$$C_n(h) = 2\frac{(n-1)\omega_{n-1}}{n\omega_n}\int_0^1 (1 - t^2)^{\frac{n-3}{2}} C_1(th)dt \qquad (3)$$

where ω_n stands for n-dimensional volume of the unit ball in R^n. If $n = 3$, the formula is reduced to:

$$C_3(h) = \int_0^1 C_1(th)dt \qquad (4)$$

which can be written also as:

$$C_1(h) = \frac{d}{dh}(hC_3(h)) \ . \qquad (5)$$

Knowledge of semivariogram and variance (i.e. $C(0)$) of a stationary random function $Z(x)$ is sufficient to determine its covariance; therefore, in this paper, for any vector h empirical variograms were calculated as:

$$\gamma(h) = \frac{1}{2|N(h)|}\sum(Z(x) - Z(x+h))^2 \qquad (6)$$

where $N(h)$ is the set of all pairwise Euclidean distances, $|N(h)|$ is the number of distinct pairs in $N(h)$, and $Z(x)$ and $Z(x+h)$ are data values at locations x and $x + h$ respectively.

Moreover, to make analyses and forecasts of Web performance only directional variograms along the time axis (i.e. $90°$) were calculated.

4   Data Analysis

As it has been already mentioned, for the purpose of this paper, only the measurements collected by the Wrocław agent were used. These measurements were taken every Monday, at 6:00 am, 12:00 pm and 6:00 pm at two time periods: the first, between 1 February 2009 and 28 February 2009, and the second, between 1 December 2012 and 24 December 2012. Table 1 presents basic statistics of Web performance for considered servers. The largest span of data occurred in February, at 6:00 pm where the difference between minimum and maximum value equaled 16.62 s. High values of variability coefficient and standard deviation, especially for 6:00 am and 6:00 pm in February, indicate the variability of the examined process. What is more, high values of kurtosis and skewness coefficient for these hours show right asymmetry of the examined phenomenon. The highest average and median occurred at 12:00 pm in both considered time periods. This can be easily justified; namely, this is an ultimate heyday – people are at work/school and by surfing the Web, they generate a lot of network traffic. Finally, it may be noted that for every considered hour, the average times are greater in December, 2012 which may indicate a slight deterioration in Web performance.

Table 1. Basic statistics of download times from evaluated web servers; measurements taken by the Wrocław agent

Statistical parameter	February, 2009			December, 2012		
	6:00 am	12:00 pm	6:00 pm	6:00 am	12:00 pm	6:00 pm
Minimum value X_{min} [s]	0.12	0.16	0.13	0.10	0.11	0.10
Maximum value X_{max} [s]	7.68	5.78	16.75	4.99	4.65	4.75
Average value $\overline{X}$ [s]	0.77	1.18	1.04	1.33	1.41	1.41
Median [s] M	0.36	0.55	0.47	0.57	0.85	0.72
Standard deviation S [s]	1.11	1.38	2.12	1.47	1.45	1.50
Variability coefficient V [%]	144.16	112.71	203.85	170.43	107.99	106.58
Skewness coefficient G	4.37	1.98	6.03	1.58	1.46	1.39
Kurtosis coefficient K	24.85	6.12	43.34	1.16	0.85	0.54

Calculation of Gaussian anamorphosis is the first step after making the preliminary data analysis. To calculate Gaussian transformation frequency, the inversion model was used and the number of adopted Hermite polynomials was equal to 100. Due to its extensiveness, Gaussian anamorphosis is not discussed in this paper. One may find, however, more information in [15]. For all considered hours, the directional variograms along the time axis were calculated. The distance classes for these variograms equaled 1.31 ° and 0.89 ° for February and December respectively. The next step in structural data analysis is modeling of a theoretical variogram function. Table 2 presents all models that were used to approximate the directional variograms and Fig. 1 shows two examples of theoretical variogram models that have been used in this paper.

Table 2. Models used to approximate the directional variograms

February, 2009		December, 2012	
Hour	Theoretical model	Hour	Theoretical model
6:00 am	Nuggets effect	6:00 am	Gamma and J-Bessel
12:00 pm	Nuggets effect and J-Bessel	12:00 pm	J-Bessel
6:00 pm	Nuggets effect and J-Bessel	6:00 pm	K-Bessel

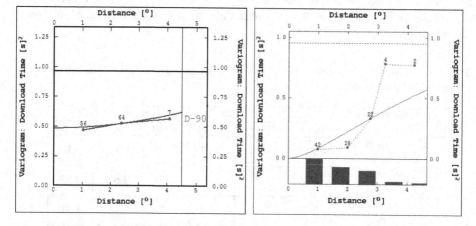

Fig. 1. Directional variograms of download times calculated along the time axis, and their theoretical models: (left) 6:00 pm, measurements taken in 2009; (right) 6:00 pm, measurements taken in 2012

5   Web Performance Evaluation

The aforementioned theoretical variogram models together with simulation type (block or punctual) as well as neighbourhood type (moving or unique) were used to create forecasts models. For every considered case, 100 simulations were conducted. What is more, for both, February and December, punctual type of simulation and moving neighbourhood were used. Additionally, in case of February, the search ellipsoids for the moving neighbourhood equaled 23.15°, 11.36° and 22.92° for 6:00 am, 12:00 pm and 6:00 pm respectively. In case of December, the search ellipsoids equaled 22.85° for 6:00 am and 16.89° for both 12:00 pm and 6:00 pm. Subsequently, these forecasts models were used to make spatio-temporal analyses and forecasts of Web performance. Forecasts were made for two consecutive Mondays, between 1 March 2009 and 14 March 2009 (by using the data collected by Wrocław agent in February) and for two consecutive

Mondays, between 25 December 2012 and 7 January 2013 (by using the measurements taken in December).

Table 3 presents global statistics of conducted forecasts. In case of February, the highest values of mean, maximum and mode were surprisingly forecasted for 6:00 pm (compare with the Table 1). In case of December, situation is slightly different; namely, the lowest Web performance was observed at 12:00 pm. Additionally, by analysing differences in mean value and mode, one may notice a slight decrease in Web performance in the last three years; once again, the case is somewhat different at 12:00 pm.

Table 3. Global statistics for the forecasted download times

Parameter Mean Forecasted value $\overline{Z}$	Min. value $Z_{min}[s]$	Max. value $Z_{max}[s]$	Mean value $Z[s]$	Mode **M** [s]	Median M [s]	Variance S^2 $[s]^2$	Standard deviation [s]	Variance coefficient $V[\%]$
Wrocław agent; 01.03.2009-04.03.2009								
for 6:00 am	0.26	2.56	0.59	0.34	0.53	0.05	0.21	35.59
for 12:00 pm	0.26	4.47	1.51	0.68	1.38	0.47	0.69	45.70
for 6:00 pm	0.21	6.93	1.83	0.9	1.58	1.17	1.08	59.02
Wrocław agent; 25.12.2012-31.12.2012								
for 6:00 am	0.13	4.90	1.60	0.64	0.99	1.70	1.30	81.25
for 12:00 pm	0.11	4.65	1.57	4.65	1.03	1.67	1.29	82.16
for 6:00 pm	0.12	4.79	1.56	0.32	0.98	1.73	1.31	83.97

Table 4. Sample results of forecasts for web servers in Budapest, Hungary and Eindhoven, the Netherlands

Evaluated web server	Date	Actual download time [s]	Forecasted download time [s]	Ex-post forecast error [%]
Budapest, Hungary	02 Mar 2009, 6:00 am	0.31	0.33	6.71
Budapest, Hungary	09 Mar 2009, 6:00 am	0.45	0.34	24.44
Budapest, Hungary	31 Dec 2012, 6:00 am	0.31	0.24	22.58
Budapest, Hungary	07 Jan 2013, 6:00 am	0.27	0.23	14.81
Eindhoven, the Netherlands	02 Mar 2009, 6:00 pm	0.44	0.43	1.97
Eindhoven, the Netherlands	09 Mar 2009, 6:00 pm	0.52	0.44	15.38
Eindhoven, the Netherlands	31 Dec 2012, 6:00 pm	0.75	0.83	10.67
Eindhoven, the Netherlands	07 Jan 2013, 6:00 pm	0.69	0.85	23.19

Table 4 presents sample forecasts results for web servers in Budapest and Eindhoven. One can observe that forecasts made using TB are fairly accurate. Nevertheless, due to the large spread between evaluated Web servers in the examined area and high dispersion of the input data not all results can be described with unerring or at least acceptable accuracy[3].

---

[3] It is assumed that acceptable forecast error is within 20%.

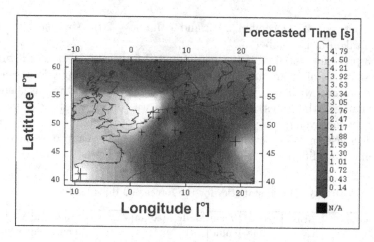

Fig. 2. Sample raster map showing forecasted download times on Monday, January 7, 2013, at 6:00 pm

Sample forecasts results for the whole considered area are presented as a raster map in Fig. 2. Crosses shown on the map represent the examined servers and the size of these crosses corresponds to the actual times needed to obtain the rfc file from evaluated web servers. The server with the lowest performance (the largest download time) is located in Porto, Portugal; forecasted download time for this server equaled 4.79 s.

6   Conclusions

This paper presented TB and MWING system as invaluable tools for analysing and forecasting Web performance. The presented herein innovative approach may be helpful in analysing and forecasting Web performance (or volume of network traffic), especially when one has to simultaneously deal with the temporal and spatial aspects of the studied phenomenon. In the future, TB may not only prove to be an indispensable way for discovering knowledge by network administrators, but it may also help them to improve the quality of the services they provide.

TB and MWING can also be successfully used to to compare Web performance in different time periods. In this paper, it has been shown that in general one may notice a slight decrease in Web performance for the considered web servers. Generally, by making forecasts with an acceptable accuracy, TB seems to be a potentially useful method for forecasting. Nevertheless, for a few web servers forecast error turned out to be unacceptable, so a need to improve accuracy of forecasts still exists. This improvement could be achieved by making forecasts for different scenarios, varying not only in the length of time horizons or the number of evaluated servers, but also in the size of considered area.

References

1. Mirza, M., Sommers, J., Barford, P., Zhu, X.: A machine learning approach to TCP throughput prediction. IEEE ACM T. Network 18(4), 1026–1039 (2010)
2. Karrer, R.: TCP prediction for adaptive applications. In: Proc. 32nd IEEE Conference on Local Computer Networks, pp. 989–996 (2007)
3. He, Q., Dovrolis, C., Ammar, M.: On the predictability of large transfer TCP throughput. Comput. Netw. 51(14), 3959–3977 (2007)
4. Yin, D., Yildirim, E., Kulasekaran, S., Ross, B., Kosar, T.: A data throughput prediction and optimization service for widely distributed many-task computing. IEEE Trans. Parall. Distr. 22(6), 899–909 (2011)
5. Kamińska-Chuchmała, A., Wilczyński, A.: Application Simulation Methods to Spatial Electric Load Forecasting. Rynek Energii 1(80), 2–9 (2009)
6. Borzemski, L., Kamińska-Chuchmała, A.: Client-Perceived Web Performance Knowledge Discovery through Turning Bands Method. Cybern. Syst. 43(4), 354–368 (2012)
7. Borzemski, L., Kamińska-Chuchmała, A.: Distributed Web Systems Performance Forecasting Using Turning Bands Method. IEEE. Trans. Ind. Inform. 9(1), 254–261 (2013)
8. Borzemski, L., Cichocki, Ł., Fraś, M., Kliber, M., Nowak, Z.: MWING: a multi-agent system for Web site measurements. In: Nguyen, N.T., Grzech, A., Howlett, R.J., Jain, L.C. (eds.) KES-AMSTA 2007. LNCS (LNAI), vol. 4496, pp. 278–287. Springer, Heidelberg (2007)
9. Borzemski, L., Cichocki, Ł., Kliber, M.: Architecture of Multiagent Internet Measurement System MWING Release 2. In: Håkansson, A., Nguyen, N.T., Hartung, R.L., Howlett, R.J., Jain, L.C. (eds.) KES-AMSTA 2009. LNCS, vol. 5559, pp. 410–419. Springer, Heidelberg (2009)
10. Borzemski, L.: The experimental design for data mining to discover web performance issues in a Wide Area Network. Cybern. Syst. 41(1), 31–45 (2010)
11. Lantuejoul, C.: Geostatistical Simulation: Models and Algorithms. Springer (2002)
12. Matheron, G.: Quelques aspects de la montée. Internal Report N-271, Centre de Morphologie Mathematique, Fontainebleau (1972)
13. Matheron, G.: The intrinsic random functions and their applications. JSTOR Advances in Applied Probability 5, 439–468 (1973)
14. Borzemski, L., Danielak, M., Kaminska-Chuchmala, A.: Short-Term Spatio-temporal Forecasts of Web Performance by Means of Turning Bands Method. In: Nguyen, N.-T., Hoang, K., Jędrzejowicz, P. (eds.) ICCCI 2012, Part II. LNCS, vol. 7654, pp. 132–141. Springer, Heidelberg (2012)
15. Wackernagel, H.: Multivariate Geostatistics: an Introduction with Applications. Springer, Berlin (2003)

A Note on the Local Minimum Problem
in Wireless Sensor Networks

Adam Czubak

Institute of Mathematics and Computer Science,
Opole University, Opole, Poland
adam.czubak@math.uni.opole.pl

Abstract. The Local Minimum Problem occurs in geographic routing
scenarios. In this paper two solutions to this problem for certain network topologies are proposed. By using the notion of virtual coordinates
a theoretical and a practical constructions are presented. A distributed
algorithm for the practical approach is proposed.

Keywords: WSN, ad-hoc networks, local minimum.

1 Introduction

Geographical routing algorithms use geographical coordinates for routing purposes. These are the most scalable propagation algorithms available for computer
networks [1].

The greatest advantage of the above greedy algorithms is the fact, that these
require only the coordinates of neighboring nodes in order to perform routing
process. This characteristic limits the amount of routing messages immensely.

Nevertheless, there are network topologies in which the neighboring node chosen for propagation has no neighbor located closer to the destination than itself[1].
The case when this greedy criterion is not satisfied is named the local minimum
problem [2]. A visualization of this very problem is shown in Fig. 1.

There are many theoretical geographical routing algorithms proposed in the
literature [3–7]. But it turns out, that in practice some of these solutions do not
perform as expected as it was shown in [2]. Most of the solutions to the local
minimum problem are based on performing planarization of the WSN graph.
This process, in theory, is relatively simple and easy to implement in a distributed
fashion at the nodes. There are some important factors which are not taken under
consideration like [2]:

- In order for planarization to work, the nodes must have accurate information
  regarding their and their neighbors' localization.
- Any mobility of the nodes is not allowed.
- Every disappearance and appearance of a new node requires replanarization.

---

[1] There are many criteria, other than the Cartesian distance, according to which a node
is *closer* or *further* to the destination.

A. Kwiecień, P. Gaj, and P. Stera (Eds.): CN 2013, CCIS 370, pp. 64–75, 2013.
© Springer-Verlag Berlin Heidelberg 2013

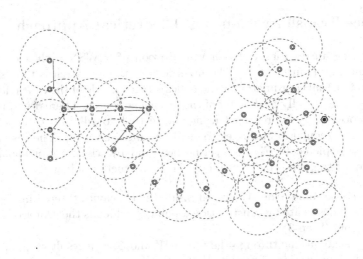

Fig. 1. Visualisation of the Local Minimum Problem

- The nodes must have a range in the shape of a perfect circle, which is never the case in WSNs.

Other solutions not involving troublesome planarization use notions of hyperbolic geometry [8] or differential geometry [9]. These two are highly theoretical, very advanced and not tested in real-world scenarios.

The aim of this paper is to give some insight into a solution to the local minimum problem which does not require planarization of the WSN [10] graph. It does not solve the local minimum problem for all network topologies, in particular non-regular, spiral and tunnel-like concavities, but gives valuable insight to the process of generation of virtual coordinates. Solutions applied in real-world scenarios cannot be over-complicated, to allow to perform a successful troubleshooting in case of an issue. The main goal of this paper is to propose a solution to the local minimum problem, which would also be relatively simple for a network engineer to troubleshoot.

2   The Applied Model

In the considered geographical routing scenario the accepted global criterion is the Cartesian distance between any two nodes. An addition restriction applies, which does not allow the nodes to change their locations. The definition follows.

For every node i there is a submodel, that consists of the neighboring nodes set $N(i)$:

$$fl_i = \sum_{i=1}^{n} \sum_{j=1}^{n} d(i,j)\alpha_{i,j}\left((x_i - x_j)^2 + (y_i - y_j)^2\right) \to \min \ . \tag{1}$$

3 Space Transformation – A Theoretical Approach

The first diagram in Figure 2 is a visualization of a WSN with a concavity and a local minimum inside it. The main idea of the approach is to designate a new set of virtual coordinates for the nodes located inside of the concavity. The nodes would be literally *pushed out* of the problematic area. Thereafter a regular routing process may be resumed using the virtual coordinates to choose the next hop and regular coordinates to send the message itself. The instructing steps below depicted in Fig. 2 describe how and by which factor should the coordinates of the nodes be pushed out. The construction follows:

1. Let's choose some node W located inside the concavity (Step 2 in Fig. 2).
2. Next let's designate by construction a point S, which is the center of gravity of the void (Step 3).
3. Draw a straight line through the points W and S, as a result we get a point P at the intersection of the line WS with the concavity (Step 4).
4. Now let's construct a straight line tangent to the void, thus closing the concavity (Step 5).
5. Point P' is the intersection of the constructed straight lines. Its coordinates are as follows:

$$P' = S + \overrightarrow{SP}\frac{|\overrightarrow{SP'}|}{|\overrightarrow{SP}|} \ . \tag{2}$$

The coordinates of the virtual point W' are the coordinates of the point W moved in relation to the center of gravity of the void by the vector $\overrightarrow{SP}$ scaled down by the ratio of the lengths of two vectors: $\frac{|\overrightarrow{SP'}|}{|\overrightarrow{SP}|}$.

6. Now let's designate virtual coordinates W' for the node W as follows:

$$W' = S + \overrightarrow{SW}\frac{|\overrightarrow{SP'}|}{|\overrightarrow{SP}|} \ . \tag{3}$$

The coordinates of every point in the concavity are moved in relation to the center of gravity of the void by the vector $\overrightarrow{SP}$ scaled down by the relation of the lengths of two vectors: $\frac{|\overrightarrow{SP'}|}{|\overrightarrow{SP}|}$.

It is important to mention, that the above construction applies to every point inside the concavity and behind it as well. The whole space behind the concavity has to be *pushed* as well.

The Figure 3 shows this transformation performed for a larger number of nodes and in Table 1 the values of node shifts from the figure. The columns of the Table 1 contain:

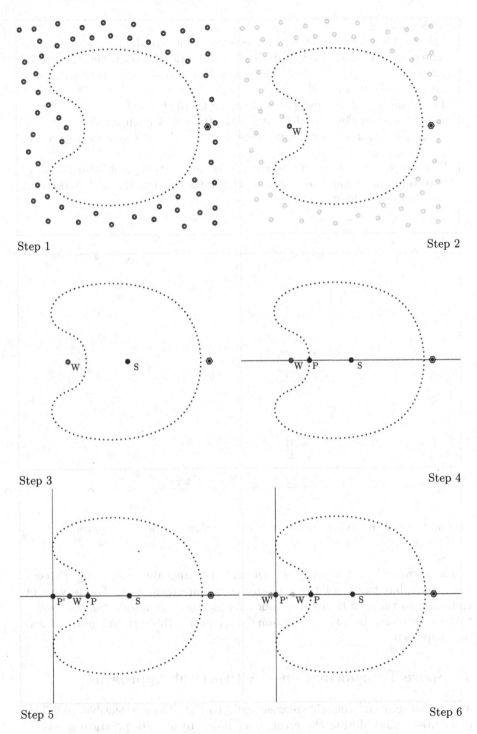

Fig. 2. Visualization of the theoretical approach to space transformation

1. Nodes' names.
2. Value of the coefficient by which the real distances of the points v_i from the center of the void will be multiplied. It is the ratio of two lengths: the stretch between center point S to border point P and the stretch between S and the constructed point P'.
3. The distance between point v_i and the center of the void.
4. The distance of the virtual point v_i' from the center of the void.
5. Value of the shift, ie. a difference between the real distance to the center and the virtual one.
6. Percentage value, stating by how much the new virtual point is further away from the center than the real one., ie. a ratio between the real distance to the center and the virtual one.

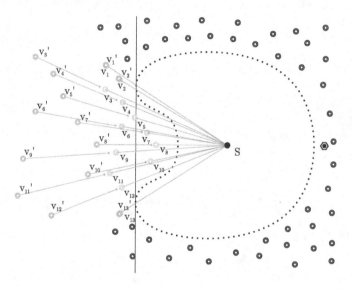

Fig. 3. Solution to the local minimum problem by the means of virtual coordinates

The above solution is a correct one and very appealing regarding it's continuousness but turns out to be insensible in implementation. The process of transforming the area behind the concavity may be very message intensive and take a rather long time. For this reason the solution in this section is a theoretical only approach.

4   Space Transformation – A Practical Approach

The above solution, thought geometrically correct is not a feasible one. The observation made during the creation of it led to a more practical solution, described in this section.

Table 1. The summary of the transformation of nodes from Fig. 3

Nodes	$\frac{|SP'|}{|SP|}$	$|Sv_i|$	$|Sv_i'|$	$||Sv|-|Sv'||$	$\frac{||Sv|-|Sv'||}{|Sv_i|}$
v_1	1.004	115.9	116.43	0.53	0.46 %
v_2	1.007	100.97	101.68	0.71	0.70 %
v_3	1.58	108.13	170.87	62.73	58.02 %
v_4	1.672	91.28	152.61	61.33	67.18 %
v_5	1.769	78.14	138.28	60.14	76.97 %
v_6	1.825	86.86	158.52	71.66	82.50 %
v_7	1.831	65.7	120.29	54.59	83.10 %
v_8	1.849	57.84	106.97	49.13	84.95 %
v_9	1.845	90.39	166.82	76.43	84.56 %
v_{10}	1.854	63.51	117.76	54.24	85.41 %
v_{11}	1.798	97.49	175.26	77.77	79.77 %
v_{12}	1.67	91.91	153.51	61.6	67.03 %
v_{13}	1.01	101.64	102.90	1.24	1.23 %

The goal was, to limit the number of generated virtual coordinates only to the nodes actually placed in the concavity and not the ones behind. The main idea is to place the nodes from the concavity on the circumference of a circle. It cannot be a straight line, because a line has a local minimum as well. The proposed solution is depicted in Fig. 4.

The construction is as follows:

1. Let's choose some node W which is located inside the concavity (Step 2 in Fig. 4).
2. Next let's mark the destination node (ie. a sink) B (Step 3).
3. As we draw a straight line through the points W and B, we get a point P on the intersection with the concavity (Step 4).
4. Now we set a point K located at the end of the concavity and construct a circle with the center P and a radius PK. Point K is actually the node which, as a next node to send data, will not use a node placed inside the concavity (Step 5).
5. Draw a straight line through points B' and W. The point W' of intersection with the circle is new virtual node W.

Figures 5, 6, 7 and 8 depict the above transformtion for a larger amount of nodes:

1. Figure 5 is the starting point.
2. Figure 6 shows the construction of a crircle.
3. Figure 7 shows how the new virtual coordinates are generated.
4. Figure 8 shows how the nodes are being pushed out of the concavity.

The above solution is feasible to implement in real-life applications thanks to the reduction of the area undergoing the transformation. The nodes are placed on the circumference of a circle, since a simple straight line would also pose a local minimum (in accordance to the model from Sect. 2). Furthermore, it

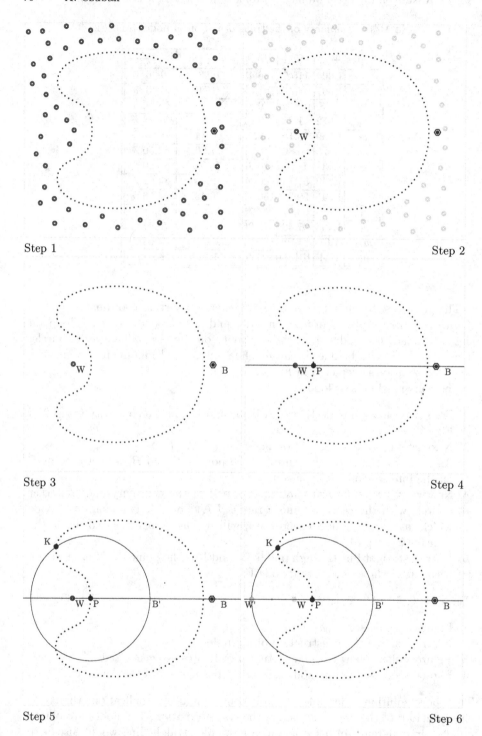

Fig. 4. Visualization of the practical approach to space transformation

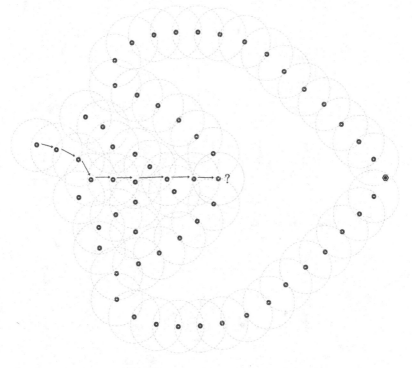

Fig. 5. Practical approach – network topology

is worth mentioning, that the area of the circle cannot contain the destination node for the same reason. The result would be a local minimum formed on the circumference.

5   A Distributed Algorithm for the Practical Approach to Space Transformation

From the implementation point of view again both, the virtual and the real sets of coordinates are required. Every node, which is located inside of the concavity in order to determine its virtual coordinates requires the location of the points P, K (or the radius of the constructed circle) and of the destination node B, which is known. Points P and K are determined as follows:

1. The node which is the local minimum sends two messages: one traverses the concavity according to the right hand rule and the second one according to the left hand rule.
2. Both messages keep track of the visited nodes.
3. When either of the two a messages encounters a node, which as a next hop does not have the node from which the message was last propagated it returns to the local minimum node. It is a probable point K.

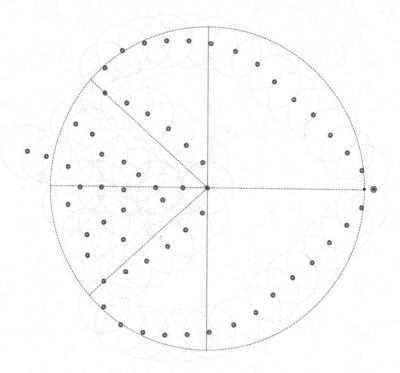

Fig. 6. Practical approach – constructing the circle

4. Now the local minimum node:
   (a) Upon receiving the returning messages chooses from the two possible K points the one located further away from itself.
   (b) Prepares a new message containing: coordinates of the point P, radius of the circle.
   (c) Floods the message to its neighbors.
5. The nodes located inside of the constructed circle calculate their virtual coordinates and propagate the message.
6. The nodes located outside of the constructed circle do not calculate new coordinates, but propagate the message to those nodes in their neighborhood that lie within the circle.
7. The routing process resumes with the help of virtual coordinates.

6 Communication Complexity, Time Complexity and Space Complexity of the Distributed Algorithm for Practical Approach

In this section the complexities of the solution proposed in Sect. 5 are given. The process that pushed the nodes onto a circumference is performed locally

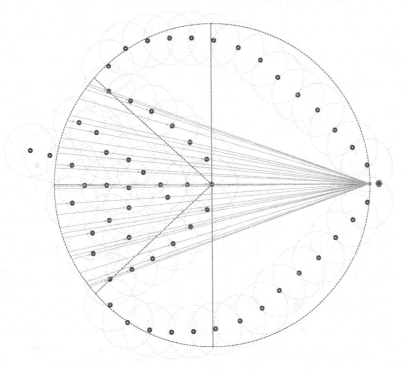

Fig. 7. Practical approach – intermediate point construction

and requires only local intersection between nodes within the concavity. For that reason the complexities refer not to the whole network $G = (V, E)$ but to the subgraph G^* containing the nodes and edges within the concavity.

Theorem 1. *For a given graph $G^* = (V^*, E^*)$ the algorithm has the communication complexity of $\mathcal{O}(|E^*|)$.*

Proof. The local minimum node sends two messages alongside the concavity, so no more than $|E^*|$. Next the messages return to the source giving $2 \times |E^*|$ messages. After that the flooding starts resulting in $2 \times |E^*| + |E^*|$ messages. Which asymptotically equals to $\mathcal{O}(|E^*|)$.

Theorem 2. *For a given graph $G^* = (V^*, E^*)$ the algorithm has the time complexity of $\mathcal{O}(|E^*| + diam(G^*))$.*

Proof. The local minimum node sends two messages alongside the concavity, to reach potential points K. These will reach the points no later than after $|E^*|$. Next the messages return to the source after $2 \times |E^*|$. After that with the flooding we get $2 \times |E^*| + diam(G^*)$. Which asymptotically equals to $\mathcal{O}(|E^*| + diam(G^*))$.

Theorem 3. *For a given graph $G^* = (V^*, E^*)$ the algorithm has the space complexity of $\mathcal{O}(1)$.*

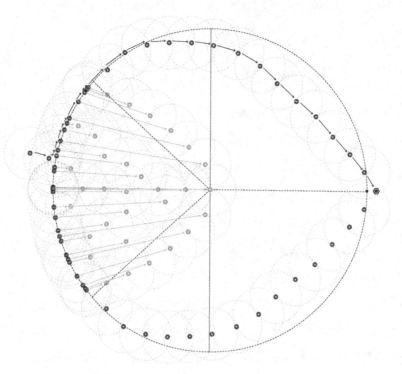

Fig. 8. Practical approach – designating virtual coordinates

Proof. Every node keeps track of its real and virtual coordinates, which gives the space complexity of $\mathcal{O}(1)$.

7   Summary

In this paper two solutions to the local minimum problem were proposed. The first one is a theoretical construction, based on the idea of transforming, *squeezing* the space inside the problematic area and the whole space behind it.

The second construction moves the nodes out of the concavity placing them onto the circumference of a circle. This method allows for limitation of the transformation to the area of concavity. A distributed algorithm was proposed and its complexities proven. It has the following characteristics:

- Requires only local interaction within the concavity;
- It is not required to assign virtual coordinates to every node in the network, only to those within the concavity;
- It is a relatively simple solution, allowing for a troubleshooting process;
- It does not solve the local minimum problem for all network topologies, in particular non-regular, spiral and tunnel-like concavities, but gives valuable insight to the process of generation of virtual coordinates.

References

1. Kranakis, E., Singh, H., Urrutia, J.: Compass Routing on Geometric Networks. In: Proceedings 11th Canadian Conference on Computational Geometry, pp. 51–54 (1999)
2. Kim, Y.-J., Govindan, R., Karp, B., Shenker, S.: On the pitfalls of geographic face routing. In: Proceedings of the 2005 Joint Workshop on Foundations of Mobile Computing, DIALM-POMC 2005, pp. 34–43. ACM, NY (2005)
3. Takagi, H., Kleinrock, L.: Optimal Transmission Ranges for Randomly Distributed Packet Radio Terminals. IEEE Transactions on Communications COM-32(3), 246–257 (1984)
4. Hou, T.C., Li, V.O.K.: Transmission Range Control in Multihop Packet Radio Networks. IEEE Transactions on Communications COM-34(1), 38–44 (1986)
5. Stojmenovic, I., Lin, X.: Power-Aware Localized Routing in Wireless Networks. IEEE Transactions on Parallel And Distributed Systems 12(10), 1122–1133 (2001)
6. Stojmenovic, I., Lin, X.: Loop-Free Hybrid Single-Path/Flooding Routing Algorithms with Guaranteed Delivery for Wireless Networks. IEEE Transactions on Parallel And Distributed Systems 12(10), 1023–1032 (2001)
7. Finn, G.: Routing and addressing problems in large metropolitan-scale internetworks, University of Southern California. Information Sciences Institute, Marina del Rey (1987)
8. Kleinberg, R.: Geographic Routing Using Hyperbolic Space. In: INFOCOM, pp. 1902–1909 (2007)
9. Sarkar, R., Yin, X., Gao, J., Luo, F., Gu, X.D.: Greedy routing with guaranteed delivery using Ricci flows. In: Proceedings of the 2009 International Conference on Information Processing in Sensor Networks, IPSN 2009, pp. 121–132. IEEE (2009)
10. Czubak, A., Wojtanowski, J.: On applications of wireless sensor networks. In: Tkacz, E., Kapczynski, A. (eds.) Internet – Technical Development and Applications. AISC, vol. 64, pp. 91–99. Springer, Heidelberg (2009)

Lessons Learned from the Deployment of Wireless Sensor Networks

Tomasz Surmacz, Mariusz Słabicki,
Bartosz Wojciechowski, and Maciej Nikodem

Wrocław University of Technology
Institute of Computer Engineering, Control and Robotics
Wybrzeże Wyspiańskiego 27, 50-370 Wrocław
tomasz.surmacz@pwr.wroc.pl

Abstract. Many theoretical works focus on maximizing the lifetime of a measurement-gathering sensor networks by researching different aspects of energy conservation or details of self-organizing network algorithms. In our practical deployment of such a network we learned that software and hardware reliability, as well as anticipation of worst-case scenarios, are the equally important factors for successful experiments. We describe our experiences with implementing a WSN for long-time unattended operation in a greenhouse data-collecting application.

Keywords: wireless sensor networks, greenhouse monitoring, measurements.

1 Introduction

Due to their intrinsic properties Wireless Sensor Networks (WSNs) have a wide range of practical applications in which they outperform traditional cable networks in terms of cost, deployment time, and flexibility. This is one of the reasons for high interest in research on WSN design and architecture. Nevertheless, much work is still needed to bridge the gap between theoretical or simulation-based works and practical tests in real-world conditions. In this paper we provide insights and conclusions drawn from our preliminary deployment of WSN network.

One typical application of WSNs is climate monitoring and control inside a greenhouse. Greenhouse is a structure isolated from the outside world and adverse influence of external environmental factors. This isolation allows controlling climate parameters inside and optimising agricultural production – lowering production cost and increasing yield at the same time [1]. In industrial greenhouses the problem of maximising crop growth is reduced to two separate problems – controlling the climate and *fertirrigation* (fertilisers and irrigation). Climate control focuses on ensuring optimal conditions for plant growth and photosynthesis. Photosynthesis is stimulated by fragment of light spectrum (wavelengths from 400 to 700 nm) but its rate depends also on the proper temperature level. Therefore light and temperature control dominate the climate control in the greenhouses. Additional parameters monitored include humidity and CO_2 levels,

A. Kwiecień, P. Gaj, and P. Stera (Eds.): CN 2013, CCIS 370, pp. 76–85, 2013.
© Springer-Verlag Berlin Heidelberg 2013

which are tightly related to temperature and the amount of sunlight radiation. Unfortunately, a greenhouse is not perfectly isolated and is under adverse influence of external factors: temperature, humidity, wind and solar radiation [2]. Consequently, precise monitoring and control of the greenhouse climate allows to minimise resource utilisation and reducing the operation costs while maximising the yield.

The ultimate goal of our research is to deploy WSN-based monitoring system to provide greenhouse owners with detailed measurement data and assist in cultivation of plants. From the computer scientist's perspective this is also an opportunity to deploy and verify various WSN architectures, protocols and algorithms in real life application and environmental conditions as well as gain experience in deciding on proper network architecture.

2   Related Work

There is a lot of research in recent years regarding different applications in which WSNs are useful. Networks are developed and tested in military, healthcare, road monitoring, and number of other applications. One of the most popular WSN application is *precision agriculture*, both in greenhouse or open field [2]. Unfortunately, developing WSN for real-world precision agriculture application is not a trivial problem. Therefore, despite the extensive research there are still many open issues which need to be resolved. Consequently, the knowledge about "how to deploy a network" is still incomplete among researchers.

In [3] authors addressed many problems with developing and testing network composed of more than 100 nodes. To avoid test-phase problems, off-the-shelf and verified components were used (i.e. MicaZ nodes and the TinyOS software from Berkeley University). Even so, the network design required a lot of corrections, e.g. achieving deep sleep modes proved to be far from trivial. In the experiment aftermath, only 2 % of all messages were collected by the base station. Authors conclude that it happened because the new deployment was not sufficiently tested but only large-scale and long lasting tests allowed to find some of the problems. The authors also note that debugging complex stack of network protocols requires well prepared infrastructure (packet sniffing and reports from each protocol layer) as well as deep understanding of protocol internal operation.

Some recently published works address the practical deployment problems [4,5,7,6]. Paper by Jelicic et al. [4] describes monitoring system for olive grove. The proposed deployment is based on ZigBee network but uses complex nodes (e.g. equipped with camera) and external power supply (e.g. solar panels). Consequently the network is not a low-power and requires high throughput to accommodate large volumes of data. Paper by Xia et al. [5] addresses problems of poor real-time data acquisition, small monitoring area and involvement of manpower. They propose a complex monitoring system based on JN5121 nodes and proprietary data frames. The proposed solution was verified in a network of 9 nodes only and tests were focused on ensuring data acquisition rather than network operation. Sugar cane farm monitoring was analysed in [6] in order to investigate how irrigation practices affect the environment and cultivation. Authors used 915 MHz radio transceivers to cover large area with a few nodes only. Literature review reveals that in some cases network deployments were successful, but most of the presented examples can be just considered an important lesson. A deeper survey may be found in [7] where a number of real-life WSNs are analysed and conclusions are drawn. The article provides rules how to deploy networks to overcome reliability and prototype testing issues. The general conclusion is that even though propagation and energy consumptions models are usually known and carefully selected, the actual deployment still poses a lot of difficulties. These in turn affect the effectiveness of the selected solutions and there is a gap between what developers expect and what reality provides [8].

3   Testing Setup

For our test deployment we used the well known and widely used TelosB nodes. They use protocol stack based on ZigBee and IEEE 802.15.4 standards and operate in 2.4 GHz wireless band (shared also by WiFi). We used four variants of the nodes, developed by Advanticsys (XM1000, CM3300, CM4000 and CM5000) that differ in antenna type (external 5 dBi or PCB), on-board sensor configuration and integrated programmer/USB interface. An example of two different node configurations can be seen in Fig. 1(a) and 1(b) Since all the nodes share the same programming environment, we find it beneficial to be able to compare e.g. communication ranges between different antennae types. Each wireless node is equipped with a set of sensors for measuring temperature, humidity and luminance in visible light spectrum (with 560 nm peak sensitivity wavelength), which are the base for environment monitoring in greenhouses [2]. The nodes operate on two AA rechargeable batteries with typical capacity of 2000 mAh. A few selected nodes were also equipped with CO_2 sensors, however their use was limited by the fact that the sensors required a separate 9 V energy source.

Our test network was deployed in a relatively small greenhouse complex that is used for both teaching and research by the Wrocław University of Environmental and Life Sciences. It consists of a set of adjacent greenhouses, measuring 30×10 m each (Fig. 2). Each wireless node (T05, S42, etc.) was fixed to the support wires located approx. 3 m above the ground (see Fig. 1(a)). One base station (PI07) was placed in greenhouse B6, approx. 0.2 m above the ground. The other one (PI06) was located in the corridor, approximately 0.6 m above the ground. This resulted in better communication conditions between relatively distant nodes (as they maintained the Line of Sight (LoS)), than between the nodes and the BS.

Each node was programmed with our custom *GreenhouseMonitor* application running under TinyOS. We used the standard Medium Access Control (MAC) implemented in TinyOS with the Low Power Listening (LPL) protocol as a way to conserve energy. Using LPL in our setup allowed to extend useful unattended operating time of the network from 4 to over 18 days. Methods of further network lifetime extension (e.g. using deep sleep modes, turning off LEDs or using more efficient routing) did not serve the goal of learning the network operation at the

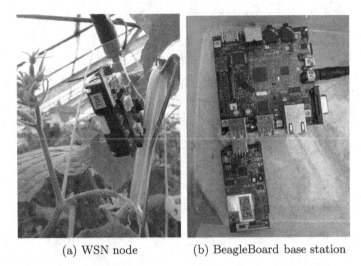

(a) WSN node   (b) BeagleBoard base station

Fig. 1. Elements of the WSN network installed in a greenhouse

target site and will be used in future tests. To avoid making wrong assumptions about network operation we used flooding as message routing method. This also provided an opportunity to analyse which routing paths were used the most. Each node was programmed to take measurements from all of its sensors and generate a message every 30 s. This is much more frequent than necessary, considering the rate of change of conditions in the greenhouse, but allowed us to gather large amount of data for analysis.

4 Datasink Architecture

The goal for the experiments described below was to create a setup for a data-gathering system that could be deployed in the remote greenhouse and would allow unattended data acquisition from wireless sensor nodes for at least 2–3 weeks (or possibly much longer). For collecting the measurement data (i.e. the data sink) we have tested different setups, including laptop computers, however, for unattended operation, the best choice is a small embedded system (Fig. 3).

In our research we have experimented with BeagleBoard and Raspberry Pi architectures. The BeagleBoard xM is a 3-year old embedded system platform running on OMAP ARM Cortex-A8 processor and costing around $125. The Revision-B board includes an ethernet port and 4 USB ports, allowing connection of external harddrives or WiFi dongles and it runs the OMAP version of the Linux operating system. Raspberry Pi is a fairly recent development board including the ARM11 processor, 512 MB of RAM, ethernet port and 2 USB ports, running Raspian Linux. It costs around $35.

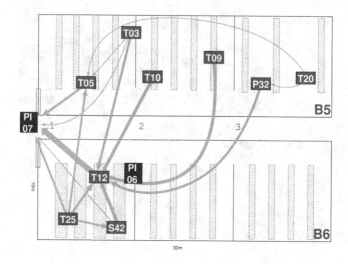

Fig. 2. Node placement at the greenhouses and the main paths to base station Pi07

Fig. 3. Architecture of the masurement system

One of the most important problems with our system running on BeagleBoard was the filesystem instability. Although the general practice with all Unix systems is to never power down the system abruptly, without the proper shutdown, sometimes such power-downs are unavoidable. In our case we have twice lost a week worth of measurement when the whole filesystem was corrupted beyond repair. Hard resets or power-downs are hard to avoid in unattended system operation and may be caused by many reasons – power outages, greenhouse workers temporarily disconnecting the system, or finally – inability to properly shutdown the system with no network connection and no input devices. Inability to ensure data safety in such cases disqualified BeagleBoard systems for our use.

With Raspberry Pi platform, our biggest problem was the USB support of the Raspian Linux distribution – the system was freezing as soon as the wireless node was accessed through the USB port. Research of various discussion fora revealed that it is a known problem, with the general consensus of blaming it on the outdated chip in the RS232-to-USB converter. Since the FT232BM chip is an integral part of the TelosB programmer, the only resolution to this problem was to change the booting parameters of the Raspian system to slow down the transfer rate on the USB bus. The proper solution however would be to rewrite the kernel driver for the FT232 chip to behave properly and not to cause the system panic. The conclusion from this experience is that even though the Raspberry Pi platform is a fairly modern one and with a great potential for growth, there is a general lack of understanding in the open systems developers community, that not only the newest and shiniest gadgets should get their support, but also some older hardware which may still be in use.

Another problem was the lack of permanent internet connection in the greenhouse installation. All modern distributions of Linux depend heavily on internet access for system updates and software installation, but it should be possible to run them as isolated systems once we have a desired system version. However, it took a lot of effort to configure the system for such a way of operation. The symptoms of malfunction included losing the networking card setup if the DHCP server was not available (thus, making the system inaccessible even over the peer-to-peer crossed-over cable connection with a laptop computer), inability to boot properly without the network access, or losing the local time during operation, even if it was properly set at the booting time. With the lack of a proper RTC hardware, this posed another category of problems as the data gathered has to be properly timestamped.

5 Observations

We have run several experiments from July 2012 until now (January 2013), but in this paper we focus on results obtained from the data collected between 4th and 22nd of January 2013 in two different greenhouses (Fig. 2). During that time succulents and herbs (section 1 and 2) and tropical plants (section 3) grew in greenhouse B5. At the end of 2012, section 1 of the greenhouse B6 was filled with tomatoes seedlings while other sections were not used for cultivation.

One aim of deploying wireless sensor network was to estimate its reliability, range and routing paths of radio communication. For that purpose, every node appended routing path information to the messages it sent. Table 1 presents ratios of messages that originated from each node and reached the base station with direct transmissions (when packets were received at the base station directly from the source node) and indirect transmissions (when packets were received only when retransmitted by other nodes). We also counted percentage of missing packets based on sequential numbers. Since broadcast communication was used, the same packet could reach the base station both directly and indirectly, therefore the sum of percentages may exceed 100 % – this is especially true for nodes located close to the base station (i.e. T25, S42, T3).

Routing path analysis clearly shows that no packet from nodes located in greenhouse B5 reached the base station PI06 directly. This is caused by the fact that heating and control equipment is located in-between both greenhouses and effectively blocks radio communication to the PI06 located 0.2 m above the ground. However, there are almost no obstacles between nodes located in greenhouses B5 and B6 as they are located 3 m above the ground. Consequently,

Table 1. Percentage of messages that reached each of the base stations

| Source node | BS 06 | | | BS 07 | | |
| | Percent of messages | | | Percent of messages | | |
	direct	indirect	missing	direct	indirect	missing
T03	0.00	84.66	15.34	29.01	69.12	24.24
T05	0.00	0.36	99.64	62.30	0.30	37.69
T09	0.00	61.14	38.86	14.18	54.84	39.67
T10	0.00	85.78	14.22	11.63	54.12	43.12
T20	0.00	0.26	99.74	0.00	0.50	99.50
P32	0.00	34.69	65.31	0.08	21.20	78.80
T12	71.73	0.09	28.27	65.15	7.92	34.62
T25	99.50	99.48	0.10	89.22	89.39	7.92
S42	71.78	71.64	28.11	54.93	62.51	35.19

nodes T12 and T25 receive packets from nodes in B5 and retransmit them to the base station PI06. This explains the big number of indirect transmissions to PI06. Also, only 0.26 % of packets from node T20 reached the PI06 (through either T12 or T25). Communication link reliability between T20 and T12, T25 is low due to distance and obstacles, so only a few hundreds of packets from node T20 were successfully received over the whole period of 18 days. Figure 2 presents node placement and main routing paths from all the nodes to base station PI07 with thickness of arrows corresponding to the number of messages. Most of the messages received at PI07 were actually transmitted through node T12. This is a consequence of two facts: first, nodes in B6 greenhouse used transmission power that was 15 dB higher than power used by nodes in B5 greenhouse; second, node T12 was the only node that had LoS visibility to the base station PI07. Higher transmission powers also enabled nodes T12, T25 and S42 to directly reach the base station more often then in case of nodes T03 and T05. For the same reason, significantly more packets from distant nodes (T09, T10, T20 and P32) were routed successfully through node T12 (routing through other nodes, e.g. P03 and P05 failed mostly due to lower transmission power).

Figure 4 presents received signal strength indicator value (RSSI) of packets received at the base station PI07. RSSI parameter estimates radio signal power at receiver and depends mostly on transmission power, propagation conditions and distance. We expected that RSSI will change periodically with daily and weekly cycles due to people working in the greenhouses, but we could not find any correlation. Although RSSI values differ for each node (e.g. as a result of different transmission paths, distances and obstacles) radio propagation changes in similar way for all nodes. When using RSSI measurements to adjust transmission power, one needs to disregard small oscillations (of 1 or 2 dBm) that most likely result from inaccuracy of measurement and not the radio propagation itself.

We have also used nodes to measure environmental conditions – temperature, relative humidity and light level. Figure 5 presents light level and temperature variations over three day period in two different greenhouses. It can be seen that temperature values exceeding 23 °C on 5th and 7th of January correlate with

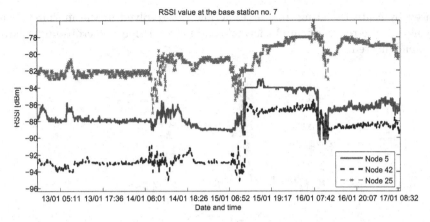

Fig. 4. Received signal strength indicator at base station PI07 from selected nodes

high levels of light. In fact these two days were sunny and direct solar radiation heated the sensors. On the contrary, there is no increase in temperature during January 6th and the light level is two orders of magnitude smaller as weather was cloudy on that day. Oscillations of temperature level (especially during the night) result from cyclic operation of central heating system. Small difference (1–1.5 °C) between temperatures in both greenhouses results from different setup of heating system – increased temperature is provided to seedlings in B6 greenhouse.

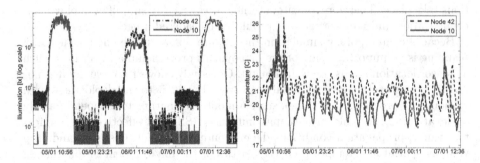

Fig. 5. Light level (left) and temperature (right) measured in two different greenhouses over the same period of time. Note that light level is in log scale.

The light level plot also reveals increased light level between midnight and morning in greenhouse B6 (node S42). This is due to additional illumination provided during the night in this greenhouse for stimulated grow of plants. Another interesting observation was that constant environment monitoring can be used for alarming in critical situations. Figure 6 shows a part of temperature profile collected during four days in November 2012. Normally the automated heating was supposed to keep the temperature around 23 °C. However, a significant drop in temperature can be observed during the second night. It was caused by central heating unit malfunction and resolved in the morning. If it was not for the relatively mild thermal conditions outside, the crops might have been endangered.

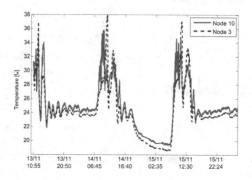

Fig. 6. Temperature in the greenhouse in a 4 days period

6   Conclusions

When deploying a network that should run unattended for long time, reliability of the whole solution is the key factor. In most cases, remote access to network is difficult, relatively expensive and often non trivial. When available, remote access allows debugging the network and taking actions when unexpected situations take place. Therefore, on-site internet access greatly simplifies deployment, configuration and monitoring, and should be considered a necessity.

Because contemporary simulation models are not accurate enough, network planning is an important part of the deployment process and all decisions (such as sensor locations) must be well thought. Our study shows that usually it is not possible to setup the network in laboratory and move it to the deployment area without any problems. This is so, since a number of parameters depend heavily on location, e.g. heights of antennas influence RSSI and effective communication range, propagation conditions depend on environment, obstacles and crop volume, spatial and temporal characteristics of the environment is different in laboratory and inside a greenhouse. We have observed some of these issues in our deployment even though it was thoroughly tested in laboratory: increased packet loss, difficulty in predicting routing paths, adverse interference from external environmental factors and also some seemingly trivial problems, such as the lack of power outlets to power the base stations.

Our experiments show that there is an increasing need for improved tools for network planning. Such tools should not only use mathematical models of typical environments but also capture a variety of different aspects of radio transmission and wireless network deployment. Until then networks need to be planned for worst case scenario and thoroughly verified in the field before they are put in operation. This is crucial in precision agriculture applications but also in any application where failures may lead to tremendous and irreparable losses.

During the deployment many typical engineering problems and challenges appeared (e.g. weak OS support for embedded hardware) which added extra workload to the team. Such problems are not possible to predict based only on experiences with simulation-based research. It also follows from our study that WSN debugging and testing is difficult. Mainly due to the distributed nature of the system and constrained capabilities of a single node (limited memory, computational power, lack of remote access). Valuable simplification in this process was achieved by turning selected nodes into promiscuous mode. This allowed us to use dedicated software to sniff and debug all nearby communication. This gave us an insight into intra-node communication and operation.

The experiences described here will allow for more complex deployment with dedicated and reliable routing. Our aim is to run the network for long period of time starting from early spring until the end of crop's vegetation cycle.

Acknowledgment. This work was supported by National Science Centre grant no. N 516 483740.

The authors would like to thank Dr Piotr Chohura of Wrocław University of Environmental and Life Sciences for providing access to the greenhouses and valuable insights in the field of precision agriculture.

References

1. Blackmore, S.: Precision farming: an introduction. Outlook on Agriculture 23(4), 275–280 (1994)
2. Berezowski, K.: The landscape of wireless sensing in greenhouse monitoring and control. International Journal of Wireless & Mobile Networks (IJWMN) 4(4), 141–154 (2012)
3. Langendoen, K., Baggio, A., Visser, O.: Murphy loves potatoes: experiences from a pilot sensor network deployment in precision agriculture. In: 20th International Parallel and Distributed Processing Symposium IPDPS 2006 (April 2006)
4. Jelicic, V., Razov, T., Oletic, D., Kuri, M., Bilas, V.: MasliNET: A wireless sensor network based environmental monitoring system. In: 2011 Proceedings of the 34th International Convention on MIPRO, pp. 150–155 (May 2011)
5. Xia, J., Tang, Z., Shi, X., Fan, L., Li, H.: An environment monitoring system for precise agriculture based on wireless sensor networks. In: 2011 Seventh International Conference on Mobile Ad-hoc and Sensor Networks (MSN), pp. 28–35 (December 2011)
6. Hu, W., Dinh, T.L., Corke, P., Jha, S.: Outdoor sensornet design and deployment: Experiences from a sugar farm. IEEE Pervasive Computing 11(2), 82–91 (2012)
7. Laukkarinen, T., Suhonen, J., Hamalainen, T., Hannikainen, M.: Pilot studies of wireless sensor networks: Practical experiences. In: 2011 Conference on Design and Architectures for Signal and Image Processing (DASIP), pp. 1–8 (November 2011)
8. Di Martino, C., Cinque, M., Cotroneo, D.: Automated generation of performance and dependability models for the assessment of wireless sensor networks. IEEE Transactions on Computers 61(6), 870–884 (2012)

Analysis and Optimization of LEACH Protocol for Wireless Sensor Networks

Błażej Adamczyk

Silesian University of Technology,
Institute of Computer Sciences
Akademicka 16, 44-100 Gliwice, Poland
Blazej.Adamczyk@polsl.pl
http://www.polsl.pl

Abstract. The growing popularity of wireless sensor networks increases the need for optimal and energy efficient routing protocols. Among many hierarchical routing protocols, the one which is really worth noticing is the Low Energy Adaptive Clustering Hierarchy (LEACH) protocol. Because of its simplicity and distributed operation it has been used as a base for several further protocols such as PEGASIS, HEED, TEEN and APTEEN. This article presents an analysis of the LEACH protocol and suggests a modification to improve its effectiveness by far.

Keywords: wireless sensor networks, LEACH, LEACH-S, routing protocol, energy-efficient.

1 Introduction

Wireless sensor networks often contain hundreds or even thousands of distributed sensors, which using the built-in battery collect data and send them to the base station. Such solution introduces one very important restriction which is the network lifetime. Usually the most energy demanding operation of such networks is the radio transmission and operation. Thus, a very effective way of optimizing the energy consumption can be the use of an optimal routing protocol. Different routing protocols for wireless sensor networks are divided into flat and hierarchical protocols (for more details see [1]). In flat protocols all nodes work in the same manner and the optimization of energy consumption is mainly based on data aggregation and elimination of data duplication. There is no division or grouping of sensors in such protocols what may not be optimal in some scenarios.

Hierarchical routing protocols, on the other hand, allow to decrease the energy consumption by creating clusters. Thanks to the division, majority of sensors send the information only to the closest cluster head[1]. Then, after collecting the information from all sensors within a group, the cluster head performs the data aggregation and finally sends it to the base station. Such protocols can also be nested creating a multi-hierarchical structure where a group of lower level clusters can form one higher level cluster and so on.

---

[1] Cluster Head – main sensor in a cluster.

A. Kwiecień, P. Gaj, and P. Stera (Eds.): CN 2013, CCIS 370, pp. 86–94, 2013.
© Springer-Verlag Berlin Heidelberg 2013

This article describes LEACH [2, 3] – one of the most popular hierarchical routing protocols. The performed analysis revealed its deficiencies and a possible improvement direction. This protocol has been analysed and extended in many other studies like [4–8] but none of them focuses on the problem with its randomness as it is done herein. Section 2 presents the way the LEACH protocol operates. Next, in Sect. 3, the problem related with a random nature of this protocol is described – LEACH choses cluster heads basing on random number generation done independently on each node what allows to create certain optimal number of clusters on average but not always. Further, Section 4 shows the proposed modification to overcome the mentioned problem. Section 5 and 6 contain results of performed simulations. Additionally the latter focuses on the influence of the number of clusters on the effectiveness of applied protocol. Finally, Section 7 gathers the conclusions.

2   LEACH Protocol

The first and one of the most popular hierarchical routing protocols for wireless sensor networks optimized for energy efficiency is the LEACH protocol. It divides the set of all sensors into groups, called clusters, which are organized around main nodes, called cluster heads. It then chooses different cluster heads in a cyclic manner to split the power consumption. All nodes belonging to the same group, are sending information to the cluster head which aggregates and sends the data to the base station afterwards.

LEACH operates in rounds, where each round consists of two phases: setup and working phase. In the setup phase each node decides whether it should be the cluster head or not and all cluster heads send advertment messages to non-cluster-head nodes to allow them to choose the appropriate head. Finally, the non-cluster-head nodes send messages to chosen heads in order to allow them to create the transmission schedule.

When all cluster heads have created the transmission schedule and distributed it accross all nodes belonging to their group the working phase begins. Thanks to the schedule all non-cluster-head nodes know exactly when to transmit and thus can sleep in the meantime to safe as much energy as possible. Finally, after recieving data from all the nodes within a schedule the cluster head aggregates it and sends it to the base station. The schedule repeats for defined time and then a new round begins.

The main adventage of LEACH protocol is the fact, it is completely distributed and all the nodes work independently of each other. No control information is required from the base station, and the nodes do not need any knowledge of the global network in order for LEACH to operate correctly.

The number of expected clusters in each round is specified as one of predefined parameters. As shown in [2] and [3] depending on the number and distribution of nodes the optimal number of clusters for the LEACH protocol might be different. As the algorithm is distributed the decision of becoming a cluster head has to be done indepedently. Thus, the authors have proposed a probability based approach. A node becomes a cluster head if a generated random number between 0 and 1 is less than a threshold $T(n)$ defined as:

$$T(n) = \begin{cases} \frac{P}{1-P(r \bmod \frac{1}{P})} & \text{if} \quad n \in G \\ 0 & \text{otherwise} \end{cases}$$

where P – the desired percentage of cluster heads in a round, r – current round number, and G is the set of nodes that have not been cluster heads in last $\frac{1}{P}$ rounds. As it can be observed such solution assures that all the nodes will become a cluster head once per $\frac{1}{P}$ rounds because the threshold grows with each round a node has not been a cluster head.

It is worth noticing that in border cases (i.e., when there are no cluster heads or when every node is a cluster head) the LEACH protocol works as there is no routing and all the nodes send information directly to the base station (so called direct communication).

3   LEACH Deficiency

As it was already stated, the energy efficiency of LEACH protocol strongly depends on the expected number of clusters (obviously, the same applies also to other hierarchical protocols). Thus, it is necessary to determine this number a priori. Unfortunately, because the cluster head decision is based on probability it does not guarantee creation of exactly the expected number of cluster heads in each round. This means that even if the expected number of cluster is set to the optimal value, the protocol will quite often choose different number of cluster heads decreasing the network energy efficiency.

The simulations in Sect. 5 will present cases where this effect is very noticable. It appears that the results of LEACH protocol would be better if it could guarantee that exactly the specified amount of clusters will be formed in each round. In the next section of this article, such modification will be proposed and further the results of an experiment will be presented to prove this assumption is correct.

There exist several other modifications of LEACH protocol like PEGASIS [4], HEED [5], TEEN [6], APTEEN [7] to only name a few. The Power Efficient Gathering in Sensor Information Systems (i.e., PEGASIS) protocol extends LEACH functionality by forming chains of sensors so that each node communicates only with its neighbors. Only one node in a chain is allowed to send data to the base station. Hybrid, Energy-Efficient Distributed Clustering (HEED) achieves the power balance by using residual energy and density as a metric for cluster selection. Such approach creates a well-distributed clusters and the decision is made according to network structure minimizing the communication costs. Finally the Threshold Sensitive Energy Efficient Sensor Network Protocol (TEEN) and later also APTEEN (Adaptive Periodic TEEN) introduce layered cluster structure. Similarily to LEACH nodes send data to their cluster heads, however the lower level cluster heads send the data further to higher level heads and so on. All these modifications improve the energy efficiency of LEACH by using additional information about nodes or network. In this paper we present a modification which allows to achieve better results but still maintain the LEACH simplicity.

4 Synchronous LEACH

The modification should fulfill two main requirements: guarantee the expected number of clusters in each round, and sustain all the mentioned LEACH protocol adventages like distributed and independent node operation. In this paper, we propose a modification of the cluster head decision proces as shown in Algorithm 1 and we called it – synchronous LEACH (LEACH-S).

Algorithm 1. LEACH-S

---

  function INITIALIZE

    ...

    Initialization of the random number generator with tha same seed on all nodes

    ...

  end function

  function DECIDECLUSTERHEAD

    for $i = 1 \rightarrow number\_of\_clusters$ **do**

      if $length(previous\_heads) = number\_of\_nodes$ **then**

        Clear $previous\_heads$

      end if

      repeat

        Draw id from $< 1; number\_of\_nodes >$

      until $ID \notin previous\_heads$

      Add $ID \rightarrow previous\_heads$

      if $node\_ID = ID$ **then**

        Mark current node as cluster head

      end if

    end for

  end function

The main idea behind this modification is to initialize the random number generator of all nodes with the same value and then perform the draws exactly in the same manner on all nodes. Such solution allows the nodes to operate independently but yet work in a synchronized manner. This way the expected number of cluster heads will be drawn in every round. The random number generator seed could be determined and distributed before the first round or could be defined statically at the time of compilation. The algorithm itself assures that each node will be cluster head the same number of times by maintaining the list of already choosen CHs. If the list contains all the available nodes it is cleared and all nodes can become cluster heads once again. This is important to distribute the energy consumption equally between the nodes as the cluster head has to aggregate and send data to the base station what costs more.

5   Simulation

To verify the correctness of the LEACH-S modification we have performed an experiment similar to the one presented in [2] and [3]. We used the network simulator 2 (ns-2 [9]) version 2.35 to perform the simulations. We have reused the energy model, the distribution of nodes and the location of base station. The simulation parameters are summarized in Table 1.

Table 1. Simulation parameters

Parameter	Value
Nodes	100
Area	100 m × 100 m
Base station location	(50, 175)
Initial node energy	2 J
Round time (for LEACH, LEACH-C and LEACH-S)	20 s
Radio speed	1 Mbps
Data size	2000 B

Figure 1 presents the network topology used for the simulation. It is exactly the same topology used in other LEACH related articles. In the first experiment the expected number of clusters was set to 5 according to [2]. It was repeated for MTE [10] (Minimum Transmission Energy protocol), LEACH [2], LEACH-C [3] (centralized LEACH), and finally for the modified LEACH-S. The relation between number of nodes still alive and number of data messages recieved at the base station is shown in Fig. 2.

In the above experiment the measurements were repeated 100 times for each protocol to see the variance and precision of the algoritm. The results are presented with 0.95 confidence intervals.

The results confirm that indeed the LEACH protocol has much better energy efficiency than the MTE routing. However the proposed LEACH-S protocol on average allows to gather almost $\frac{1}{3}$ more data during the execution than LEACH. This effect is mainly caused by the fact the LEACH protocol is based on probability and does not guarantee to draw the expected number of nodes in each round.

We have analyzed all the results and verified that indeed for expected number of clusters set to 5 there were round when the protocol did not draw any cluster heads at all or in contradiction has drawn much more cluster heads than expected. Figure 3 presents such an example where the LEACH protocol has chosen 12 different cluster heads in the same round. On the other hand, the LEACH-S protocol guarantees that exactly the expected number of clusters will be created in each round (obviously, the only exception is when some nodes have already died).

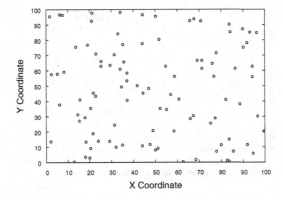

Fig. 1. Testing network consisting of 100 nodes distributed on 100 m × 100 m area. The base station has coordinates (50, 175) and is not shown.

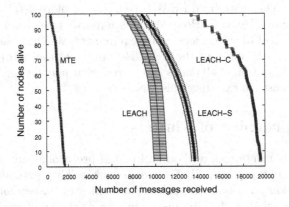

Fig. 2. Number of nodes still alive as a function of number of messages received by the base station. Number of clusters set to 5.

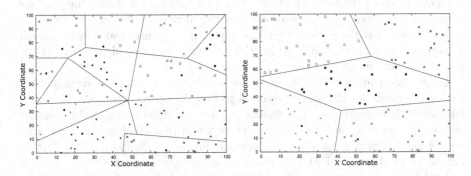

Fig. 3. An example of cluster division in LEACH (left) and LEACH-S (right) protocols where the expected number of cluster is 5

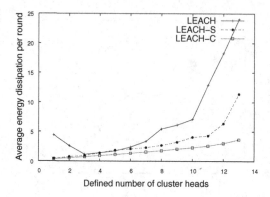

Fig. 4. Average energy consumption per round for different number of clusters

Additionally, the centralized LEACH energy efficiency is also compared in this experiment. The centralized LEACH (see [11] for more detail) outperforms both LEACH and LEACH-S. However, it is important to remember that the centralized version of LEACH has several requirements which are hard to satisfy. For example, in each round of the LEACH-C protocol the information about energy state and location of all nodes is required what may be hardly achievable and brings new costs to exchange this data.

6   Optimal Number of Clusters

As it was already mentioned, authors of LEACH protocol estimate the optimal number of clusters experimentally for given scenario. They evaluated a curve of average dissipation per round with respect to number of clusters for the LEACH protocol. It appears that the curve is decreasing as the number of clusters decreases and finally when the number of clusters approaches 0 the curve is starting to raise creating a minimum value (at around 3–5 clusters see Fig. 4).

Knowing about the random nature of LEACH protocol and that the protocol behaves like a direct communication when the number of chosen cluster heads is 0, suggests that the reason the curve is starting to raise near 0 is because it is more probable that LEACH will not choose any cluster heads at all what has a great impact on energy efficiency.

Taking into consideration the above we decided to evaluate the average dissipation per round for the other protocols to find the optimal number of clusters for each of them separately. The results are also presented in Fig. 4. As it can be seen it appears that our intuition was correct and that elimination of randomness of number of clusters allows to decrease the energy dissipation even further for the LEACH-S and LEACH-C protocols in the given scenario.

As it can be seen on Fig. 4 the optimal number of clusters for the considered scenario is equal to one. It depends mostly on the node distribution and data aggregation costs. The higher the aggregation cost the more clusters are to be created. For the sake of clarity and to allow the reader to compare our results with the results of LEACH authors we decided to use exactly the same scenario – other studies also seem to make the comparison using the same topology and parameters.

To present how this effect can influence the lifetime of whole network we have performed the first experiment one more time and set the expected number of clusters to the optimal value specifically for each protocol (for LEACH – 3, for LEACH-C – 1 and for LEACH-S – 1). The results are presented in Fig. 5.

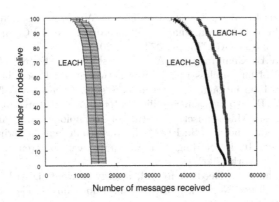

Fig. 5. Number of nodes still alive as a function of number of messages received by the base station. Number of clusters set to optimal for each protocol.

The impact of number of clusters on the energy efficiency is very noticable. For 5 clusters the best – LEACH-C protocol allowed to receive almost 20 000 messages and the same algorithm for 1 cluster achieved more than 50 000 messages. What is more, choosing optimal value for expected number of clusters improves the LEACH-S protocol efficiency by far. This time it is almost as efficient as the centralized LEACH version.

7   Conclusions

In this article we have analysed the operation of LEACH protocol. We have summarized its adventages but also presented one defficeincy which influences its energy efficiency. Further, we have proposed a modification called LEACH-S which allows to overcome the mentioned problem. The simulations confirmed that LEACH-S modification is correct and can be used to increase the network lifetime. Hopefully, this modification can also be applied to other variations of LEACH.

Possibly this modification of LEACH protocol could additionally use the information about energy state of the nodes in the routing decision as it is done in many other power efficient routing protocols concepts. For example the LEACH-C uses the global energy state information to calculate the best routes.

This topic has been approached many times and considered in detail not only for wireless sensor networks but also for ad-hoc mobile networks applications (e.g., see [12–14]).

Acknowledgement. This material is based upon work supported by the Polish National Science Centre under Grant No. N N516 479240.

References

1. Goyal, D., Tripathy, M.R.: Routing Protocols in Wireless Sensor Networks: A Survey. In: 2012 Second International Conference on Advanced Computing Communication Technologies (ACCT), pp. 474–480 (2012)
2. Heinzelman, W.B., Chandrakasan, A., Balakrishnan, H.: Energy-efficient communication protocol for wireless microsensor networks. In: Proceedings of the 33rd Annual Hawaii International Conference on System Sciences, vol. 2, p. 10 (2000)
3. Heinzelman, W.B.: Application-Specific Protocol Architectures for Wireless Networks. PhD thesis, Massachusetts Institute of Technology (June 2000)
4. Lindsey, S., Raghavendra, C.S.: PEGASIS: power efficient gathering in sensor information system. In: Proceedings of IEEE Aerospace Conference, pp. 1125–1130. IEEE Press, New York (2002)
5. Younis, O., Fahmy, S.: Heed: A hybrid, Energy-efficient, Distributed Clustering Approach for Ad-hoc Networks. IEEE Transactions on Mobile Computing 3(4), 366–369 (2004)
6. Manjeshwar, A., et al.: TEEN: A routing protocol for enhanced efficiency in wireless sensor networks. In: Proceedings of the 15th Parallel and Distributed Processing Symposium, pp. 2009–2015. IEEE Computer Society, San Francisco (2001)
7. Manjeshwar, A., Agrawal, D.P.: APTEEN: a hybrid protocol for efficient routing and comprehensive information retrieval in wireless sensor networks. In: Proceedings of the International Parallel and Distributed Processing Symposium, IPDPS 2002, pp. 195–202 (2002)
8. Wang, J., Yang, G., Chen, S., Sun, Y.: Secure LEACH routing protocol based on low-power cluster-head selection algorithm for wireless sensor networks. In: International Symposium on Intelligent Signal Processing and Communication Systems, ISPACS 2007, pp. 341–344 (2007)
9. The Network Simulator – ns-2 (February 01, 2013), http://www.isi.edu/nsnam/ns/
10. Shepard, T.: A channel access scheme for large dense packet radio networks. In: Proc. ACM SIGCOMM, Stanford, CA, pp. 219–230 (August 1996)
11. Xinhua, W., Sheng, W.: Performance Comparison of LEACH and LEACH-C Protocols by NS2. In: Ninth International Symposium on Distributed Computing and Applications to Business Engineering and Science (DCABES), pp. 254–258 (2010)
12. Gelenbe, E., Lent, R.: Power-aware ad hoc cognitive packet networks. Ad Hoc Networks 2(3), 205–216 (2004)
13. Rabiner, W., Chandrakasan, A., Balakrishnan, H.: Energy-Efficient Communication Protocol for Wireless Microsensor Networks. In: Hawaii International Conference on System Sciences, Maui, pp. 10–19 (January 2000)
14. Muruganathan, S.D., Ma, D.C.F., Bhasin, R.I., Fapojuwo, A.O.: A centralized energy-efficient routing protocol for wireless sensor networks. IEEE Communications Magazine 43(3), 8–13 (2005)

Performance Analysis of IEEE 802.11 EDCA for a Different Number of Access Categories and Comparison with DCF

Olga Leontyeva and Kvitoslava Obelovska

Lviv Polytechnic National University, 12/806 Bandera St., Lviv-13, 79013, Ukraine
olya_leon@rambler.ru, obelyovska@gmail.com

Abstract. Performance analysis of the 802.11 EDCA and DCF using simulation and analytical model is presented. An impact of different number of ACs in the network performance is studied under nonsaturation and saturation network condition. Additionally, it is shown that the EDCA doesn't provide a good prioritized access in contrast to the DCF, when only one traffic type is being transmitted through the wireless network.

Keywords: wireless networks, 802.11, medium access control, DCF, EDCA, throughput analysis.

1 Introduction

IEEE 802.11 [1] standard is one of the most popular standards of wireless LANs. The wireless LAN's performance strongly depends on the transmission rate and on the design of the medium access control (MAC) protocol. Technology innovations have significantly increased the transmission rate through wireless environment. At the same time almost all wireless networks still use the same MAC protocol based on the Carrier Sense Multiple Access with Collision Avoidance (CSMA/CA) scheme with slotted Binary Exponential Backoff (BEB) algorithm. This MAC protocol provides contention-based access to the physical environment. As it has been shown in [2] the throughput of the 802.11 wireless LANs is bounded by the overhead of the MAC protocol. Therefore, there are many studies focused on the MAC protocol improvement and extension to maximize channel capacity and utilization.

The CSMA/CA scheme, implemented in the early versions of 802.11 MAC, is referred to as Distributed Coordination Function (DCF). Later, with significantly growing multimedia traffic and number of real-time applications that require supporting of a Quality of Service (QoS), the DCF has been improved by Enhanced Distributed Channel Access (EDCA) scheme. The EDCA scheme defines four access categories (ACs) that provide support for the delivery of prioritized traffic.

This paper is focused on the performance analysis of the EDCA scheme with a different number of ACs compared to the performance of the DCF scheme under saturation and nonsaturation network conditions.

A. Kwiecień, P. Gaj, and P. Stera (Eds.): CN 2013, CCIS 370, pp. 95–104, 2013.
© Springer-Verlag Berlin Heidelberg 2013

2 Overview of 802.11 DCF and EDCA Schemes

The distributed scheme DCF, based on the CSMA/CA, provides the basic asynchronous and contention-based shared access to the physical environment. Before a station starts transmission, it senses the wireless medium. If a station senses no transmission on the channel, it considers the channel state as idle; otherwise it considers the channel state as busy. If the medium is idle for the Distributed Interframe Space (DIFS) the station waits for a random backoff interval. Here backoff counter is uniformly and randomly chosen in the range [0, CW], where CW – Contention Window. If during and after backoff the medium is still sensed idle, then the node is permitted to begin the transmission process. If the medium is busy, then station postpones its transmission for a random period of time. In case of successful transmission a receiver after the Short Interframe Space (SIFS) immediately transmits a positive Acknowledgement (ACK). Thus, only one station can successfully transmit in the network at a given time. The DCF has not been developed to support prioritized traffic; therefore packets with a different priority are being processed identically.

As opposed to DCF scheme the EDCA scheme has been designed to support QoS. In EDCA, frames from the upper layers are mapped onto one of the four ACs according to their priority: background (AC_BK), best effort (AC_BE), video (AC_VI) and voice (AC_VO). Each AC has a transmission queue and the access parameters: minimal and maximal values of CW, Arbitration Interframe Space (AIFS), which is larger or equal to the DIFS. Each AC independently executes DCF scheme in order to resolve internal collision between ACs. If waiting time reaches zero simultaneously for two or more frames in different AC queues, then the frame with the higher priority gets the opportunity to be transmitted. Minimal and maximal CW as well as AIFS are shorter for higher-priority ACs. Therefore, higher-priority frames have a better chance to get transmission opportunity than lower-priority ones.

3 Performance Analysis

It is assumed that EDCA scheme can have a different number of ACs (not only four as it is defined in IEEE 802.11e). Our goal is to study the throughput variations of a wireless network for different number of ACs as a function of the number of stations in the case of saturation network condition, or as a function of the offered load in the case of nonsaturation network condition.

3.1 Simulation Setup

To perform the simulation experiments, we have developed a wireless simulator [3]. Additionally, we use an Engelstad's analytical model [4] (we call the model by the first author's name). Engelstad's model allows to simulate EDCA and DCF schemes under both saturation and nonsaturation network condition.

A simulated wireless network consists of several wireless stations and an access point, which are located within the Basic Service Set (BSS), i.e., every station is able to detect a transmission from any other station. The wireless network works in the Infrastructure mode, when all stations send and receive traffic via an access point. The channel condition is assumed to be ideal and each station operates at the transmission rate of 54 Mbit/s. The other parameter settings for MAC and physical layers are shown in Table 1.

Table 1. Parameter settings for MAC and physical layers

Frame size	2312 bytes
MAC-header	34 bytes
PHY-header	32 bytes
ACK	14 bytes
Slot time	20 μs
SIFS	10 μs
Retry limit	16

3.2 Simulation Scenarios

In order to study the impact of the number of ACs on the wireless network performance we define five scenarios:

Scenario 1. There is only one AC and the wireless network works using DCF scheme.

Scenario 2. There are two different ACs for video (AC_VI) and background (AC_BK) traffic.

Scenario 3. There are four ACs, as defined in the IEEE 802.11e standard.

Scenario 4. There are eight ACs that correspond to the user priorities defined in the IEEE 802.1D.

Scenario 5 is similar to Scenario 3, but only one of the four traffic types is present in the wireless network.

The AC parameters for each scenario are shown in Table 2. The higher number the AC has the higher-priority traffic it corresponds to. Traffics for each AC type are equally generated by station.

3.3 EDCA Performance under Saturation Network Condition

Simulations using the simulator and the analytical model have been done for each scenario where the number of the ACs changes from 1 to 8.

Our goal is to simulate the throughput variations of the wireless network as a function of the number of stations. Additionally, we compare them with the results obtained from the analytical model.

Table 2. AC parameters for five scenarios

Scenario	1	2		3 and 5				4			
AC	0	0	1	0 (AC_BK)	1 (AC_BE)	2 (AC_VI)	3 (AC_VO)	0,1	2,3	4,5	6,7
CWmin	31	31	15	31	31	15	7	31	31	15	7
CWmax	1023	1023	31	1023	1023	31	15	1023	1023	31	15
AIFS (μs)	50	150	50	150	70	50	50	150	70	50	50

Fig 1 shows the total normalized throughput of the saturated wireless network, when a station always has frames, which are ready to be sent, for Scenarios 1–4 is shown.

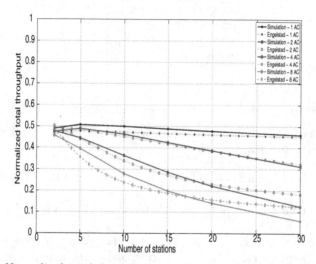

Fig. 1. Normalized total throughput as a function of the number of stations

In each scenario the total normalized throughput decreases with increasing the number of stations in the network. For a small network (number of stations is less than 5) the normalized throughput varies in the range 0.4–0.5. For larger networks the difference between the throughputs increases with increasing the number of stations. For example, in the case of 15 stations the throughput difference between network with 1 and 2 ACs is 0.06. For the network with 4 ACs the total normalized throughput is approximately one half of the total normalized throughput for the network with 1 AC. And in the case of the network with 8 ACs the total normalized throughput is bounded by the value 0.2. If the number of stations in the network is increased to 30 then the total normalized throughput for the networks with 2, 4 and 8 ACs is 1.5, 3.7 and 7.9, respectively, i.e., less than the normalized throughput for the network with 1 AC, that is DCF scheme.

Hence, we can conclude that an increase in the number of ACs in the saturated network may significantly decrease the total throughput. In the worst case – up to 8 times, and in the case of the 802.11e standard – up to 3.7 times. Also, the total throughput dramatically decreases with increasing the number of stations in the network. One of the reasons of such dependency may be the use of collision avoidance mechanism. Fig 2 shows collision probability as a function of the number of stations and the number of ACs in the network.

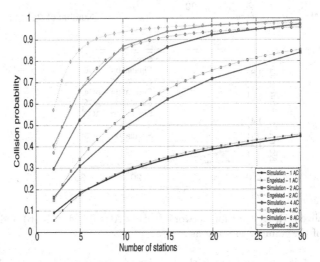

Fig. 2. Collision probability as a function of the number of stations

The collision probability for the network with 1 AC is approximately one half of the collision probability for the networks with 4 and 8 ACs. For the large networks with 4 and 8 ACs the collision probability tends to 1. Hence, a station wastes time applying the mechanism of collision avoidance instead of successful transmitting.

If we analyze the normalized throughput for each AC (Fig. 3), then we can see that the network with 2 ACs provides quality of services for the higher-priority frames under saturation condition. Even for the large network the normalized throughput is approximately 0.3, as opposed to the networks with 4 and 8 ACs. The network with 4 ACs provides good priority-access for the higher-priority frames only in small networks. The performance efficiency decreases with increasing the number of stations. For 30 stations the normalized throughput of the highest-priority traffic is bounded by the value 0.1, whereas for the network with 2 ACs this value is 0.3. For the network with 8 ACs the best throughput for the higher-priority frames (approximately 0.14–0.16) is provided if the network size is small. Otherwise the throughput is significantly low for each type of traffics.

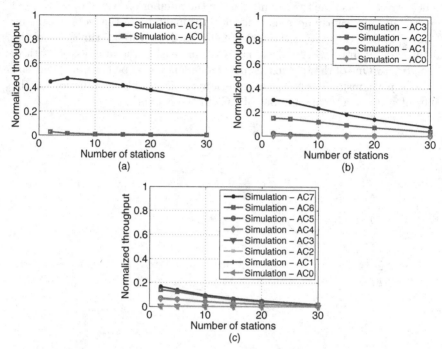

Fig. 3. Normalized throughput for networks with 2 ACs (a), 4 ACs (b) and 8 ACs (c) as a function of the number of stations

For Scenario 5 it is assumed that the saturated network uses the EDCA scheme with 4 ACs as it is defined in the 802.11e standard. Each station transmits only one of the four traffic types. Normalized throughputs for each traffic type in comparison to throughput of the DCF scheme are shown in Fig. 4.

Hence, we can see that under the saturation condition the throughput for the EDCA scheme in the presence of only one of the four traffic types is lower than the throughput for the DCF scheme. The highest throughput difference is observed when only voice traffic is transmitted in the network. In the case of 30 stations the throughput for the EDCA scheme is one half of the throughput for the DCF scheme. It is because under the saturation condition frames collide more frequently, and at the same time for the higher-priority frames the CWmax is half or equal to the CWmin of the lower-priority frames. Thus, after applying the collision avoidance mechanism, a probability of two or more stations to generate equal backoff and the probability for higher-priority frames to collide again increases. Because CWmin equals to CWmax for the background and best effort traffics, the only reason for the throughput difference is to apply the AIFS-differentiation. Hence, an increase in the number of AIFS from 70 to 150 μs decreases the throughput by about 0.05. For the saturated wireless network, when stations always have ready-to-be-sent frames, the EDCA priority-access scheme in general is less efficient (provides a lower total throughput) than the DCF scheme that doesn't provide priority access. Using of the EDCA with 2 ACs

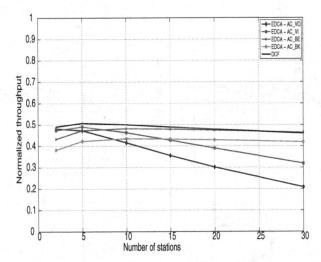

Fig. 4. Normalized throughput for each type of traffic as a function of the number of stations

allows to provide a good prioritized access for the higher-priority traffic. And the total throughput for the EDCA with 2 ACs is lower than the throughput for the DCF by just about 0.05–0.1.

3.4   EDCA Performance under Nonsaturation Network Condition

The saturation network condition allows to study a limiting behaviour of the 802.11 standard and to obtain a bound value for the network throughput and delay. However, realistic networks are not fully saturated due to bursty data traffic. Therefore, in this section we study network throughput in a full load range from a nonsaturated to a saturated input traffic.

Simulations for the variable-offered load have been done taking into account the network size. The network consisting of 5 stations corresponds to a small network, 15 stations – a midsized network, 30 stations – a large network. Different network size allows to analyse its influence on the network characteristics.

The total normalized throughput as a function of the offered load is shown in Fig. 5.

The total normalized throughput dependence increases for both the simulator and the analytical model for the offered load between 0 and 0.5. Then it saturates. The network transmits almost all offered load that is lower than 0.4 for all scenarios. Only in the case of the large network (Fig. 5c) for Scenarios 3 and 4 the total normalized throughput becomes saturated for the offered load higher than 0.3. At the same time, the total throughput saturates faster for more stations and more ACs in the network.

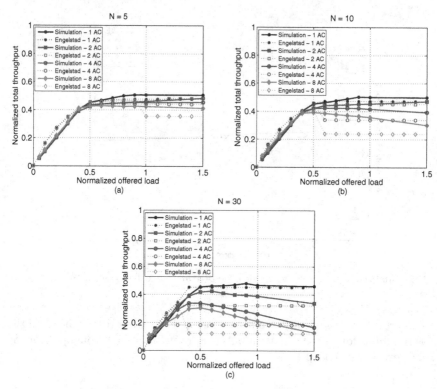

Fig. 5. Normalized total throughput for small (a), midsized (b) and large (c) networks as a function of the offered load

For the offered load in the range 0.1–0.4 the total normalized throughput obtained by the analytical model is higher than the offered load by about 0.07. It means that the station transmits more traffic than it generates. These discrepancies have also been noted in [4]. There it has been suggested that probably the AIFS-differentiation is done rough. For the higher offered load than 0.4–0.5 there are also noticeable discrepancies between the results of the analytical model and the simulator due to numerical errors.

The efficiency of the EDCA prioritized scheme in the full load range can be estimated by analysing the throughput for each AC. Fig. 6 shows a normalized throughput for each AC of the midsized network that has 2, 4 or 8 ACs as a function of the offered load. As it is shown analysing the total throughput the offered load lower than 0.4 is almost all transmitted through the network. As the offered load increases, the higher-priority traffics achieve higher throughput than the lower-priority traffics. Even for a highly loaded network the EDCA provides a prioritized access for the higher-priority frames. With an increase of the number of ACs the effectiveness of the differentiated service decreases.

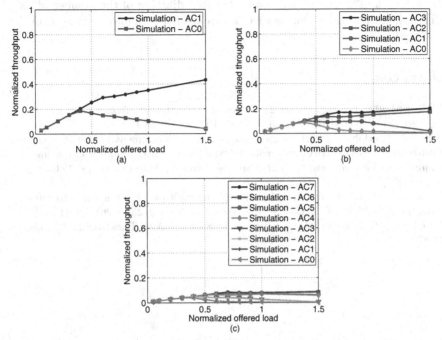

Fig. 6. Normalized total throughput for midsized networks with 2 ACs (a), 4 ACs (b) or 8 ACs (c) as a function of the offered load

Also, it can be observed that for the traffic AC1 in case of network with 4 ACs (Fig. 6b) the normalized throughput in the range 0.4–1 is about one half of the throughputs of the traffics AC2 and AC3. This is revealed by means of an increase of the offered load that leads to more collisions between higher-priority frames, and gives the opportunity to be transmitted for the traffic AC1.

We conclude that the EDCA efficiency strongly depends on the offered load, the number of stations and on the number of ACs. The higher these parameters are the less efficient EDCA is.

4   Conclusions

We presented a simulation and analytical analysis of the EDCA efficiency for the network with a different number of the ACs. The prioritized channel access provided by the 802.11 standard is sensitive to the size of the network, the network load, and the number of ACs. The higher these parameters are the less efficient EDCA is. It is shown that decreasing of the number of ACs can improve the total throughput, even under highly load network condition. Use of the EDCA with 2 ACs provides a good prioritized access for the higher-priority traffic in the full load range (from nonsaturation to saturation). Although the EDCA provides the prioritized access, for larger networks EDCA is inefficient in contrast to the DCF, especially when only higher-priority traffic is present in the network. The possibility of an adaptive adjusting of the number of ACs for increasing the total performance of a wireless network may be studied in the future works.

References

1. IEEE Std 802.11^{TM}-2007, Part 11: Wireless LAN Medium Access Control (MAC) and Physical Layer (PHY) Specifications. IEEE Std. (2007)
2. Xiao, Y., Rosdahl, J.: Performance analysis and enhancement for the current and future IEEE 802.11 MAC protocols. ACM SIGMOBILE Mobile Computing and Communications Review (MC2R), Special Issue on Wireless Home Networks 7, 6–19 (2003)
3. Leontyeva, O., Obelovska, K.: Modeling of the multiple access method to physical environment of wireless networks. The Technical News 1(25)-2(26), 78–81 (2007)
4. Engelstad, P., Østerbø, O.: Analysis of QoS in WLAN. Telektronikk 1, 132–147 (2005)

Simulation Comparison of LEACH-Based Routing Protocols for Wireless Sensor Networks

Agnieszka Brachman

Silesian University of Technology, Gliwice, Poland
agnieszka.brachman@polsl.pl
http://www.ztipsk.aei.polsl.pl

Abstract. Battery-powered nodes have limited energy reserves therefore applications and protocols used for WSNs, should be designed, concerning the optimized energy consumption in order to prolong the network lifetime. Data reception and transmission are the main energy consuming operations and they are regulated by the network layer, hence the routing protocol plays very important role in network optimization.

In this paper information concerning the LEACH routing protocol is gathered and the classification of LEACH-based modifications is presented. Furthermore this paper focuses on improvements to the LEACH protocol that address problems of the cluster head selection, load balancing and lifetime enhancement as well as presents the simulation results for the selected group of LEACH-based protocols.

Keywords: LEACH, routing, sensor network, simulation.

1 Introduction

Wireless sensor networks (WSNs) are believed to be one of the fundamental technology that will have and actually already has an enormous impact on our everyday life. The potential field of application is unlimited. WSNs play an important role in many industrial, commercial and domestic applications concerning asset tracking systems, controlling and monitoring the buildings' equipment (lighting, ventilation, security systems, fire systems), security systems, habitat monitoring, environment monitoring, vehicular tracking, medical applications, Automated Meter Reading (AMR) for water, electricity, heat and gas and many many more. This dynamic and rapid development is possible thanks to the revolution in wireless technologies as well as the introduction of smaller and more effective electronic devices.

Typical WSN consists of many sensor nodes (SN) which are usually small and inexpensive devices. Nodes may be equipped with one or more, different kinds of sensors, embedded processors, memory, radio transmitter and are normally operated with a battery. Nodes of a WSN communicate with each other by establishing multi hop, wireless network. Each SN is responsible for sensing a desired parameters, some of SNs may also perform some data preprocessing

A. Kwiecień, P. Gaj, and P. Stera (Eds.): CN 2013, CCIS 370, pp. 105–113, 2013.
© Springer-Verlag Berlin Heidelberg 2013

or data aggregation. Furthermore SNs relay data to the same location usually called the WSN sink or the Base Station (BS).

Battery-powered nodes have limited energy reserves therefore applications and protocols used for WSNs, should be designed, concerning the optimized energy consumption in order to prolong the network lifetime. Data reception and transmission are the main energy consuming operations and they are regulated by the network layer, hence the routing protocol plays very important role in network optimization. Energy efficient routing protocols may reduce the number of transmitted packets as well as optimize the selection of traces and nodes for data relaying.

Mainly because of the large number of nodes, deficiency of global addresses, scarce energy resources and synchronization problems existing routing protocols used in IP and cellular networks are not applicable in WSNs. A WSN may consist of hundreds or thousands of SNs. The deployment of nodes in WSN may be deterministic or random, dense or spacious. Data transmission to the BS may be continuous, event-driven or query-driven. Some nodes may stop working in time course, therefore the routing protocol must be able to adapt to the topology changes and always find the optimal (according to defined criteria) route.

There are three main communication schemes in WSNs: direct, flat and hierarchical. In direct scheme, each SN communicates directly with the BS. The solution is rather useless, because not every node is in the sufficient proximity. Routing protocols using the flat communication scheme treat all nodes equally and they all take part in routing. The main flaw is that nodes close to the BS more often take part in data forwarding than farther nodes. To provision the efficient energy consumption, WSNs use clustering. The network is divided into a cluster, in each cluster one node is selected as the Cluster Head (CH). Nodes within one cluster communicate with the CH and the CH communicates directly with BS. This is the hierarchical communication scheme.

One of the first and most common hierarchical routing protocol is LEACH (Low-Energy Adaptive Clustering Hierarchy) [1]. Since it was described over twelve years ago, many modifications were proposed to elevated some of its limitations. LEACH and its derivatives are described in Sect. 2

The main purpose of this paper was to gather information concerning LEACH and present classification of LEACH-based modifications. Furthermore this paper focuses on improvements to the LEACH protocol that address problems of the cluster head selection, load balancing and lifetime enhancement as well as presents the simulation results for the selected group of LEACH-based protocols.

The rest of the paper is organized as follows. As mentioned before, Sect. 2 presents the fundamentals of LEACH protocol as well as description of the newly proposed LEACH modification is covered. Related work is referenced in Sect. 3. Sections 4 and 5 contain simulation details and its results respectively. The concluding remarks and future work suggestions are provided in Sect. 6.

2   LEACH and LEACH-Based Protocols

LEACH [1] is one of the first hierarchical routing algorithms proposed for the WSNs. The routing is two hop, according to the following rules. Each node may act as a cluster head (CH) or a regular sensor node. Communication to the sink goes through the CHs. Every time interval (round), a node declares himself as a CH with the certain probability. The node selects a CH which is closest to him. The node itself makes a decision, if or not become a CH. It selects a random number between 0 and 1, if the chosen number is less than the threshold $T(n)$, the node starts being the CH. The $T(n)$ threshold is defined as follows:

$$T(n) = \begin{cases} \frac{P}{1-P(r \bmod \frac{1}{P})} & \text{if } n \in G \\ 0 & \text{if } n \notin G \end{cases}. \tag{1}$$

Where P is the desired percentage of CHs (usually 0.05), r is the number of the current round, G is the set of nodes that have not been cluster heads for the last $1/P$ rounds.

Sensors organize themselves in clusters. Every round reorganization is performed. Only the CHs can communicate with the BS, nodes use CHs as a route to pass data. The CH collects, aggregates, sometimes compresses and transmits received data. The steady state phase starts after selecting CHs. In the state phase nodes transmit data to the sink, during allocated time slots, otherwise they remain asleep.

2.1   LEACH Limitations

There are several problems with LEACH, that lead to the rapid battery drain. All nodes are assumed to have the same capabilities and the same residual energy level, which may not be correct.

The main LEACH limitations, as depicted in [2,3,4], are as follows :

- Two-hop routing – some CHs may be far from the BS, therefore transmission may use considerable amount of energy.
- Number of cluster heads is predefined, however the selected number, depends on the node distribution; the number may not be sufficient and the cluster formation may be suboptimal.
- The threshold $T(n)$, defined in Equation (1), doesn't take into account the residual energy level while selecting CHs.
- The cluster size may differ significantly every round when selected randomly.
- In each round, all nodes take part in network reconstruction, which consumes their energy.

2.2   LEACH Improvements

The modifications of LEACH protocol, concern mainly several parameters. Their categorization, presented in [4], is as follows:

- The cluster head selection;
- Multihop Data Transmission;
- Heterogeneous – support heterogeneity among the nodes;
- Chain Based – focused on the construction of chains among the nodes;
- Others: Mobility, Security, Spare Management, Application Specific, Clusters Radius Fixation.

Most modifications, introduced to the LEACH algorithm, depict how the cluster head is selected and/or add the multihop transmission.

In the original LEACH, cluster heads are selected randomly. The intuitively better methods, take into consideration, that different nodes have different energy level, especially in the time course, and use it to increase the probability of becoming a CH, basing on the energy level e.g. HEED [5], PEACH [6], PEGASIS [7] and more [8].

There is a whole group of the LEACH modifications, that change the original LEACH into the multi hop protocol [3,9,10,11,12,13,14]. The multi hop versions are usually designed with one of two assumptions: reducing power consumption [3,10,12], reducing the amount of traffic or hybrid [11].

Furthermore this paper focuses on two LEACH improvements, described further in this section, that address problem of the proper selection of the cluster heads. The advantage of the first version is the constant percentage of the cluster heads throughout the network lifetime; the second protocol considers the residual energy level for all nodes.

2.3   LEACH-Balanced

The protocol called LEACH Balanced (LEACH-B) was presented in [15]. At each round, after the selection of CHs, according to the original LEACH procedure, the second selection is performed, that leads to the fixed number of cluster heads. If too much cluster heads are selected, the ones having the lowest energy level, are eliminated from the CH list; on the other hand, if too few are selected, some additional nodes are converted into the cluster heads. The nodes with the highest energy level, have the highest probability of being selected as the additional cluster heads. The improvements provide, that the required number of CHs is always assured, moreover nodes, that have the highest energy level, are selected as the CHs in the first place.

2.4   Energy LEACH

Xiangning et al. [14] proposed two improvements to the original LEACH. The Energy-LEACH modification redefines the cluster head selection procedure. The residual energy of a node is the main indicator whether or not turn the node into the cluster head. The second modification, proposed in aforementioned paper, concerns the multihop routing and, as out of the scope of this paper, is not discussed. The Energy-LEACH although proposed in 2007, represents major trend in LEACH development, therefore is a good representative of its group.

3   Related Work

The detailed survey of then routing protocols, along with their description and discussion, is presented in [16], with their classification to three main categories: data-centric, hierarchical and location-based. The comparative analysis of LEACH and its variants, however without simulation results, is presented in [17]. Another survey is presented in [18]. Through Matlab simulation, the authors compare LEACH and few LEACH-based modifications, both centralized and distributed, multi-hop and dedicated for mobile nodes. The comprehensive survey of all main classical and swarm routing protocols in WSNs is presented in [19]. The analytical and simulation comparison of the selected algorithms is also covered.

There are many papers, that present detailed survey of the routing algorithms, however there is lack of papers, that present comparison based on simulation or emulation results. There are also few papers, that are focused on the LEACH based protocols and, according to the author's best knowledge, no simulation comparison among numerous improvements described in the literature is covered. The purposed of this paper, is to compare the LEACH-based modifications, that improve the procedure of the cluster head selection. Two base algorithms were selected, as mentioned in the previous section, which represent the main trends for LEACH development.

4   Simulation Scenario

To compare the depicted LEACH modifications, the network simulator NS-2.34 [20] was used with LEACH model developed by [21], with own modifications for the presented algorithms. The parameters were set according to the directives given in [22,21], details are presented in Table 1.

Table 1. Simulation parameters

Simulation area	100*100
Simulation time	1000 s
Number of nodes	100
Initial energy of node	2 J
Round time	20 s
Radio speed	1 Mbps
Data size	2000 B

Several simplifications were assumed, i.e. all nodes are identical, static and have the same initial energy. Nodes are displaced randomly, however they stay within transmission range. Nodes always have some data to send. All LEACH protocols were configured to select 5 % of nodes as the cluster heads. Simulation results and discussion are presented in the following section.

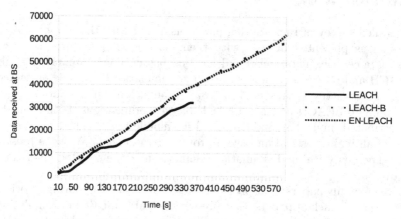

Fig. 1. Time vs no of data signals received at BS

5   Result Analysis

The network lifetime for each LEACH version is as follows:

– LEACH – 363 s,
– Balanced LEACH – 604 s,
– Energy LEACH – 3600 s.

The obtained results, depicting the number of data signals, total energy dissipation and number of nodes alive are presented in the Figs. 1, 2, 3.

From Fig. 1 it can be concluded, that LEACH-B and Energy LEACH are able to deliver more data to the BS, comparing to the original LEACH. In LEACH-B the number of the cluster heads is constant, some CHs are selected with regard

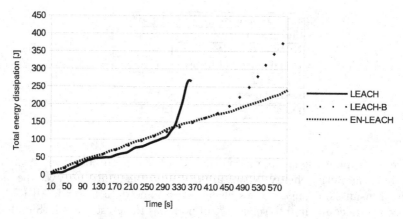

Fig. 2. Time vs total energy dissipation

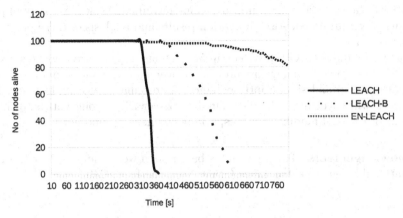

Fig. 3. Time vs no of nodes alive

to the residual energy of a node, there are no rounds with an extra number of the CHs, hence the clustering is more effective and consequently more data is delivered. Moreover, during the network lifetime, the number of data transmitted with LEACH-B is comparable with the amount delivered with Energy LEACH, which has the longest network lifetime.

Figure 2 depicts, how the energy is dissipated for the evaluated protocols. To some point, for all LEACH versions the total energy dissipation linearly increases. For LEACH-B and Energy LEACH this tendency last longer, due to the higher number of nodes alive. For original LEACH we can see, that at some point, where too much nodes are dead, the energy usage significantly increases, which leads to the network death. At the end of every WSN life, similar increase can be observed; it strictly depends on the number of nodes alive and the distance between nodes and cluster heads.

From Fig. 3 we can see, that the number of nodes alive decrease much slowly in Energy LEACH, than in any other version. Therefore the network lifetime is significantly prolonged. For pure LEACH, uneven distribution and variable number of the cluster heads, significantly reduces the number of nodes alive in short time course. The faster some of the nodes die, the shorter the remaining nodes live, due to the higher extend of network scattering, longer distances and therefore higher transmission power for each connection. From this figure, it can be observed how important balancing the energy usage is.

6   Conclusions and Future Work

The cluster head selection and the energy consumption are the most important factors when discussing the hierarchical, clustering routing algorithms for WSNs. Low Energy Adaptive Clustering Hierarchy (LEACH) is the fundamental clustering protocol for WSN and is taken as a benchmark solution – basis for the newly proposed findings. In this paper, detailed discussion, concerning ongoing work is provided. Brief description of chosen LEACH modifications is presented, along with the classification of improvements introduced, since the original protocol has been proposed. Also simulation results and analysis of these protocols are presented.

The presented LEACH protocol improvements represent major trends in LEACH development. They are proved to overcome the shortcomings of the original protocol and significantly enhance the original protocol efficiency. Future work will cover more detailed simulation scenarios, along with additional versions of LEACH modifications, especially the newly proposed.

Acknowledgements. This material is based upon work supported by the Polish National Science Centre under Grant No. N N516 479240.

References

1. Heinzelman, W.R., Chandrakasan, A., Balakrishnan, H.: Energy-efficient communication protocol for wireless microsensor networks. In: Proceedings of the 33rd Annual Hawaii International Conference on System Sciences, vol. 2, pp. 3005–3014 (January 2000)
2. Jianyin, L.: Simulation of improved routing protocols LEACH of wireless sensor network. In: 2012 7th International Conference on Computer Science & Education (ICCSE), pp. 662–666 (2012)
3. Yan, J.-F., Liu, Y.-L.: Improved LEACH routing protocol for large scale wireless sensor networks routing. In: 2011 International Conference on Electronics, Communications and Control (ICECC), pp. 3754–3757 (2011)
4. Tyagi, S., Kumar, N.: A systematic review on clustering and routing techniques based upon LEACH protocol for wireless sensor networks. Journal of Network and Computer Applications (2013), doi:10.1016/j.jnca.2012.12.001
5. Younis, O., Fahmy, S.: HEED: a hybrid, energy-efficient, distributed clustering approach for ad hoc sensor networks. IEEE Transactions on Mobile Computing 3(4), 366–379 (2004)
6. Yi, S., Heo, J., Cho, Y., Hong, J.: PEACH: power-efficient and adaptive clustering hierarchy protocol for wireless sensor networks. Network Coverage and Routing Schemes for Wireless Sensor Networks 30(14-15), 2842–2852 (2007)
7. Lindsey, S., Raghavendra, C.: Pegasis: Power-efficient gathering in sensor information systems. In: Aerospace Conference Proceedings, vol. 3, pp. 3-1125–3-1130. IEEE (2002)
8. Pawar, S., Kasliwal, P.: Design and evaluation of en-LEACH routing protocol for wireless sensor network. In: 2012 International Conference on Cyber-Enabled Distributed Computing and Knowledge Discovery (CyberC), pp. 489–492 (2012)
9. Yektaparast, A., Nabavi, F.H., Sarmast, A.: An improvement on LEACH protocol (cell-LEACH). In: 2012 14th International Conference on Advanced Communication Technology (ICACT), pp. 992–996 (2012)
10. Bo, W., Han-Ying, H., Wen, F.: An improved LEACH protocol for data gathering and aggregation in wireless sensor networks. In: International Conference on Computer and Electrical Engineering, ICCEE 2008, pp. 398–401 (2008)
11. Abdulla Ahmed, E.A.A., Nishiyama, H., Kato, N.: Extending the lifetime of wireless sensor networks: A hybrid routing algorithm. Special Issue: Wireless Sensor and Robot Networks: Algorithms and Experiments 35(9), 1056–1063 (2012)

12. Kumar, N., Sandeep, Bhutani, P., Mishra, P.: U-LEACH: a novel routing protocol for heterogeneous wireless sensor networks. In: 2012 International Conference on Communication, Information & Computing Technology (ICCICT), pp. 1–4 (2011)

13. Xu, J., Jin, N., Lou, X., Peng, T., Zhou, Q., Chen, Y.-M.: Improvement of LEACH protocol for WSN. In: 2012 9th International Conference on Fuzzy Systems and Knowledge Discovery (FSKD), pp. 2174–2177 (2012)

14. Xiangning, F., Yulin, S.: Improvement on LEACH protocol of wireless sensor network. In: International Conference on Sensor Technologies and Applications, SensorComm 2007, pp. 260–264 (October 2007)

15. Tong, M., Tang, M.: LEACH-B: An improved LEACH protocol for wireless sensor network. In: 2010 6th International Conference on Wireless Communications Networking and Mobile Computing (WiCOM), pp. 1–4 (September 2010)

16. Akkaya, K., Younis, M.: A survey on routing protocols for wireless sensor networks. Ad Hoc Networks 3(3), 325–349 (2005)

17. Haneef, M., Deng, Z.: Comparative analysis of classical routing protocol LEACH and its updated variants that improved network life time by addressing shortcomings in wireless sensor network. In: 2011 Seventh International Conference on Mobile Ad-hoc and Sensor Networks (MSN), pp. 361–363 (2011)

18. Aslam, M., Javaid, N., Rahim, A., Nazir, U., Bibi, A., Khan, Z.: Survey of extended LEACH-Based clustering routing protocols for wireless sensor networks. In: 2012 IEEE 14th International Conference on High Performance Computing and Communication & 2012 IEEE 9th International Conference on Embedded Software and Systems (HPCC-ICESS), pp. 1232–1238 (2012)

19. Zungeru, A.M., Ang, L.-M., Seng, K.P.: Classical and swarm intelligence based routing protocols for wireless sensor networks: A survey and comparison. Service Delivery Management in Broadband Networks 35(5), 1508–1536 (2012)

20. The Network Simulator ns-2: Documentation, http://www.isi.edu/nsnam/ns/doc/index.html

21. μAMPS ns Code Extensions, http://www.isi.edu/nsnam/ns/doc/index.html

22. Geetha, V., Kallapur, P.V., Tellajeera, Sushma: Clustering in wireless sensor networks: Performance comparison of LEACH & LEACH-C protocols using NS2. In: 2nd International Conference on Computer, Communication, Control and Information Technology (C3IT-2012), vol. 4, pp. 163–170 (February 2012)

Routing Protocols for Border Surveillance Using ZigBee-Based Wireless Sensor Networks

Hoda Sharei-Amarghan[1], Alireza Keshavarz-Haddad[2], and Gaëtan Garraux[1]

[1] Centre de Recherche du Cyclotron, Université de Liege, Belgium
[2] School of Electrical and Computer Engineering, Shiraz Universtity, Iran
{h.sharei,ggarraux}@ulg.ac.be, keshavarz@shirazu.ac.ir

Abstract. In this paper, we study proper routing protocols for border surveillance missions using wireless sensor networks (WSNs). We assume that the sensor nodes are equipped with ZigBee transceivers for wireless communications. Three well known routing algorithms (AODV, DSR, and OLSR) are simulated in a WSN surveillance scenario. The performances of these routing algorithms are compared in terms of traffic load, delay, packet loss, and energy consumption. Our results indicate that DSR performs better than other algorithms for border surveillance applications. Moreover, a novel algorithm called "DSR_OP" is proposed for improving DSR routing in terms energy management in the network and extending the network life time. However Comparisons of WSN routing protocols (DSR, AODV, OLSR and others) are presented since many years ago, but there is no simulation with OPNET, so our novelty is that the validity of proposed method and the comparisons are confirmed by simulations in OPNET.

Keywords: WSN, routing protocols, OPNET, border surveillance.

1   Introduction

The technology of sensor networks has paved the way for an accurate and intangible monitoring of an environment or a process in a large physical space. Such networks are comprised of many sensor nodes. These nodes can cover a very large physical environment and gather information. Besides, nodes can help each other to gather the information in a centralized unit for decision making.

Therefore today, world is witnessing a growing interest on the topic of Wireless Sensor Networks (WSNs) and their applications in different fields. Some of these applications include monitoring environment, detection and identification of vehicles, hacker detection machine, sanitary and medical care, environmental control, monitoring the quality of agricultural products, etc [1]. The nodes of a WSN, encounter many limitations in energy consumption and processing power. Besides, the technology of WSN is not reached the required maturity yet. Therefore, these networks are facing many challenges such as: energy consumption, latency, scalability, lower cost, communication security, robustness against technical problems, and optimal routing algorithms.

A. Kwiecień, P. Gaj, and P. Stera (Eds.): CN 2013, CCIS 370, pp. 114–123, 2013.
© Springer-Verlag Berlin Heidelberg 2013

In this paper, we study proper routing protocols for border surveillance missions using wireless sensor networks (WSNs). Border is defined as any physical region which we want to monitor it and dependent on its applications, the entrance or exit of an intruder should be detected. We assume that the sensor nodes are equipped with ZigBee transceivers for wireless communications. Using Zigbee protocol is due to its good feature in comparison with other protocols such as Wi-Fi.

There are various types of routing protocol. To have a good conception of two Proactive and Reactive catagory of routing, three well known routing algorithms (AODV, DSR, and OLSR) are simulated in a WSN surveillance scenario. The performances of these routing algorithms are compared in terms of traffic load, delay, packet loss, and energy consumption. Our results indicate that DSR performs better than other algorithms for border surveillance applications. Moreover, a novel algorithm called "DSR_OP" is proposed for improving DSR routing in terms of energy management in the network and extending the network life time. The validity of proposed method and the comparisons are confirmed by simulations in OPNET software.

The rest of this paper is organized as follows: in Sect. 2, border surveillance mission is briefly described. In Section 3, preliminaries on ZigBee standard and routing protocols are presented. A novel optimized routing protocol is proposed in Sect. 4, and simulation results and comparisons of different routing algorithms are presented in Sect. 5. Finally, conclusion and future work are included in Sect. 6.

2 Realated Works on Border Surveillance

The nodes of a WSN can be set up on the ground, in the air, under water, on a vehicle or even in a human's body [2]. In a border surveillance mission, the border is defined a physical region which should be monitored accurately, and dependent on its applications, the entrance or exit of an intruder should be detected. Basically, a linear structure of sensors can be used for coverage of a marginal region for detection of unauthorized activities and crossings [3,4]. A sample structure is shown in Fig. 1. As it shows, the network topology is a collection of nodes, randomly distributed throughout the area. Traffic is sent from all the sensor nodes to the sink. The region may be hundreds of kilometers, so it can be handled by partitioning the area into multiple parts and putting a sink for each area, then all sinks can send their information to a central sink [4].

The existing works that address the border surveillance problem, consider different aspects of it. One of the challenging subjects, which should be considered, is routing.

In [5] the quality of deployment issue is surveyed and analyzed. Suitable measures are discussed for the assessment of the deployment quality. Also some simulation results evaluate the impact of the node density on the detection ratio and on the time-to-detect an intruder. In [6], a method is proposed which specifies the breach paths and the deployment quality is defined as the minimum of

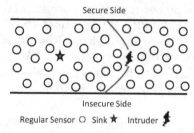

Secure Side

Insecure Side

Regular Sensor ○   Sink ★   Intruder ⚡

Fig. 1. Intruder detection in border surveillance scenario

the maximum detection probabilities on the breach paths in the presence of obstacles. In [7], two protocols are discussed to provide secure detection of trespass within the monitored area and also node failures. The sensor type which is used in this paper, is a simple Passive Infra-Red sensor (PIR sensor).

This paper focuses on routing w.r.t. the detection of intruder across an area monitored by a sensor network. The paper presents a novel algorithm called "DSR_OP" for improving DSR routing in terms of energy management in the network and extending the network life time, and evaluate the system by means of simulation.

3   Preliminaries

In this section, we present a brief introduction on ZigBee standard and three well-known routing protocols are described.

3.1   ZigBee Standard

In 2000, IEEE Standards introduced a low rate wireless personal area network standard, called 802.15.4. In 2003, Zigbee Alliance introduced Zigbee standard protocol. In a technical view, the stack of the ZigBee protocol has four main layers [8]: Physical layer, Media Access Control layer, Network layer, Application layer.

In particular, the first two layers are defined by IEEE 802.15.4, and the other two layers are defined by ZigBee alliance [9]. Network layer provides routing.

The main specifications which have concluded to the vast development of ZigBee standard include but are not limited to: low cost, security, self-healing, flexibility, and high potential for further developments, low power consumption, cheap and easy placement, using free radio bounds.

3.2   Routing Protocols

Routing is one of the key issues in WSN, so a lot of routing protocols has already been proposed [10,11]. Since our simulation region is two-dimensional, so it's better to apply two-dimensional routing protocols in order to increase efficiency. On the other hand, from the point of topological view, two main routing protocols are defined: Proactive, and Reactive. So, we use the AODV, and DSR protocols from Reactive category, and OLSR from Proactive categories and all are two-dimensional. In this subsection, we have a quick overview on these protocols.

Ad hoc On-Demand Distance Vector (AODV) is an on-demand routing algorithm in that it establishs a route to a destination only when a node wants to send a packet to that destination. Such behavior is very useful in networks with low traffic load to keep the routing overhead small. In AODV, every node maintains a table, which stores the information about the next hop to the destination and a sequence number which is received from the destination and ensures the freshness of routes. It is one of the key features of AODV, to avoid counting to infinity that is why it is loop free [12,13]. There are three AODV messages: Route Request (RREQs) which is sent when the host does not know the route to the needed destination host or the existed route is expired, Route Replies (RREPs) which is sent by a node when it has a route to the destination or to a node which has a route to destination, and Route Errors (RERRs) which is sent when the link breakage happens [14,15]. The route discovery is used by broadcasting the RREQ message to the neighbours with the requested destination sequence number, which prevents the old information to be replied to the request and also prevents looping problem. Each passed host makes update in their own routing table about the requested host. The route reply use RREP message that can be only generated by the destination host or the hosts who have the information that the destination host is alive and the connection is fresh [13].

Dynamic Source Routing (DSR) protocol is a reactive routing protocol and like AODV, is known as an on demand routing protocol. It is a source routing protocol which means that the originator of each packet determines an ordered list of nodes through which the packet must pass while traveling to the destination [16]. Each node along the route forwards the packet to the next hop. If, after a limited number of retransmissions of the packet, nest node doesn't recieve the packet, it returns a ROUTE ERROR to the original source of the packet and it means the link from itself to the next node was broken. The sender then removes this link from its Route Cache and tries to discover another route to this destination.

The DSR network is totally self organizing and self configuring. This protocol is comprised of two mechanisms: *Routh Discovery* and *Routh Maintanance*:

- Route discovery is used by a source node S, when it aims to find a route to a sink node D. This process is used just when no route from S to D is known in advance.
- Route maintenance is used to maintain and rebuild the routes which are already known. Therefore, when a path between S and D is know, due to some topological changes, this route might change. At this point, the maintenance algorithm might prefer to use replace another path from its database,

or start a discovering process. However, the Maintenance mechanism is only used when a package is already sent from S to D [14,15].

Optimized Link State Routing (OLSR) is a proactive routing protocol and is also called as table driven protocol because it permanently stores and updates its routing table and so the routes are always immediately available when needed [17,18]. It is an optimization of pure link state protocols in that it reduces the message overhead in the network by using MPR. MPRs are an arbitrary subset of one-hop neighbors of a node N while they could cover all the nodes that are two hops away. Each node in the network keeps a list of MPR nodes. Information is rebroadcast only by MPRs, Whenever a packet is received, a node checks its sender; if it is MPR, the packet is forwarded, otherwise the packet is discarded [14,15].

The performances of these routing algorithms are compared in terms of delay, traffic load, packet loss, and energy consumption. Therefore, a brief description of these parameters are presented as following [15]:

Delay – the time which is needed for packets to go from source to sink. This time is expressed in seconds, and have different kinds such as *processing delay, transmission delay,* and *propagation delay.*

Network Load. The network load shows the overall load (bps) of every node in a wireless network. In other words, the network load is the sent packets of the network in each second.

Packet loss which could occur for different reasons, such as the distance, battery depletion, collision and etc.

Energy consumption is the total amount of the consumed energy during simulation runtime.

Throughput. Throughput is defined as the ratio of the total data which reaches a receiver from the sender. Different factors affect this power, such as various changes in network topology, non-reliable links between nodes, limited bandwidth, and energy limitation. In every network, the highest throughput is the optimal one.

4   DSR_OP: A Novel Routing Protocol

In [19], Bashyal and Venayagamoorthy stated that knowing the number of alive nodes in a sensor network does not reveal how effective the system is, except when all, or none of the sensor nodes are alive. What should also be known is the distribution of the surviving nodes in the sensor network so that the area that is being monitored could be estimated. Figure 2 and the discussion following it show the importance of sensor node distribution for effectiveness of the wireless sensor network.

As we explained earlier in this paper, in DSR the source S might receive several routing replies from the network. In DSR protocol, it chooses the best route which is the shortest one, and saves the other routes in its table. Whenever a link breaks down, then it uses the stored data for choosing another route.

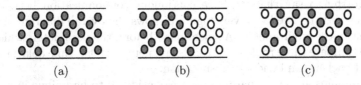

Fig. 2. Different possible sensor node distribution: (a) initial network with all surviving nodes, (b) uneven distribution of surviving sensor nodes, (c) more uniform distribution of surviving sensor nodes

Therefore, the nodes used for transferring data, are used as much as possible until they are impaired of any reason, including the battery loss, and at this time another route is replaced. So the effectiveness of the system is decreased.

The main objective of this novel optimized DSR algorithm (DSR-OP) is to address this problem and suggest a way for preventing the complete break down of nodes. In a normal situation the nodes will randomly get involved in a route and in case of failure of some nodes, the remaining nodes shall take the place. In our method, we suggest that when one node reaches a predefined threshold energy, it leaves the network for a random time period. In this way, other nodes of the network get involved, and the energy consumption is uniformly dispersed in the network.

5 Results

In this section, we first present detailed comparisons between three well known routing protocols, and then simulation results of the novel DSR_OP algorithm proposed in this paper are illustrated.

In all simulations of this subsection, following parameters are considered:

- Simulation region is 10 km × 4 km, it is selected a rectangular area to be similar to border.
- One sink node, one source node which generate traffic, and 28 normal nodes.
- Time of the simulation is 1:00 hour.

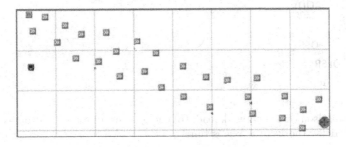

Fig. 3. Network topology

It's noteworthy to say that we have continuous communication between the fixed sender and the fixed receiver. Figure 3 shows the overall diagram of nodes deployment in the environment. As mentioned before, the parameters which will be used to compare three routing protocols are: network load, time delay, number of dropped packets, and energy consumption.

For all simulations, every parameters are considered to be unique in order to make the protocols comparable.

5.1 Comparison on DSR, AODV, and OLSR Protocols

In Figure 4a, the network load in terms of bit-per-second is illustrated for all three protocols. As expected, the OLSR protocol which is a proactive protocol, has the highest network load. After that, we have AODV, and finally DSR with less network loads.

In Figure 4b, the time averaged energy consumption is depicted in term of Joule. We can see that the highest energy consumption belongs to OLSR algorithm for its being proactive, and then AODV, and DSR are the next ones.

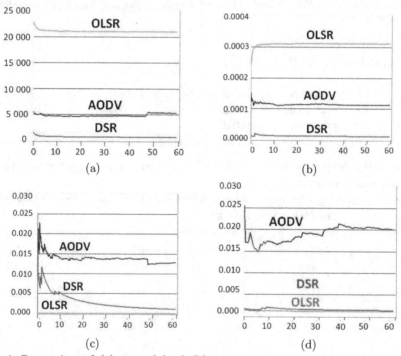

Fig. 4. Comparison of: (a) network load, (b) energy consumption, (c) dropped packets, (d) network delay (horizontal axis is time)

In Figure 4c, the time averaged dropped packets of the network are shown. Generally, this packet drop is a result of TTL time finishing. If one protocol, uses much time for choosing a route to sink, then the packets with limited life length are possible to be dropped. However, the effective protocols can decrease the packet dropping rate intelligently. Therefore, we anticipate that DSR has a lower packet drop rate than AODV [20]. In OLSR, almost no packet is dropped which is a result of predetermined routes for the source-sink connection. Besides, the AODV has the most packet dropping rate.

In Figure 4d, the time averaged overall delay of the network is shown. In AODV and DSR, based on its specific conditions, it takes a fixed time for the route from source to sink to be established. Therefore, the starting delay of the network is initialized by a large value [21]. Again, we see that OLSR has the least delay among other methods, and DSR has a less delay than AODV which are both of the on-demand kind.

Based on the presented simulation results in this subsection, it is intuitively clear that the DSR protocol has the best responses for a network with limited number of nodes.

5.2   Simulation Results for DSR_OP Protocol

As shown in previous subsection, the DSR routing technique has a higher performance than AODV and OLSR techniques. Therefore, in this subsection, we analyze performance of the proposed DSR_OP technique. The conditions for simulations of this subsection are considered as following:

- Simulation region is $10 \, \text{km} \times 4 \, \text{km}$.
- One sink node, two source node which generate traffic, and 4 normal node.
- Time of simulation is 20:00 hours.

In Figure 5a, the $n - 3$ node is the bottleneck and both $N - 1$ and $N - 2$ sources communicate with the sink via $n - 3$. In our proposed DSR_OP algorithm, in order to prevent $n - 3$ from being died, after consumption of 30 % of the node battery, it is departed from the $n - 4$ route for a random time. However, the link $n - 3$ and $n - 5$ remains connected. Therefore, $n - 4$ is forced to maintain its connection with the sink (in this example, via $n - 6$). The new links of the network are depicted in Fig. 5b.

After about 15 hours, with 30 % of energy consumption in node $n - 3$, this node is withdrawn of the communication link and $n - 6$ is used for establishing the connection. The energy consumption of node $n - 3$ is depicted in Fig. 5c, which shows that its energy consumption is decreased after the 15th hour.

In Figure 5d, the routing traffic received in the nodes $n - 3$ and $n - 6$ are shown. After the 15th hour, traffic of $n - 3$ is decreased and instead traffic of $n - 6$ is increased.

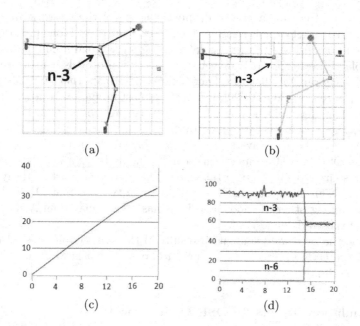

(a)   (b)

(c)   (d)

Fig. 5. Simulation of DSR_OP: (a) routes before leaving $n-3$, (b) routes after leaving $n-3$, (c) energy consumption in $n-3$, (d) routing traffic received by nodes $n-3$ and $n-6$ (horizontal axis is time)

6   Conclusion

In this paper, the routing protocols for ZigBee-based WSNs were studied, where the WSN was designed for a border surveillance scenario. Three well-known routing protocol were compared in detail which showes that DSR is suitable for border surveillance applications and outperforms in all specified scopes (i.e. traffic load, delay, packet loss, and energy consumption). Next, the AODV protocol has better result in comparison with OLSR. Finally, the OP-DSR was suggested as an improved version of the DSR in terms energy management in the network and extending the network life time. The validity checks were all confirmed by simulations in OPNET.

References

1. García-Hernández, F., Ibargüengoytia-González, H., García-Hernández, J., Pérez-Díaz, A.: Wireless Sensor Networks and Applications: a Survey. IJCSNS International Journal of Computer Science and Network Security 7(3) (March 2007)
2. Arora, A., et al.: A line in the sand: A wireless sensor network for target detection, classification, and tracking. Computer Networks Special Edition 46(5), 605–634 (2004)

3. Chen, Y., Chuah, C.-N., Zhao, Q.: Network Configuration For Optimal Utilization Efficiency of Wireless Sensor Networks. In: IEEE MILCOM 2005, Atlantic City, NJ, USA, pp. 17–20 (2005)
4. Jawhar, I., Mohamed, N.: A Hierarchical and Topological Classification of Linear Sensor Networks. In: Wireless Telecommunications Symposium, Prague, pp. 1–8 (2009)
5. Onur, E., et al.: Surveillance Wireless Sensor Networks: Deployment Quality Analysis. IEEE 21(6), 48–53 (2007)
6. Onur, E., et al.: Surveillance with wireless sensor networks in obstruction: Breach paths as watershed contours. Computer Networks Special Edition 50(3), 428–441 (2010)
7. Dudek, D., et al.: FleGSens – Secure Area Monitoring Using Wireless Sensor Networks. World Academy of Science, Engineering and Technology (2009)
8. http://www.zigbee.org/
9. Severino, R.: On the use of IEEE 802.15.4/ZigBee for Time-Sensitive Wireless Sensor Network Applications. In: School of Engineering. Polytechnic Institute of Porto (2008)
10. Al-Karaki, J.N., Kamal, A.E.: Routing techniques in wireless sensor networks: A survey. IEEE Wireless Communications 11(6), 6–28 (2004)
11. Akkaya, K., Younis, M.: A survey on routing protocols for wireless sensor networks. Ad Hoc Networks 3(3), 325–349 (2005)
12. Perkins, C., Belding-Royer, E., Das, S.: Ad hoc On-Demand Distance Vector (AODV) Routing (February 2003)
13. Huhtonen, A.: Comparing AODV and OLSR Routing Protocols. In: Seminar on Internetworking, Sjökulla (2004)
14. Holter, K.: Comparing AODV and OLSR. Helsinki University of Technology (2005)
15. Ali, S., Ali, A.: Performance Analysis of AODV, DSR and OLSR in MANET. Department of Computing Systems. School of Engineering Blekinge Institute of Technology, Sweden (2009)
16. Poonia, R., Sanghi, A.K., Singh, D.: DSR Routing Protocol in Wireless Ad-hoc tworks: Drop Analysis. International Journal of Computer Applications 14(7), 975–8887 (2011)
17. Clausen, T., Jacquet, P.: Optimized Link State Routing Protocol (OLSR). Project Hipercom, INRIA (2003)
18. Jacquet, P., et al.: Optimized link state routing protocol for ad hoc networks. In: IEEE INMIC 2001 (2001)
19. Bashyal, S., Venayagamoorthy, G.K.: Collaborative Routing Algorithm for Wireless Sensor Network Longevity. In: 3rd International Conference on Intelligent Sensors, Sensor Networks and Information (ISSNIP) (2007)
20. Narsimha, V.B.: Comparison of Routing Protocols in Mobile Ad Hoc Networks. Global Journal of Advanced Engineering Technologies (2012)
21. Ali, A., Akbar, Z.: Evaluation of AODV and DSR Routing Protocols of Wireless Sensor Networks for Monitoring Applications. Department of Electrical Engineering Blekinge Institute of Technology, Sweden (2009)

Quantitative Risk Analysis for Data Storage Systems

Tomasz Bilski

Poznań University of Technology
tomasz.bilski@put.poznan.pl

Abstract. In IT systems quantitative risk analysis is a method for security risk assessment used as a part of risk management process which in turn is a part of security policy. The main advantage of the method (in comparison to other techniques) is its accuracy – a better start point for security policy definition. Obviously the accuracy is directly related to input data dependability. The basic problem of the method is related to acquisition of input data necessary to perform the analysis. Data should be complete and reliable. The main purpose of the paper is to define comprehensive set of data necessary for the quantitative risk analysis for data storage systems (including magnetic disks and solid state disks) and to discuss trustworthiness of the data. Such comprehensive set of data should include data related to storage technology, features of storage processes (e.g. compression, deduplication), security events probabilities. Some of the parameters are dynamic, they change in time, they are related to environmental conditions. Different sources, different means of data acquisition are presented together with discussion on trustworthiness and dependability of the acquired data in the second part of the paper[1].

Keywords: data storage, security, risk analysis.

1 Introduction

One of many legal requirements related to IT systems is the obligation to prepare and implement information security policy. The process of policy development is composite and multistage. An important part of the process is risk analysis. Risk in computer system security domain is defined as a function (e.g. arithmetic product) of negative event occurrence probability and level of the event impact on IT resources (data, services, hardware) [1]. Complete risk analysis process should include each security requirement (confidentiality, integrity, availability), all categories of security violations (human/adversarial and non-adversarial), all contexts/states (parts of IT system) in which data may be violated. From data context point of view there are three circumstances: data storage, data processing and data transmission. In the next sections of the paper we will focus on risk

---

[1] The research project is scheduled for years 2010–2013 and partly supported by scientific grant from the Polish Ministry of Education and Science.

A. Kwiecień, P. Gaj, and P. Stera (Eds.): CN 2013, CCIS 370, pp. 124–135, 2013.
© Springer-Verlag Berlin Heidelberg 2013

analysis related to non-adversarial threats in data storage part of IT system, including networked storage systems.

2 Risk Analysis

In general, risk analysis is a process used to evaluate negative impacts of all security related events (intentional and accidental) on IT system resources. Risk analysis is based on many facts and parameters related to: threats, vulnerabilities, incidents, events, assets, business impact. The process is a part of risk management, consisting of: risk framing, risk analysis, risk treatment and risk monitoring. The process is often an initial part of security policy establishment [2]. The main goal of risk analysis is to categorize all security problems in order to optimize security policy.

Risk analysis methods may be classified with a use of many different criteria. For example from the point of view of utilized tools we have: functional dependency network analysis, attack tree analysis for adversarial threats, fault tree analysis for other types of threats. From the point of view of input and output data type we have two common methods for risk analysis: qualitative and quantitative. Qualitative risk analysis is based on some arbitrary levels of threats, events initiation or occurrence probability, likelihood threat events result in adverse impact and negative impacts extent (e.g. low, medium, high; but the number of levels is also arbitrary). Qualitative risk analysis is relatively simple and easy but the results are imprecise. For example one has to assess risk qualitative level for an event that has very small probability and high impact – it is not possible to compute arithmetic product of adjectives.

The solution is semiquantitative risk analysis. The method is based on numbers but the numbers do not have any meaning outside the analysis. For example one may assign values for levels of probability and levels of impact: low=2, medium=5, high=10. Using the values one may perform some arithmetic operations for risk assessment. For example multiplication of low probability (2) and high impact (10) gives risk level value equal to 20.

Quantitative risk analysis is based on numbers (risk assessment metrics) related to threats, probabilities and impact extent. The numbers used in quantitative methods have their meaning also outside the analysis. Quantitative method is time consuming. It requires trustworthy data that may be difficult to obtain (economic value of a given file, database or service is relatively hard to establish). The advantage of the method is that the results of the analysis are more precise than the results of qualitative analysis – therefore quantitative results are good starting point for security policy definition [3].

A compromise (relatively fast and precise) is a two-phase process. In the first coarse phase qualitative risk analysis is done for all threats in a system. The phase lets to categorize and select resources and threats. Selected resources (e.g. most valuable) and selected threats are included in the second phase with detailed quantitative analysis.

3 Data for Quantitative Risk Analysis

3.1 Categories of Data

Quantitative risk analysis is based on real metrics. Risk is defined as arithmetic product of security violation event probability and business impact of the violation measured e.g. in money. The second part of the product is dependent on data and services values in a given IT system. The economic value of a given file lost due to hardware failure or malicious attack may not be assumed universally. In general economic value of data is related to quality and quantity of information pieces. Such costs should be estimated using mission/business impact analyses and asset inventories. On the other hand, probability of some security (non-adversarial) violation events may be estimated or calculated with a use of a particular IT hardware characteristic.

The numbers in quantitative methods have their meaning also outside the analysis and may be associated with some measurable features related to adversarial and non-adversarial threats (e.g. storage hardware error or failure probability). In the case of data storage system one has to include, in analysis, the following metrics:

- transmission BER (Bit Error Rate) for communication channel,
- information BER for communication channel,
- lost packet ratio,
- BER for storage device,
- UBER (Uncorrectable Bit Error Rate) for storage device,
- MTBF (Mean Time Between Failures),
- number of read/write cycles,
- risk data related to particular processing techniques (e.g. deduplication),
- adversarial/human threats quantitative metrics.

Network reliability metrics are necessary in the case of networked storage systems such as cloud storage, which are extensively used in contemporary information systems.

3.2 MTBF

MTBF is reliability measure commonly used by disk (magnetic as well as semiconductor) manufacturers. MTBF is defined as number of hours between two consecutive failures of a given device. There are some pitfalls. MTBF only applies to the aggregate analysis of large numbers of devices; it says nothing about a particular unit. It is defined for recommended operating temperature. MTBF doesn't take into account such factors as: project errors, production errors, human errors and environmental factors.

Furthermore, the parameter does not directly give probability (or risk level) of device failure. Nevertheless, it is possible to compute failure rate (e.g. probability of failure over the course of 1 year) with a use of MTBF. Number of failures per hour is simply equal to 1/MTBF. Number of failures per year AFR (Annualized

Failure Rate) is equal to product of failures per hour and number of working hours per year. For example, assuming MTBF is equal to 1 200 000 hours (representative value for contemporary hard disks), disk is working 24 hours a day (8 760 hours per year) we have failure per year rate $(1/1\,200\,000) \times 8\,760 = 0.0073$ (0.73 %).

3.3   BER

Storage device and communication channel reliability is usually measured with a use of BER.

BER in Communication Channels. In communication channels two distinct BERs are defined. First of all, the transmission BER equal to the number of incorrect bits received by receiver divided by the total number of all transferred bits. Another type of BER is called information BER equal to the number of bits, that remain incorrect after the error correction processes performed by the receiver, divided by the total number of decoded bits. Transmission BER is usually larger than information BER – the ratio is related to EFR (Error Correction Code Failure Rate).

Transmission BER is dependent on transmission medium and protocols (Table 1). It may be affected by such phenomena as: transmission channel noise, interference, distortion, bit synchronization problems, attenuation, wireless multipath fading, etc.

Table 1. Nominal transmission BER for different channels specifications

Communication channel	Transmission BER
IEEE 802.11	10^{-5}
Ethernet 100 BASE-TX	10^{-10}
Ethernet 1 Gbit	10^{-12}
Fibre Channel	10^{-12}
OBSAI RP3	10^{-15}

From risk analysis point of view lost packet ratio is more important. According to test data [4] it may be assumed that globally lost packet ratio is below 1 %. It is obvious that risk level related to the lost packet parameter is dependent on protocols in higher (from 4 to 7) layers of ISO/OSI model. TCP (Transmission Control Protocol) retransmission is common solution to lost or incorrect packets problem.

BER in Storage Systems. Similarly, in storage media and storage devices two distinct error rates are defined: BER and UBER. BER is calculated after data read and before error correction and UBER (Uncorrectable Bit Error Rate) is equal to the number of erroneously read bits that could not be corrected by error correction mechanisms in read module (Table 2).

Table 2. Representative UBER for storage media

Storage medium	UBER
Flash SLC	$10^{-9} - 10^{-11}$
Flash MLC (2 bits/cell)	$10^{-5} - 10^{-7}$
Flash TLC (3 bits/cell)	$10^{-3} - 10^{-4}$
SCSI hard disk	$10^{-15} - 10^{-16}$
ATA hard disk	$10^{-13} - 10^{-15}$
DLT tape	$10^{-11} - 10^{-15}$
CD Mode 1	10^{-14}
CD Mode 2	10^{-9}
DVD	10^{-15}

UBER for Magnetic Media. Specifications (provided by manufacturers) of contemporary disks include UBER in very wide range from 10^{-11} to 10^{-18}. Ordinary PATA (Parallel Advanced Technology Attachment) and SATA (Serial Advanced Technology Attachment) disks have UBER from 10^{-13} to 10^{-15} (one uncorrectable error in every 10 TB to 1000 TB read). Ordinary SCSI (Small Computer System Interface) or SAS (Serially Attached SCSI) disks have UBER 10^{-15} to 10^{-16} (one uncorrectable error in every 1 PB to 10 PB). High end disks have UBER as low as 10^{-18}.

It must be added that UBER in hard disk specifications is related to drive error rate. In order to calculate total error rate we have to take account of error rate of disk controller, cables, PCI (Peripheral Component Interconnect) bus, memory, processor. Only some of the errors may be masked by the controller or operating system. BERs and UBERs for hard disks seem relatively small but in large IT systems such services as: data pipeline processing, data mining, and backup/restore routinely read tens of terabytes per day. So, impact of bit error risk becomes significant.

UBER for Flash. UBER for flash memory media (including solid state disks) is much greater than in magnetic media. The highest UBER ($10^{-3} - 10^{-7}$) is characteristic for Multi Level Cell (MLC) NAND technology.

Flash is a matrix of cells (transistors) and bits of data are represented by electrical charge accumulated in floating gate. In Single Level Cell (SLC) NAND technology each cell stores single bit of data, in MLC each cell stores many bits of data (3 bits in TLC (Triple Level Cell) technology). The separation between adjacent voltage levels in MLC and TLC is much smaller than in SLC flash. This reduced separation affects data reliability and performance, due to the following causes: electrons stored in adjacent levels can be disturbed with electrical noise and can shift from one level to another, causing a higher bit error rate. The risk related to relatively high BER should be carefully evaluated especially in the case of flash memories, that are used very frequently, e.g. solid state disks.

BER in flash memory is dynamic. In [5] it was shown that the average BER is a function of the number of program/erase cycles – for example after about 3500 program/erase cycles BER in TLC may drop from 10^{-4} to 10^{-3}. Finally, it must be added that semiconductor storage media are very resistive to such factors as magnetic fields, shocks, high temperature, high humidity but they are sensitive to electrostatic field.

UBER for Optical Media. BER for optical media is relatively high, usually between 10^{-5} and 10^{-6}. UBER is reduced by special data encoding and redundancy and is dependent on optical disk mode of operation. For example in the case of CD-ROM mode 1 (which is protected by RSPC (Reed Solomon Product Code) and CIRC (Cross Interleaved Reed Solomon Code)) the typical UBER is about 10^{-14}. On the other hand, CD-ROM mode 2 (protected only by CIRC) has UBER equal to 10^{-9}. DVD uses only one level of error detection and correction RSPC code. RSCP in DVD is more advanced, more parity bits are generated by the RSPC in DVD than in CD-ROM. With only one level of error detection and correction code a DVD UBER of 10^{-15} can be achieved. For Blue-ray disks (BD) BER is at the level of 10^{-6} [6].

Given above BER and UBER data are correct if disks are manufactured according to standard specifications (e.g. standards specifying details of internal disk structure). Nevertheless there are some non-standard disks, manufactured to increase the capacity of optical readout system. One of such methods is based on reduction of the track pitch. It was proved that UBER for such non-standard disks is much worse [7].

3.4   Read/Write Cycles

Each storage device has limited number of operations (or read/write cycles) which may be successfully terminated. It may be assumed that malfunction probability and data loss risk increase quickly after a given number of operations. This seems obvious in the case of mechanical devices such as hard disks and streamers. Nevertheless, the same phenomena is observable in flash memory.

Internal structure of the flash transistor and mode of operation cause the transistor properties deteriorate during normal use. Each block of a flash storage device can only be erased (and therefore written) a limited number of times

Table 3. Read/write cycles

Storage medium	Number of read/write cycles
Flash SLC	100 000 – 1 000 000
Flash MLC 2 bits/cell	10 000
Flash MLC 4 bits/cell	1 000
Hard disk	1 000 000
DLT tape	10 000 load cycles

before it fails. The problem is especially visible in MLC flash technology. MLC chips with 4 bits per cell become useless after just 1000 write/erase cycles. MLC flash lifetime may be slightly extended by controllers and special block load balancing algorithms (Table 3).

3.5 Hash Collision Risk in Deduplication

Contemporary backup systems use some methods to reduce volume of data. Lossless data compression is an example of such method. Another way to reduce data volume is deduplication – removal of non-unique data units from backup system.

Deduplication process is based on hashes (e.g. MD5, SHA) – hashes are computed for each unit (or chunk) of data and compared in order to detect identical units. The deduplication unit size (deduplication granularity) is usually between 4 KB to 128 KB (the lowest unit size is equal to allocation unit size on most NTFS (New Technology File System) and EXT3 volumes and is an important feature which has to be optimized.

Smaller unit size saves more storage, results in more granular fingerprints and in identifying more duplicates. However, smaller chunks have additional costs in terms of database size and deduplication time. Higher fragmentation normally results in more file system metadata and hence can require more storage. The space consumed by the deduplication database and the increased file system metadata can reduce the savings achieved via deduplication. Additionally, fragmentation can also have a negative effect on performance.

Larger unit sizes normally result in a smaller deduplication database size, faster deduplication and less fragmentation. These benefits come at the cost of less storage savings.

The same hashes that are used for block comparison are used as indices to lookup tables. Only one unit of identical data is stored entirely, all subsequent units with the same data (in fact units with the same hash values) are replaced by pointers to lookup table. Deduplication is in opposition to redundancy and encryption. Deduplication risk is associated to hash collisions – two (or more) different data units may give the same hash value and one (or more) of the data units is lost due to deduplication based on hash comparison. Risk level related to data availability is much higher in backup systems integrated with deduplication.

Furthermore, the general requirement for encryption systems says that two encrypted units of the same data should be different – so deduplication of such encrypted units could not be performed. Here, we have an old dilemma – confidentiality vs. functionality.

From the quantitative risk analysis point of view we have to analyze risk related to hash function collisions. Statistically, there are multiple possible units that have the same hash. If two different units of data give the same hash values then one of the units will be mistakenly removed from backup system and data availability will be lost [8]. Probability of collision in hash functions is dependent on hash length and hash algorithm.

Let's analyze collision probability with a use of birthday paradox. In general the expected number of N-bit hashes that can be generated before getting a collision is not 2N, but rather only 2N/2.

More generally, number n of different data units such that the collision probability is equal p for hash function with d different outputs may be approximated with a use of the following equation:

$$n(p, d) \approx \sqrt{2d \times \left(\ln \left(\frac{1}{1 - p} \right) \right)} .$$

Approximately (from the equation), if the number of data units is equal to square root of the number of all hash values for the units, then collision probability is about 50 % [9]. Approximate volumes of data storage system for 0.5 collision probability in deduplication system (it is assumed that deduplication unit size is equal to 4 KB) and two hash functions are given in Table 4. In the second table (Table 5) very small collision probability equal to 10^{-18} has been chosen since this is also the lowest level of UBER of high end disk.

Table 4. Approximate volume of data for 0.5 collision probability

Length of hash	Approx. number of 4 KB units	Approximate volume of data
128 bits (MD5)	2.2×10^{19}	9×10^{10} TB
512 bits (SHA-512)	1.4×10^{77}	6×10^{68} TB

Table 5. Approximate volume of data for very small 10^{-18} collision probability

Length of hash	Approx. number of 4 KB units	Approximate volume of data
128 bits (MD5)	2.6×10^{10}	9×10^{1} TB
512 bits (SHA-512)	1.6×10^{68}	6×10^{59} TB

Collision probability of hashes of data units in deduplication system is related to hash length and number of data units. So, in order to decrease collision probability we may use longer hashes or/and greater data units for deduplication. However, in both cases, probability reduction is at the expense of performance and data volume reduction factor (longer hashes mean longer pointers to data units) – longer pointers to units of data have to be stored. For example four times longer hash output means approximately 8 times longer hash function processing time [10] and four times longer pointers to units of deduplicated data.

Similarly, probability may be reduced by increasing data unit size (e.g. from 4 KB up to 128 KB). Larger unit sizes normally result in a smaller deduplication database size, faster deduplication and less fragmentation of files. However in the case of small files deduplication unit size may be greater than file and instead of storing small file we have to store large unit – this can reduce the savings achieved via deduplication.

4 Sources of Data

In order to use quantitative analysis one has to collect quantitative data on all vulnerabilities, security threats, probabilities of negative events and impact levels in a given system. Sources of vulnerabilities and threat probabilities can be either internal or external to organizations. Internal sources can provide insights into specific threats to particular data storage system and can include, for example: incident reports, security logs and monitoring and test results. External sources can include manufacturers, standardization institutes, cross-community organizations and scientific teams. The sources provide, for example: hardware technical specifications, hardware test results, security assessment reports and statistics, vulnerability assessment reports, risk assessment reports, incident reports. In each case trustworthiness of the sources and data should be evaluated.

4.1 Standardization Bodies

There are hundreds of standards related to data storage media and devices. Most important storage standardization institutions are: ISO, IEC, ECMA, ANSI. A number of standardization documents provide qualitative and quantitative requirements for storage media and storage devices. The requirements include numerical data on reliability features such as BER and UBER. Such data may be treated as dependable, it is assumed that devices and media built according to standards will fulfill the requirements.

It must be noted that storage media reliability parameters described in standards and specifications are usually presented for a given set of conditions (temperature, humidity and other working conditions). For example magnetic disk UBER is related to internal temperature, work load, mode of operation. It is estimated that UBER increases by 2–3 % with 1 degree Centigrade increase of operating temperature [11]. Furthermore, each hardware reliability parameter deteriorate with time.

4.2 Manufacturers

First of all, reliability parameters of IT hardware and software may be obtained directly or indirectly from hardware and software instruction manuals and technical specifications. The main problem is that the trustworthiness of manufacturers' specifications is relatively low. Results of reliability tests carried on by a number of third parties [12,13,14] proved that manufacturers' MTBF values are undependable, they are much greater than experimental, test data. Furthermore, failure rates are not constant. There are differences across suppliers and differences also within, for example, specific disk family from a single supplier [15].

4.3 Internal Monitoring

An important method for data acquisition is based on tests, monitoring and audits performed regularly inside the system. One may monitor storage devices as well as network reliability.

Storage device performance may be monitored with such tools and methods as PFA (Predictive Failure Analysis) and SMART (Self-Monitoring Analysis and Reporting Technology). PFA and SMART are monitoring performance of the disk drive, logging and analyzing data from internal measurements, and recommending replacement when specific thresholds are exceeded. At given intervals, PFA's Generalized Error Measurement (GEM) performs a suite of self-diagnostic tests, which measure changes in the drive's characteristics. GEM measures various magnetic parameters of the head and disk, as well as figures of merit for the channel electronics. The analysis of the error log information is performed periodically during idle periods. Data collected by PFA or SMART are used internally by operating system but they are also available to users. For example, smartmontools, on GPLv2 license, contains utility programs (smartctl, smartd) to control/monitor storage systems.

Data related to a given network reliability may be obtained with a use of such popular tools, services and protocols as: ping, SNMP, syslog, Nmap.

4.4 Network Monitoring, Statistical Data

A lot of historical data related to tests of Internet performance (with, for example, lost packet ratio) are available in some public databases. An example is PingER administered by SLAC (Stanford Linear Accelerator Center) [4]. The problem is that such databases provide historical data and simple extrapolation of the data for potential risk assessment is not possible.

In [16] it was shown that predicting average network parameters (such as lost packet ratio, RTT (Round Trip Time), throughput) with a use of historical, statistical data is very difficult. In long term perspective it is hardly possible, even with a use of great number of statistical data. Approximating statistical data with regression trends is scarcely possible. The best approximation for some sets of older data is linear regression but this regression type is useless if more recent data will be taken into account. In short-term perspective predicting (with historical data) future QoS (Quality of Service) level for normal WAN (Wide Area Network) operation is much simpler. Nevertheless irregular disasters e.g. with submarine cable faults (caused by earthquakes) make this predictability more complicated. Significant QoS deterioration together with high fluctuations of performance parameters may last weeks and even months after disaster.

4.5 Adversarial Threats, Statistical Data

Many different data sources need to be analyzed in order to quantifiably capture the elements of human (external as well as internal; it must be noted that risk related to insiders is usually greater than risk related to outsiders) threats for data storage system. Objective and accurate statistical data for such aspects as human threat factors may be collected from outside institutions, trade or industry groups, insurance companies and scientific teams [17]. Capabilities and intents of adversaries, targeting of organizational operations, assets, or individuals are constantly under control. Some statistical data (e.g. on adversarial threats' probabilities) are available from third parties monitoring and evaluating data security such as: CSI (Computer Security Institute), Symantec, McAfee, SecurityFocus.com, SecurityWatch.com, SecurityPortal.com, SANS.org. Statistical data on adversarial attacks indicate that risk level is related to the organization profile. Some organizations (e.g. financial institutions) are attacked more frequently than others.

Trustworthiness of third party statistical data is related to data collection methods. Some institutions provide details on methods (e.g. sample size) used to accumulate and process data. In general, dependability of data from such third parties should be evaluated using status and reputation of a given institution.

5 Conclusion

There are many parameters that should be included in quantitative risk analysis for data storage system. The values of the parameters may come from many different sources. In general the dependability of the sources is limited. For example manufacturers' MTBF values are unreliable, they are much greater than experimental data. There are differences across suppliers and differences also within specific disk family from a single supplier. Reliability parameters are related to working conditions. Failure rates in storage media and storage devices are not constant, they deteriorate with time and number of operations. Relatively important is risk evaluation in the case of innovative technologies such as: flash memory and deduplication processes.

Probabilities of some dangerous events related to data storage systems may seem very small (e.g. UBER $= 10^{-15}$). Nevertheless, one has to keep in mind that data centres store and transfer enormous volumes of bytes, often reaching many petabytes (petabyte is equal to 10^{12} bytes or about 10^{13} bits). So, actual risk level becomes much greater than it may seem at first look.

Furthermore, one has to keep in mind, that there are many risks and their combined effect is much greater than simple addition of particular threats probabilities. Similarly, as in many other situations, we have many trade-offs which have to be carefully evaluated. All such observations should be taken into account in quantitative risk analysis.

References

1. ISO/IEC 27005 Information technology – Security techniques – Information security risk management. ISO/IEC (2011)
2. ISO/IEC 17799 Information Technology – Code of practice for information security management. ISO/IEC (2005)

3. NIST Special Publication 800-30 (2011) – Guide for Conducting Risk Assessments. NIST (2011)
4. Cottrell, L., Matthews, W., Logg, C.: Tutorial on Internet Monitoring and PingER at SLAC. SLAC (2007), http://www.slac.stanford.edu/comp/net/wanmon/tutorial.html
5. Yaakobi, E., Grupp, L., Siegel, P.H., Swanson, S., Wolf, J.K.: Characterization and Error-Correcting Codes for TLC Flash Memories. In: International Conference on Computing, Networking & Communications, Maui Hawaii (February 2012), http://cseweb.ucsd.edu/users/swanson/papers/ICNC2012TLC.pdf
6. Blu-ray Disc Format A Physical Format Specifications for BD-RE, 3rd edn. Blu-ray Disc Association (October 2010), http://www.blu-raydisc.com/Assets/Downloadablefile/ BD-RE-physical-format-specifications-18325.pdf
7. Huang, J., Lo, F.: Effect of reducing track pitch in DVD-ROM. IEEE Transactions on Magnetics 41(2), 1073–1075 (2005)
8. Cannon, D.: Data Deduplication and Tivoli Storage Manager. IBM Corporation 2009 (2009), http://www.ibm.com/developerworks/wikis/download/ attachments/106987789/TSMDataDeduplication.pdf?version=1
9. Mathis, F.H.: A Generalized Birthday Problem. SIAM Review 33(2), 265–270 (1991)
10. Nakajima, J., Matsui, M.: Performance Analysis and Parallel Implementation of Dedicated Hash Functions. In: Knudsen, L.R. (ed.) EUROCRYPT 2002. LNCS, vol. 2332, pp. 165–180. Springer, Heidelberg (2002)
11. Bilski, T.: Storage media security. In: NATO Regional Conference on Military Communications and Information Systems 2001. Partnership for CIS Interoperability, Wojskowy Instytut Lacznosci, Zegrze (October 2001)
12. Schroeder, B., Gibson, G.: Disk failures in the real world: What does an MTTF of 1,000,000 hours mean to you? In: Proceedings of the Fifth Usenix Conference on File and Storage Technologies FAST (February 2007)
13. Sun, F., Zhang, S.: Does hard-disk drive failure rate enter steady-state after one year? In: Proceedings of the Annual Reliability and Maintainability Symposium. IEEE (January 2007)
14. Shah, S., Elerath, J.G.: Disk drive vintage and its effect on reliability. In: Proceedings of the Annual Reliability and Maintainability Symposium, pp. 163–167 (January 2004)
15. Elerath, J.G., Pecht, M.: Enhanced reliability modeling of RAID storage systems. In: 37th Annual IEEE/IFIP International Conference on Dependable Systems and Networks, Edinburgh, UK (June 2007)
16. Bilski, T.: QoS Predictability of Internet Services. In: Kwiecień, A., Gaj, P., Stera, P. (eds.) CN 2010. CCIS, vol. 79, pp. 163–172. Springer, Heidelberg (2010)
17. McCumber, J.: Assessing and Managing Security Risk in IT Systems. A Structured Methodology. Auerbach Publications (2004)

Onion Routing Efficiency for Web Anonymization in Various Configurations

Tomas Sochor

University of Ostrava, Ostrava, Czech Republic
tomas.sochor@osu.cz
http://www1.osu.cz/home/sochor/en/

Abstract. Web traffic anonymization is hiding the originator of the web request. Although is required in some circumstances, such feature is not included in the WWW service. Anonymization can be reached using various commercial and public domain tools. The obvious adverse aspect of anonymization is the significant slowing down of the anonymized traffic compared to the normal traffic. This study focuses on measurement of the parameters of such slowing down (in terms of transmission speed and latency). This study is based on previous studies concluding that TOR (The Onion Routing) is the best available free web anonymization tool. Nevertheless the TOR's cost of operation in terms of response time increase is still too high so the main goal of this paper is to find ways how to make them lower. The extensive set of WWW pages and files was formed and the latency and response time during their download was measured repeatedly both with anonymization in various configurations and without it. The result showed that there are some ways how to improve the TOR efficiency but despite them the cost still remains too high for ordinary use.

Keywords: web anonymization, TOR, WWW, I2P, latency, round-trip time, transmission speed, Mozilla Firefox, AdBlock Plus, Firebug, NoScript, Torbutton, iMacros for Firefox.

1 Web Anonymization Using TOR

There are situations when a user could desire to hide details about his/her activity in the Internet, especially on a web site. Certain part of the task could be done by encryption (e.g. VPN) but the traces of the activity still remain at the user's computer and at the web server as well. If a user desires to eliminate them he/she needs a tool performing so-called *anonymization*. TOR (The Onion Routing) is one of the best known free anonymization tools. Recently also other tools (e.g. JAP, I2P) have emerged. The author's recent study [1] focused on the comparison of TOR with JAP and I2P concluding that TOR is the best present anonymization tool. More thoroughgoing study focusing on TOR and especially its capability to improve its behavior was performed and the results are presented below. The primary aim of this study was to measure the latency

A. Kwiecień, P. Gaj, and P. Stera (Eds.): CN 2013, CCIS 370, pp. 136–145, 2013.
© Springer-Verlag Berlin Heidelberg 2013

and response time inside web browsers using TOR-anonymized WWW traffic having various configuration.

As explained above only TOR is studied here in details. The majority of anonymization tools including TOR rely on concealing the IP address of the originator. The operation of public anonymization tools is possible thanks to the fact that many people throughout the world allow using their computers as anonymization nodes (in case of TOR called *onion routers*). The TOR client installed on the user's computer initiates the forming of the network (called *TOR circuit*) composing 3 onion routers. The initiation of the TOR circuit is the most vulnerable phase of anonymized communication. The compromising of this phase is avoided by authentication of the list of available TOR nodes. The communication between these TOR routers is encrypted so that every onion router can directly communicate only with their 2 neighbors. The result is that the other (non-neighboring) nodes do not know anything about the node (including IP address). More detailed information about TOR operation can be found in [2].

An obvious drawback of anonymization tools is significantly increased latency and total response time (i.e. the decrease of transmission speed). Such significant worsening of transmission quality is obviously not only due to adding more nodes into the path between the client and the target but due to encryption of data traffic between intermediate nodes as well. The main aim of the study described here was to quantify such decrease for various TOR configurations so as to allow the assessment of efficiency of using TOR. Some subsequent experiments have been performed using I2P too but they demonstrated insufficient stability and limited usability of I2P. More details about I2P are available in [3]. The comparison between TOR and I2P can be seen in the Fig. 1.

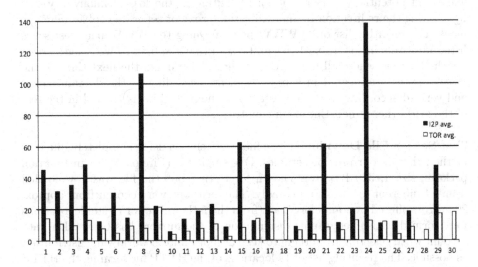

Fig. 1. Load time of web pages from the Test set 1 compared between TOR (averages from 15 measurements) and I2P (averages from 3 measurements, missing I2P column means repeatedly faulty measurement)

2   Measurement Methods

2.1   Selection of Model Traffic

Similarly as in the case of the previous work [1] it was decided to limit the scope of the study to WWW traffic. The WWW traffic was chosen not only because of its significant and stable share in total Internet traffic but also because of the nature of WWW traffic causing it is much more likely to require anonymization. The most important reason was the limitation of the most anonymization tools to WWW (in some cases even only to selected WWW browsers). This was the reason why our previous studies were focused on WWW traffic so we decided to foliow this model to achieve comparability with the older results.

The experiments were intended to measure latency and total response (or download) time in various TOR configurations so the important part of the study was proper selection of the test sets. Two separate test sets were formed, the first test set was the set of WWW pages and the other was the set of locations of a single file available via HTTP. The test sets were built based on the corresponding sets used for previous measurements. The original test sets were unable to use in the same composition because the significant part of the sets changed since the last measurements were performed.

Test Set of WWW Pages. The web pages to test were chosen so that they are stable in time. Multimedia-rich or active-contents web pages were avoided as well as pages requiring active plugins (e.g. Silverlight, Flash, Java etc.) to display properly. It should be noted that many plug-ins could pose a problem in respect to anonymity because web browser plugins could operate independently and could potentially perform actions resulting in the loss of anonymity, e.g. by ignoring the redirection to anonymization proxy or separate cookies management. The resulting list of 30 WWW pages forming the WWW page test set is listed in details in the Table 1. It can be seen that the first 10 web pages are in Czech Republic where all measurements have been done, the next third of the WWW pages are located in Europe (at least according to the domain names and verified according to IP addresses) and the final third is located in the rest of the world (i.e. in non-European countries).

Test Set for File Download. The files were chosen using completely different method than in our previous studies. One single file (Firefox 16.0.1 installation package, size 19 329 KB) was chosen for testing but the file was downloaded from 14 different locations repeatedly. The locations were chosen from approx. 20 mirrors offered by sourceforge.net (from them the subset of 14 stable locations were formed). Also geographic distribution of the mirrors has been taken into account when selecting from them to keep as mot uniform worldwide distribution as possible. The resulting list of 14 locations of the HTTP file forming the HTTP file test set is listed in the Table 2 below.

Table 1. Details of WWW pages test set

Item	URL	Size [KB]	Country	HTML	CSS	JS	Images
1	www.databazeknih.cz/zajimavosti-knihy/kmotr-34837	406.58	Czech	1	5	12	28
2	http://www.pepak.net/dalsi-projekty/	197.83	Czech	1	8	6	31
3	https://ispforum.cz/	289.24	Czech	1	5	3	21
4	http://www.fcviktoria.cz/cs/historie	494.65	Czech	1	6	16	59
5	http://instalateri-ostrava.cz/	331.22	Czech	1	3	4	18
6	http://www1.osu.cz/home/sochor/cz/info.html	230.17	Czech	1	1	0	3
7	http://pocasi.divoch.cz/	148.32	Czech	4	6	4	21
8	http://insolvencni-zakon.justice.cz/	306.14	Czech	1	1	2	30
9	http://www.csfd.cz/tvurce/3698-kar-wai-wong/	549.58	Czech	6	2	3	19
10	http://www.nm.cz/snm/matice.htm	125.59	Czech	1	0	0	4
11	http://www.flliferrari.it/ff2/jsp/en/whereweare.jsp	233.28	Italy	1	2	1	8
12	http://www.luomus.fi/	390.89	Finland	2	5	0	23
13	http://www.loxon.de/index.php?id=5	275.02	Germany	1	8	7	9
14	http://www.pps.univ-paris-diderot.fr/~jch/software/polipo/	43.37	France	1	1	0	2
15	https://www.wikipedia.org/	80.42	Netherlands	1	1	1	12
16	http://www.letrasymusica.es/grupos/a/1/	81.14	Spain	3	1	2	18
17	http://www.cityofathens.gr/en/cityofathens	332.2	Greece	1	29	8	47
18	http://rt.com/about/partners/	508.19	Russia	1	4	10	39
19	http://www.rnc.ro/new/cinfo.shtml	46.73	Romania	1	2	0	15
20	http://www.eenet.ee/EENet/Kprojekt.html	65.77	Estonia	1	1	1	9
21	http://web.up.ac.za/default.asp?ipkCategoryID=17610&articleID=13614	299.85	South Africa	2	1	4	14
22	http://www.nic.ly/contactus.php	35.5	Libya	1	1	0	16
23	http://about.anu.edu.au/profile/history	298.8	Australia	1	3	2	13
24	http://bodhizazen.net/Tutorials/TOR	505.35	USA	1	4	2	16
25	http://jam.canoe.ca/Music/Artists/D/Daft_Punk/	107.82	Canada	1	2	16	38
26	http://www.nic.cl/anuncios/2012-06-19.html	120.85	Chile	1	1	0	5
27	www.bestbrazil.org.br/cgi/cgilua.exe/sys/start.htm?sid=1	281.97	Brazil	1	2	3	16
28	http://jprs.co.jp/en/regist.html	49.9	Japan	1	6	1	16
29	http://www.india.gov.in/abouttheportal.php	376.88	India	1	4	5	22
30	http://www.medee.mn/main.php?eid=21066	517.31	Mongolia	3	3	2	16

Table 2. Locations of the HTTP test file

No.	Location	Abbr.	IP address	Provider
1	Zurich, Switzerland	CHE	130.59.138.21	SWITCH
2	Ancona, Italy	ITA	193.206.140.34	garr.it
3	Praha, Czech Rep.	CZE	62.109.128.11	Ignum
4	Ireland	IRL	192.1.193.66	HEAnet
5	Koln, Germany	DEU	78.35.24.46	NetCologne
6	Paris, France	FRA	158.255.96.7	Free France
7	Moscow, Russia	RUS	212.118.44.106	CityLan
8	Nomi, Japan	JAP	150.65.7.130	Japan Adv. Inst. of Science and Technology
9	Tchao-jüan, Tchaj-wan	TWN	140.115.17.45	National Central University
10	Melbourne, Australia	AUS	202.158.214.107	AARNet
11	Wynberg, South Africa	ZAF	155.232.191.245	TENET
12	Tampa, USA	USA	74.50.101.106	HiVelocity
13	Montréal, Canada	CAN	70.138.0.134	iWeb
14	Curitiba, Brazil	BRA	200.236.31.2	Centro de Comp. Cientifica e Software Livre

2.2   Latency and Transmission Speed Measurement

Parameters like latency (to be more precise *round-trip latency* because all measurements were made locally) and response time are quite difficult to measure exactly because of their *soft* (i.e. insufficiently exact) definition. The definition itself (the time difference between sending the request and the first/last byte of the response) seems exact but the problem lies in the question what time to accept as the end of the measured period. Moreover also the effect of DNS resolution had to be taken into account. If DNS requests are not eliminated the latency measurement would be affected. The exact determination of the end of the load period is particularly complicated due to the nature of WWW (HTTP) communication where the traffic consists of downloading multiple files consecutively initiated by the WWW client. Therefore our study assumed certain additional limitations. The main such limitation was that only a single web client software has been used (namely Mozilla Firefox, see details below) so as to eliminate extrinsic fluctuations in measured data due to different implementation of client HTTP communication functions.

All the measurements were performed solely on the application layer despite the fact that in some cases the L3 measurement tools could be more precise. The limitation of the measurement to L7 is due to the nature of the desired comparison. The main obstacle making the L3 measurement useless was the fact that all anonymization tools studied here perform the anonymization on the application layer so the measurement of L3 parameters would not produce any results applicable to this study.

Client Software for Measurement. As explained above only single WWW client was used for measurement, i.e. Mozilla Firefox. It was chosen due to the extensive support for TOR (e.g. Firefox configurability, available plugins). All measurements were done using the following software:

– Mozilla Firefox 16.0.2,
– TOR 0.2.2.39 with Vidalia 0.2.20 Control panel,
– cURL 7.28.0 (used for file download only),
– and following Mozilla Firefox plugins: AdBlock Plus 2.1.2, Firebug 1.10.5, NoScript 2.6, Torbutton 1.4.6.3, iMacros for Firefox 7.6.0.2.

Also Wireshark 1.8.3 was used for verification of results where necessary.

Measurement of WWW Pages. All measurements were performed on the application layer. i.e. inside the client application because the anonymization effect can be observed at L7 only. Both latency and response time had been measured. To eliminate undesired fluctuations in results due to caching of previously downloaded data the use of the WWW client's cache was completely deactivated (using Firebug (see [4]) plugin). Latency (RTT) is defined as time from the end of establishing the TCP connection (i.e. sending the first HTTP request) to the target till the beginning of the data reception from the target. These times were measured by Firebug and the RTT is then defined as follows:

$$latency(RTT) = Sending + Waiting \ . \tag{1}$$

The time required for DNS lookup and establishing the TCP connection are not part of the RTT. The total response time was defined as the time from sending the first request till the reception of the complete web page (all its files). In order to obtain comparable results the time reported by Firebug was compared to the time measured by Fasterfox and Lori used in previous studies. It was observed (and verified using multilayer measurement using Wireshark too) that data measured by Fasterfox and Lori are systematically incorrect (longer) than results from Firebug verified by Wireshark. Therefore the previous results presented in [1] in terms of transmission speed contain a systematic error. Due to the systematic nature of the error it is obvious that all measured times were slightly longer and it means that comparison of transmission speed between normal and anonymized traffic gave slightly lower speed ratio between normal and anonymized traffic. The results presented here do not contain any such error. Certain measurement error in respect to the DNS resolution was observed in Firebug but it can be neglected due to the fact that DNS resolution time is not a part of the measured times.

The WWW page download has been measured in 15 sets. Each set of measurements comprising gradual displaying of each of the WWW page from the set described in the Table 1 was performed in different weekday and time of the day to avoid systematic circadian and hebdomadal errors.

Measurement of File Download. File download was measured using cURL instead of Firefox because of easier automation of the measurements. The latency (RTT) measured in the case of file download was defined the same as in the case of web pages. The cURL values $time\_starttransfer$ and $time\_conect$ were used for RTT calculation. The file response time was calculated using $time\_total$ cURL value. The response time was calculated as follows:

$$response\_time = time\_total - time\_starttransfer \ . \tag{2}$$

Thanks to the subtraction of the $time\_starttransfer$ value the resulting time could be directly used for transmission speed estimation as follows.

$$transmission\_speed = file\_size/response\_time \ . \tag{3}$$

Similarly like in the case of WWW pages each set consisted of subsequent downloading of the test file from all locations described in the Table 2. The precision of the cURL program measurement was tested too and the time difference between data from cURL and Wireshark were approx. 0.01 % so it was considered as neglectable.

3   Results

The measured data were preprocessed by computing average and standard deviation values. The extreme values at the 5 % significance level were eliminated, i.e. all values with the distance from average exceeded 1.96 times standard deviation were excluded from further processing. This elimination was applied for 140 measurements of RTT and 112 values of response time from total 2 250 measurements for web pages (i.e. approx. 11 % measurement eliminated) while 40 measurements of RTT and 34 ones of transmission speed from total 630 measurements for file download measurement (i.e. approx. 12 % measurements eliminated).

WWW Page Download Results. Web pages were downloaded in Firefox 16 with all plugins inactive (except Firebug for measurement and iMacros for Firefox for loading all required DNS records). The DNS records loading also caused establishing the TOR circuit so that it is not necessary to initiate the establishing the TOR circuit after the first measurement started. In the case of multiple sets of subsequent measurements always a new Vidalia identity was used (causing to form a new TOR circuit). It should be noted that TOR efficiency depends not only on TOR configuration settings but on the WWW client (Mozilla Firefox) detailed configuration as well. It means that the aim of the measurement was not only to find the ways how to configure TOR but the optional configuration of Firefox too. Total 15 sets of measurements were performed, each of them with all 30 web pages and with 5 different configurations, namely:

- A1 – WWW without anonymization,
- A2 – WWW with TOR in default setting,
- A3 – WWW with TOR in configured setting,
- A4 – WWW with TOR in default setting an Firefox configured,
- A5 – WWW with TOR in configured setting an Firefox configured.

The comparison of the cases above demonstrate how much the default TOR anonymization is worse (slower) compared to the WWW traffic without anonymization. This comparison as well as the comparison with other cases in terms of latency (RTT) is shown in the Fig. 2 while the response times are shown in the Fig. 3.

Also a simple comparison of TOR and I2P was performed. It was impossible to perform the same set of measurements with I2P for web pages (file download is not supported by I2P) because of I2P instability. Also the offer of tunnels (roughly corresponding to entry TOR node) is too narrow (at present the only tunnel entry point is publicly available). Therefore only 3 sets of measurements were done. The comparison of load time between I2P and TOR is shown in Fig. 1. The missing columns for I2P (web pages 18, 28 and 30) mean that the web page was permanently unavailable when accessing to it via I2P. The web page load time (excluding latency) is significantly longer for I2P than for TOR in most cases and it is never significantly shorter. It confirms the conclusion presented in

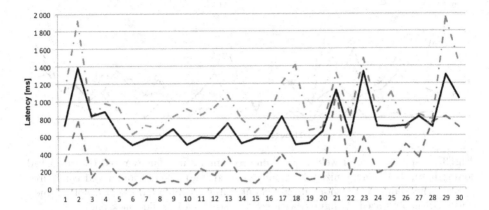

Fig. 2. Latency of web pages (RTT) compared among non-anonymized (dashed line), TOR-anonymized in default (chain-dotted line) and TOR-anonymized in configured setting (solid line)

the study [1] that I2P is less efficient anonymization tool than TOR, disregarding its much worse stability and overall usability.

TOR and Firefox Configuration Parameters. TOR configuration parameters for fine tuning are accessible by edition of torrc file. This is a common text file where each parameter can be set individually. The most influential parameter was CircuitBuildTimeout with default value of 30 but the optimum value found during experiments was 3.

There are also several configuration parameters of Firefox that can have an effect on the total TOR operation efficiency (i.e. on the efficiency of web page loading when using TOR). These parameters are accessible via Firefox GUI after entering *about:config* into the address line. There are variety of configuration parameters. The parameter that plays the most significant role in cooperation with TOR anonymization is, as our experimental results showed, the network.http.proxy.pipelining. (Boolean FALSE default parameter). Its setting to TRUE shortens the load time because it allows simultaneous loading of more files. Elimination of advertisements using the AdBlock Plus plugin and deactivation of scripts using NoScript was used too. The size of web pages after ad and script deactivation has been reduced up to almost 90 % in some cases (while it had no or neglectable effect in others).

File Download Results. File download was measured in the following three modes of operation as describer in the caption of the Fig. 4, The results of file download in terms of latency (RTT) are shown in Fig. 4 and the transmission speed in the Fig. 5.

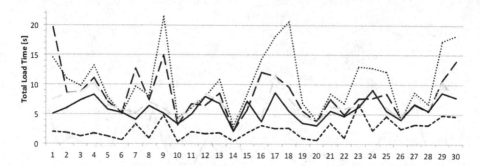

Fig. 3. Response time of web pages compared among non-anonymized (short-dashed lowest line), TOR-anonymized in default (dotted line), TOR-anonymized configured (long-dashed line), TOR-anonymized default, Firefox configured (dashed-dotted line), and TOR-anonymized configured with configured Firefox (solid black line)

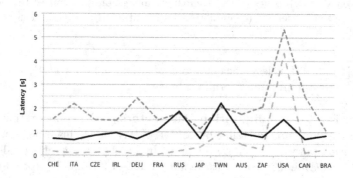

Fig. 4. Latency (RTT) of file download compared among non-anonymized (long-dashed line), TOR-anonymized in default (short-dashed line) and TOR-anonymized in configured setting (solid line)

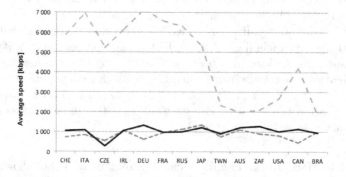

Fig. 5. Transmission speed of file download compared among non-anonymized (long-dashed line), TOR-anonymized in default (short–dashed line) and TOR-anonymized in configured setting (solid line)

4   Conclusions

In accordance with the results presented in [1], the presented measurements confirmed that using TOR causes significant slowing down comparing normal WWW use but experiential data presented here show that the efficiency of TOR could be improved much by fine-tuning of both TOR and the WWW client configuration. This was demonstrated on Firefox with results that e.g. latency increase due to TOR was more than 3 times but using proper configuration the increase factor was reduced to approx. 2.5. The similar conclusion is true for response-time too. The average increase of the file download time using TOR is 81%. Thanks to fine-tuning of TOR parameters the increase ratio is slightly smaller, approx. 78%. Despite the fact that the effect of configuring parameters is not significant in case of the transmission speed the proper configuration seems to have a potential to play an important role in the latency of web browsing. Bearing permanently changing nature of HTTP communication it is difficult to conclude with permanent configuration recommendations. Therefore some type of automatic configuration mechanism could be considered as a potential way ahead.

References

1. Sochor, T.: Anonymization of Web Client Traffic Efficiency Study. In: Kwiecień, A., Gaj, P., Stera, P. (eds.) CN 2012. CCIS, vol. 291, pp. 237–246. Springer, Heidelberg (2012)
2. Dingledine, R., Mathewson, N., Syverson, P.: Tor: The Second-Generation Onion Router (2013),
   https://svn.torproject.org/svn/projects/design-paper/tor-design.pdf
   (quot. January 22, 2013).
3. I2P Anonymous Network – I2P, http://www.i2p2.de/index.html
   (quot. January 22, 2013)
4. Odvarko, J.: Firebug Net Panel Timings. Software is hard (2012),
   http://www.softwareishard.com/blog/firebug/firebug-net-panel-timings/
   (quot. August 01, 2012)

Multi-agent Based Approach for Botnet Detection in a Corporate Area Network Using Fuzzy Logic

Oksana Pomorova, Oleg Savenko, Sergii Lysenko, and Andrii Kryshchuk

Department of System Programming, Khmelnitsky National University,
Instytutska, 11, Khmelnitsky, Ukraine
o.pomorova@gmail.com, sirogyk@ukr.net, rtandrey@rambler.ru
spr.khnu.km.ua

Abstract. A new botnet technique based on multi-agent system with the use of fuzzy logic is proposed. The analysis of the botnets' actions demonstrations in the situation of the intentionally computer system reconnection with the use of fuzzy logic is performed. Fuzzy expert system for making conclusion of botnet presence degree in computer systems is developed. It takes into account the demonstration degree of reconnected computer system, demonstration degree of probably infected computer systems and demonstration degree of other computer systems available in the corporate area network that probably weren't infected.

Keywords: botnet, multi-agent system, botnet detection, agent, sensor, fuzzy logic, infected computer system.

1 Introduction

The analysis of malware development shows dynamic growth of its quantity. The most numerous and danger malware during the last years are Trojans and worm-viruses that spread and penetrate into computer system (CS) for the purpose of information plunder, anonymous access to network, DDoS attacks, spamming etc. Such techniques as signature-based, code emulators, encryption, statistical analysis, heuristic analysis and behavioral blocking are used in modern antiviruses for botnet detection [1] show the decreasing of its efficiency for new malware detection. The efficiency of new malware detection in recent years is decreasing [2]. One of the main reasons of the low efficiency of detection is the spreading of a new malware class – botnet.

Bot-nets are the most serious cyber-threats today. They are the main base for such danger acts as distributed denial of service attacks, malware distribution, phishing, theft of confidential corporate data, organization of anonymous proxy servers etc. The peculiarity of botnet is the using of specialized commands and controlled channels of interaction that provides the updating of functional bots' parts of and actions features. The term Botnet denotes a network of compromised end hosts (bots) under the remote command of a botmaster. After botnet

A. Kwiecień, P. Gaj, and P. Stera (Eds.): CN 2013, CCIS 370, pp. 146–156, 2013.
© Springer-Verlag Berlin Heidelberg 2013

construction they are controlled autonomously and automatically. Sometimes they perform some illicit monetary activities [1,2].

That is why the actual problem of computer systems safety is a development of a new more perfect technique for new botnet detection. One of possible way to increase the detection efficiency is a developing of multi-agent system for new botnet detection in computer systems.

2   Related Works

The new approaches approach of botnet detection are developing in different directions. Authors in [1] used machine learning techniques to identify the command and control traffic of IRC-based botnets. They split this task into two stages: (I) distinguishing between IRC and non-IRC traffic, and (II) distinguishing between botnet and real IRC traffic. Results of research indicated that the proposed labeling criterion may not be representative of botnet traffic and that more accurate labeling, either through more extensive botnet testbed traffic, or by using more accurate botnet telltales, is crucial for this stage of botnet traffic identification. In [2] the analysis shows that random network models (either direct Erdös-Rényi models or structured P2P systems) give botnets considerable resilience. Such formations resist both random and targeted responses. The analysis also showed that targeted removals on scale free botnets offer the best response. Authors have demonstrated the utility of this taxonomy by selecting a class of botnets to remediate. The analysis suggested that by removing command and control nodes, targeted removal was an effective response to scale-free botnets. authors measured the impact of such responses in simulations, and using a real botnet. In [3,4,5] authors tried to shed light on the transmission methods used by current spamming botnets. The idea is that measures at the network level can be very effective in neutralizing spambots. The first case is when spambots reside inside a network. Spam relay and delivery attempts can be prevented when email traffic is managed according to MAAWG recommendation. In cases where this cannot be adopted, monitoring outgoing email traffic can give an indication of spamming activities. The main disadvantage of mentioned techniques is the impossibility of new botnet detection.

3   Previous Work

In order to increase the efficiency of botnet detection the multi-agent system that allows us to make antivirus diagnosis via agents' communication within corporate area network was offered [6]. It uses the set of agents. Each agent implements antivirus diagnosis via a set of sensors $A = \langle S_1, S_2, S_3, S_4, S_5, S_6 \rangle$, where S_1 – agent sensor of signature-based analysis; S_2 – checksum sensor; S_3 – sensor of heuristics analysis; S_4 – behavioral analysis; S_5 – sensor of comparative analysis through application programming interface API and driver disk subsystem via IOS (API sensor); S_6 – sensor – "virtual bait". Also agent includes a set of effectors that effect the computer system with purpose of blocking suspicious programs and then notify the other agents in the network about the infection in order to launch the suspicious programs detection with similar behavior. Agent has the CPU which processes the input data and determines the level of risk of specified object in the computer system based on some knowledge. In situation when agent cannot communicate with other agent it is as autonomous unit and is able to detect different malware relying on knowledge of the latest updates and corrections in the trusted software base.

The main disadvantage of this system is the decrease of the efficiency of antivirus detection by the recent period. Thus the efficiency of detection is 67 % (January, 2013) versus 70 % (February, 2012). Other problem is the comparatively high level of the false detection which is about 7–10 % (January, 2013) versus 3–7 % (February, 2012).

To overcome mentioned problems the new techniques and methods are to be developed for the high efficiency botnet detection based on proposed multi-agent antivirus system.

4 Multi-agent Based Approach for Botnet Detection in a Corporate Area Network Using Fuzzy Logic

The first step of the botnet detection is a construction of a schematic map of connections which is formed by corresponding records in each antiviral agent of multi-agent systems for some corporate area network. All agents based on this information can perform communicative exchange data to each other. Botnet detection process can be presented as a scheme shown in Fig. 1.

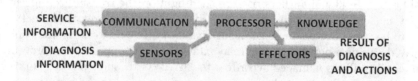

Fig. 1. The scheme of antiviral agent multi-agent system operation

In order to overcome the problem of reducing the reliability of new botnet detection, a new method for determining the degree of presence of botnet is proposed. Offered method is based on analyzing the bots actions demonstration in situations of intentional change of connection type of probably infected CS. This approach is performed in the case of insufficient (low) values of suspicion software, but this suspicion is present in a definite amount of computer systems of the corporate area network.

During computer system functioning the antivirus detection via sensors available in an each agent is performed. The antivirus diagnosis results are analyzed in order to define which of sensors have triggered and what suspicion degree it has produced. If triggering sensors are signature S_1, checksum S_2 analyzers or API sensor S_5 the results R_{S_1}, R_{S_2} or R_{S_5} are interpreted as a 100 % malware detection. In this situation, the blocking of software implementation and its subsequent removal are performed.

For situations when the sensors of heuristic S_3, behavioral S_4 analyzers or "virtual bait" S_6 have triggered, the suspicion degrees R_{S_3}, R_{S_4} and R_{S_6} are analyzed, and in the case of overcoming of the defined certain threshold n, $n \leq \max(R_{S_3}, R_{S_4}, R_{S_6}) \leq 100$, the blocking of software implementation and its subsequent removal are performed. If the specified threshold hasn't overcome the results R_{S_3}, R_{S_4}, R_{S_6} are analyzed whether they belong to range $m \leq \max(R_{S_3}, R_{S_4}, R_{S_6}) < n$ in order to make the final decision about malware presence in CS. If the value is $\max(R_{S_3}, R_{S_4}, R_{S_6}) < m$ than the new antivirus results from sensors are expected. In all cases the antiviral agents information of infection or suspicion software behavior in CS must be sent out to other agents.

The important point of this approach is to research the situation where the results of antivirus diagnosis belong to range $m \leq \max(R_{S_3}, R_{S_4}, R_{S_6}) < n$. In this case, the antiviral agent of CS asks other agents in the corporate area network about the similarity of suspicion behavior of some software that is similar to the botnet. If the interrogated agent receives information from one or more agents about the similar of software suspicious behavior, then the probably infected computer systems are marked and map reconstruction is implemented (Fig. 2). From the set of "marked" computer systems some CS must be chosen for the changing of network connection type (reconnection) – specific network settings prevent the network functioning of the bot in the computer system (DNS change, non-standard port connection to network, etc).

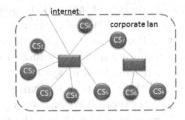

Fig. 2. The scheme of antiviral multi-agent system operation

The means of choosing the one computer system from the "marked" is the expert system. It contains a set of rules that are present in the knowledge of each antiviral agent. This computer system must meet the defined criteria.

After the reconnection of the chosen CS, the analysis of botnet demonstrations on reconnected computer system, on "marked" computer systems and other computer systems of the corporate area network and the definition of the degree of a new botnet presence in the network must be determined.

The presence of botnet in the corporate area network is concluded by the fuzzy expert system that confirms or disproves this fact. The determining of the botnet presence degree in computer system in situation of changed connection is changed is shown in the Algorithm 1.

Algorithm 1. Botnet Detection Algorithm

---

for $i=1$ *to* k *of* CS_i **do while** CS_i *is_ on* **do**
   if $R_{S_1} = true \cap R_{S_2} = true \cap R_{S_5} = true$ **then**
   | block and delete malware;
   else
      if $R_{S_3} = true \cap R_{S_4} = true \cap R_{S_6} = true$ *and*
      $n \leq \max(R_{S_3}, R_{S_4}, R_{S_6}) \leq 100$ **then**
      | block and delete malware;
      else
         if $R_{S_3} = true \cap R_{S_4} = true \cap R_{S_6} = true$ *and*
         $m \leq \max(R_{S_3}, R_{S_4}, R_{S_6}) < n$ **then**
         | communicate with other agents; analyze the degree of botnet
         | demonstration in corporate network;
         else
            if $R < m$ **then**
               wait for results $R_{S_1}, R_{S_2}, R_{S_3}, R_{S_4}, R_{S_5}, R_{S_6}$;

;

4.1 Choosing the Computer System to Change Its Type Connection in Corporate Area Network

Determination of the presence of botnet network is possible due to the fact that when we change the type of connection of some computer system, bots can demonstrate itself in some way (bots can try to communicate with other bots, update lists of active bots, reconfigure itself according to the new lists, etc.).

Note. We must pay attention to computer system place in the topology of the corporate area network. If the computer system is a unifying node with neighboring computer systems in corporate area network (e.g. CS_7 Fig. 3), which can be a server or a firewall, we cannot not change the type connection of this computer system.

In order to choose some CS we must analyze the features and properties of probably infected computer systems with botnet. For this purpose let take the concept of "suitability" of some computer system, which takes into account such fuzzy concepts as: antivirus diagnosis result – number R, produced by one of the antivirus agent's sensors; antivirus base relevance. There is a probability of delayed virus updates, which reduces the degree of botnet detection; computer system uptime. This characteristic affects the probability that the heuristic or behavioral analyzers can identify the behavior of potentially malicious software as well as a demonstration of malware actions which are directly proportional to the computer system uptime; operating system vulnerability. Taking into account the type of operating system we can distinguishe them by their degree

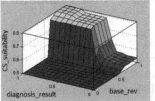

if (base_relevance is low) and
(OS_vulnerability is high) and (uptime is now)
and (diagnosis_result is low) then
(CS_suitability is not_suitable);

if (base_relevance is high) and
(OS_vulnerability is more_then_medium) and
(uptime is long_time_ago) and
(diagnosis_result is high) then (CS_suitability is
maybe_suitable);

Fig. 3. Fuzzy inference system results and rules for the choosing the most "suitable" CS for changing the type of network connection

of vulnerability. According to reports [7] the most vulnerable operating system today is the MS Windows XP, and the least vulnerable – Windows Server 2008.

Thus, we are interested in the computer system with the most relevant antivirus databases, with the highest uptime duration, with the lowest vulnerability degree of the operating system and the best result of virus diagnosis. Determination of computer system "suitability" is performed with the use of a fuzzy inference system which is present in the agent structure. Determination is based on the input linguistic variable names and terms which are given in the Table 1.

Table 1. Linguistic variables, terms and its values for determination of the most "suitable" CS which must be reconnected

♯	Linguistic variable name	Linguistic variable terms	Values
1	antivirus base relevance	not relevant	more than week
		more less relevant	from day to week
		relevant	within a day
2	duration CS is on	low	more than 6 hours
		medium	during last 6 hours
		high	during last hour
3	operating system vulnerability	high	Windows XP
		more than medium	Windows Vista
		medium	Windows 7
		more than low	Windows Server 2003
		low	Server 2008
4	Antivirus diagnosis result	low	is to be determined
		medium	experimentally in the range
		high	$m \le \max(R_{S_3}, R_{S_4}, R_{S_6}) < n$

Example of fuzzy inference system results and rules is presented in Fig. 3. Each agent of probably infected CS calculates the rate of its "suitability" and then communicates with other agents in order to choose CS as the most "suitable" one for the changing the type of network connection.

4.2   The Demonstrations of the Botnet After Computer System Reconnection

When the reconnection of CS is performed we must monitor all the actions both locally in CS (some malicious actions) and in the network (requests, DDoS, etc). All this events can be the demonstrations of the botnet activity [1].

Examples of the different botnets that belong to different types with its actions that can be the demonstrations of the botnet presence in the corporate area network are given in the Table 2.

Table 2. Botnets, its characteristics and demonstrations

Group of Bots	Ports	Actions	Example
Agobot/ Phatbot/ Forbot/ XtremBot/	21-23,25,53, 80,81,88,110,113, 119,135,137,139, 143,443,445, 3306,3389,5000, 6667,8000,8080	sniff traffic; sort traffic; Rootkit capabilities; debugers' detection; virtual machines' detection; initiate a DDoS attack; sending Spam	W32/Gaobot.worm W32/Agobot-Fam Worm_Agobot.Gen Backdoor.Agobot.gen
SDBot/ Rbot/ UrBot/ UrXBot/	6667,7000, 113,139,445,80, 135,1025,1433, 5000,6129,42 2745,3127,3410 903,17300,27347	launch the mIRC chat-client; collect system information; download files; execute files; connect to a specific server; initiate a DDoS attack; sending Spam	IRC-SDBot, Backdoor.Sdbot, Mydoom worm, Backdoor.IRC.Sdbot, Worm_Agobot.Gen, RBot. Bagle worm
mIRC-based Bots/ GT-Bots	113,139,445, 27374,53,80, 1000 – 6669	launch the mIRC chat-client; used HideWindow; used NetBIOS; scan sockets; initiate a DDoS attack; sending Spam	GT Bot Anti_Net_Bus, GT Bot HideWindow, GT Bot Aurora.d , GT Bot Bachir, GT Bot B0rg Bot

4.3   The Analysis of the Botnet Demonstrations and the Conclusion about Computer System Infection

For the determination of the presence degree of botnet in CS we must analyze botnet's demonstrations when some CS was reconnected. For this purpose all demonstrations are divided into three categories and the degrees, each of them must be determined: demonstration degree of reconnected CS, demonstration degree of probably infected computer systems and demonstration degree of other computer systems belonging to the corporate area network that probably weren't infected. To determine the possibility of the botnet presence in CS, the estimation of the demonstration degree for each of the three categories is performed. Demonstrations' degrees of three categories are presented as the fuzzy linguistic variables "demonstration degree" with terms "low", "medium" and "high".

The task of determination of membership function for input variable "demonstration degree" of reconnected CS we will consider as the task of the ranking for each of mechanisms f_i of penetration ports p_j with the set of indications of danger Z and a choice of the most possible p_j with activation of some function m_i. Then we generate a matrix of advantage $M_{adv} = |\gamma_{ij}|$. Elements of given matrix γ_{ij} are positive numbers: $\gamma_{ij} = \gamma_i/\gamma_j$, $0 < \gamma_{ij} < \infty$, $\gamma_{ji} = 1/\gamma_{ij}$, $\gamma_{ii} = 1$, $i, j = \overline{1, l}$, l – amount of possible results. Elements γ_{ij} of matrix M_{adv} are defined by calculation of values of pair advantages to each indication separately taking into account their scales $Z = \{z_k\}$; $k = \overline{1, r}$ with usage of formula

$$\gamma_{ij} = \sum_{k=1}^{r} \gamma_{ij}^k \cdot p_k / \sum_{k=1}^{r} \gamma_{ik}^k \cdot p_k \ . \tag{1}$$

Using the matrix of advantage M_{adv}, in which γ_{ij} are defined according to (1) the eigenvector $\prod = (\pi_1, \dots, \pi_i)$ is defied $\prod = (\pi_1, \dots, \pi_i)$. This eigenvector corresponds to maximum positive radical λ of characteristic polynomial $|M_{adv} - \lambda \cdot E| = 0$. $M_{adv} \cdot \prod = \lambda \cdot \prod$, where is an identity matrix. Elements of vector $\prod (\sum \pi_i = 1)$ are identified with an estimation of experts who consider the accepted indications of danger. The same procedure is performed for all f_i. As a result we receive a matrix of relationship $V_p = |f_i, p_j|$, in which each pair (relationship) f_i, p_j value $0 \leq \pi \leq 1$ responds. Using matrix $V_p = |f_i, p_j|$, we build matrix $V_p^* = |f_i, p_j|$ in which the relationship (f_i, p_j) is used and the elements of this relationship have value $\pi_{max}(0 \leq \pi_{max} \leq 1)$. Using matrix $V_p^* = |f_i, p_j|$, we build normalized curve for membership function $\mu_{xp}(R)$ of an input variable.

The task of determination of membership function for input variables "demonstration degree" of "marked" computers and common (not infected) computer systems are considered as the calculating the botnet demonstration degree. We must take into account the botnet action danger, the number of computer systems and where the demonstrations took place.

Let accept ω_η^u, $0 \leq \omega_\eta^u \leq 1$ – one of the signs of the demonstration, $j = \overline{1, x}$, $u = \overline{1, y}$, where y – number of botnet demonstration, b – number of computer systems in corporate area network. The estimation of each CS can be performed with the use of formula:

$$\omega^1 = \sum_{u=1}^{y} \alpha_u^1 \omega_u^1/y \ , \quad \omega^2 = \sum_{u=1}^{y} \alpha_u^2 \omega_u^1/y \ , \quad \dots \ , \quad \omega^\eta = \sum_{u=1}^{y} \alpha_u^y \omega_u^\eta/y \ , \tag{2}$$

where α_u – coefficients of the danger of some demonstration, $\alpha_1 + \alpha_2 + \dots + \alpha_y = 1$, $0 \leq \omega^u \leq 1$.

Thus if we choose some threshold value for each computer system with the estimation ω^η, for example $\tau \in (0; 1]$, then we can select some group g of "suspicious" computer systems if $\omega^\eta > \tau$. Then we calculate d_u – number of nonzero demonstrations of d_u^η in each computer system and average value ω_u with nonzero demonstrations ω_u^η (Fig. 4).

If number of nonzero demonstrations $d_u \neq 0$ then number of nonzero demonstrations is calculated with the use of formula:

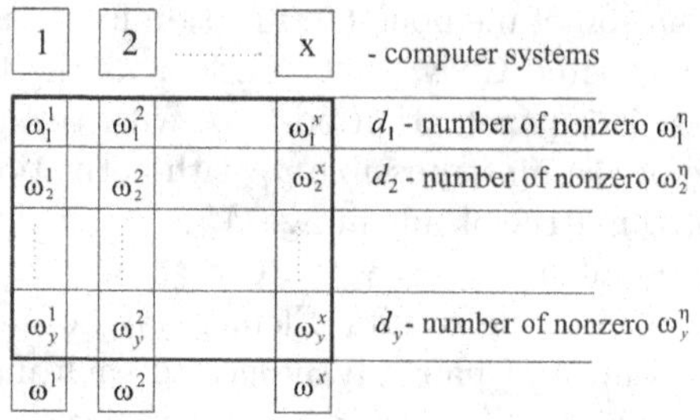

1	2	....	x	- computer systems
ω_1^1	ω_1^2		ω_1^x	d_1 - number of nonzero ω_1^η
ω_2^1	ω_2^2		ω_2^x	d_2 - number of nonzero ω_2^η
ω_y^1	ω_y^2		ω_y^x	d_y - number of nonzero ω_y^η
ω^1	ω^2		ω^x	

Fig. 4. Counting of demonstrations in each computer system

$$\omega^u = \sum_{\eta=1}^{x} \omega_\eta^u / d_u \, , \quad d = \sum_{u=1}^{y} d_u \leq y \cdot b \, . \tag{3}$$

We have to normalize the number ω_u, $u = \overline{1,y}$ so that $\omega_1 + \omega_2 + \cdots + \omega_y = 1$. Then general demonstration degree of botnet presence in "marked" CSs is:

$$P_d(d_1, d_2, \ldots, d_y) = \frac{d!}{(d_1!, d_2!, \ldots, d_y!)} \cdot \omega_1^{d_1} \cdot \omega_2^{d_2} \cdot \cdots \cdot \omega_y^{d_y} \, . \tag{4}$$

Let b', $b' \leq b$ – number of "marked" as infected computer systems. Then the arithmetic middling $\overline{\omega}$ of its correspondent ω_η must be calculated. After that the number P_d is determined and is interpreted as degree of botnet demonstration in "marked" computer systems.

The resulting conclusion of botnet presence degree in computer systems is performed by fuzzy inference system. It operates on determined demonstration degrees for three categories of computer systems (reconnected, "marked", and other computer systems of the network). The results of fuzzy inference system are presented in Fig. 5.

So, the usage of fuzzy logic enables the estimation the botnet presence degree in computer systems by determining botnet's demonstration degrees.

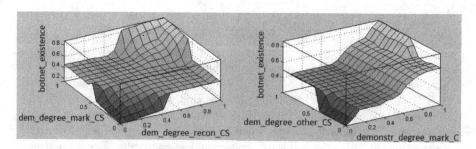

Fig. 5. The results of fuzzy inference system for calculating of the botnet presence degree in computer systems

5 Experiments

To validate the proposed method, software was developed and series of experiments were held. The research have been conducting for 6 months and such results have been obtained: the dependencies of the reconnection number and choosing the range $m \leq \max(R_{S_3}, R_{S_4}, R_{S_6}) \leq n$ on false positives and new botnet detection number were found.

Note. For the implementation of an experiment 25 programs with the botnet properties (Agobot, SDBot and GT-Bot) were generated. During each experiment (24 hours) computer systems in the network were infected only by one botnet. The results of the experiment is shown in Table 3.

Table 3. The results of the experiment

Reconection number	50–80 %		40–70 %		40–80 %	
	False positives	detection, %	False positives	detection, %	False positives	detection, %
1	5	84	7	88	6	88
2	5	88	7	92	6	92
3	5	88	7	96	6	96
0	5	82	7	84	6	84

Experimentally proved that the degree of botnet presence in computer system produced by fuzzy inference system should be 0.75 in order to make conclusion that "marked" CS is infected with botnet.

As we can see in Table 3 the decrease of the lower threshold m from the range $m \leq R < n$ increases the false positives and detection number. At the same time false positives significantly do not depend on the number of CS reconnection. The increase of the range $m \leq R < n$ and the number of reconnection increases the botnet detection number. The decreasing of the upper threshold n from the range $m \leq R < n$ leads to the increasing of the false positives and significantly does not depend on the number of botnet detection number.

Experiment results prove the efficiency of the multi-agent botnet detection in comparison with the use of the proposed technique and without it. The increasing of the efficiency is about 6–8 % and there is no increasing of the false positives.

6 Conclusions

The new botnet technique based on multi-agent system with the use of fuzzy logic is proposed. The detection is performed in the situations of priori uncertainty of the botnet presence in the corporate area network with taking into account the botnet demonstrations in the several computer systems available in the network.

With the usage of fuzzy logic, the analysis of the botnets' actions demonstrations in the situation of the intentionally computer system reconnection is performed. Fuzzy expert system for making conclusion about botnet presence degree in computer systems is developed. Fuzzy expert system takes into account the demonstration degree of reconnected computer system, demonstration degree of probably infected computer systems and demonstration degree of other computer systems available in the corporate area network that probably weren't infected.

The involvement of the developed method proves the effectiveness of the botnet detection with its growth which is about 6–8 %. At the same time the increase of false positives hasn't observed.

The consistency of agents in order to improve the efficiency of botnet detection is the direction of the further research.

References

1. Buxbaum, P.: Battling Botnets. Military Information Technology (MIT) vol. 12 (2008)
2. Zhaosheng, Z., Guohan, L., Yan, C., Fu, Z.J., Roberts, P., Keesook, H.: Botnet Research Survey. In: 32nd Annual IEEE International Computer Software and Applications, COMPSAC 2008, pp. 967–972 (2008)
3. Livadas, C., Walsh, R., Lapsley, D., Strayer, W.T.: Using Machine Learning Techniques to Identify Botnet Traffic. In: 31st IEEE Conference on Local Computer Networks, pp. 967–974 (2006)
4. Lee, W., Wang, C., Dagon, D.: A Taxonomy of Botnet Structures. In: Botnet Detection. Countering the Largest Security Threat, pp. 143–164. Springer, US (2008)
5. Stern, H.: A Survey of Modern Spam Tools. In: Proceedings of the Fifth Conference on Email and Anti-Spam (CEAS), Mountain View, CA (2008)
6. Savenko, O., Lysenko, S., Kryschuk, A.: Multi-agent Based Approach of Botnet Detection in Computer Systems. In: Kwiecień, A., Gaj, P., Stera, P. (eds.) CN 2012. CCIS, vol. 291, pp. 171–180. Springer, Heidelberg (2012)
7. Florian, C.: The Most Vulnerable Operating Systems and Applications in 2011 (2012), http://www.gfi.com/

Monitoring System's Network Activity for Rootkit Malware Detection

Mirosław Skrzewski

Silesian University of Technology, Institute of Computer Science,
Akademicka 16, 44-100 Gliwice, Poland
miroslaw.skrzewski@polsl.pl

Abstract. Contemporary malware authors attempt many ways to make its products "invisible" for antymalware programs, and after infection deeply conceal its operation from users sight. The presence of concealed malware can be detected many ways. Most of them operate "on demand" and provides high scanning overload of the system, blocking the chances for normal users operation. The paper presents new method of rootkit operation detection, suitable for continuous operation, based on the analysis of network activity pictures viewed from two sources (internal and external to system), along with the results of method tests on virtual machines infected with the selected rootkits code samples.

Keywords: network flows monitoring, network auditing, rootkit traffic detection.

1 Introduction

Detecting the presence of contemporary malware on user systems becomes over time more and more difficult task. Malware authors attempt many ways to make its program (through meta- and polymorphism modified code, many levels of obfuscation, generaton of multiple short series) "invisible" for antymalware programs, and after infection deeply conceal its operation from users sight.

Unix environment has known for many years rootkits – the methods of hiding malicious programs on the system via modifications of system tools, actively filtering system state informations from users, masking the presence of malware files, services and communication channels. This technology was also transferred to the Windows environment, and from about 2000 begin to appear malware programs [1] that use it to hide its (and other downloaded programs) presence in the system, and according to Microsoft researches [2], in 2006 nearly 20 % of malware detected on Windows XP SP2 systems contained elements of rootkits technology.

As rootkits technology is known from many years, there exists also many anty-rootkit technologies [3], attempting to counteract rootkit installation and to detect and remove installed rootkit from systems. Most of them operate "on demand" and provides high scanning system overhead, effectively ruling out the

A. Kwiecień, P. Gaj, and P. Stera (Eds.): CN 2013, CCIS 370, pp. 157–165, 2013.
© Springer-Verlag Berlin Heidelberg 2013

possibility of using the system by the user when searching for rootkits, and the results of their work are not always guaranteed to detect threats.

The presence of concealed malware can be detected from outside of the infected system by detecting hidden (invisible to the system) elements of network communication – such as additional open ports or network connections, but this also requires the use of external special tools (port scanners, network sniffers) to manualy check the state of system's network communication.

The paper presents a new method of rootkit operation detection based on the analysis of network activity pictures from the local activity monitoring program and from external monitoring system, which allows for continuous monitoring of the system behavior without compromising the work ability of its users in contrary to the most of current rootkit detection methods.

Remaining parts of the paper are organized as follows: Sect. 2 describes elements of rootkit and anty-rootkit technology, network activity monitoring tools are presented in Sect. 3, Sect. 4 presents details of the rootkit detection algorithm and its tests on research network on virtual machines and final conclusions are presented in Sect. 5.

2   Rootkit Technology

Rootkits technology was established in order to allow legitimate applications or malware programs for permanent, undetectable presence and the ability to perform actions on a computer system. This task includes the ability to hide in the presence of used resources such as processes, files, registry keys, and open communication ports. These possibilities can be used with different intentions, both legal (protection of compliance monitoring programs for DRM and copyright [4,5], compliance with security policies) or not (concealment of an intruder operation or malware program).

Initial rootkits implementations [19] were based on code modification of unix system tools (*ls, ps, lsof, find, ...*), and filtration of the information provided system users. Modifications to the code were easy to detect, so the next generation of rootkits implements filtration mechanisms deeper within the operating system, by modifying the data supplied to the unmodified system tools. Now hiding mechanisms may be found implemented at different levels and places of the operating system, from the application level, through system libraries, kernel structures, virtualization platforms, to the hardware (firmware level), and may have the form of various modification of kernel objects, system tables, or direct memory structures modification [6].

There are emerging constantly new ideas for possible new rootkits technologies [7,8]. Common feature of proposed solutions is tendency to move deeper and deeper in system with the mechanisms of information filtering, closer to the system hardware, and modify at lower-levels information provided for tools operating at higher levels.

Rootkit detection methods are aimed at detection of differences between the image of the system resources (files, processes, services, open ports) presented to the user and their real state. They are usually carried out by scanning with various methods the system resources, and by comparing the information provided by the system tools to the user with the information about the state of system resources provided by the low-level tools (such as the kernel API functions, system drivers, etc.), or by detecting changes in the content of key kernel objects.

Another, rarely used alone method for detecting the presence of rootkits in the system is analysis of the image of network communication. Comparison of the image of network communication channels visible in the system with the image visible from the outside requires the use of external port scanners like nmap. It allows for the detection of concealed open communication ports, but like the other methods, is not very suitable for continuous monitoring of system status.

Most of the methods of rootkits detection has the form of a single searching inside the system, usually performed without user activity due to the generated load. With new types of rootkits such exploration does not always have a chance to detect them. A more reliable method to disclose the presence of rootkits appears to be a continuous monitoring of the system network communications, both inside and outside, and the detection of concealed activity before the system tools.

3   Network Activity Monitoring

In Windows, as in other systems, there are tools to monitor network connections, open ports such as netstat, PortReporter but they are not suitable for continuous operation. For continuous monitoring it is required the ability of accurately recording the moments of changes in network connectivity, the associated processes, and information on the amount of data transferred in each connection. For greater efficiency, the registration mechanism should not rely on cyclic polling of system interfaces, but on capturing of events generated by changes in the state of system connections.

Figure 1 shows the architecture of the software interfaces and communication protocol stack of Windows. Of the system's kernel-level interfaces for recording operation of the network best suited is NDIS interface through which passes all network traffic.

Events generated at this level, however, do not contain information about the programs using them or addresses inside the upper layer protocols to which they are directed. This information is available on the transport layer interface TDI (Transport Driver Interface), but does not include auxiliary protocols such as ICMP and ARP.

According to these requirements, a tool for logging on the interface TDI – *tdilog* was developed. The program records the launch of processes, the moments of opening and closing ports, the establishment and completion of connections, with an indication of the direction to or from the system.

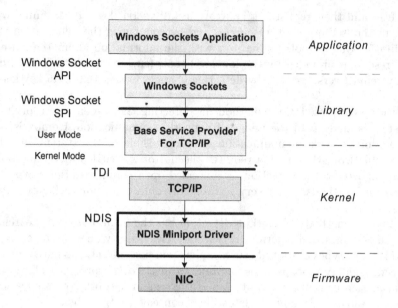

Fig. 1. Windows network protocols stack

From the point of view of rootkits detection it is important to assess the possibility of hiding information about events against the program. In Figure 1 indicated the possible levels of placement of rootkits of different classes. It seems that *tdilog* is likely to provide accurate information on the application layer rootkits activity, and perhaps also on rootkits modifying the system libraries, but rootkits acting from the kernel and firmware can modify the data recorded by the program. For this reason, with simpler rootkits the expected differences in information about the network connections may not arise.

4   Rootkit Detection Algorithm

Proposed algorithm for detection signs of rootkit operation is based on the possibility of the comparison of the images of network communication in the form of flows list, coming from two different sources – local recorder *tdilog* and from external auditor – *argus* [20]. Comparison covers the number of registered flows initiated in a certain direction, the number of destination ports, and in the presence of quantitative differences are compared also destination addresses.

Despite the same description of the communication model – the flow, described by protocol, source IP address and port, destination IP and port, there exists small differences in program records. In *tdilog* single flow is stored in the form of two records – the opening and completion of connection, the latter contains information about the amount of transferred data. Argus internally writes to the database a lot of information, which manner of presentation differs depending on the client program – *ra* (*read argus*), *ramon*, *rasort*, *racluster* and others. The primary client – *ra* program presents a list of arranged chronologically subsequent connections.

For fast detection of differences a more useful is an aggregate picture, grouping all the connections to the external system in one entry – a session, allowing to easily notice the lack of certain ports or IP addresses in one of the images. To obtain such form of information, the text logs of *tdilog* were saved to the postgress database, and then using the SQL command group by sport, d_ip, dport order by time the list of sessions was created. From *argus* the data were read by racluster program, which groups in a similar way connection records. To simplify the analysis the incoming and outgoing sessions were compared separately for TCP and UDP protocols.

To test the algorithm, network communication of some Windows XP systems were monitored during websites browsing and in online activities, and were obtained full compliance in quantity and in addresses of registered session. There were slight differences of the session start time from 20 to 80 ms, presumably due to delays in the operation of the network infrastructure. Below, few lines of session log from *tdilog* database:

```
    time          oper      prot   dir    sport       dstIP       dport
11:57:30.401   CONNECT    TCP    OUT    1035   74.125.136.94      80
11:57:30.602   CONNECT    TCP    OUT    1036   74.125.136.94      80
11:57:31.152   CONNECT    TCP    OUT    1037   74.125.136.120     80
11:57:31.243   CONNECT    TCP    OUT    1038   74.125.136.94      80
11:58:17.058   CONNECT    TCP    OUT    1040   62.149.142.34      80
```

And for comparison, from *argus* using racluster client program:

```
racluster -M dsrs=time,flow,metric -n -p3 -r argus.out - tcp and
           src host 172.17.114.215

StartTime    Proto Sport Dir    DstAddr.Dport    Pkts   Bytes  State
11:57:30.455  tcp  1035  <->   74.125.136.94.80   108   89243  CON
11:57:30.662  tcp  1036  <->   74.125.136.94.80    55   40296  CON
11:57:31.207  tcp  1037  <->   74.125.136.120.80   16    8375  CON
11:57:31.307  tcp  1038  <->   74.125.136.94.80     8    1082  CON
11:58:17.085  tcp  1040  <->   62.149.142.34.80    11    1930  CON
```

4.1 Detection Algorithm Verification

In order to verify the proposed method of detecting the presence of rootkits on the systems a series of experiments on the research network was planned, covering the infection of the virtual machines Windows XP SP3 known copies of the malware programs. Obtaining copies of active programs that use rootkit mechanisms was quite embarrassing. On Internet there are websites posting the informations about domain names of malware infected systems [9,10], monitoring the activity of systems that make up botnets [11], or offering access to archived copies of the malware programs.

The validity of this information, however, changed quickly over time, and often indicated on the sites services were not active or systems were restored after the removal of malware. Part of the already obtained copies had been inactive – program was looking for hard-coded addresses of C&C system, that have already been turned off, and did not takes of any other activity.

Most copies of malware recorded by honeypot systems (e.g. Dionaea [12]) showed no evidence of masking of any information about the state of network communication. After many attempts were collected, mainly from [13], and [11] and used to perform a number of tests a samples of programs classified as rootkits.

4.2   Test Environment

All tests were run on separated research subnetwork. On main test system with installed xen virtualization platform and Centos 5.8 system as dom0, virtual Windows XP SP3 instances were run. Another linux system was charged with task of monitoring the network communication. Both systems were connected via ethernet hub to the remaining part of unfiltered, internet connected, malware monitoring network.

Launched malware samples were monitored by the installed on Windows XP tools – *tdilog*, procmonitor, procexplorer [14], capturebat [15] recorded the internal workings of malware on the system, changes to the file system and computer's network communication. On the monitoring system runs programs *argus* monitor, tcpdump configured to register DNS queries and etherape [16], visualizing the current picture of network communication.

Depending on the activity of the tested malware infected system operate from a few minutes (in case of very active worms scanning the network) to a few hours for not showing greater network activity programs. Then virtual machine was turn off, external recording was stopped and registered logs were exported for analysis. Each of the selected malware programs was tested independently.

4.3   Malware Samples

A well-known problem in the examination of malware samples are malware names. Particular copies of the malware occur in a variety of AV software manufacturers under a very different names. The tests used copies of programs described as:

```
MaxRootkit    Md5:  392ddf0d2ee5049da11afa4668e9c98f
DarkMegi      Md5:  dd313b92f60bb66d3d613bc49c1ef35e
PhazeTDL      Md5:  4a052246c5551e83d2d55f80e72f03eb
ZeroAccess    Md5:  251a2c7eff890c58a9d9eda5b1391082
Ramnit        Md5:  607b2219fbcfbfe8e6ac9d7f3fb8d50e
Rustock-23    Md5:  1a713083a0bc21be19f1ec496df4e651
Zeus          Md5:  a7772183d2650d9d4f26ffa02fd41d64
```

On the web one can find many different descriptions of the same malware, but do not by the end of them can figure out what type of rootkits given malware represents. Some occur under many different names used interchangeably, and often are classified into different categories (trojan, backdoor, downloader or virus).

From selected for tests malware MaxRootkit, ZeroAccess, Rustock and Zeus are referred to as kernel mode rootkits, DarkMegi and Ramnit as user mode and PhazeTDL (TDL-4) as a boot mode rootkit.

4.4 Tests Results

Network activity of tested malware during the tests was varied. Some of them behaved like active Internet worms (Ramnit, MaxRootkit, ZeroAccess) and strongly scanned the network environment, trying to make connections to a number of selected systems, others (e.g. Zeus) scanned the surroundings less intensively, at 200–250 ms trying to connect to the 8–10 systems, and others (Rustock, PhazeTDL) did not show excessive activity.

For part of tested malware were no differences in the number of registered connections via *tdilog* and the *argus*. This was the case for rootkits Ramnit, DarkMegi, PhazeTDL and Rustock. In other cases, in the logs recorded by *argus* there are connections that do not exist in the data of *tdilog*. The differences relate to individual ports to or from which communication takes place in hidden form.

In case of ZeroAccess rootkit these were connection on port 16464, both TCP and UDP, incoming and outgoing. In the course of a few minutes of test *argus* registered a lot of udp scans and 6 successful tcp connections:

StartTime	SrcAddr.Sport	Dir	DstAddr.Dport	Pkts	State
11:00:10.587	96.239.109.92.1167	<->	172.17.114.215.16464	21	CON
11:00:12.960	76.172.46.93.3932	<->	172.17.114.215.16464	3	CON
11:04:01.975	75.109.6.239.1598	<->	172.17.114.215.16464	9	CON
11:04:25.003	50.193.131.1.29578	<->	172.17.114.215.16464	2	CON
10:58:51.741	172.17.114.215.1030	<->	68.50.148.254.16464	10	CON
11:04:22.065	172.17.114.215.1032	<->	61.67.19.54.16464	10	CON

Similarly, in case of rootkit MaxRootkit they were hidden connection on port 13620, *argus* registered 11 additional TCP connections in about 5 minutes. For Zeus rootkit the situation was a little different, were hidden some connections to various IP and ports (13606, 14869 and 10333 in this case), and on 10 TCP connections registered during test by *tdilog argus* registered four more:

StartTime	SrcAddr.Sport	Dir	DstAddr.Dport	Pkts	State
16:29:24.027	172.17.114.193.1055	<->	195.169.125.228.13606	6	CON
16:30:07.384	172.17.114.193.1057	<->	99.148.69.143.14869	6	CON
16:33:41.144	172.17.114.193.1068	<->	110.168.24.157.10333	6	CON
16:34:14.478	172.17.114.193.1077	<->	110.168.24.157.10333	6	CON

5 Conclusions

The proposed algorithm for rootkits detection on the basis of analysis of information about the network communication is working properly as intended, and indicates the presence of rootkits modifying the user reported information. In conducted tests the hiding of information was not always present and it is not clear how this can be explained.

After starting rootkit's code in the system there was some program activity – there were changes in the registry, were created new files, often malware file were removed, but there was no expected communication in networks to the C&C servers, even after a few hours of the program operation. Perhaps the reason might be an incomplete rootkit code, the detection of the virtual machine environment and the cessation of activity, or another than the required version of the operating system.

Adopted model of detection assumes that not in every case may go register the expected differences in communication, e.g. for rootkits modifying the information in the application layer the locally and externally recorded image of communication can be the same. The proposed algorithm is aimed at the detection of symptoms of infection and on prevention of persistence of its effects.

The main advantage of the algorithm is the possibility of it continuous operation, without affecting the normal working conditions of the system users. In conjunction with the threats detection algorithms based on the system's network activity monitoring it allows for detection of malware in the system, preventing its permanent presence in the corporate network.

Nowadays, increasingly larger and more complex systems are created [17] requiring the security and confidentiality of information processed and stored, as well as high performance and reliability. Such systems require extensive protection against threats [18], including not only the level of preventing infection (the first line of defense), as well as continuous monitoring and detection of signs of the possible presence of active malware programs, represent a threat to the confidentiality and integrity of information processed. A need to develop such a second line of defense is becoming increasingly obvious. Presented algorithm for rootkit's traffic detection can be an important part of the such solution.

References

1. Shields, T.: Survey of Rootkit Technologies and Their Impact on Digital Forensics, http://www.donkeyonawaffle.org/misc/txs-rootkits_and_digital_forensics.pdf
2. Naraine, R.: Microsoft: Stealth Rootkits Are Bombarding XP SP2 Boxes, http://www.eweek.com/c/a/Security/Microsoft-Stealth-Rootkits-Are-Bombarding-XP-SP2-Boxes/
3. Josse, S.: Rootkit detection from outside the Matrix. Journal in Computer Virology 3(2), 113–123 (2007)
4. Geist, M.: Sony Rootkit Redux: Canadian Business Groups Lobby For Right To Install Spyware on Your Computer, http://www.michaelgeist.ca/content/view/6777/125/
5. Brown, B.: Sony BMG rootkit scandal: 5 years later, http://www.networkworld.com/news/2010/110110-sonybmg-rootkit-fsecure-drm.html

6. Rozas, C., Khosravi, H., Sunder, D.K., Bulygin, Y.: Enhanced detection of malware. Intel Technology Journal 13(2) (2009)

7. King, S.T., Chen, P.M.: SubVirt: implementing malware with virtual machines. In: 2006 IEEE Symposium on: Security and Privacy, pp. 315–327 (May 2006)

8. Tsaur, W.-J.: Strengthening digital rights management using a new driver-hidden rootkit. IEEE Transactions on Consumer Electronics 58(2), 479–483 (2012)

9. http://www.malwaredomainlist.com/mdl.php

10. https://secure.mayhemiclabs.com/malhosts/malhosts.txt

11. https://zeustracker.abuse.ch/monitor.php?browse=binaries

12. http://dionaea.carnivore.it/

13. http://contagiodump.blogspot.com/search/label/rootkit

14. http://technet.microsoft.com/en-US/sysinternals

15. http://www.honeynet.org/project/CaptureBAT

16. http://etherape.sourceforge.net/

17. Gorawski, M., Marks, P.: Towards Reliability and Fault-Tolerance of Distributed Stream Processing System. In: International Conference on Dependability of Computer Systems (DepCoS – RELCOMEX 2007), pp. 246–253. IEEE, Szklarska (2007)

18. Gorawski, M., Marks, P.: Checkpoint-based resumption in data warehouses. In: Sacha, K. (ed.) IFIP Software Engineering Techniques, Design for Quality, vol. 227, pp. 313–323. Springer, Boston (2006)

19. McAfee: Rootkits. Part 1 of 3: The growing threat,
http://download.nai.com/Products/
mcafee-avert/whitepapers/akapoor_rootkits1.pdf

20. ARGUS – Auditing Network Activity, http://www.qosient.com/argus

Research of Failure Detection Algorithms
of Transmission Line and Equipment
in a Communication System with a Dual Bus

Błażej Kwiecień[1] and Marcin Sidzina[2]

[1] Silesian University of Technology, Institute of Informatics
[2] University of Bielsko-Biala, Department of Mechanical Engineering Fundamentals
blazej.kwiecen@gmail.com,
msidzina@ath.bielsko.pl

Abstract. The paper presents the results of empirical tests, which examine the algorithm of failure detection in transmission line or network node in dual bus system. The implemented algorithm is the base that use the second bus for parallel data transmission. It also presents the research results, which refer to measuring the duration of basic transaction in the network system, duration of the bus failure detection A(B) and slave station failure.

Keywords: PLC, distributed real time system, industrial computer network, time cycle of exchange data, PLC programming, avalanche of events, dual bus, network redundancy.

1 Introduction

This paper is the continuation of research [1,2,3], which presents the concept of using the redundant communication system based on Master-Slave model to reduce the duration of data exchange cycle. There are various forms of redundancy in communication systems [4]. The main goal of redundancy is to ensure the correctness of system work, and in case of system failure, its appropriate activation in order to prevent the loss of important information (e.g. skipping the object state in real-time system). The possibility of failure in the redundant system enforces the use of certain mechanisms, which enable a smooth transition from one to another state. The majority of redundant mechanisms generate the *idle transactions* during the normal operation of the system, however the incurred expenses are profitable only when the system failure occurs rarely [5,6,7,8]. The increase of reliability of communication system is the principle aim for using redundant system. Therefore, the idea of transmitting data via the unused bus (different then via primary bus), obtained by dividing tasks between the redundant buses seems to be an interesting issue. The aim of suggested modifications from the point of view of distributed real-time system is to improve the basic parameters, which define the RT system, such as: useful throughput P_U and efficiency η_U of the computer network. The main task is to reduce duration of

A. Kwiecień, P. Gaj, and P. Stera (Eds.): CN 2013, CCIS 370, pp. 166–176, 2013.
© Springer-Verlag Berlin Heidelberg 2013

data exchange, which may increase the system throughput, despite of the fact that the throughput of the data bus has not been increased. The application of the second bus as an additional transmission medium in proper communication system, change the primary function of redundancy. Thus, the unused bus does not exist anymore in such system, because both of them are used for data transmission [9]. In case of system malfunction, the main goal of the system is to detect the reason of failure and ensure the proper system operation by using proposed mechanism.

2 Analysis of Redundant Communication System According to the Master-Slave Access to the Link Method

Section two presents typical cases of analysis of monomaster systems with a single and redundant buses.

2.1 The Duration of the Network Cycle in Case of a Single Bus Analysis

The discussed communication system is based on two independent communication buses (Fig. 1). In the primary analysis of data exchange time, each of the buses can be treated as a separate system. Communication is based on a monomaster model with one Master station and n Slave stations [3].

For each bus the duration of executing transaction *request/command-response* is as follows:

$$T_{\mathrm{W_{ZPO}}} = T_{\mathrm{ZM}} + T_{\mathrm{PM}} + 2\left(T_{\mathrm{PR}} + T_{\mathrm{TR}} + T_{\mathrm{DT}} + T_{\mathrm{AR}}\right) + T_{\mathrm{AS}} \qquad (1)$$

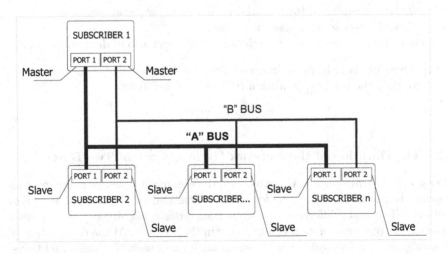

Fig. 1. Model of redundant communication system [2]

where:

$T_{W_{ZPO}}$ – time of data exchange *request/command-response*,
T_{ZM} – time measured from the moment of exchange request during the T_{AP}
phase till the start of coommunication phase,
T_{AS} – time measured from the moment of transmission request during the automatic machine cycle of Slave station till the start of communication phase,
T_{PM} – time measured from the moment of decoding the information by the Master station coprocessor, till the data are ready to use by the application in Master station,
T_{PR} – time of frame preparation,
T_{TR} – time of frame transmission,
T_{DR} – time of frame detection,
T_{AR} – time of frame analysis.

Times T_{AM} and T_{AS} presented in Formula (1) refer to the waiting time for the signal of the cycle end in Master and Slave stations, where in an extreme case, they stand for the maximum cycle duration of a controller. Times of frame preparation, detection and analysis are constant for Master and Slave stations. Transmission time of a single frame T_{TR} is the quotient between the number of transmitted bits and data transfer rate. Taking into consideration the fact that transmission scenario includes N *request-response* transmissions, then the total time of all data transmission in the scenario is as follows:

$$T_{CW} = \sum_{i=1}^{N} T_{wi} \qquad (2)$$

where:

T_{CW} – total transmission time of all data in the exchange scenario,
T_{wi} – time of i *request-response* transaction,
N – number of *request-response* exchanges in the scenario of data transmission.

If the times of data transmission are constant ($T_W = const$) in transaction scenario, then the total transmission time will be as follows:

$$T_{CW} = N * T_w \ . \qquad (3)$$

2.2   The Duration of the Network Cycle in Case of Two Buses

In case of data transmission via two buses, the execution time of a single *request-response* transaction is the same for each of them and can be defined as in Formula (1). The main differences between transmission via single and two buses concern the total time of data transmission in the scenario. At the design stage of transmission scenario which is realized through two buses, the individual transmissions time for each bus should be similar to each other. Moreover, every single station should be queried only by one of the buses. In other words, it is required that the total transmission time of all data included in the scenario for each of buses would be similar:

$$T_{CW} \cong T_{CW2} \tag{4}$$

where:

T_{CW} – total transmission time of all data in scenario for the first bus,
T_{CW2} – total transmission time of all data in scenario for the second bus.

If there are M *request-response* transmissions in transaction scenario of the first bus and R *request-response* transmissions in transaction scenario of the second bus, then the time of data transmission for each buses is as follows:

$$T_{CW1} = \sum_{i=1}^{M} T_{wi} \ , \quad T_{CW2} = \sum_{i=1}^{R} T_{wi} \tag{5}$$

where: T_{wi} – time of i *request-response* transaction.

The total time of all data transmission in the scenario is as follows:

$$T_{CW} = \max\left(T_{CW1}, T_{CW2}\right) \tag{6}$$

where: max – function which returns the maximum value of the input parameters.

3   The Algorithm of Transmission and Device Failure Detection

During the normal operation of communication system, the doubled Master stations execute *request/command-response* transaction in the arranged order. Data are exchanged among network subscribers using two communication buses. In order to provide the basic functionality of redundant systems, such as security and data integrity a special procedures were prepared. The main goal of these procedures is to ensure the protection of communication system, in case of one of the bus or subscriber error. The algorithm predicts the failure detection of communication medium (single interface), subscriber error or entire communication failure. Presented algorithm (Fig. 2), is used for permanent verification of the network accuracy. The system collects the information each time, when any of transactions included in the exchange scenario is executed. The executed transactions are basic and the only source of information for the algorithm, which enable to detect the failure of communication medium. The first symptom informing about system error is so called timeout, which occurs when the time for response is exceeded. It should be noted, that this is one of the most important parameters, which determine the response time of each communication system in case of bus or subscriber error. The timeout itself is not the main goal of this study. However, it is the basic indicator informing about the failure of a bus, interface or subscriber. Therefore, the wrong estimation of timeout may cause unnecessary delays [1,10,11]. In presented algorithm the timeout is a fundamental parameter. It appears once during the bus failure, and twice during detection of subscriber error.

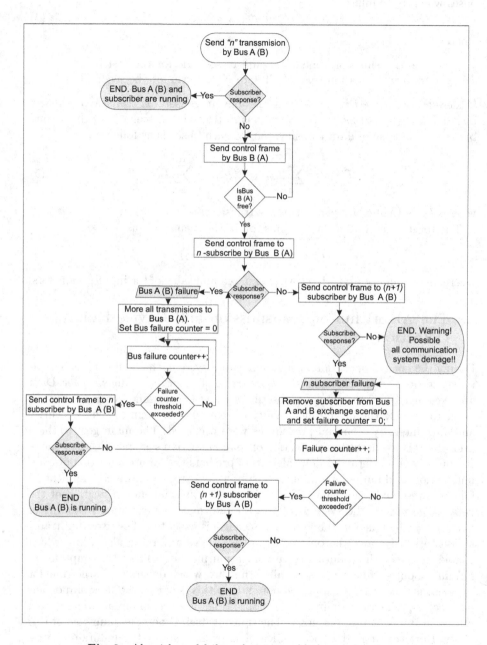

Fig. 2. Algorithm of failure detection of link or subscriber

The main goal of presented algorithm is to detect the failure of bus A or B and also Slave subscriber. The efficient detection of system failures allows for permanent updating the status, which helps to make following decisions connected with transactions in distributed system. The source code below, presents the point of decision related to the bus choice in three following cases: normal work, bus failure, Slave error.

```
if(ERROR_BUS_1) { //or ERROR_BUS_2
  count_error_Bus_1++;
  send_Bus_2(i);
  Send_Exchange_Bus_1 = false;

  if(count_error_Bus_1 == limit) { //number of cycles skipped
    count_error_Bus_1 = 0;
    STEP_Bus_1 = 0;
    STEP_Bus_2 = 0;
    ERROR_BUS_1 = 0;
    ERROR_BUS_2 = 0;
  }
}
else {
  if(error_Slave) {
    count_error_Slave++;
    send_Bus_1(i+1);

    if(count_error_Slave == limit) { //number of cycles skipped
      count_error_Slave = 0;
      error_Slave = false;
      STEP_Bus_1 = 0;
      STEP_Bus_2 = 0;
    }

  }
  else {
    send_Bus_1(i);
  }
}
```

- In case of failure of one of the buses, it is necessary to change the exchange scenario so that the information would be sent only via one bus. The exchange cycle during which the failure occurred, can be considered in two ways:
  - Repeat the entire exchange scenario, during which a failure occurred.
  - Sent only the exchanges (transactions) which had not been sent before.

  It is possible to determine programmatically during the failure, which exchanges had been sent, and estimate the time necessary for preparing a new scenario (including those already sent) and the time necessary to send the entire scenario. Comparing these times it is possible to answer which is more profitable: sending the entire scenario, or sending only those data which had not been sent before? In case of one bus failure, the total transmission time of all information included in scenario, will be increased in relation to the work of one bus by:

T_{OD} – the maximum response time for request frame,

T_{PR2} – time of preparing frame for another subscriber,

T_{TR2} – time of frame transmission to the next subscriber,

T_{SCE} – time of changing transmission scenario from two into one bus,

T_{ALG1} – time of executing the algorithm used for bus failure detection.

And it will be as follows:

$$T_{CW} = \sum_{i=1}^{N} T_{Wi} + T_{OD} + T_{PR2} + T_{TR2} + T_{SCE} + T_{ALG1} \ . \tag{7}$$

– In case of the subscriber failure it is not necessary to change the exchange scenario from two *streams* into one, but only to specify the number of network cycles, in which the subscriber will not be queried (hole in the network). Having executed a particular number of cycles, the control frame will be sent to the subscriber in order to verify its accessibility. If the subscriber is available, then it is again added into the exchange scenario. In the above case the total time of transmission of all information in the scenario will be extended in relation to correct work of the system by:

T_{OD} – the maximum response time for request frame,

T_{PR2} – time of preparing frame for another subscriber,

T_{TR2} – time of frame transmission to the next subscriber,

T_{ALG2} – time of executing the algorithm used for subscriber failure detection.

And it will be as follows:

$$T_{CW} = \max\left(T_{CW1} + T_{CW2}\right) + T_{PR2} + T_{TR2} + T_{OD} + T_{ALG2} \ . \tag{8}$$

4   Test Bench and Research Procedure

In order to conduct the basic empiric test, a test bench consisted of two devices was prepared. The first device is Master station, which executes the exchange scenario and implements the algorithm of failure detection of link or subscribers. Slave station, on the other hand, was implemented into the second device, in such a way to make possible to simulate the work of many subscribers, link failure and subscriber error. Compared to [1] the concept of test bench has been changed. Instead of test bench based on PLC controllers, the current model of network node uses Arduino Mega 2560 platform. This platform is equipped with 8 bits AVR MEGA 2560 microcontroller by ATMEL with four UART circuits for asynchronous transmitting and receiving serial data. Two UART circuits of Master station were connected to two UART circuits of Slave station *Master Tx → Slave Rx; Slave Tx → Master Rx*. The presented test bench gives an opportunity to check and test the algorithm. Contrary to the popular solutions used in PLC, in presented test bench it is necessary to implement a mechanism of communication systems, which helps to control fully the sequence of events.

The test bench presented in Fig. 3 is the base for network system, which focuses on the concept of dual bus in order to increase the frequency of message exchanges in distributed control systems (Fig. 1) [2]. This concept in contrast to the one presented in [12], uses two hardware interfaces (Fig. 4), and the exchange scenario is divided between each of them. Most of current redundant systems transmit the same data packet via both buses [8]. It means that the system executes the same work twice. The discussed conception of the redundant system enables to reduce the time of data exchange cycle, but at the same time it does not change the fundamental features of redundancy, which prevent the system from data loosing.

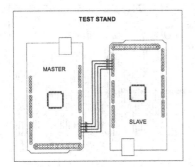

Fig. 3. The test bench   **Fig. 4.** The conception of double TTL converter to RS-485 interface

5   Research Results

The principal research of algorithm was conducted on test bench presented in Sect. 4. The aim of the research was to examine the effectiveness of the algorithm and failure detection time of subscriber, as well as the data bus. Measurements consisted of three phases (research procedure):

- measurement of duration of query and response transactions for 10 bytes control frame,
- measurement of duration of the failure detection algorithm of bus A (B) – for various Timeout parameters: 1 s, 0.5 s, 0.1 s,
- measurement of duration of the failure detection algorithm of the subscriber – for various Timeout parameters: 1 s, 0.5 s, 0.1 s,
- measurement of the entire failure detection (failure of bus A and B or Master station).

The results are presented in Figs. 5–8.

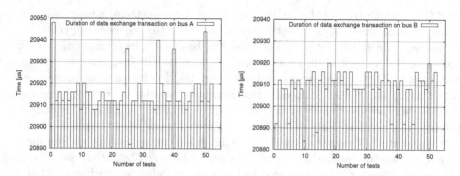

Fig. 5. Measurement of data transaction time (a) in bus A, (b) in bus B

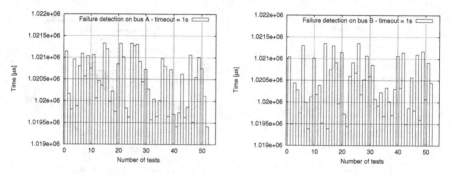

Fig. 6. Measurement of failure detection time of bus (a) A, (b) B

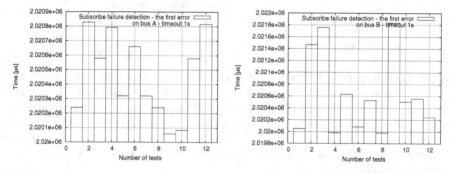

Fig. 7. Measurement of subscriber failure detection time, the first query (a) bus A, (b) bus B

The presented research clearly define the usefulness of proposed solutions. The algorithm can detect the failure of interface, data link or subscriber in the redundant network, which is based on Master-Slave exchange model. During the test, the algorithm in each case identified correctly the type of the error introduced into the system. The most important parameter, which causes the greatest delays is Timeout. As the results shows, the Timeout parameter in Master-Slave

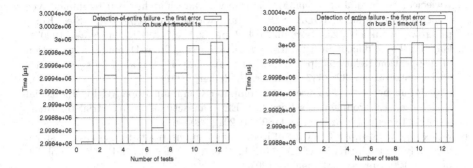

Fig. 8. Measurement of the entire failure detection time on (a) bus A, (b) bus B

network is responsible for the biggest delays during switching communication between the buses in redundant system. The other parameters are comparable with the duration of a single data exchange.

6   Conclusion

The presented algorithm is proved to be an effective tool used to detect failures occurring in one of the buses of communication redundant system. The paper presents a solution based on Master-Slave communication model. The developed mechanism is effective, especially in case of failure detection of a single bus, as well as failure detection (disconnection) of Slave device. The mechanism, which checks whether the bus or a subscriber restarted working properly, was also taken into consideration. Recheck the correct operation of the subscriber or the bus is a separate parameter in the proposed system, and can be freely determined depending on the tasks of the communication system. The paper presents the basic principles for test position, which is currently being under development. Another advantage of the proposed solution is the fact that it will be able to be applied in the redundant systems which already exist. The possibility of testing or applying the algorithm requires installing an extra software procedures in existed Master station program. The research which are being currently conducted refers to the software of presented communication system, in order to verify such features as increased reliability and effectiveness. The analysis of the basic time assumptions indicates that this algorithm is a useful tool for extending the work of certain communication systems with *monomaster* architecture.

References

1. Sidzina, M., Kwiecień, B.z.: The Algorithms of Transmission Failure Detection in Master-Slave Networks. In: Kwiecień, A., Gaj, P., Stera, P. (eds.) CN 2012. CCIS, vol. 291, pp. 289–298. Springer, Heidelberg (2012)
2. Kwiecień, A., Sidzina, M.: Dual Bus as a Method for Data Interchange Transaction Acceleration in Distributed Real Time Systems. In: Kwiecień, A., Gaj, P., Stera, P. (eds.) CN 2009. CCIS, vol. 39, pp. 252–263. Springer, Heidelberg (2009)

3. Kwiecień, A., Stój, J., Sidzina, M.: Analiza wybranych architektur redundantnych z zastosowaniem sieci MODBUS/RTU. In: Kwiecień, A., et al. (eds.) Sieci Komputerowe. Aplikacje i zastosowania, vol. 2, pp. 359–367. WKiŁ, Warszawa (2007)

4. Gaj, P.: The Concept of a Multi-network Approach for a Dynamic Distribution of Application Relationships. In: Kwiecień, A., Gaj, P., Stera, P. (eds.) CN 2011. CCIS, vol. 160, pp. 328–337. Springer, Heidelberg (2011)

5. Gaj, P., Jasperneite, J., Felser, M.: Computer Communication Within Industrial Distributed Environment – a Survey. IEEE Transactions on Industrial Informatics 9(1), 182–189 (2013), doi:10.1109/TII.2012.2209668

6. Kwiecień, A., Stój, J.: The Cost of Redundancy in Distributed Real-Time Systems in Steady State. In: Kwiecień, A., Gaj, P., Stera, P. (eds.) CN 2010. CCIS, vol. 79, pp. 106–120. Springer, Heidelberg (2010)

7. Stój, J., Kwiecień, B.: Real-time System Node Model – Some Research and Analysis. In: Contemporary Aspects of Computer Networks, vol. 2, pp. 231–240. WKiŁ, Warszawa (2008)

8. Stój, J., Kwiecień, A.: The Response Time of a Control System with Communication Link Redundancy. In: Contemporary Aspects of Computer Networks, vol. 2, pp. 195–202. WKiŁ, Warszawa (2008)

9. Kwiecień, B., Stój, J.: Network Integration on the Control Level. In: Kwiecień, A., Gaj, P., Stera, P. (eds.) CN 2011. CCIS, vol. 160, pp. 322–327. Springer, Heidelberg (2011)

10. Jestratjew, A., Kwiecień, A.: Performance of HTTP Protocol in Networked Control Systems. IEEE Transactions on Industrial Informatics 9(1), 271–276 (2013)

11. Kwiecień, A.: The improvement of working parameters of the industrial computer networks with cyclic transactions of data exchange by simulation in the physical model. In: The 29th Annual Conference of the IEEE Industrial Electronics Society, Roanoke, Virginia, USA, vol. 2, pp. 1282–1289 (November 2003) (curr. ver. April 2004)

12. Stój, J.: Real-Time Communication Network Concept Based on Frequency Division Multiplexing. In: Kwiecień, A., Gaj, P., Stera, P. (eds.) CN 2012. CCIS, vol. 291, pp. 247–260. Springer, Heidelberg (2012)

The Concept of Using Multi-protocol Nodes in Real-Time Distributed Systems for Increasing Communication Reliability

Andrzej Kwiecień[1], Marcin Sidzina[2], and Michał Maćkowski[1]

[1] Silesian University of Technology, Institute of Informatics
[2] University of Bielsko-Biala, Department of Mechanical Engineering Fundamentals
andrzej.kwiecien@polsl.pl,
msidzina@ath.bielsko.pl,
michal.mackowski@polsl.pl

Abstract. The paper presents the considerations on the method which enables to accelerate the exchange of data in distributed control system, using multi-protocol nodes. The method presented in this paper is related to hardware and software, and based on new original communication protocol, which is used by the node for supporting various communication protocols to simultaneous data exchange via different buses. This method will enable to increase not only network bandwidth, but also the security of data transmission.

Keywords: distributed control system, redundancy, network node, industrial networks, distributed real-time systems, industrial protocols, medium access, acceleration of data exchange.

1 Introdution

Industrial real-time distributed systems are built on the basis of several models of access ([1–6]) to the communication link:

- Master-Slave,
- Token-Bus, Token-Ring,
- PDC (Producer-Distributor-Consumer),

or other being a combination of the above. There are many commercial products, which implement partly or entirely models mentioned above ([7–9]). It is also necessary to mention that there are many attempts and many works which refers to the use of Ethernet in industrial communication, such as in [10–13]. However the mentioned scope, is out of the authors' interest. Thus it was not discussed in the paper. Taking into consideration the reliability of the system (which can be increased thanks to the redundancy) or network overload (permanent or temporary) there is a problem how to increase the flexibility of network connections in order to improve the time of data exchange. The mentioned redundancy can also be used ([14, 15]) as a way of improving the transmission time parameters [14]

A. Kwiecień, P. Gaj, and P. Stera (Eds.): CN 2013, CCIS 370, pp. 177–188, 2013.
© Springer-Verlag Berlin Heidelberg 2013

and it also may increase the bandwidth. However, it should be noted that the greatest advantages of using redundant communication link to increase the performance of transmission time, occur when the redundancy already exists in the system (for example because of the reliability) ([16, 17]). This is due to the high cost of system with redundancy. Based on the above discussion some questions arise. First of all, whether there are other possibilities capable to unload intensive data traffic, especially when temporary overload occurs ([18–20])? Next, is there a possibility to take advantage of multi-protocol communication that uses specialized nodes of a distributed system (maintaining the rules of real-time system) in case of system failure, such as: damage of transmission system (cable, optical fiber, network coprocessor, etc.), events in industrial area (communication failure) or other reasons, the security of transmission can be ensured? It would be interesting to use multiprotocol communication [20] using specialized nodes of distributed system according to rules of real time system.

The rules of real-time system means that the global time response to any event or serious of events cannot be exceed.

$$T_{RE} < T_G \tag{1}$$

where:

T_{RE} – the result of design and is the maximum time of system reaction to an event or the sequence of events,

T_G – is the maximum of acceptable response time of system to the sequence of events and it is usually force by the technology of system.

2   The Concept of Multi-protocol Node

Figure 1 presents the concept of multi-protocol network with one highlighted node $A0$, which additionally includes three communication interfaces with A, B and C networks. Subscribers $B1$ and $B4$ are connected to network A, subscribers $B2$, $B3$ and $B4$ to networks B as well as C. It is assumed that node $A0$ is the master station to networks A, B and C. The idea is that each node of network S may be a node for other network. Obviously, the networks A, B and C may use various protocols and transmission media. For these considerations some simplifications have been made, which in the authors opinion do not limit the scope of current discussion.

Figure 1 presents network S ("ring") which can be defined as follows:

$$S\{N\langle *A0\rangle \gg N\langle A1\rangle \gg N\langle A2\rangle \gg N\langle A3\rangle \gg N\langle A4\rangle \gg N\langle A5\rangle \gg N\langle *A0\rangle\}$$
$$\tag{2}$$

where S stands for network name, and $N\langle NAME\_OF\_NODE\rangle$ is node name $NAME\_OF\_NODE$.

Notation (2) should be then interpreted as follows: Network S (here "ring") consists of six nodes called: $A0$, $A1$, $A2$, $A3$, $A4$, $A5$, which are connected together in the following way: $A1$ with $A2$; $A2$ with $A3$; $A3$ with $A4$; $A4$ with $A5$; $A5$ with $A0$ and $A0$ node is additionally the node of other or others networks.

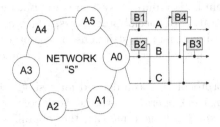

Fig. 1. The idea of network with multi-protocol nodes

The connections between the nodes are arranged as an example, because there are many possible ways to connect the nodes. A given example is only to keep order. If the node name is preceded by * it means that a particular node is at the same time a node for other networks.

As the above notation indicates, this is the network of "ring" type with undefined protocol. The only requirement is that it should be a deterministic protocol that guarantees to each subscriber the access to the communication medium within a maximum time T_{ACC}. For further discussion network S will not be taken into account. The only node taken under discussion is node $A0$ in order to describe the phenomena which can be reproduced on other nodes in network S. However, it is important to remember that networks A, B and C do not have to be isolated from network S, which means that any subscriber of network S can transmit data to/from networks A, B and C via node $A0$. As already mentioned, at the current stage of considerations these data transfers are not taken into account.

The considered network is a Master-Slave network in which the node $A0$ is the Master station. Choosing the Master-Slave model to these considerations was intentional, but it in no way limit the scope of discussion. It intends only to make a practical study designed to answer a number of questions concerning the possibility to reduce transmission time and increase its reliability. Thus, the network with node $A0$, can be described as follows:

$$Master A0 \underbrace{\{N\langle B1\rangle N\langle B4\rangle\}}_{Network\,A} \underbrace{\{N\langle B2\rangle N\langle B3\rangle N\langle B4\rangle\}}_{Network\,B} \underbrace{\{N\langle B2\rangle N\langle B3\rangle N\langle B4\rangle\}}_{Network\,C} \quad (3)$$

where $A0$ is simultaneously master station for networks A, B and C.

And node $B1$, $B2$, $B3$ and $B4$ are slave stations in networks A, B and C.

It should be stated that Master-Slave networks are characterized by the fact that Slave devices are of secondary nature and they cannot make a communication in network system. This is due to the fact that all exchange scenario is supervised by Master station. The only one station which is able to establish network communication is Master. It sends a request for the answer to other nodes and request to write data into memory address space of Slave station. Triggering the communication by the Slave station, is not very often discussed in literature. However if this problem is mentioned then it is usually limited to the possibility of system node failure or system crash but outside the computer system. The implementation of special procedures, which are prepared for such situations is constrained with many restrictions. These procedures are only performed in critical situations, for example when the communication system tends to fall.

The Master-Slave protocol has been defined for a single communication bus. The device coprocessor which operates in the network normally does not support two communication buses. In order to increase the reliability and security of the network by introducing a redundant system it is necessary to equip the nodes in this system with additional interfaces (communication coprocessors). It is also possible to use for example "sleep" Master stations which take control over the network on case of failure of communication for the main Master station [14, 15].

Thus $A0$ node is Master station. According to the idea of such network, the so called exchange scenario is placed in Master station. It determines the order of exchanges between Master and Slave stations. In a classic Master-Slave network, where Master is equipped with one communication interface, the analysis of data flow is relatively simple. The concept contains also one exchange scenario and the Master station has to choose via which bus the request should be sent in order to minimize the exchange time.

Figure 2 presents the exchange scenario. It requires the data transfer to/from specific nodes ($B1$, $B2$, $B3$, $B4$). This transfer may occur regardless of situation in network S. However, the data transmitted by a node $A0$ may not be local value essential for networks A, B and C. Then, the essential will be "time stamp" or in other words "recent indicator", which will be transmitted by $A0$ node:

- At the time of receiving them from network S (data from the network S will be forwarded to the network A, B and C).
- At the time of receiving them from any A, B or C network (data will be transmitted to the network S).

Therefore, assume the exchange scenario as in Fig. 2 (realized from top).

The first data to be transmitted are data to/from node $B1$. It can be realized only with the use of network A. Next transaction is to/from node $B2$. This can be done in three ways:

- with network B (if the network is not busy),
- with network C if the network B is busy, or
- waiting until one of them is not busy.

The next data to be transmitted are to/from the node $B3$. Three cases can be considered:

- the network B will be used when it is not busy, or
- data will be transmitted with network C, in case when network B is busy and C is free, or
- waiting until one of them is not busy.

In the following step, data to be transmitted are data to/from node $B4$. In this case, the transfer can be accomplished in the following way:

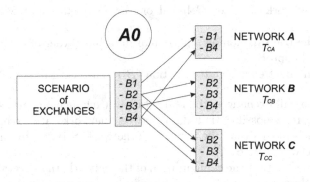

Where T_{CA}, T_{CB}, T_{CC} are network cycle time, respectively A, B and C

Fig. 2. The exchange scenario in networks A, B and C

- with the network A (if free), or
- with network B, if A is busy, or
- with network C, if the both above are busy, or
- waiting until one of them is not busy.

3 Criterion of Network Selection

With this proposed sequence of exchanges, there are some problems with choosing, which of the available network will be used for data transmission. This will depend on a number of factors that are associated with both transmission parameters as well as differences in transmission protocols. A network designer will be faced to the problem related to the criterion of network selection. The more differences in time cycle of particular network (A, B or C), the more important the problem will be. Such situation occurs when the protocols are significantly different from each other because of using various transmission media. The criterion of network selection may be described in several steps as follows:

1. Determining the fastest network based on the test data, (T_{TR}).
   In this point the static calculation of T_{TR} parameter is done, which corresponds to the transmission time of a particular data block or even a single frame as a function of data transfer rate, protocol overhead (number of bits per single transmitted character, numbers of control bit, etc.) and the size of data block. Having calculated T_{TR} it is possible to order the priority of using particular network as a function of its speed.
   The aim of the basic research was to determine the duration of data transmission depending of the type of transmission protocol, duration of application of master station and coprocessor type (Sect. 4).
2. Preparing the information exchange.
3. Checking the busy flag for each network FL_i.

4. Selecting network A, B or C based on the T_{TR} value and network busy flag Fl_i.
5. Running the transmission on a particular data bus (order of transmission to the selected coprocessor).
6. Fetching the next exchange transaction (B_i) and return to step 2.

After completing all exchanges, the realization of exchange scenario starts from the beginning. It is possible that data between networks are exchange asynchronous. Which means that the flow in the network S is totally independent from flow in networks A, B, C.

It should be noted that the configuration of the network, the selection of nodes and assigning them to particular networks will be a time consuming process. A necessary condition to obtain the proper effects is to provide the nodes with a sufficient and independent communication interfaces (network coprocessors). Additionally, some problems appears which are the object of the research. The above discussion assumed that the network operates without any failure. This group may consists of:

− If at least one of the networks (A, B, C) will be a token ring network, it will be necessary then to calculate the cycle time of the network in order to eliminate the phenomenon of monopolizing the communication link;
− The necessity of estimating the maximum time of exchange in each node of A type, in order to calculate the bandwidth of entire network;
− If the protocols will vary significantly, then the algorithms of network failure detection will be getting more complicated.

4   The Initial Research

The method discussed in the paper, or in other words a schema of operations, has been developed on the basis of many tests, which aimed at shortening the duration of data exchange in distributed real time communication system. The research results confirm a lot of relationships:

− Duration of the basic cycle of a node has a significant influence on duration of data exchange (Fig. 3–6).
− There is a possibility to shorten effectively the duration of automation cycle, both static and dynamic, in order to reduce the duration of data exchange, e.g. [21].
− Generally, each type of coprocessor and its localization (e.g. embedded in central unit, installed in PLC slot) is characterized by different exchange durations. It is advised to conduct separate tests for each device, which check the duration of information exchange (comparison of Figures 3 with 4).

As it was mentioned in Sect. 3, the main goal of research was to determine the duration of data transmission (T_{TR}), depending on the type of transmission protocol, the duration of application of Master station and type of coprocessor

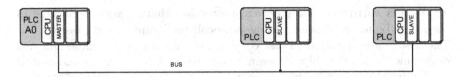

Fig. 3. The schema of test bench for Modbus/RTU (separate network coprocessor)

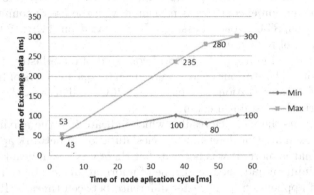

Fig. 4. Test bench 1. Coprocessor of Modbus/RTU protocol placed in PLCV Reading of ten words for automat cycle is as follows: 3.7 ms, 37.6 ms, 46.5 ms, and 55.8 ms.

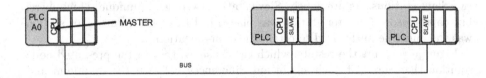

Fig. 5. The schema of test bench for MODBUS/RTU (network coprocessor embedded in central unit)

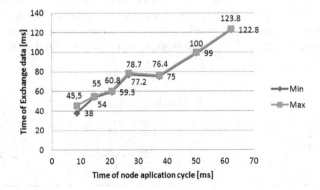

Fig. 6. Test bench 2. Coprocessor for Modbus/RTU placed in the central unit. Reading of ten words for automation cycle is as follows: 8.6 ms, 14.6 ms, 20.7 ms, 26.6 ms, 37.2 ms, 50.2 ms and 62.3 ms.

(Sect. 3). It is worth to mention once again the fact that networks connected to A0 node can be networks with various protocols (including Token-passing).

The first phase of empirical test consisted of measurements whose aim was to determine the relationships between various network coprocessors and central unit. The basic parameters, which were taken into account included: duration of cycle of application node and the type of used transmission protocol. Therefore, there were four test benches prepared for measuring those parameters. The duration of information exchange was measured with the use of communication protocols: Modbus/RTU, Token-Passing, SRTP (based on TCP), for different durations of cycle of node application.

Test bench 1 for Modbus/RTU protocol consisted of one Master station and two Slave stations. The stations (Slave) were given unique numbers of ID station. Coprocessor of Master station was placed in PLC, and it was the only one stations authorized to establish communications.

Figure 4 presents the research results, which indicate that the maximum time of information exchange for determined cycles' duration of network applications was 300 ms. What is more, it was noticed that there was a significant difference between the minimum and maximum time of information exchange. The longer the duration of application, the greater difference between times is. It is a consequence of network structure, in which a coprocessor is an independent device that works on the same bus as the rest I/O modules of PLC controller.

Test bench 2 for Modbus/RTU protocol consisted of one Master station and two Slave stations. As previously, Slave stations were given unique ID numbers, but coprocessor of Master station was placed in PLC central unit. This station was the only one authorized to establish communication.

Figure 6 presents the results, which clearly show, that in the presented node solution there are not such significant differences between the minimum and maximum time for information exchange during the realization of node application. In addition, it can be seen that with the adopted node architecture the time for data exchange is much shorter than in the previous solution.

Test bench 3 (Fig. 7) used Toke-Passing protocol (GENIUS). All nodes were given unique SBA address (Serial Bus Address) and were authorized to transmit and receive data in the network. All network coprocessors were placed on basic cassettes of particular PLC controllers, and they realized only periodical transactions.

GENIUS

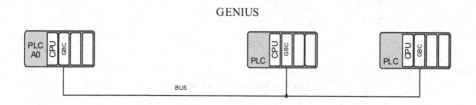

Fig. 7. The schema of test bench for Token-Passing – GENIUS network

Figure 8 presents the results, which indicate that the maximum time of information exchange for determined cycles durations of node application was 46.2 ms, with the duration of node application 15.8 ms. As in case of test bench 1, it was also noticed that there was a significant difference between the minimum and maximum time of information exchange. The longer the duration of application, the greater difference between times is.

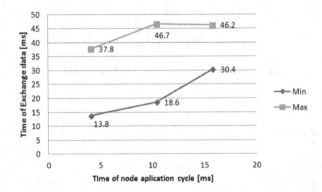

Fig. 8. Test bench 3. Coprocessor for Token-Passing protocol placed in the main cassette of controller. The reading of eight words for automation cycle is as follows: 4.1 ms, 10.4 ms and 15.8 ms.

Test bench 4 for SRTP protocol (Service Request Transfer Protocol) consisted of three PLC equipped with modules of Ethernet communication network connected to switch in star architecture. Each node was given unique IP number. In such configuration each node may be a "client". However, because of the measurement of time for data exchanges in a system with a deterministic access to the link, only the one node can be a "client" station, being authorized to establish communication with the "server" stations.

Figure 9 illustrates the research results, which indicate that the maximum time of information exchange for determined cycles durations of node application was 177.4 ms (Fig. 10), with the duration of node application 66.5 ms. Such as in case of test bench 2, there is a difference between the minimum and maximum time of information exchange.

The empirical research confirmed the thesis, which refers to the necessity of determining the duration of cycles for information exchange in the network supervised by node $A0$ (see also Sect. 3.1). Determining the time T_{TR} is necessary before the network installation is turned on, because only then the algorithm of network controlled by node $A0$ will be work correctly. A very interesting is the fact, that during the research there were the differences between minimum and maximum time of information exchange during the realization of node application. It is also advised to take into account the network architecture in case of choosing a network controlled by node $A0$ (which means position of coprocessor, model of protocol, etc.). One of the factor which increase duration of node

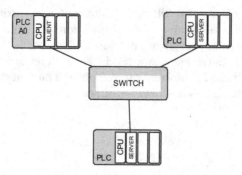

Fig. 9. The schema of test bench for SRTP

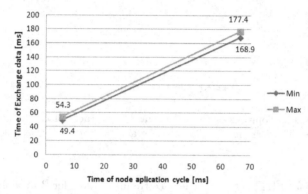

Fig. 10. Test bench 4. Coprocessor of SRTP protocol placed in main cassette of controller. The reading of eight words for automat cycle is as follows: 5.6 ms and 66.5 ms.

cycle is the number of network coprocessors. The bigger number of coprocessors means that the expenditure for system diagnostic are increased, and it influences on the duration of automat cycle. The research results are very important base for further research of shortening data exchange time, and yet they are used to determine the reliability and effectiveness of presented method.

5   Conclusions

This paper presents the concept of creating complex network, which are based on a model of multiprotocol nodes. If the presented idea turn out to be verified by positive results, then it will be the beginning of work on formal structure and definition of a new communication protocol. What is more, it will be possible to use such solution in designing the industrial computing real time systems. It was created on the basis of already tested redundant structures ([22]) with the use of existing MODBUS RTU protocol. The idea of using this protocol aimed at creating a new model of protocol in order to test and determine the practical usefulness of a suggested solution for designing industrial applications.

The already obtained results of practical verification and theoretical background described in Sect. 3, tend to encourage further tests and research. The expected results will be the base of new controller prototype equipped with new integrated software which realizes an algorithm of new protocol. The authors intended to build a new controller which will be able to work with existing hardware solutions. An additional advantage of the proposed solution is the fact that it will be able to work in existing installations, only by installing additional specialized software procedures into the existing system. Currently, the research on developing new software for the presented communication system are being conducted. The aim of this research is to verify the reliability and efficiency of the system. Thanks to the analysis of fundamental time assumptions, it is possible to extend the work of some communication systems with various architecture.

Acknowledgments. This work was supported by the European Union from the European Social Fund (grant agreement number: UDA-POKL.04.01.01-00-106/09).

References

1. Kwiecień, A.: The improvement of working parameters of the industrial computer networks with cyclic transactions of data exchange by simulation in the physical model. In: The 29th Annual Conference of the IEEE Industrial Electronics Society. Conference Center, Roanoke, Virginia, USA (November 2003)
2. Modbus-IDA. Modbus Application Protocol Specification V1.1b3. (December 2006), http://modbus.org/docs/Modbus_Application_Protocol_V1_1b3.pdf
3. Modbus-IDA. Modbus Messaging on TCP/IP Implementation Guide V1.0b. (October 2006), http://modbus.org/docs/Modbus_Messaging_Implementation_Guide_V1_0b.pdf
4. Miorandi, D., Vitturi, S.: Analysis of master-slave protocols for real-time-industrial communications over IEEE802.11 WLAN. In: 2nd IEEE International Conference on Industrial Informatics 2004, pp. 143–148 (June 2004)
5. Conti, M., Donatiello, L., Furini, M.: Design and Analysis of RT-Ring: a protocol for supporting real-time communications. IEEE Transactions on Industrial Electronics 49(6), 1214–1226 (2002)
6. Raja, P., Ruiz, L., Decotignie, J.D.: On the necessary real-time conditions for the producer-distributor-consumer model. In: IEEE International Workshop on Factory Communication Systems, WFCS 1995, pp. 125–133 (October 1995)
7. Genius® I/O System and Communications. GE Fanuc Automation, doc. no: GEK-90486f1 (November 1994)
8. PACSystemsTM Hot Standby CPU Redundancy, GE Fanuc Intelligent Platforms. doc. no: GFK-2308C (March 2009)
9. Siemens. Simatic Profinet description of the system. Siemens, doc. no: A5E00298288-04, Warsaw (2009)
10. Decotignie, J.-D.: Ethernet-Based Real-Time and Industrial Communications. Proceedings of the IEEE 93(6), 1102–1117 (2005)
11. Felser, M.: Real-Time Ethernet-Industry Prospective. Proceedings of the IEEE 93(6) (June 2005)

12. Jestratjew, A., Kwiecień, A.: Performance of HTTP Protocol in Networked Control Systems. IEEE Transactions on Industrial Informatics 9(1), 271–276 (2013)
13. Jestratjew, A., Kwiecień, A.: Using Cloud Storage in Production Monitoring Systems. In: Kwiecień, A., Gaj, P., Stera, P. (eds.) CN 2010. CCIS, vol. 79, pp. 226–235. Springer, Heidelberg (2010)
14. Sidzina, M., Kwiecień, B.: The Algorithms of Transmission Failure Detection in Master-Slave Networks. In: Kwiecień, A., Gaj, P., Stera, P. (eds.) CN 2012. CCIS, vol. 291, pp. 289–298. Springer, Heidelberg (2012)
15. Kwiecień, A., Sidzina, M.: Dual Bus as a Method for Data Interchange Transaction Acceleration in Distributed Real Time Systems. In: Kwiecień, A., Gaj, P., Stera, P. (eds.) CN 2009. CCIS, vol. 39, pp. 252–263. Springer, Heidelberg (2009)
16. IEC 62439, Committee Draft for Vote (CDV): Industrial communication networks: high availability automation networks. Chap. 6: entitled Parallel Redundancy Protocol (April 2007)
17. IEC 62439, Committee Draft for Vote (CDV): Industrial communication networks: high availability automation networks. Chap. 5, entitled Media Redundancy Protocol based on a ring topology (April 2007)
18. Gaj, P., Jasperneite, J., Felser, M.: Computer Communication Within Industrial Distributed Environment – a Survey. IEEE Transactions on Industrial Informatics 9(1), 182–189 (2013), doi:10.1109/TII.2012.2209668
19. Stój, J.: Real-Time Communication Network Concept Based on Frequency Division Multiplexing. In: Kwiecień, A., Gaj, P., Stera, P. (eds.) CN 2012. CCIS, vol. 291, pp. 247–260. Springer, Heidelberg (2012)
20. Gaj, P.: The Concept of a Multi-Network Approach for a Dynamic Distribution of Application Relationships. In: Kwiecień, A., Gaj, P., Stera, P. (eds.) CN 2011. CCIS, vol. 160, pp. 328–337. Springer, Heidelberg (2011)
21. Kwiecień, A., Sidzina, M.: The Method of Reducing the Cycle of Programmable Logic Controller (PLC) Vulnerable "to Avalanche of Events". In: Kwiecień, A., Gaj, P., Stera, P. (eds.) CN 2011. CCIS, vol. 160, pp. 379–385. Springer, Heidelberg (2011)
22. Kwiecień, A., Stój, J.: The Cost of Redundancy in Distributed Real-Time Systems in Steady State. In: Kwiecień, A., Gaj, P., Stera, P. (eds.) CN 2010. CCIS, vol. 79, pp. 106–120. Springer, Heidelberg (2010)

Bandwidth Optimization Method
for Non-critical Data Transmission
in Real-Time Communication Systems

Rafał Cupek, Kamil Folkert, and Mateusz Starzyk

Silesian University of Technology, Institute of Informatics
{rcupek,kamil.folkert}@polsl.pl, matt.starzyk@gmail.com

Abstract. Manufacturing Execution Systems (MES) require the communication services, which allow for easy access to information processed in industrial control systems. Contemporary distributed control systems are based on the IEC 61131 and IEC 61158 standards. This situation causes a bottleneck in communication between control and MES systems. This paper deals with the problem of large data block acquisition between the distributed control system and MES. Authors have put a special focus on Master PLC – Station PLC communication realised in a distributed control system. The limited bandwidth of existing fieldbus network was used. Presented approach has been developed to support data collection process in existing distributed control system used in automotive production. Considered MES functions are related to energy efficiency analysis on machine and production line level. The OPC UA (IEC 62541) standard has been taken into account during large data block size estimation.

Keywords: PLC, MES, PROFIBUS, bandwidth optimization, real-time communication systems.

1 Introduction

Contemporary industrial computer systems use advanced data processing methods like Data Mining (DM) or Artificial Intelligence (AI) for manufacturing processes optimisation, anomaly detection, hazards detection or search for technology dependencies used in the production system engineering. DM and AI services need an access to large data blocks available on the control system level. Unfortunately, industrial computer systems are characterised by relatively large inertia related with long life of installations they support. New services introduced in the higher system layers require their integration with existing control systems [1]. This paper presents data integration approach which was implemented in a distributed control system realised accordingly to IEC 61131 and IEC 61158 standards in order to provide information necessary for MES based on OPC UA – IEC 62541 solution.

Although, new communication solutions dedicated for communication between MES (Manufacturing Execution System) and distributed control systems

A. Kwiecień, P. Gaj, and P. Stera (Eds.): CN 2013, CCIS 370, pp. 189–199, 2013.
© Springer-Verlag Berlin Heidelberg 2013

are developed [2] but in the case of existing industrial installation such possibilities are very limited [3,4]. The adaptation of existing real-time networks to communication with new MES system is very complicated both by the limitations of real-time network technology, as well as by RT (real-time) network functional limitations which are focused on control system [5]. The aim of this study is to ensure access to data processed by distributed control system in order to supply information required by MES. The proposed data block division algorithm for limited sub-bandwidth transmission was optimised according to requirements related with OPC UA – IEC 62541 communication services [6].

The reason for undertaking the described problem was a real need for access to data processed in the automotive engine production line in order to support energy efficiency analysis on the MES level. An important limitation was the underlying control system topology, presented in Fig. 1. Particular production stations were realised on SIMATIC S7-300 PLCs and were interconnected by PROFIBUS real-time network. The production control was synchronised by Master PLC which was also responsible for transport control. PROFIBUS was used both for control and synchronisation functions.

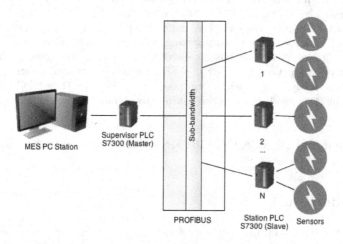

Fig. 1. Existing control system architecture

The designers of the considered distributed control system didn't plan an additional communication necessary for large data blocks transferring. Due to limited bandwidth of PROFIBUS [7] network such opportunities are not available. However, designers predicted the possibility of a future communication system expansion and they reserved a few communication channels not used by distributed control system. The characteristic of available communication channels corresponds to the control system needs but not to MES communication requirements. PROFIBUS network allows for realising cyclic small data buffer transmissions between Master's PLC CPU (Central Processing Unit) and stations' PLC CPUs. External MES system can access data from Master PLC CPU using direct Ethernet connection. This structure causes that all necessary information should be transmitted through existing reserved communication channels from line stations to the master station. In our case, information is necessary for machine and production line energy efficiency analysis but presented algorithm and its effectiveness evaluation are generic and may be useful for other MES applications.

The basic idea under presented solution is data recording in station's PLC CPUs (accordingly to conditions defined by MES) and then data block transfer to Master CPU by a small communication window implemented on the basis of available real-time cyclic communication channels. Additional mechanisms of data block integrity checking and communication status verification were implemented. Simulation results proves the earlier analysis of the dependence between data block size and total transmission time. Experiments take into account the impact of the fieldbus communication parameters (window size, exchange cycle and random communication errors). Analysed range of data block size fulfil both the simple exchange of a small amount of data requirements but also can be used for metadata transfer analysis in respect to in OPC UA objects, types, variables, or even entire address space description [8]. Section 2 presents the generic data flow and communication time analysis. It includes both communication system limitations and data transmission errors which may occur during normal exploitation. Section 3 presents some details of data blocks division algorithm and presents selected use cases defined for OPC UA data transmissions. Section 4 presents simulation studies which may be useful to estimate data transmission time in the real OPC UA communication. Conclusions and future works are presented in Sect. 5.

2 Data Flow Analysis

According to assumptions described in Sect. 1, the data transmission methodology must fit to existing system architecture. The data flow must be organised in a way that won't give any influence on control algorithms and its crucial parameters, especially communication cycle duration must remain unchanged [9]. The Master-Slave paradigm was selected for data block transmissions. Master-Slave application level protocol uses a limited sub-bandwidth of the existing PROFIBUS [10,11] communication network. The energy efficiency data is transferred from Slave PLCs to Master PLC on Master's demand. As presented in Fig. 2, the Slave PLC acquires process data directly from sensors (analogue and binary signals).

The data is stored in PLC CPU's memory data blocks (DB) and then sent on MES demand. These transmissions include timestamp, station's PLC status, state of the underlying sensors and some process data, which is needed to recognise the state of the process during energy efficiency analysis in higher layers. In order to send required data block of size D using communication buffer of size B we require minimum N number of transmissions (minimum is in the case when all transmissions are successful) as shown in Formula (1):

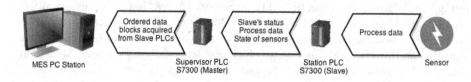

Fig. 2. Data flow between process and MES

$$N = \left\lceil \frac{D}{B} \right\rceil . \tag{1}$$

The maximum data block transmission time (T) outcomes directly from a modified Master-Slave transmission algorithm [12]. We use doubled fieldbus network transition period (T_N) instead of timeout used in classic Master-Slave network analysis. Because transfers are prepared and confirmed on application level we have to consider Master CPU cycle (T_M) and Station CPU cycle (T_S). To initialize data block transfer Master PLC sends a data block transmission request. Then Station PLC starts to realise N confirmed communication buffer transfers. To estimate the maximum data block transmission time Formula (2) can be used

$$T \leqslant T_N + N \cdot (2\,T_N + T_M + T_S) \tag{2}$$

where:

T is the maximum data block transmission time,
T_N is a fieldbus network transition period,
T_M is a maximum Master CPU cycle,
T_S is a maximum Station CPU cycle,
N is a transmission number as calculated in Formula (1).

In the case of communication errors the value of N has to be increased by the number of errors E. Hence, the final formula for estimated maximum transmission time is presented in Formula (3):

$$T \leqslant T_N + (N + E) \cdot (2\,T_N + T_M + T_S) \tag{3}$$

where E is the number of occurred transmission errors. If we assume the constant communication error occurrence probability (equals P_E) we can calculate the total probability P that N frames will be transmitted:

– without errors ($E = 0$):

$$P_{E=0}(N) = (1 - P_E)^{2N} \tag{4}$$

N was multiplied by 2 because both Master-Slave transmission and Slave-Master confirmation may be unsuccessful.

– with minimum one repetition ($E = 0$ or $E = 1$):

$$P_{E \leqslant 1}(N) = (1 - P_E)^{2N} + 2N \cdot \left[P_E \cdot (1 - P_E)^{2N-1} \cdot (1 - P_E) \right] =$$
$$= (1 - P_E)^{2N} + 2N \cdot \left[P_E \cdot (1 - P_E)^{2N} \right] . \tag{5}$$

The second part of Formula (5) expresses the probability of successful first ($E = 1$) repetition in the case of 1 error and $2N - 1$ successes.

– with minimum R repetitions ($E \leqslant R$):

$$P_{E \leqslant R}(N) = (1 - P_E)^{2N} + \sum_{i=1}^{R} \frac{N!}{(N - i)!} \cdot P_E^i \cdot (1 - P_E)^{2N-i} \tag{6}$$

where $\frac{N!}{(N-i)!}$ is the number of variations for i errors which can occur during N transmissions.

The sample probability calculation results for $P_{E<R}(N)$ assuming $P_E = 10^{-2}$ are presented in Table 1. However, as we can see, in the case of small N the probability of retransmission is relatively small, but in the case of larger N this probability will increase significantly. If we consider higher N the $\frac{E}{N}$ ratio seems do not have important impact on total transmission time. Otherwise in the case of low N value transmission error is not very high probable but when it occurs it can have a significant impact on the total transmission time.

Table 1. Success data block transmission probability assuming N transmissions with maximum R repetitions and constant error rate $P_E = 10^{-2}$

$2N$	$E = 0$	$E \leqslant 1$	$E \leqslant 2$	$E \leqslant 4$	$E \leqslant 8$
1	0.99	0.9999	0.9999	0.9999	0.9999
2	0.9801	0.999702	0.999801	0.999801	0.999801
4	0.960596	0.99902	0.99960203	0.99960596	0.99960596
8	0.922745	0.996564	0.999174053	0.999227442	0.999227447
16	0.851458	0.987691	0.998011715	0.998514183	0.998514578
32	0.72498	0.956974	0.993296291	0.997233888	0.997249803
64	0.525596	0.861978	0.969008797	0.994793613	0.995255948
128	0.276252	0.629854	0.856659212	0.983252314	0.992756041
256	0.076315	0.271681	0.523289532	0.875945754	0.989530654
512	0.005824	0.035643	0.112599237	0.414603049	0.915613353
1024	3.39E-05	0.000381	0.002175772	0.024272032	0.302195921

3 Timestamp-Based Complex Data Blocks Division Algorithm for Limited Sub-bandwidth Transmission

The core feature of investigated system is an acquisition of data which describes energy efficiency. That information will be used for further analysis in MES, so it should be time-consistent. For that reason, every data package acquired by higher level from Slave PLC is a snapshot of monitored data values, in accuracy of single PLC cycle. This process should be dynamic. MES system should be free to decide which data set will be necessary and when. This will allow to extend system's functionality in future, e.g. using artificial intelligence supervisor module or even complex multi-agent system which will be responsible for deciding which data should be available for monitoring on the field level.

To ensure maximum system's flexibility the list of pointers contains all possible references to PLC's CPU memory. It refers to process data and meta data that are expected for monitoring (list S). As soon as the list is created, the word of status is changed and the Slave PLC is ready to communicate with master. Just after that, Master PLC is able to fetch the S list of pointers supplemented by variable ID numbers and S list version number. Master PLC creates the actual list of values that will be monitored (M list), which by default contains all of the values available on the Slave side. However, this also can be configured so not every value from S list has to occur in M list. The M list is created for every Slave. Then, the M list has also its version number, which is sent to Slave PLC. Basing on that, Slave creates new list (list S') as a multiplication of S and M lists, so the S' list contains only pointers that occurred in both S and M lists. The Slave PLC's memory structures supporting described pointers creation operations are presented in Fig. 3.

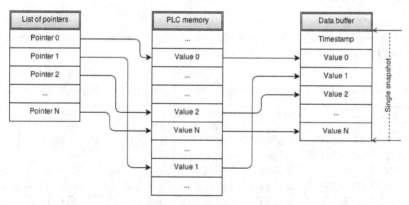

Fig. 3. Slave PLC – data block memory organisation

After successful S' list creation, Slave exposes that fact by changing its status word. Next, Master PLC sends to Slave a configuration package, which describes the parameters of data acquisition:

1. Time interval (in milliseconds) between data snapshots.
2. Acquisition termination method (given number of production cycles or as long as the sufficient memory is available).

After the operations described above the Master-Slave pair is ready to exchange the data. However, the first exchange will be started on Master's demand, given by specific order. Moreover, the Master and Slave system clocks are synchronised during exchange preparation procedure. Afterwards, Slave starts data acquisition and buffering according to configuration. When the task is completed it is exposed to Master by changing the status word.

Master-Slaves data exchange is realised using data exchange buffer. This buffer is then transmitted by fieldbus network communication channel. The minimal buffer size is 4 bytes, which comes from aggregated size of protocol instruction (2 bytes) and parameter (2 bytes). The reserved 4 bytes is an optimal number as a minimum size; the empirical tests indicated that further minimisation of the buffer size would significantly increase the time of data blocks transmission.

The optimal buffer size depends on the actual use case. As the example, if we would like to transfer OPC UA data model in a standard notation (based on OPC Foundation's compressed XSD format) [13] for a simple AddressSpace (26 nodes with data types and references) it would fit in a data package of size 1.7 KB. With the PROFIBUS network exchange cycle given as 100 ms, the whole model would be transferred in approximately 43.5 s according to Formula (2).

The data transfer is initiated by Master PLC. Before the actual energy efficiency data blocks are transferred, the Master PLC asks the Slave PLC for the configuration data: status and S and S' lists, as well as current buffering configuration. If the configuration is correct (has not changed from last check), Master initiates transmission of data blocks containing time-stamped snapshots. In the contrary, if the configuration is to be updated, Master asks for the S list from Slave, updates the M list according to the available values, sends back the M list to Slave (S' is automatically updated) and sends buffering configuration (which is also updated on Slave). When the configuration is updated on Slave, Master initiates transmission of data blocks.

The time-stamped snapshots are transferred iteratively in packages of size equal to buffer size (4 bytes by default). After the package is sent, Slave PLC is waiting for confirmation frame from receiver, which is supposed to contain the number of next package prepared for transmission. If the confirmation frame contains another number, it means that the package of received number was the last successfully transferred. If there will be more than 10 lost packages in a row, the whole transmission process is aborted. As soon as the data is transferred, Master sets proper bit in Slave PLC status. The Slave becomes idle, waiting for Master order. In the same time, Master is waiting for next command.

4   Simulation Studies

To examine efficiency of proposed communication protocol's several simulations were carried out. The measurements were performed using Master's and Slave's applications, implemented in Ladder Diagram for Siemens S7-300 PLC. The applications were running in a simulation environment, provided by SIMATIC S7PLCSIM Professional 2010 V5.4 + SP4. The simulation was carried out on a single PC (Intel Core 2 Duo T7300 2 GHz, 2 GB DDR2, Windows XP Professional with Service Pack 3).

Network delays were simulated with programmable timers support. The random transmission time and random transmission error generation functions were used in order to express reliability parameters observed in real industrial environment. The tests of transmission time as a function of data block size were conducted against influence of two system parameters: data exchange buffer size and network cycle time. For both of them four series of data were measured in domain of data block size values form 16 B to 16 KB. If the single measurement took more than 60 s, the experiment was aborted. The network cycle was simulated with 10 % accuracy. The probability of packet loss was equal 1 %. In the Fig. 4 we can see transmission time which was obtained for network cycle equal to 1 ms (value typical for PROFINET I/O network) [7]. In the Fig. 5 we can analyse results of similar experiment but with network cycle equal to 100 ms (value typical for PROFIBUS network [14]). For each of them four values of data exchange buffer size were considered (4 B, 16 B, 64 B and 256 B). As expected in Formula (3), the transmission time is linearly dependent on data exchange buffer size (the horizontal axis scaled exponential). However, we can observe anomaly for buffer size 16 B and data block size 4096 B which reflects the fact that transmission was repeated in result of more than 10 transmission errors. The single packet retransmissions due to low $\frac{E}{N}$ factor can't be observed on trends directly.

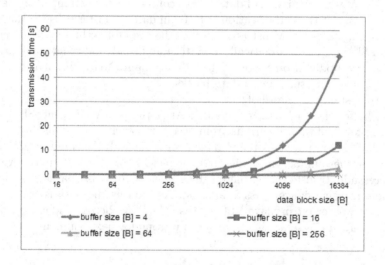

Fig. 4. Total data block transmission time for 1 ms network cycle

In the Fig. 6 we can analyse the situation when buffer size was fixed to 4 B but different network cycles were compared(1 ms, 10 ms, 100 ms and 1 s). According to expected results, the linear dependency of transmission time is visible for all network cycles. Please notice that for the biggest network cycle value (1000 ms),

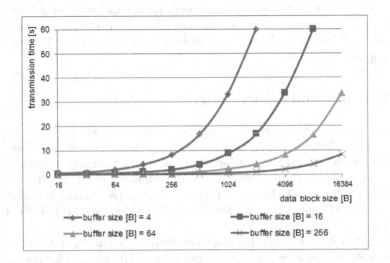

Fig. 5. Total data block transmission time for 100 ms network cycle

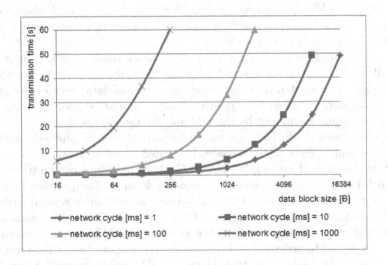

Fig. 6. Total data block transmission time for various network cycles

even the first measurement for data block size 16 B took more than 6 s. This is caused by the 3-step protocol message exchange mechanism:

1. start transmission,
2. get first package,
3. send confirmation (package number).

Each of these steps is followed by the confirmation message. Thus, the accumulative time t satisfies the following inequality: $6 \cdot (network\ cycle) \leqslant t \leqslant 6 \cdot (network\ cycle) + 3 \cdot [(Master\ PLC\ cycle) + (Slave\ PLC\ cycle)]$, as expected according to Formula (3).

5 Summary and Conclusions

Presented solution allows for data acquisition between MES and distributed control system. Authors shown that limited bandwidth of existing fieldbus communication network can be used for large data block transfers. Proposed solution was analysed on the example of control system based on the S7-300 PLCs and PROFIBUS network. However, proposed scheme can be easy transferred to other distributed control systems built according to IEC 61131 and IEC 61158 standards.

Presented in Sect. 4 simulation results confirm the linear dependence between total transmission time and size of the transmitted data block. This dependency was not significantly disturbed in the event of a communication errors occurrence as was envisaged in Sect. 2. Simulations show the impact of the communication buffer size and fieldbus cycle time on data block transfer abilities. Unfortunately, above parameters depend on the the existing communication system and in most cases it is not possible to change them.

More open question is related with the data block size selection. Larger data block allows for more accurate and more sophisticated analysis performed on the MES system level. On the other hand bigger size of data block causes longer time of information exchange. The actual value should be chosen taking into account the requirements put by the DM and AI algorithms.

The next issue was related to process and control system parameters modelling as object oriented data address space. Presented in Sect. 3 values prove that meta data model used for in OPC UA (IEC 62541) standard can be transferred between MES and control system in a reasonable time. The proposed solution may be a step towards mapping OPC UA services directly on existing distributed control systems created accordingly to IEC 61131 and IEC 61158 standards.

Presented work does not address the issue of how MES system determines which data should be collected and when it should happen. One of the promising solutions is use of multi-agent technology for DM and AI support. Considered distributed control system consists of many hundreds of distributed PLC based nodes which are involved in the different stages of the production. The above premise causes that presented approach may be useful in future research on applying multi-agent technology for DM and AI tasks.

Acknowledgement. This work was supported by the European Union from the European Social Fund (grant agreement number: UDA-POKL.04.01.01-00-106/09).

References

1. Cupek, R., Huczala, L.: Passive PROFINET I/O OPC DA Server. In: 14th IEEE International Conference on Emerging Technologies and Factory Automation, ETFA 2009 (2009)
2. Sauter, T.: The continuing evolution of integration in factory automation. IEEE Industrial Electronic Magazine 1(1), 10–19 (2007)
3. Tan, V., Yi, M.-J.: Development of an OPC Client-Server Framework for Monitoring and Control Systems. Journal of Information Processing Systems 7(2), 321–340 (2011)
4. Felser, M.: Real-time ethernet-industry prospective. Proceedings of the IEEE 93(6), 1118–1129 (2005)
5. Gaj, P., Jasperneite, J., Felser, M.: Computer Communication Within Industrial Distributed Environment – a Survey. IEEE Transactions on Industrial Informatics 9(1), 182–189
6. Lange, J., Iwanitz, F., Burke, T.J.: OPC – From Data Access to Unified Architecture, pp. 196–201. VDE Verlag (2010)
7. Kleines, H., Detert, S., Drochner, M., Suxdorf, F.: Performance Aspects of PROFINET IO. IEEE Transactions on Nuclear Science 55(1) (February 2008)
8. Cupek, R., Fojcik, M., Sande, O.: Object oriented vertical communication in distributed industrial systems. In: Kwiecień, A., Gaj, P., Stera, P. (eds.) CN 2009. CCIS, vol. 39, pp. 72–78. Springer, Heidelberg (2009)
9. Risso, F., Deogianni, L., Varenni, G.: Profiling and Optimization of Software-Based Network Analysis Application (2011)
10. Popp, M., Weber, K.: The Rapid Way to PROFINET. PROFIBUS Nutzerorganisation e.V., Karlsruhe (2004)
11. PROFINET Technology and Application. Siemens information materials, Karlsruhe (2005)
12. Robert, J., Georges, F., Rondeau, E., Divoux, T.: Minimum cycle time analysis of Ethernet-based real-time protocols. International Journal of Computers, Communications and Control 7(4), 743–757 (2012)
13. Mahnke, W., Leitner, S.-H., Damm, M.: OPC Unified Architecture. Springer, Heilderberg (2009)
14. GSDML Specification for Profinet IO. Version 2.16, PROFIBUS Nutzerorganisation e.V., Karlsruhe (2008)

Communication Performance Tests in Distributed Control Systems

Marcin Jamro, Dariusz Rzońca, and Bartosz Trybus

Rzeszow University of Technology, Department of Computer and Control Engineering,
al. Powstancow Warszawy 12, 35-959 Rzeszow, Poland
{mjamro,drzonca,btrybus}@kia.prz.edu.pl
http://kia.prz.edu.pl

Abstract. The paper presents a concept and implementation of Communication Performance Tests (CPT) for small distributed control systems. Requirements for the communication performance are specified using SysML notation. Test cases included in the specification are translated into a dedicated test definition language CPTest+. System implementation is then verified by executing the tests generated from the specification and analyzing results of test runs. The procedure is supported by specialized tools integrated with IEC 61131-3 development environment, including SysML model editor and CPTest testing environment.

Keywords: control systems, communication, performance, testing.

1 Introduction

Process automation involves either stand-alone controllers or distributed control systems (DCS). In the latter case, devices such as controllers, I/O modules, HMI panels, or measurement appliances communicate to exchange values used in control programs [1]. Distributed control is applied not only in large scale applications, but become more common also in small and medium systems. Typically, mini-DCS involves communication between the controllers and external I/O modules [2], between monitoring and engineering stations with SCADA software, or between an HMI panel and a controller. Usually, traditional field communication protocol is used, e.g. defined in IEC 61158 standard [3], however nonstandard solutions are also described, like field protocol based on HTTP [4]. The protocols are based on different models of communication, among them master-slave type is fairly common.

The paper concerns master-slave communication between a controller and slave I/O modules. A single *communication task* is defined as data exchange with one device and is performed by a periodical execution of *transactions*. Each transaction involves a *request* and *response* (Fig. 1). A request is sent from the controller to the slave, while a response is sent in an opposite direction.

Testing is an important and complex part of software design and can be classified using various criteria [5]. Real-time systems work in a different way than traditional IT projects [6] and deal with some hardware-related issues. Thus,

A. Kwiecień, P. Gaj, and P. Stera (Eds.): CN 2013, CCIS 370, pp. 200–209, 2013.
© Springer-Verlag Berlin Heidelberg 2013

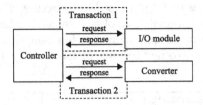

Fig. 1. Communication transactions between the controller and external devices

dedicated testing solutions are introduced for this kind of applications, including the framework using Test-First Development approach [7], as well as specialized communication tests. Failures and inaccuracy of communication can cause risk by performing calculations and control with not up-to-date or untrue values. To meet requirements for functional safety covered in IEC 61508-1 standard [8], the authors propose specification and execution of *Communication Performance Test* (CPT). While different kinds of communication tests are commonly used in large scale networks [9], an appropriate equivalence for industrial fieldbuses needs special consideration.

The proposed procedure is as follows (Fig. 2). First, requirements are specified during design stage using SysML language [10] to indicate performance assumptions for the communication tasks. Then, a set of test cases is automatically generated in textual CPTest+ language (see Sect. 2.2). The tests are executed by a specialized test execution engine (see Sect. 2.3) to verify performance of communication tasks in shorter and longer periods and to check if they are finished in a specified amount of time. Results of the test runs are presented to the developer (see Sect. 2.3).

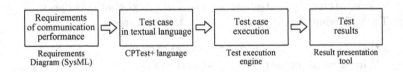

Fig. 2. Process of CPT specification, implementation, execution, and result analysis

2 CPT Specification and Execution

2.1 Requirements Specification

System Modeling Language (SysML) [10] is a graphical specification language based on UML (Unified Modeling Language), but provides better support for systems that are not only oriented to objects. SysML consists of several diagram types. One of them is the Requirements Diagram, applied in the proposed solution to specify communication performance assumptions (Fig. 3). The stereotype «requirement» is used to specify suitable requirements. The diagram may contain other stereotypes to form a tree-like structure. The top requirement is related to the whole distributed control system, while the others to particular communication tasks. Each requirement can have multiple test cases, specified by «testcase» stereotypes, to perform one or more verification checks.

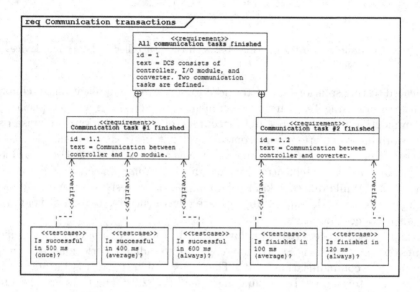

Fig. 3. Model of communication performance tests

Each «requirement» contains an unique identifier and a description text. The identifier is used in other «requirement» stereotypes connected by containment relations. Stereotypes «testcase» consist of a text that specifies verification checks. The following checks can be used in the CPT tests:

- Is [not] successful in **number unit (recurrence)**?
- Is [not] finished in **number unit (recurrence)**?

where **number** is an integer value that indicates amount of time in specified **unit** (ms or s) with given **recurrence** (**once**, **always**, **average**). The recurrence indicator **once** is used for a single run test, and two other – for tests run in longer period of time.

2.2   Test Cases in CPTest Language

Requirements are used to automatically generate test cases in CPTest+ textual language. It is dedicated to creation of unit tests for POUs in control systems and contains extensions for the communication performance tests. CPTest+ supports the following basic instructions:

- SET – sets value of a logical variable (or variables) to TRUE,
- RESET – sets value of a logical variable (or variables) to FALSE,
- ASSIGN – sets value of a variable,
- WAIT – holds execution of a test for specified period of time,
- LOG – saves information to logs,
- ASSERT – checks whether a condition is met.

ASSERT is the most complex instruction and has the following syntax: ASSERT [OPERATOR] [VARIABLE] [VALUE] or ASSERT [ISTRUE|ISFALSE] [VARIABLE], e.g. ASSERT EQ PACKETS_NUMBER 5 or ASSERT ISTRUE PACKET_RECEIVED. The operators which can be used in the ASSERT clause are: EQ (equal), NEQ (not equal), LT (less than), LTE (less than or equal), GT (greater than), GTE (greater than or equal), ISTRUE (is equal TRUE), ISFALSE (is equal FALSE).

The following instructions are dedicated to perform CPT:

- ASSERT {CT_SUCCESSFUL|CT_UNSUCCESSFUL} [ID] [TIME] [UNIT]   – checks whether a communication transaction is finished successfully or unsuccessfully in the given time,
- ASSERT {CT_FINISHED|CT_UNFINISHED} [ID] [TIME] [UNIT] – checks if a communication transaction is ended or not in the given time (without checking correctness),
- GET CT_SUCCESSFUL [ID] [VARIABLE] – stores to VARIABLE a logical value indicating whether a specified communication transaction is already finished successfully,
- GET CT_FINISHED [ID] [VARIABLE] – stores to VARIABLE a logical value indicating whether a specified communication transaction is already ended (without checking correctness),
- GET {CT_LAST|CT_MAX|CT_AVG} [ID] [VARIABLE] – stores to VARIABLE an integer value of the last, maximum or average time (in ms) of the specified communication transaction.

ID is a number of communication transaction and TIME indicates a period in specified UNIT (ms or s). Passing * as ID means that the test will check all defined communication transactions.

For example, a test verifying that a transaction is finished successfully in 500 ms can be represented in CPTest+ language as ASSERT CT_SUCCESSFUL 1 500 MS. It can be created with a set of instructions as well:

```
WAIT 500 MS
GET CT_SUCCESSFUL 1 CT_RESULT
ASSERT ISTRUE CT_RESULT
```

Long running tests can be specified in CPTest+ using CT_LENGTH and CT_MODE special variables. CT_MODE can be set to ONCE, ALWAYS, and AVERAGE. A test running for 30 seconds which checks whether a transaction is finished always within 300 ms is represented by the following CPTest+ code:

```
DEFINE CT_LENGTH T#30s
DEFINE CT_MODE ALWAYS
ASSERT CT_SUCCESSFUL 2 300 MS
```

2.3 Test Execution

Communication performance tests are run by the dedicated test execution environment which interprets CPTest+ code, reads current values from the controller, and checks test conditions, e.g. whether a specified communication transaction is already finished, or whether all communication transactions ended without timeout in the last attempt. Data from the controller are fetched via a master-slave communication protocol (typically Modbus). Tests are executed sequentially and each test runs a set of instructions (Fig. 4). The results indicates whether the test run is finished successfully. Additional result data can be used to observe progress of the test run and analyze the time required for communication.

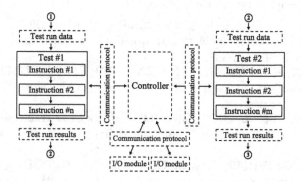

Fig. 4. Execution of communication performance tests

3 Modelling and Testing Tools

The CPT tests have been implemented in a set of dedicated tools supporting design, execution, and result analysis. CPModel (Control Program Modelling) and CPTest (Control Program Tester) are the main modules. They have been integrated in the CPDev engineering environment [11,12] which is used to develop control systems according to IEC 61131-3 standard [13]. CPDev has been applied in some industrial applications including small DCS systems [12,2].

3.1 Control Program Modelling Tool

CPModel is an editor of graphical SysML diagrams supporting Requirements, Block Definition, and State Machine Diagrams [10]. The first type of diagrams specifies functional and non-functional requirements for the designed control system. A structure of the system software can also be modelled, including definition of IEC 61131-3 Program Organization Units (POUs, i.e. programs, functions, and function blocks) and test cases dedicated to them.

CPModel main window (Fig. 5) consists of two parts. Project tree is on the left with information about the diagrams involved. The remaining space is used as a workspace containing diagram design windows. Fig. 5 contains the sample requirements diagram discussed earlier. To make diagram creation easier, mechanism of finding connections between stereotypes (e.g. «requirement» and «testcase») automatically selects a suitable path between the elements and updates it whenever one of the connected elements is moved [14]. The user can adjust settings of various elements, including relations between stereotypes and diagram itself. To simplify design process, the editor supports showing an auxiliary grid and snaps elements to it.

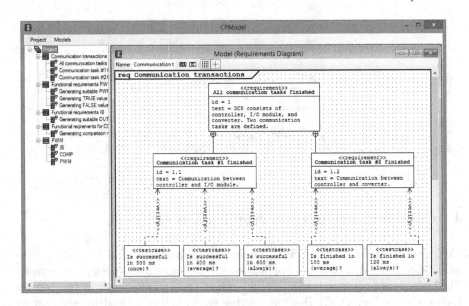

Fig. 5. CPModel tool with a set of diagrams

3.2 Control Program Tester Tool

CPTest is an integrated tool which allows to edit test cases, execute them and analyze the results. Its main window (Fig. 6) is split into the tests tree (on the left) and design area. In the approach presented in the paper, communication performance tests are automatically generated accordingly to the SysML model. Other tests can be added manually in the CPTest tool, e.g. unit and table tests for POUs.

CPTest contains a test execution engine working in on-line mode to obtain CPT performance data from the target controller. The user can see the progress and results of test runs in the main window (bottom part of Fig. 6). Each test run can be in the following states: not started, running, passed, and failed. In case of failure, CPTest tries to execute the next test. CPTest provides details about test runs (like exact execution time and date) for reporting purposes.

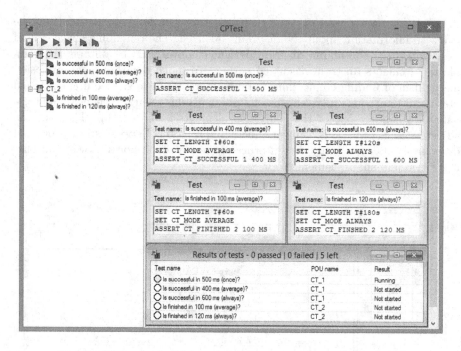

Fig. 6. CPTest tool with a set of CPT tests and execution results

4   Communication Tests Application Example

SMC controller from Lumel S.A., Poland [15], operating as a central node in mini-DCS, is used here as an application of communication performance tests. The SMC is not equipped with internal inputs and outputs, but involves remote I/O modules (Fig. 7). CPDev engineering environment is used to program its control functions.

Function Block Diagram (FBD) shown in Fig. 8 is an example of SMC program created in CPDev. It consists of six blocks. The logical blocks OR, AND control ENGINE according to rule ENGINE := (ENGINE OR START) AND NOT STOP AND NOT ALARM. The blocks (DELAY_ON and DELAY_OFF) delay PUMP activation and deactivation by 5 and 10 seconds, respectively. COM_OUT and COM_IN are instances of communication blocks defining communication with SM4 and SM5 I/O modules (Fig. 7). START, STOP, and ALARM signals are read from IN1–IN3 binary inputs of SM4 module, ENGINE and PUMP are written to OUT1–OUT2 outputs of SM5 module. Communication timeout is set to 100 ms in both cases.

The communication blocks from FBD define two communication tasks shown in Fig. 9. The first row in the table represents the task which periodically communicates with slave 1 (SM4 here) using Modbus function FC3 to read 8 registers from remote address 4003. The obtained values are converted from 16 to 8 bits and stored at local address 63. The task has normal priority and its timeout is set to 100 ms. The last two columns are used by CPTest tool for checking

CPDev
PC
USB
SMC
RS-485
SM5
SM4
START
STOP
ALARM
ENGINE
PUMP

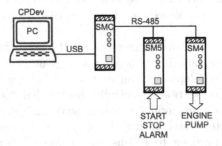

Fig. 7. Small distributed control system with SMC controller

START
OR
1
STOP
ALARM
AND
2
ENGINE
DELAY_ON
TON
IN
Q
T#5s
PT
ET
3
DELAY_OFF
TOF
IN
Q
PT
ET
4
PUMP
T#10s
COM_OUT
COM_SM4
TRUE
EN
C
BYTE#2
NO
T#100ms
TOUT
ENGINE
OUT1
PUMP
OUT2
OUT3
OUT4
OUT5
OUT6
OUT7
OUT8
5
COM_IN
COM_SM5
TRUE
EN
C
BYTE#1
NO
IN1
START
T#100ms
TOUT
IN2
STOP
IN3
ALARM
IN4
IN5
IN6
IN7
IN8
6

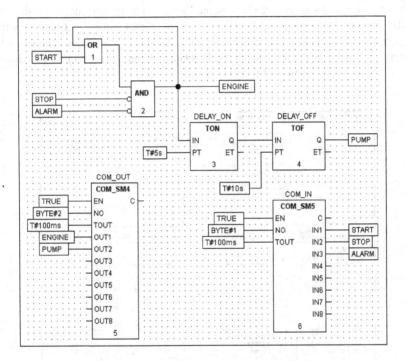

Fig. 8. Sample FBD control application

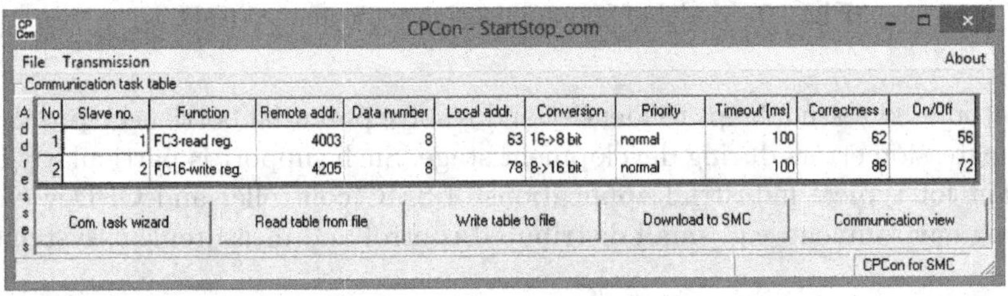

CPCon - StartStop_com
File Transmission
About
Communication task table
No Slave no. Function Remote addr. Data number Local addr. Conversion Priority Timeout [ms] Correctness On/Off
1 1 FC3-read reg. 4003 8 63 16->8 bit normal 100 62 56
2 2 FC16-write reg. 4205 8 78 8->16 bit normal 100 86 72
Com. task wizard
Read table from file
Write table to file
Download to SMC
Communication view
CPCon for SMC

Fig. 9. Communication tasks in CPCon

correctness of communication. The second row represents communication with the other module i.e. SM5. As seen, each of the tasks involve a single communication transaction.

The modules SM4, SM5 are connected via RS-485 with baud rate set to 115.2 kbps. The transmission time of a request is about 2 ms. The slave responds after 10–15 milliseconds, depending on its type and internal state. Thus, time required for a single transaction is typically shorter than 20 ms. In every working cycle the controller reads inputs, executes program and writes outputs. The cycle time is set to 50 ms and both tasks should be served to ensure validity of variables. Timeouts are set to 100 ms, so the request is retransmitted if no response is received within 100 ms.

SysML specification with CPT test cases for such requirements is shown in Fig. 10. If the tests are run once, each of the tasks must be finished within 20 ms (bottom test cases). In longer periods of time the average time of overall communications (both tasks) should be less than 50 ms (top right). If any timeout occurs (100 ms), the tests will fail (bottom).

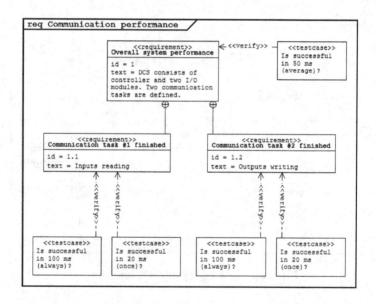

Fig. 10. Model of communication performance tests

Automated testing helps to identify and resolve potential bottleneck problems and regression errors during development stage. Such support is particularly important for typical industrial applications of SMC controller and CPDev engineering environment, e.g. small distributed control and measurement systems.

5   Conclusion

In the proposed approach, SysML language is used to specify both requirements for the system and communication test cases. According to the specification, communication tests are generated using specialized test description language called CPTest+. Dedicated tools introduce the concept into IEC 61131-3 development environment for distributed control systems. By running the tests, communication implementation can be verified against the specification.

References

1. Gaj, P., Jasperneite, J., Felser, M.: Computer Communication Within Industrial Distributed Environment – a Survey. IEEE Transactions on Industrial Informatics 9(1), 182–189 (2013)
2. Rzońca, D., Stec, A., Trybus, B.: Data Acquisition Server for Mini Distributed Control System. In: Kwiecień, A., Gaj, P., Stera, P. (eds.) CN 2011. CCIS, vol. 160, pp. 398–406. Springer, Heidelberg (2011)
3. IEC 61158 Standard: Industrial Communication Networks – Fieldbus Specifications. IEC (2007)
4. Jestratjew, A., Kwiecień, A.: Using HTTP as Field Network Transfer Protocol. In: Kwiecień, A., Gaj, P., Stera, P. (eds.) CN 2011. CCIS, vol. 160, pp. 306–313. Springer, Heidelberg (2011)
5. Vegas, S., Juristo, N., Basili, V.R.: Maturing Software Engineering Knowledge through Classifications: A Case Study on Unit Testing Techniques. IEEE Transactions on Software Engineering 35(4), 551–565 (2009)
6. Thane, H., Hansson, H.: Towards systematic testing of distributed real-time systems. In: Proc. The 20th IEEE Real-Time Systems Symposium, pp. 360–369 (1999)
7. Winkler, D., Hametner, R., Ostreicher, T., Biffl, S.: A framework for automated testing of automation systems. In: 2010 IEEE Conference on Emerging Technologies and Factory Automation (ETFA), pp. 1–4 (2010)
8. IEC 61508-1 Standard: Functional safety of electrical/electronic/programmable electronic safety-related systems – Part 1: General requirements. IEC (2010)
9. Angrisani, L., Narduzzi, C.: Testing communication and computer networks: an overview. IEEE Instrumentation & Measurement Magazine 11(5), 12–24 (2008)
10. SysML Open Source Specification Project website, http://www.sysml.org/
11. CPDev website, http://cpdev.kia.prz.edu.pl
12. Rzonca, D., Sadolewski, J., Stec, A., Swider, Z., Trybus, B., Trybus, L.: Mini-DCS system programming in IEC 61131-3 Structured Text. Journal of Automation, Mobile Robotics and Intelligent Systems 2(3), 48–54 (2008)
13. IEC 61131-3 Standard: Programmable Controllers. Part 3. Programming Languages. IEC (2003)
14. Jamro, M.: Graphics editors in CPDev environment. Journal of Theoretical and Applied Computer Science 6(1), 13–24 (2012)
15. LUMEL S.A. website,
   http://www.lumel.com.pl/en/area_of_activity/
   network_solutions/art272,smc-programmable-logic-controller.html

Some Problems of Integrating Industrial Network Control Systems Using Service Oriented Architecture

Marcin Fojcik and Joar Sande

Sogn og Fjordane University College
{marcin.fojcik,joar.sande}@hisf.no

Abstract. In this paper, the authors present methods for connection Service Oriented Architecture with OPC UA to control systems. Different disciplines such as data science, communication theory and control systems have different viewpoint on the properties of communication systems. Requirements of the control systems and requirements of information systems are presented. The integration of the two systems is analysed. Problem is formulated: how it is possible to get new functionality without losing existing features. Most important parameter is time. The control systems have to achieve fixed time of data sampling, regardless of situation. For Service Oriented Architecture time is not so important. SOA emphasizes openness, modularity and compatibility. Integration of two different systems and integration problems in distributed control systems are presented and analysed.

Keywords: OPC Unified Architecture, Network Control Systems, Service Oriented Architecture, fieldbus, network traffic.

1 Introduction

Today, information systems are designed more open and modular. The information systems consist of universal sets of interfaces and services. It simplifies creation, installation and maintenance. The reason for this is the use of one communication protocol – TCP/IP, as a standard. Standardization enables easy connecting between different applications. Integration gives possibility of using the same services, common data exchange and independence of operating systems and hardware. Similar, unified tendency is observable in automation systems. Unfortunately, one common communication standard is not available here. There are many different, incompatible standards define to data exchange in completely different situations and environments. Connection of some of these standard communication protocols in one system is possible. This action requires individual processing, human intervention is often necessary.

There are many research projects [1,2,3] with aim to define universal rules and methods for integration. The aim is to achieve an open control system. The pursuit on openness can lead to loss the characteristic properties of the

A. Kwiecień, P. Gaj, and P. Stera (Eds.): CN 2013, CCIS 370, pp. 210–221, 2013.
© Springer-Verlag Berlin Heidelberg 2013

system. In industry protocols, typically the raw data without description are sent. Without knowledge about the system and protocol configuration, these data are not usable. Adding descriptions to the protocol increases the amount of data to transfer, and therefore increases the transfer time. Other, potentially negative, situation is the mutual influence of interfaces for the control system and the information system. Both have different communication possibilities: fast transfer of small amount of data versus slow transfer of big amount of data. It is possible to send all information from the control system into the information system. Transfer in the other direction leads to increased traffic and blockage of the control system.

In this paper, the authors suggest methods to connect information systems based on Service Oriented Architecture to industrial networks. Some problems of integrations are also shown. Requirements, especially for data transfer speed, are described. Authors present description and analysis of some control system protocols and services available in Service Oriented Architecture. In particular, services used to connection SOA to the control system are considered. With analysis of existing research projects, literature and own research, authors show requirements necessary to correct integration. Nowadays, software and hardware do not allow full integration, but even partial integration gives many new possibilities to reduce time of creating software for automation systems.

This paper is divided into 6 parts. In Section 2, the authors present time problems in control systems; Sect. 3 contains mathematical equations used during parametrization of industry network. In Section 4 the ideas of the SOA based system are presented. Section 5 contains description of the OPC UA protocol and shows how it can be used to connect SOA and industry network. In Section 6, results of calculations and experiments in connecting are presented.

2 Control Systems

In a control system there are several control loops, some of them single loop, others with multiple inputs and outputs. Some of the control loops can be interconnected such that change in a control parameter in one control loop affects a different loop. The control loop calculates new parameters based on measurements from the field, which in turn are sent to an actuator. Some loops e.g. temperature control, can be slow, while others e.g. motor control, requires an immediate response. In some cases the states of a control system can be estimated based on inputs, outputs and the physical measurements that are done. This lowers the costs and the number of sensors can be minimized, without reducing the quality of the control.

A process can be described by a state space model, where x are the states, u is the input and y it the output.

$$x' = A * x + B * u, \qquad y = C * x, \qquad \text{for continuous control,}$$
$$x_{k+1} = \Phi * x_k + \Gamma * u_k, \qquad y(k) = C * x(k), \quad \text{for discrete control,}$$

where:

- $\Phi = e^{Ah}$,
- $\Gamma = B * \int_0^h e^{As} ds$,
- A, B and C are matrices describing the controlled system: A – system matrix, B – input matrix, C – output matrix,
- h is fixed sampling period.

Figure 1 shows a control schema of a process plant [4]. The measurements $y(t)$ are sampled at time t_k. A mathematical model calculates the different states of the control system. The states are combined into an output from the controller(s) to the actuator(s), controlling different plant processes involving e.g. temperature and motor speed. The sampling time t_k is crucial, it must be short enough to detect any changes in the controlled process. A slow sampling rate can lead to an unstable system. The controller will overreact; the controlled system will receive too much energy, and therefore becomes unstable. The destruction of both equipment and production batch can be the result. Communication through a network induces delays. The duration between sampling of measurement signal until it is used by the actuator, introduces an additional time delays that must be dealt with. Time delay caused by the transmission of measurements to the controller, processing and then sending data from the controller (Fig. 1) is defined [4] as:

$$t_{\text{delay}} = t_1 + t_2 + t_3 .$$

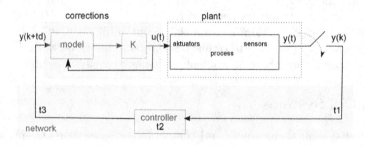

Fig. 1. Model of control system with network delays

In order to maintain correct operation of the control system, time delay tdelay should be equal or lower than fixed sampling period h.

$$t_{\text{delay}} \leq h . \tag{1}$$

Some cases of control systems with non-constant time delay can be corrected with additional mechanisms. One of them is Smith Predictor – a delay compensation algorithm. Other strategies are the Dahlin Algorithm [5] or the Kalman filter [6] which can be used to refine the output signal from imprecise input data but with reduced quality.

In [5] response time and computation time for the cases of control systems without delays are presented (Table 1). Changing of the time delay leads to serious (up to 46 times longer) extension of the response time.

Table 1. Comparison of some correction strategies for control systems

	Control system		Control + Smith Predictor		Control + Dahlin algorithm	
	computation time [ms]	response time [ms]	computation time [ms]	response time [ms]	computation time [ms]	response time [ms]
nodelay	20	2.4	40	5	40	240
delay without computation time stability	—	—	—	230	—	480
with controller optimization	—	—	40	150	40	350

3   Data Exchange in Industry Networks

For Network Control System, the performance is a function of not only the sampling period, but also the traffic load on the network [7]. The traffic load can cause variable time delays in the transmission of data to the control system. The control system requires that time delays should not be longer than specified sampling time (1) It is necessary to calculate maximum time delay for known configuration of the control system.

Several have calculated the transmission time used to send one data package. We refer to [8,9] for such calculations. This approach is often insufficient in distributed control systems, because of protocols requirements. To achieve proper control, continuously sending of many measurements and control signals in specified time is necessary. For example, in protocols based on pooling, a slave station can send data after receiving a request from the Master station, and then wait for the next request, while the Master station asks the next Slave station. To calculate the shortest time period between two transmissions of the same data, we need to know the structure of the entire distributed system (protocol model, amount of devices, processing times for each device, transmission speed, and many others). The time used to send the data between all devices is called network cycle [10]. This time can be calculated for all protocols, which guarantee constant and repeatable actions.

The traffic depends on the network protocol and the configuration. There are many protocols according to different needs. Industrial protocols based on some models: Time Division Multiple Access, token, Master-Slave or Producer-Distributor-Consumer, have precisely defined order of data exchanges. Data exchanges are executed strictly in accordance to pre-defined scenarios. This guarantees receiving all data within a fixed, calculated time. Dense traffic caused by continuous data exchange is a disadvantage in these models (Fig. 2a).

Another solution is to send data on demand only. Devices, with additional processor, send data only on events. Protocols based on this model induce much

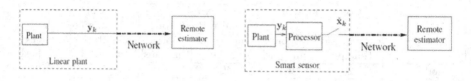

Fig. 2. Types of industry networks: (a) with constant network cycle, (b) event-driven [11]

lower network traffic, but it is not possible to plan the order of exchanges, it is therefore not possible to get fixed frequency of data exchanges (Fig. 2b). The first type of protocols requires calculation and planning of time parameters in the installation phase. In the working phase data packages are sent according to a planned scenario. Due to exchange-control mechanisms, these types of protocols guarantee regular delivery of the data. Data are sent continuously in the same time interval, independent of the actual situation. To achieve required network cycle time, the protocol generates a significant traffic on the network. Connection of a new device to an existing system, without renewing the configuration can easily lead to traffic blockage.

Event-driven data exchanges are used in protocols based on random or hierarchic access to media. Every device has a processor which chooses to send data or not. Only measurements which have changed value are transmitted. Operation of choosing data before sending reduces the network traffic. It is possible to detect, transmit and receive all, even short-term changes in process. These events cannot be detected in protocols with constant cycle time. A drawback of such protocols is that during an alarm situation many sensors suddenly detect changed values. A cascade occurs when several sensors send data almost at the same time. Such cascades can result in a blocked network. As an example, random access to media with collision detection or avoidance (CSMA/CD, CSMA/CA) can be mentioned [8,9,12]. A possible solution is to use data packets with priority. Information with lower priority has to wait until all information with higher priority is transmitted. This mechanism is used in Producer-Consumer (CAN) [13].

The network cycle time in protocols with constant time of cycle can be calculated. To simplify the problem, we assume only correct hardware condition for devices. For TDMA, the time is available as a protocol parameter. For calculations for protocols with *token* it is necessary to know the amount of devices, amount of exchanges and how the data are sampled, processed, prepared and sent by each device. In [10,14,15] the cycle time is calculated for many industrial protocols.

Minimal network cycle time for *token* protocols can be calculated by the formula:

$$T_{\text{NCT}} = time_{\text{token transmission}} + time_{\text{processing}} + time_{\text{data transmission}} +$$
$$+ time_{\text{data detection}} + time_{\text{acknowledgement transmission}}$$

$$T_{\mathrm{NCT}} = 3 * L_{\mathrm{A}} * (T_{\mathrm{TR}} + T_{\mathrm{DR}} + T_{\mathrm{AR}} + T_{\mathrm{PR}}) + \sum_{i=1}^{L_{\mathrm{A}}} T_{\mathrm{A}_i} +$$

$$+ \sum_{i=1}^{L_{\mathrm{A}}} (T_{\mathrm{PR}_i} + T_{\mathrm{TR}_i}) + 2 * L_{\mathrm{A}} * (T_{\mathrm{DR}} + T_{\mathrm{AR}}) + L_{\mathrm{A}} * T_{\mathrm{TP}} \qquad (2)$$

Respectively, calculation for Master-Slave protocols:

$$T_{\mathrm{NCT}} = time_{\text{request transmission}} + time_{\text{data detection and processing}} +$$
$$+ time_{\text{answer transmission}}$$

$$T_{\mathrm{NCT}} = \sum_{i=1}^{L_{\mathrm{A}}} (T_{\mathrm{PR}_i} + T_{\mathrm{TR}_i}) + \sum_{i=1}^{L_{\mathrm{A}}} (T_{\mathrm{AR}_i} + T_{\mathrm{A}_i} + T_{\mathrm{DR}}) +$$

$$+ \sum_{j=1}^{L_{\mathrm{A}}} (T_{\mathrm{DR}} + T_{\mathrm{A}_j} + T_{\mathrm{PR}_j} + T_{\mathrm{TR}_j} + T_{\mathrm{AR}_j}) \qquad (3)$$

where:

T_{NCT} – network cycle time,
L_{A} – number of devices,
T_{TR} – token transmission time,
T_{DR} – token detection time,
T_{AR} – time of analysis of token,
T_{PR} – time of preparation of token,
T_{PR_i} – time of preparation of data package,
T_{TR_i} – time of data package transmission,
T_{A_i} – cycle time (complete processing time for all actions in device),
T_{AR_i} – time of data package analysis,
T_{DR} – time of package detection,
T_{TP} – time of acknowledgement transmission.

Ethernet is an example of a protocol with random access. There are no constant delays and all stations transmit on demand. In this case network cycle time has to be calculated with probabilistic methods. In a normal situation, protocols are quick enough. In case of an alarm situation (cascade of events), these protocols cannot provide a reliable and efficient transmission of the data within, there are no mechanisms for controlling of exchanges. Efficiency of Ethernet protocol is dependent on network traffic, on small traffic is much better than in Master-Slave protocols but in heavy traffic is much worse [12].

4   Service Oriented Architecture

Today, SOA is one of the most successful paradigms in information systems. It gives many benefits like: openness, modularity, scalability and interoperability. These properties are also important in manufacturing systems. However, connecting SOA (which is based on Event-Driven Architecture) to the control systems requires additional software. There are many projects [1,2,3] with description of rules and methods on how to implement system for a complete automation. A complete industrial automation consists of planning, preparation, reporting, monitoring, control and regulation. For the ERP and MES, planning and reporting are the most important. Application on these levels should be universal and available on all software platforms for users at different locations. In SOCRADES [1], interfaces and procedures to create and configure control applications for distributed control system are defined. SOCRADES describes all levels of production process according to ISA 95 [16]: sensors and actuators, fieldbus, SCADA, local network, MES, ERP.

Specifications of services, interfaces, protocols on levels 2–4 (see Fig. 3) are presented precisely. For levels 0–1 there are only general guidelines such as: to use open protocol instead proprietary, if only this is possible or more versatile services needs more processing power. Due to lack of one, existing standard, (or due to too many standards) it is impossible to describe precisely integration of levels 0 and 1. Each protocol may have different way of connection. This is the reason that Service Oriented Architecture, which is proposed in this project, requires high computing power to provide proper connection.

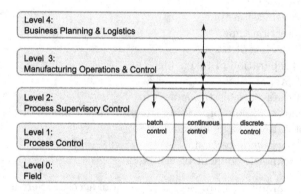

Fig. 3. ISO 95 model of production

Time is not the most important parameter of ERP or MES system. The situation is different for control processes, where time parameters (time delay and variations of time delay) are crucial. Time delay should be sufficiently short and stable. Reduction of time delay and more precisely reducing variations of time delay, increase the stability of the system and quality of control. However, increased time delay (variations of time delay) leads to loss of stability and to errors in control. For some protocols which require high speed of data transfer, using of SOA is not sufficient to achieve the desired integration goals. The authors of SOCRADES hope that, in near future, together with progresses in computer hardware, a total integration can be achieved.

The term fieldbus, used in the ISA 95 standard, can mean one of several protocols [17] on the lowest level of production. There are some methods for connecting the fieldbus level to the higher level. Because of big differences in characteristics and requirements of protocols used on these levels, there are no standard ways to connect (integrate) them yet.

One method is to use an agent – additional device connected physically to communication channels on both levels. An agent monitors all network traffic on the lower level and sends chosen, necessary data to the higher level. This method allows only a one-way communication. A different method is to use an interconnection device. It can be hardware or software device which allows to-ways communication (through conversion) between protocols. This method causes time delay from conversion and impact on industry network [6,7,18].

There are some methods for integration SOA system to the control system. There are two proposals: Devices Profile for Web Services (DPWS) and OPC Unified Architecture (OPC UA). DPWS enables UPnP which enables devices to use dynamic discovery, services description, messaging and events and subscriptions. There is only set of services for connecting and communication between separate devices, which include collection of data and security, but no standardization. OPC UA provides models for both whole system and services for proper operation. In addition OPC UA (as the only software package now) enables object oriented model. It means that not only data from manufacturing are collected and saved but relations and dependency between them too.

The differences are that DPWS was invented for direct communication between devices (levels 0 and 1 in ISO 95) [19] and OPC UA evolved to integrate different measurement and relations in a system (levels 2, 3 and 4) [20].

For further consideration OPC UA was chosen. It has the possibility to map full industrial processes with access through standard interfaces. It is possible to make independent objects with measurements and relations, services to discover, read, write, subscribe with secure channels. It can be used in interconnection of different standards such as ISO 15926 [21].

5   OPC UA

OPC UA has many implemented services. There are services between server and client such as browsing, subscription and read/write enables full control over process. A client application, without prior complicated preparation, can get list of variables, status and configuration, and even change values in process. The only condition is to have TCP/IP protocol. In theory, any client application, from any place, can assist in control.

Other services can be used in server communication to synchronize data, to improve safety by providing redundancy or use so called discovery server. The discovery server works as a special server which collects all information about other servers. This server contains information about all measurements and relations between them (ontology). It can simplify discovery services by obtaining all information at a single location (Fig. 4).

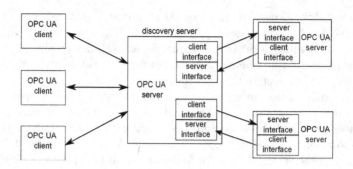

Fig. 4. Aggregating servers

A system consists of at least one OPC UA Server and many OPC UA Clients. All services and mapping are defined in standards.

6 Integration Problems

In theory, full integration of SOA and control systems is possible. In practice, the main problem is the impact of additional processing on control system time parameters. For several industry protocols, mathematical calculations can be used to find minimal network cycle time. There are no relevant equations for SOA systems. Integration has to be made with respect to:

- impact on network cycle time,
- sufficient processing power on OPC UA server.

Consider first problem. As was mentioned in Sect. 4, there are two types of integrations with fieldbus according to direction; one way, which can be made with agent device and two-ways, with specialized interconnecting device. The method with agent can be used in monitoring systems, where it is not necessary to change control system parameters. Interconnecting devices can be used in both types, but it adds additional delay. To calculate the time delay, Equation (2) and (3) can be used. Assuming that all devices acting in control are the same and processing, detection, transmission times are much shorter ($< 1\,\mathrm{ms}$) than PLC processing time (typically 5–100 ms), it is possible to evaluate network cycle time.

The simplified network cycle time depends proportionally on number of devices and device processing time.

$$T_{\mathrm{C}} = K * L_{\mathrm{A}} * T_{\mathrm{A}}$$

where:

K – constant value,
L_{A} – number of devices,
T_{A} – device processing time.

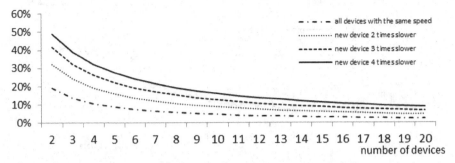

Fig. 5. The network cycle time extension after connection of the next device

The more devices are connected, the longer is cycle time, and typically PC has much longer processing time than PLC, which causes longer cycle time what is presented on Fig. 5.

The second problem concerns sufficient processing power on OPC UA Server. The server is connected to industry and to local network. In experiments [22] impact of both networks was measured. First, different methods for updating data in OPC UA server were tested. Updating data, without internal queue on server, causes heavy processor load and leads furthermore to server time parameter exceeding (Fig. 6).

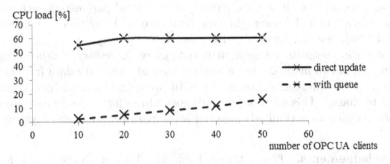

Fig. 6. Direct vs indirect updating of Servers data model [22]

In next experiment, dependency between server processor load and amount of the clients and amount of the subscribed data was measured (Fig. 7). Critical is amount of transferred data. Server has boundary for processed data amount, if data amount exceeds limitation, server cannot work properly. The boundary is dependent on hardware, operating system or application type. Solution to boundary problem can be dividing whole data processing from one OPC UA server into several OPC UA servers. The division causes reducing processor load on each server, but increases traffic load on industry network. Each of the servers has to obtain data from the control system, and thus increases network cycle time.

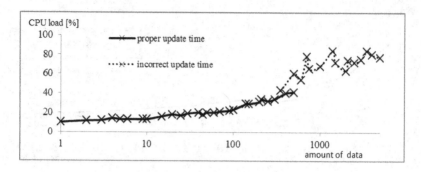

Fig. 7. Model of control system with network delays

7   Conclusions

New technologies like SOA, introduce new possibilities in information systems and maybe also in automation. Manual configuration of devices can be replaced with standard and open service sets. Due to object oriented model in OPC UA, it can be possible to get information not only about one measuring, but about all related measurements. The services should be independent of control system and hardware. It can reduce developing time and costs. But full automation is not always possible. There are two main problems: network cycle time and processing speed. Both problems depend on additional parameters such as type of protocols, amount of devices, time restrictions, etc. Today, there is no standard solution which can configure all parameters automatically.

In practice, complete integration is not always necessary. Authors suggest combination of two methods: use agent(s) to get all necessary data from control system, and one interconnection device with pre-defined parameters, which are allowed to change. This solution gives all possibilities for browsing and subscription of data and some (limited) possibilities to parametrize control system.

Acknowledgements. The authors thank Mr. Preben Gråberg Nes for his advice and help.

References

1. Taisch, M., Colombo, A.W., Karnouskos, S., Cannata, A.: SOCRADES Roadmap: The future of SOA-based factory Automation (2009)
2. Delsing, J., Eliasson, J., Kyusakov, R., Colombo, A.W., Jammes, F., Nessaether, J., Karnouskos, S., Diedrich, C.: A Migration Approach towards a SOA-based Next Generation Process Control and Monitoring. In: 37th Annual Conference of the IEEE Industrial Electronics Society (IECON 2011), Melbourne, Australia (2011)
3. SOA in Manufacturing. Guidebook, White Paper. A MESA International, IBM Corporation and Capgemini (May 2008)

4. Nilsson, J.: Real-Time Control Systems with Delays. Ph.D. Thesis, Department of Automatic Control, Lund Institute of Technology, Lund (1998)
5. Ogunnaike, B.A., Ray, W.H.: Process Dynamics, Modelling, and Control. Oxford University Press (1994)
6. Addad, B., Amari, S.: Delay Evaluation and Compensation in Ethernet-Networked Control Systems. In: 16th International Conference on Real-Time and Network Systems, pp. 139–148 (2008)
7. Lian, F.-L., Moyne, J., Tilbury, D.: Network design consideration for distributed control systems. IEEE Transactions on Control Systems Technology (2002)
8. Florescu, O., de Hoon, M., Voeten, J., Corporaal, H.: Probabilistic modelling and evaluation of soft real-time embedded systems. In: Vassiliadis, S., Wong, S., Hämäläinen, T.D. (eds.) SAMOS 2006. LNCS, vol. 4017, pp. 206–215. Springer, Heidelberg (2006)
9. DeVan, W., Hicks, S., Lawson, G., Wagner, W., Wantland, D., Williams, E.: Using a control system ethernet network as a field bus. In: Proceedings of 2005 Particle Accelerator Conference. Tennessee, Knoxville (2005)
10. Kwiecień, A., Sidzina, M.: The Method of Reducing the Cycle of Programmable Logic Controller (PLC) Vulnerable "to Avalanche of Events". In: Kwiecień, A., Gaj, P., Stera, P. (eds.) CN 2011. CCIS, vol. 160, pp. 379–385. Springer, Heidelberg (2011)
11. Hespanha, J.P., Naghshtabrizi, P., Yonggang, X.: A Survey of Recent Results in Networked Control Systems. Proceedings of the IEEE 95(1), 138–162 (2007)
12. Gaj, P., Jasperneite, J., Felser, M.: Computer Communication Within Industrial Distributed Environment – a Survey. IEEE Transactions on Industrial Informatics 9(1), 182–189 (2013)
13. CAN network – ISO 11898 standard
14. Communications Networks. Programming Manual. Cegelec, Clamart (1993)
15. Open MODBUS/TCP Specification. Schneider Electric (1999)
16. ISO 95 standard
17. IEC 61135 standard
18. Barbosa, R.R.R., Sadre, R., Pras, A.: Difficulties in Modeling SCADA Traffic: A Comparative Analysis. In: Taft, N., Ricciato, F. (eds.) PAM 2012. LNCS, vol. 7192, pp. 126–135. Springer, Heidelberg (2012)
19. OASIS Devices Profile for Web Services (DPWS) Version 1.1
20. OPC Unified Architecture standard
21. Sande, O., Fojcik, M., Cupek, R.: OPC UA Based Solutions for Integrated Operations. In: Kwiecień, A., Gaj, P., Stera, P. (eds.) CN 2010. CCIS, vol. 79, pp. 76–83. Springer, Heidelberg (2010)
22. Fojcik, M., Folkert, K.: Introduction to OPC UA Performance. In: Kwiecień, A., Gaj, P., Stera, P. (eds.) CN 2012. CCIS, vol. 291, pp. 261–270. Springer, Heidelberg (2012)

OFDM Transmission with Non-binary LDPC Coding in Wireless Networks

Grzegorz Dziwoki, Marcin Kucharczyk, and Wojciech Sulek

Silesian University of Technology, Institute of Electronics
{grzegorz.dziwoki,marcin.kucharczyk,wojciech.sulek}@polsl.pl

Abstract. High-quality information exchange between upper layers of the communication network (e.g. TCP, IP layers) requires reliable connection of communicating devices on the physical layer. Any non-corrected errors at this level force the upper layers to perform proper action to recover transmitted information. It reduces data throughput and increases delay to unacceptable level for some services. Among physical media, wireless one is the most hostile environment, due to its unpredictable behavior. In that case, OFDM (Orthogonal Frequency Division Multiplex) modulation and LDPC (Low Density Parity Check) error correction codes appear the best choice to provide high transmission quality on the physical layer. This paper presents the results of the authors' simulation of a LDPC-coded OFDM system with particular emphasis on codes over high order Galois fields (non-binary) which are not commercialized yet.

Keywords: wireless network, physical layer, OFDM modulation, LDPC code.

1 Introduction

LDPC-coded OFDM modulation becomes a popular transmission scheme on the physical layer of diverse communication networks. OFDM modulation has an established position among many existing commercial standards of wireless networks. The LDPC coding is often an option in the current specifications, although it seems to be a strong candidate as main coding method in the future releases of them. LDPC-coded OFDM transmission is successfully adopted for instance in IEEE WiFi [1], IEEE WiMAX [2], and it is also considered in the future releases of a 3GPP LTE (Long Term Evolution) specification called the LTE-Advanced.

Wireless environment, convenient from consumers' perspective, poses a serious obstacle to high-rate reliable transmission. Multipath propagation and mobility cause that the transmission channel is frequency selective and dynamic. The OFDM modulation, owing to the intrinsic subchannel orthogonality, copes with the selectivity by enabling a simple implementation of channel equalization, where each subchannel is equalized separately [3]. Additionally, MIMO (Multiple Input Multiple Output) transmission [4], different subchannel modulation and coding rules improve the system performance. Recently popular binary LDPC

A. Kwiecień, P. Gaj, and P. Stera (Eds.): CN 2013, CCIS 370, pp. 222–231, 2013.
© Springer-Verlag Berlin Heidelberg 2013

error correction coding is an appealing technique for data protection, which out-performs other one [5]. But, when length of the codeword is short to moderate, the classical binary LDPC code should be replaced with LDPC one over high order finite fields (non-binary) in order for the performance preservation [6].

Numerous scientific reports explore properties of different LDPC and OFDM combinations. For example, optimization of the binary LDPC codes for OFDM systems is investigated in [7,8], iterative channel estimation supported by the LDPC decoding (turbo equalization) is considered in [9,10], while non-binary LDPC-coded OFDM system for underwater acoustic communication is analyzed in [11].

This paper presents a simulation analysis of a general OFDM system with non-binary LDPC coding. Various configurations of the OFDM and the LDPC parameters settings are considered to evaluate system performance. The code rates were $R = 0.5$ as suggested in [11], but different system parameters including the LDPC code generation method were tested. A purpose was to further explore the influence of the relationship between the OFDM and LDPC settings on transmission quality, especially in case of constant time duration of the OFDM symbol and without restriction on relation between the field order p and the constellation size 2^s.

2   Transmission System

Evaluation of the system performance for various LDPC coding and OFDM parameters is a continuation of research presented in [12] where the focus was put on the binary codes only. The remaining elements of the transmission system like preambles, headers, training sequences, packet payload construction, etc. were omitted during simulations. There was assumed perfect synchronization and its corresponding blocks were omitted in the system model too. The kind of the transmitted information was not relevant. Concentrating on the LDPC and OFDM processing only, there was proposed a general model of the transmission system which is presented in Fig. 1. Additional elements of the model act as converters (mappers). They are responsible for matching the signal's structures between the different processing stages.

Data bits submitted for transmission are grouped in blocks of p bits each at first and then the K consecutive p-bit blocks comprise a single word. The word is encoded using a $GF(2^p)$ LDPC encoder with rate $R = K/N$. The obtained codewords of length N are matched by the word-to-symbol converter to the IFFT size L and the subchannel constellation size 2^s of the OFDM modulator. After IFFT processing, the OFDM waveform with a cyclic prefix is transmitted over a dispersive channel. The length of the cyclic-prefix is at least as long as the length of the channel impulse response. Complying with this requirement, the frequency subchannels can be equalized separately using one-tap equalizer in case of static or slow fading environment. Next, the equalized OFDM symbols are soft detected, and the obtained a posteriori probabilities are used as the soft decision metrics in the LDPC decoder. The receiver's converter located between

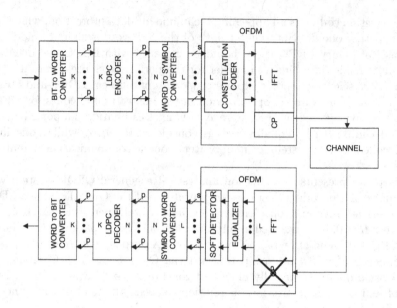

Fig. 1. LDPC-coded OFDM transmission system

the OFDM and LDPC processing units performs far more complicated tasks than its transmitter counterpart. In addition to forming the proper structure of the output signal, it must recalculate the input probabilities according to input properties of the LDPC decoder.

The key units of the transmission system are discussed a bit more in the next subsections.

2.1   OFDM Modulation

An orthogonal frequency division multiplexing (OFDM) modulation is a kind of multicarrier modulation with strong resistance to interchannel (ICI) and intersymbol (ISI) interferences caused by multipath propagation over a wireless channel. The OFDM modulation uses a lot of orthogonal narrowband subchannels with slow signalization in each of them. The subchannels can be considered as a set of AWGN channels with different gains and consequently different SNR (signal-to-noise) ratios. Overall transmission speed depends on the FFT length, the sizes of the QAM constellations in the subchannels and coding rate of the error correction code.

The OFDM symbol of length L is created in frequency domain and then it undergoes IFFT processing. A cyclic-prefix of length G (removed at the receiver side), being the exact copy of the tail of the transformed OFDM symbol, is inserted at the beginning of the symbol itself before it is transmitted. The properly chosen length of the cyclic-prefix ensures time separation of consecutive OFDM symbols at the receiver side. Reliable communication with the OFDM modulation requires good synchronization and channel equalization. The latter is mainly performed in frequency domain after the FFT transformation.

2.2 LDPC Coding

LDPC codes are linear block codes defined over the Galois field GF(q) with restriction to fields of the size being power of two ($q = 2^p$). In the case of the well known binary codes the field size is 2 (thus $p = 1$), whereas for the non-binary codes $p > 1$.

The (N, K) LDPC code with a source vector length K and a code vector length N is defined by a low density parity check matrix $\mathbf{H}_{M \times N}$ with GF(2^p) entries, where $M = N - K$ is the number of parity checks. Note that the information vectors are over GF(2^p), therefore the source block length is $K \cdot p$ bits and the code block length is $N \cdot p$ bits. We denote the entries of the parity check matrix as h_{mn}; $m = 1, \dots, M$; $n = 1, \dots, N$. The column weight d_c of $\mathbf{H}$ is the average number of non-zero entries in the columns. The row weight has the similar meaning in relation to the rows of the parity check matrix.

A row vector $\mathbf{c}$ (in GF(2^p)) of length N is a valid codeword if it satisfies the parity check equation:

$$\mathbf{H}\mathbf{c}^T = \mathbf{0}_{M \times 1} , \tag{1}$$

where the operations are performed in the Galois field arithmetic. Equation (1) can be partitioned into M checks associated with M rows of $\mathbf{H}$.

The parity check matrix $\mathbf{H}$ may be randomly constructed. But, in this paper, the codes are created with the PEG algorithm, which is preferred in the case of relative short-length block codes. The detailed description of the algorithm can be found in [13].

The goal of the decoder is to find the most probable originally transmitted vector $\mathbf{c}$ that satisfies (1), taking into account the received channel values. In the soft decision decoding system, the values initializing the decoder are likelihoods. Considering the iterative decoding algorithm, a convenient representation of the parity check matrix is the Tanner graph, which is a bipartite graph with variable nodes (VNs) and check nodes (CNs). The edges in the graph are associated with positions of the non-zero entries in the parity check matrix. The classical formulation of generalized belief propagation (BP) decoding algorithm assumes two major calculation steps performed in every iteration and in the each node: check node processing for calculation of probabilities associated with given check equation and variable node processing with tentative decoding. A brief description of the decoding procedure can be found in [6].

2.3 Wireless Channel

Radio channel is probably the most hostile environment. The transmitted signals are distorted by two major overlapping phenomenons other than background noise. They are shortly explained below:

multipath propagation causes frequency selectivity of the channel, i.e. the
channel characteristic varies in frequency and received signal may experience
deep fading;

mobility causes time variation of the received signal power. It is the result of
relative movement of transmitter, receiver, and even the whole surrounding
environment.

Because communication systems work in various frequency bands, the respective
channel models may have different number of taps of the impulse response and
different statistical properties. For simulation purposes the channels are usually
normalized according to carrier frequency and sampling period.

A model of frequency selective wireless channel for terrestrial propagation
in an urban area was considered in the investigations. It is based on COST
207 typical urban propagation profile (TU6) with parameters given in Table 1.
Among many possible instances of randomly generated channel characteristic,
the chosen one, presented in Fig. 2 was used in further simulations of the system
model.

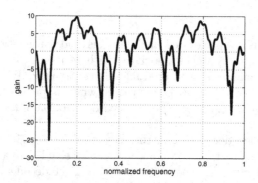

Fig. 2. The frequency response of the wireless channel (randomly generated according
to COST 207 TU6 propagation profile)

Table 1. TU6 profile

Tap number	Delay [μs]	Power [dB]
1	0	-3
2	0.2	0
3	0.5	-2
4	1.6	-6
5	2.3	-8
6	5	-10

3   Numerical Experiments

3.1   Assumptions

The simulations was performed according to the following assumptions:

- the parameters of the LDPC codes are presented in Table 2. All codes have rate $R = 0.5$ and are generated according to PEG algorithm. The maximum number of iteration of the BP decoding algorithm amounts to 40;

Table 2. LDPC codes

No.	N	K	p	column weight d_c
1	128	64	2	2.5
2	128	64	4	2.5
3	128	64	6	2
4	192	96	4	2.5
5	256	128	1	3
6	256	128	3	2.5
7	384	192	2	2.5
8	512	256	1	3
9	768	384	1	3

- time duration of the OFDM symbol is the same for all three considered cases. The number L of the subchannels is always equal to 128. The constellation sizes, assigned to the every subchannel, are 4-QAM ($s = 2$), 16-QAM ($s = 4$), 64-QAM ($s = 6$) respectively;
- the foregoing LDPC and OFDM parameters of the simulated transmission system may be combined in any way. It complements the commonly used approach presented in many science reports where constellation size, assigned to every OFDM subchannel, is often confined to two points ($s = 1$) (e.g. [8,10]) or follows the rule $s = p$ (e.g. [11]);
- the coherence time is sufficiently large to ensure a static channel conditions during transmission of a single OFDM symbol;
- synchronization and equalization are perfect. Referring to the equalization, the receiver knows the channel frequency characteristic. Although, MMSE (Minimum Mean Square Error) equalization algorithm with pilot sequences is usually used in practice, the foregoing assumption of perfect equalization is acceptable for the analyses herein, because the LDPC coding schemes and their combination with OFDM were the main concern of the research;
- WER (Word Error Rate) as a function of E_b/N_0 (bit energy to noise density) is used as a metric of system performance. The word $Word$ in the name of the metric refers to a single codeword of the LDPC code.

3.2   Results

Scenario 1. The existing practical LDPC-coded OFDM systems use the binary LDPC codes. An exemplary results of the transmission performance for general model of the system are presented in Fig 3. There were considered all assumed instances of the 128-point OFDM modulations and three binary LDPC codes with the codewords length $N = \{256, 512, 768\}$. Apart from the subchannel constellation size of the OFDM modulation, the best system performance is obtained for the longest LDPC code ($N = 768$). Among the considered codes, there is also the case of the one-to-one relation at the word/symbol level. It means that one LDPC codeword is included exactly in one OFDM symbol. Despite this correspondence, the word-to-symbol converter is still necessary in the system due to the difference between p and s values. In the case of $s = 4$ and $s = 2$, the one LDPC codeword of length $N = 768$ is transmitted by 1.5 and 3 OFDM symbols respectively. Preservation of the exact relation between the LDPC and OFDM block lengths for these values of s requires use of shorter LDPC codes (here $N = 192$ and $N = 384$ respectively), that consequently has impact on the performance reduction.

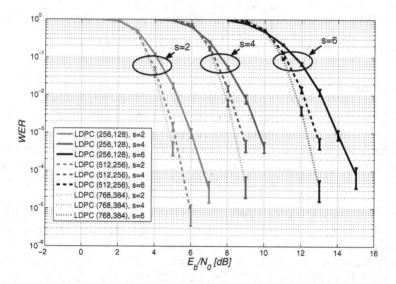

Fig. 3. WER for the binary LDPC-coded OFDM system

Scenario 2. Preservation of the direct relation between LDPC and OFDM signal structures at bit and word/symbol levels separately (i.e. $s = p$ and $N = L$) was the main assumption of the second analysis. The word-to-symbol converters are not required in case of this relationship. The system performance for this direct connection, in comparison with the best results for binary codes from Scenario 1, is presented in Fig. 4. Except for the absence of the word-to-symbol converters in the system model, another advantage is the system performance improvement. On the whole, the higher Galois field order, the better system performance. It is about 2 dB and 3 dB bit-energy reduction for $p = 4$ and $p = 6$ respectively for the investigated system. But, there is no improvement for $p = 2$. The reason is much shorter length of the non-binary LDPC code from binary point of view ($128 \cdot 2 = 256$ bits) in comparison with the binary one (768 bits). For example, the binary LDPC code with the same length in bits ($N = 256$) has slightly worse performance only (dotted curve in Fig. 4) than the non-binary one (light-gray dashed curve in Fig. 4).

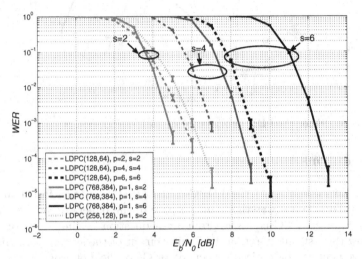

Fig. 4. WER for the nonbinary LDPC codes vs the best binary one

Scenario 3. The possibility of significant performance improvement was demonstrated in the previous scenario, when high Galois field orders were used in the code design. Unfortunately, on the other hand, an implementation complexity of the LDPC code also increases for high values of p. The third scenario explored the combination of different LDPC code parameters under assumptions $N \cdot p = \text{const}$ and the OFDM modulation with $L = 128$ and 64-QAM constellation ($s = 6$) in every subchannel. The purpose was to evaluate the differences in the system performance for various kinds of the LDPC codes, providing that the whole codeword is transmitted within a single OFDM symbol (in that case $N \cdot p = 768$).

The obtained results demonstrate a relatively good transmission quality for the code with $p = 3$ (Fig. 5). In comparison with the best one (for $p = 6$) it characterizes small reduction in performance only. Regarding practical implementation, there is a considerable improvement. The Belief Propagation decoding algorithm requires 2^p probabilities for each p-bit block of the codeword. The total amount of probabilities (for a single codeword) is 8192 for the best quality code among the considered ones. In case of the LDPC code, which is second in the performance rank ($p = 3$), it is 2048 probabilities only.

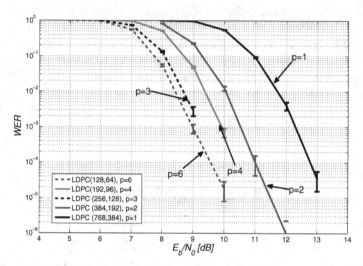

Fig. 5. WER for the LDPC codes comply with assumption $N \cdot p = \text{const}$

4   Conclusions

The numerical evaluation of the LDPC-coded OFDM systems for binary and non-binary LDPC codes was the main purpose of the paper. A commercial use of the non-binary LDPC codes is probably a question of time. The obtained results show their strong potential for improvement of the system performance. The improvement can be noticed especially for Galois field of orders higher than 4. The presented case study for the 128-point OFDM modulation and subchannel coding with 64-QAM constellation shows the attractive results for LDPC code over Galois field of order 2^3.

The certain disadvantage of the non-binary LDPC codes is more complex implementation than the binary one. But seeking for more efficient system solutions (e.g. in power consumption), a wider look at the whole system is necessary. Note that beside the LDPC coder/encoder and OFDM modulator/demodulator there are the converters (mapper/demapper) between them (except for the direct assignment), which consume resources too. A compromise is to find a combination of the LDPC and OFDM parameters to best fit implementation and performance issues.

A closer look at implementation issues is assumed in future work. It particularly concerns the non-binary LDPC decoder and its combination with the symbol-to-word converter.

Acknowledgement. This work was supported by the Polish National Science Centre under Grant number 4698/B/T02/2011/40.

References

1. IEEE 802.11-2012:IEEE Standard for Information technology-Telecommunications and information exchange between systems. Local and metropolitan area networks-Specific requirements Part 11: Wireless LAN Medium Access Control (MAC) and Physical Layer (PHY) Specifications (2012)
2. IEEE 802.16e-2006: IEEE Standard for Local and metropolitan area network. Part 16: Air Interface for Fixed and Mobile Broadband Wireless Access Systems (2006)
3. Wang, Z., Giannakis, G.: Wireless Multicarrier Communications: Where Fourier Meets Shannon. IEEE Signal Process. Mag. 47, 28–48 (2000)
4. Martyna, J.: The Least Squares SVM Approach for a Non-Linear Channel Prediction in the MIMO System. In: Kwiecień, A., Gaj, P., Stera, P. (eds.) CN 2011. CCIS, vol. 160, pp. 27–36. Springer, Heidelberg (2011)
5. MacKay, D.J.C.: Good Error-Correcting Codes Based on Very Sparse Matrices. IEEE Trans. Inf. Theory 45, 399–431 (1999)
6. Davey, M.C., MacKay, D.: Low-Density Parity Check Codes over GF(q). IEEE Commun. Lett. 2, 165–167 (1998)
7. Lu, B., Yue, G., Wang, X.: Performance Analysis and Design Optimization of LDPC-Coded MIMO OFDM Systems. IEEE Trans. Signal Process. 52, 348–361 (2004)
8. Serener, A., Natarajan, B., Gruenbacher, D.: Lowering the Error Floor of Optimized Short-Block-Length LDPC-Coded OFDM via Spreading. IEEE Trans. Veh. Technol. 57, 1646–1656 (2008)
9. Guan, W., Xiang, H.: Low-Complexity Channel Estimation Using Short-Point FFT/IFFT for an Iterative Receiver in an LDPC-Coded OFDM System. Wireless Personal Communications 64, 739–747 (2012)
10. Oh, M.K., Kwon, H., Park, D.J., Lee, Y.H.: Iterative Channel Estimation and LDPC Decoding with Encoded Pilots. IEEE Trans. Veh. Technol. 57, 273–285 (2008)
11. Huang, J., Zhou, S., Willett, P.: Nonbinary LDPC Coding for Multicarrier Underwater Acoustic Communication. IEEE J. Sel. Areas Commun. 26, 1684–1696 (2008)
12. Dziwoki, G., Kucharczyk, M., Sulek, W.: Transmission over UWB Channels with OFDM System using LDPC Coding. In: Proc. SPIE 7502, Photonics Applications in Astronomy, Communications, Industry, and High-Energy Physics Experiments, pp. 75021Q–75021Q6 (2009)
13. Hu, X.Y., Eleftheriou, E., Arnold, D.M.: Regular and Irregular Progressive Edge-Growth Tanner Graphs. IEEE Trans. Inf. Theory 51, 386–398 (2005)

Spectrum Access Game for Cognitive Radio Networks with Incomplete Information

Jerzy Martyna

Institute of Computer Science, Jagiellonian University,
ul. Prof. S. Lojasiewicza 6, 30-348 Cracow, Poland

Abstract. In this paper, the competitive interactions of radio devices dynamically accessing the radio spectrum in the cognitive radio network are studied. The dynamic spectrum access is modelled by a game with incomplete information. The notion of incomplete information means that some players do not completely know the structure of the game. This paper provides a spectrum auction to address the problem of radio channel allocation for cognitive radio networks. The VCG auction to maximise the auctioneer's revenue or maximise social welfare in the spectrum auction is also examined. A dynamic programming algorithm is then applied to solve the spectrum auction problem. Some simulation results are provided.

Keywords: cognitive radio networks, game theory, wireless communication.

1 Introduction

Cognitive radio network is believed to be an effective solution to enhance overall spectrum efficiency. The devices with cognitive radio networks are able to switch between bandwidthss to adapt to varying channel qualities, network congestion, interference, service requirements, etc. [1–3]. Keeping in mind that the Federal Communications Commission (FCC) Spectrum Policy Task Force has published a report [4] in 2002, in which it thoroughly investigates the underutilisation in the radio spectrum. Recent measurements by the FCC show that 70 % of the allocated spectrum in the United States is not utilized. Cognitive radio networks are envisioned to be able to exploit all holes in the spectrum, by means of knowledge of the environment and cognition capability, to adapt their radio parameters accordingly.

Dynamic spectrum access is based on software-defined radio technology, which is proposed to enhance the adaptability and flexibility of wireless transmission. To realize dynamic spectrum access, spectrum management, together with an appropriate model, is required. The spectrum management optimizes to fully utilise spectrum bands. The service provider wishing to increase his profit by increasing the revenue with limited spectrum bands can do so by allowing secondary users (SUs) to access the unused spectrum bands of primary users (PUs). Because SUs cannot provide possible interference under some minimum service constraint to PUs, cognitive radio devices have some constraints to utilize the

A. Kwiecień, P. Gaj, and P. Stera (Eds.): CN 2013, CCIS 370, pp. 232–239, 2013.
© Springer-Verlag Berlin Heidelberg 2013

spectrum bands of PUs. Game theory is a fundamental technology for spectrum management in these networks.

Researchers have been drawn to explore the dynamic spectrum access system in the papers [5, 6]. Traditionally, dynamic spectrum assignment is proposed in [7] as spectrum broker. Distributed spectrum allocation approaches [8, 9] have been studied to enable efficient spectrum sharing only based on local observations. F. Fu et al. [10] focuses on developing solutions for wireless secondary users to successfully compete with others in limited and time-varying spectrum opportunities based on the auction mechanism. A class of auctions, i.e. the multiunit sealed-bid auction (i.e., the Vickrey auction [11]), is suitable to execute in a deterministic time with an acceptable signalling effort in comparison to the English auction [12].

The Vickrey auction has many weaknesses. For example, the it does not allow for price discovery, meaning that the it does not allow for discovery of the market price if the buyers are unsure of their own valuations. Moreover, sellers may use skill bids to increase profit. A way to omit these weakness has been proposed by E. H. Clarke [13] and T. Groves [14]: the auction mechanism referred to as the VCG auction. However, the VCG auction is vulnerable to bidder collusion and to shill bidding with respect to the buyers. The VCG auction isn't necessarily maximise seller revenues; seller revenues may even be zero.

Recently, in a paper by T. Wysocki and A. Jamalipour [15], the portfolio theory has been applied as a spectrum management tool for QoS management and pricing in cognitive radio networks. This approach incorporates the variance of investment returns, a key measure of economic welfare, into pricing and trading strategies. In order to assess opportunistic spectrum access scenarios in cognitive radios, two oligopoly game models are reformulated by L. C. Cremene and D. Dumitrescu [16] in terms of Cournot and Stackelberg games. A new optimal auction mechanism to determine the assigned frequency bands and prices in the cognitive radio, called the generalized Branco's mechanism, was proposed by Sung Hyun Chun [17]. Nevertheless, these papers have been not devoted to the regulator rights allocation for the primary and the secondary networks.

We make the following key contributions. We have investigated the spectrum access game for cognitive radio networks with incomplete information. Spectrum auctions are suitable for selling the rights to the primary and secondary users on the radio channels. Additionally, we have proposed the VCG action that can be used to maximise the auctioneer's revenue or to maximise social welfare in the spectrum auction. Next, we considered the case in which the bids of a network are independent of the networks it shares a channel with, and provided an optimal dynamic programming algorithm for the access allocation problem. Using simulations, we provided some numerical results that the above algorithm performs optimally in a variety of scenarios.

The rest of this paper is structured as follows. Section 2 provides our system model. Section 3 presents auction framework and is followed by the formulation of the spectrum auction game and solves the auction game by use dynamic programming algorithm. Simulation results are presented in Sect. 4. Section 5 concludes the paper.

2   System Model

Let M be identical orthogonal channels in the cognitive radio network. We recall that a regulator conducts an auction to sell the rights to be the primary and secondary networks on the channels. N bidders participate in the auction, with each bidder being an independent network. Each network can evaluate utilities or valuations that are functions of the number of channels on which they get primary and secondary user rights, how many and which other networks they share these channels with, etc.

The assumption of incomplete information is here related to two different notions: imperfect information and imperfect channel state information (CSI). *Imperfect information* means that a player (device) does not know exactly what action other players take at that point in games. *Imperfect channel state information* means that a player has perfect CSI about its own channel, but it has imperfect CSI about any other device's channel.

Following the approach given by J. C. Harsanyi [18], the Bayesian game can be obtained by introducing some randomness in a strategic game. Suppose that each player knows its own the utility network function and does not know the network utility function of all the other players. In other words, each player knows that there exists a finite set of possible types $\mathcal{T}_j$ for each player. The corresponding type for each player is a random variable that follows a probability distribution known by all the players.

The dynamic spectrum access problem can be modelled as a Bayesian game. In order to give insight into the Bayesian game, we provide the description of this game.

Definition 1 (Bayesian Game). *[18]*
   A Bayesian game is completely described by the following set of parameters:

- *A set of $\mathcal{N}$ players, $\mathcal{N} = \{1, \ldots, N\}$.*
- *A finite set of T types of players $\mathcal{T} = \{1, \ldots, T\}$.*
- *A probability density function of the different types of players: $\{f(t) \in [0, 1] \mid \forall t \in \mathcal{T}\}$.*
- *A set of T finite sets of strategies: $\mathcal{S}_\infty, \mathcal{S}_\in, \ldots, \mathcal{S}_T$ each one for for each type of player.*
- *A set of T utility function $u_i : \mathcal{T} \times \mathcal{S} \to \mathcal{R}_+$ for each type of player.*

The set of types $\mathcal{T}$ corresponds to all the possible probability distributions that can model the channel realisation of each player. It is defined as follows.

Definition 2 (Strategy Set). *The strategy set in the game is defined as $\mathcal{S} = \mathcal{S}_1 \times \ldots \times \mathcal{S}_N$, where S_i is the strategy set of player i and is given by*

$$\mathcal{S}_i \left\{ x_i = (x_{i,1}, \ldots, x_{i,N}) : \quad \forall n \in \{1, \ldots, N\}, \quad x_{i,n} > 0 \quad \text{and} \quad \sum_{n=1}^{N} x_{i,n} \leq x_{\max} \right\}$$

where x_i is the valuation of a network i for a channel alocation k.

We assume that $u_i(k)$ be the i-th network utility from the channel allocation $k \in \mathcal{K}$. This value means that since network i will share channels with other networks in the allocation k, the actual utility that network i will derive from an allocation k after the networks start using the allocated channels. In other words, the term network utility $u_i(k)$ should be understood to mean valuation of network i for the channel allocation k.

The valuation of a network i for a channel allocation $k \in \mathcal{K}$ depends on the number of channels on which network i has primary and secondary rights in the allocation k and is expressed by $x_i(k)$. The network utility is given by

$$u_i(k, \tau_i, x_i) = x_i(k) - \tau_i \tag{1}$$

where τ_i is the payment that network i makes to the auctioneer. We assume that the auctioneer determines the channel allocation and each network i makes the payment τ_i to the auctioneer. The social welfare of allocation k is given by

$$w(k) = \sum_{i=1}^{N} u_i(k, \tau_i, x_i) . \tag{2}$$

From a noncooperative point of view the goal of each network as the player is to selfishly maximise its own utility function.

It could be stated that this kind of game can be solved by use a Nash equilibrium (NE). Under the assumption of incomplete information there exists a unique NE at which all players use the available channels. Therefore, we can give the following definition.

Definition 3 (Nash Equilibrium). *The NE for the game with incomplete information is the vector* $x^* = \{x_1^*, \ldots, x_N^*\}$ *where*

$$\forall i \in \mathcal{N}, \quad x_i^* = \frac{x_{\max}}{N} 1_N .$$

It means that under the assumption of incomplete information, there is a unique NE at which all the networks uniformly spread all the data between all the available channels.

3 Spectrum Access Game for Cognitive Radio Networks with Incomplete Information

A possible objective for the game should be to achieve efficiency that maximises "social welfare". It follows from the revelation principle [19] that to maximise social welfare, it is sufficient to consider mechanisms in which the payments are chosen that for each bidder i. However, truth-telling is a weakly dominant strategy. This mechanism is called *incentive compatible*.

The Vickrey-Clarke-Groves mechanism [20] is the only known general incentive compatible mechanism that can be used to maximise social welfare. Under this mechanism for the given valuation function $z_i(.)$ of the bidders, the channel allocation k_i^* is chosen as follows. Let k_{-i}^* be the channel allocation that would have maximised the social welfare if network i did not participate in the auction. That is, k_{-i}^* satisfies the declared valuation function, namely

$$\sum_{j=1,j\neq i}^{N} z_j(k_{-i}^*) \geq \sum_{j=1,j\neq i} z_j(k) \quad \forall k \in \mathcal{K} . \tag{3}$$

By use the VCG auction the payment made by network i is given by

$$\tau_i = \sum_{j=1,j\neq i} z_j(k_{-i}^*) - \sum_{j=1,j\neq i}^{N} z_j(k^*) . \tag{4}$$

To implement the VCG auction, the channel allocations k^* and and k_{-i}^*, $i = 1,\ldots,N$ must be determined. To find the allocation, k^* and k_{-i}^* can be used the algorithm given in Sect. 3.1 for the channel allocation problem and for the bidders $\{1,\ldots,N\}$.

3.1   An Algorithm in a VCG Combinatorial Auction for Spectrum Allocation

In this section, we present an algorithm for solving the problem of access allocation in the CR network. The algorithm is polynomial-time when the number of possible parameters of sets of the secondary network on a fixed-bounded channel. In our study, we generalise an approach given in [21], namely:
(1) all objects in a combinatorial auction are indivisible; (2) the allocation in the auction has to be feasible, e.g. auctioneer's revenue must be maximised. The algorithm works as follows:

Let M be the channel number, such that $M_1 + \cdots + M_n = M$. The maximum possible renevue from all participating networks is given by $T(j_0, j_1, \ldots, j_n, i)$. Let j_0 be primary parts and j_1 secondary parts of type $t, t = 1,\ldots,n$, that must be allocated to the networks $1,\ldots,i$ in the auction. In other words, let $K(j_0,\ldots,j_n,t)$ be the set of allocations $k_i = \{n_{0,t} : v = 1,\ldots,i; t == 0,\ldots,n\}$ satisfying the following conditions:

$$\sum_{v=1}^{i} n_{v,0} = j_0, \quad 0 \leq n_{v,0} \leq M, \quad v = 1,\ldots,i \tag{5}$$

$$\sum_{v=1}^{i} n_{v,t} = j_t, \quad t = 1,\ldots,n; \quad 0 \leq n_{v,t} \leq M_t, \quad v = 1,\ldots,i . \tag{6}$$

Then

$$T(j_0, j_1, \ldots, j_n, i) = \max \left\{ \sum_{v=1}^{i} z_v(n_{v,0}, n_{v.1}, \ldots, n_{v,n}), \ k_i \in K(j_0, j_1, \ldots, j_n, t) \right\} . \tag{7}$$

For finding the values of $T(j_0, j_1, \ldots, j_n, 1)$ is used the following equation:

$$T(j_0, j_1, \ldots, j_n, 1) = \begin{cases} z_1(j_0, j_1, \ldots, j_n) & \text{if } j_0 \leq M, \ j \leq M_t, \ t = 1, \ldots, n \\ -\infty & \text{otherwise} \end{cases} \quad (8)$$

For the single network ($i \geq 1$) by use the Equation (8) is allocated all parts of channel to network 1. However, if $j_0 > M$, then $n_{1,0} > M$, which violates condition $0 \leq n_{v,0} \leq M$, $v = 1, \ldots, i$. Analogously, if $j_1 > M_t$, then $n_{1,t} > M_t$ ($v = 1, \ldots, i$). Hence, if $j_0 > M$ or $j_t > M_t$ then $T(.)$ is set to $-\infty$.

For the two networks the following recursion is used:

$$T(j_0, j_1, \ldots, j_n, i) = \max \{T(j_0 - l_0, j_1 - l_1, \ldots, j_n - l_n, i-1) + z_i(l_0, l_1, \ldots, l_n)\},$$

$$l_0 \in \{0, 1, \ldots, \min(j_0, M)\}, \quad l_v \in \{0, 1, \ldots, \min(j_0, M)\},$$
$$l_v \in \{0, 1, \ldots, \min(j_v, M_v)\}, \quad v = 1, \ldots, n \ . \quad (9)$$

In the Equation (9), if the primary parts, l_0, and the secondary parts l_1 of type v ($v = 1, \ldots, n$) are allocated to network i, then the pay $z_i(l_0, \ldots, l_n)$, and the maximum revenue from the networks $1, \ldots, i-1$ are allocated for the remaining parts, namely $T(j_0 - l_0, j_1 - l_1, \ldots, j_n - l_n, i-1)$. Thus, $l_v \leq \min(j_v, M_v)$ for $v = 1, \ldots, n$, and $l_0 \leq \min(j_0, M)$. Summarizing, the Equation (9) maximizes the renevue from networks $1, \ldots, i-1$ over all possible values of $l_0, \ldots, l_n$.

The Equation (9) is recursively repeated for all sets $M_1, \ldots, M_n$ such that $M_1 + \cdots + M_n = M$. Then, the optimal set $(M_1^*, \ldots, M_n^*)$ is found as follows:

$$(M_1^*, \ldots, M_n^*) = \arg \max_{M_1 + \cdots + M_n = M} T(M, m_1 M_1, \ldots, m_n M_n, N) \ . \quad (10)$$

It is obvious that the allocation with $M_1 = M_1^*, M_2 = M_2^*, \ldots, M_n = M_n^*$ maximises revenue over all channels.

4 Empirical Results

In this section, we report on computational results to prove that the given methodology solves the access allocation problem in cognitive radio networks.

In our simulation, we assumed that the number of secondary networks on channel can be selected from a possibly large set. Moreover, the set of secondary networks on each channel can be an arbitrary subset of the set of all secondary networks. We simulated the case in which the bid function of every network is different and is linearly approximated by the quadratic function.

We found the optimal revenue by using the dynamic programming algorithm given in previous section. The bid functions of each secondary network are approximated by the following functions respectively:

$$y_i(x) = A_1 \left(1 - \frac{e^{-a_1 x}}{x}\right), \quad i = 1, \ldots, N_1$$

$$y_i(x) = A_2 \left(1 - \frac{e^{-a_2 x}}{x} \right), \quad i = N_1 + 1, \ldots, N$$

where A_1, A_2, a_1, a_2 are the parameters of networks, N is the number of networks N_1 is the number of primary networks. We studied the auction revenues of the used algorithm to find the channel allocation that maximises revenue. The revenue for the number of channels varied from 3 to 20. Next, we found the value of revenue that maximisea the auctioneer's revenue and allows us to compute all players' bids.

5   Conclusion

In this paper, we have proposed a new methodology for spectrum allocation games with incomplete information. Although the given approach looks quite different from the traditional solutions, it is a sufficient for implementation. We demonstrated ways to apply the Bayesian game and the VCG auction mechanism to reach an effective solution to the problem. This approach is a step forward compared to recent approaches that only consider users having complete information, which may not be realistic in many practical applications.

References

1. Mitola, J.: Cognitive Radio an Integrated Agent Architecture for Software Defined Radio. Ph.D. Thesis, Royal Institute of Technology (KTH), Stockholm, Sweden (May 2000)
2. Haykin, S.: Cognitive Radio: Brain-empowered Wireless Communications. IEEE Journal on Selected Areas in Communications 23(2), 201–220 (2005)
3. Akyildiz, I.F., Lee, W.Y., Vuran, M.C., Mohanty, S.: Next Generation/dynamic Spectrum Access/cognitive Radio Wireless Networks: A Survey. Computer Networks 50(13), 2127–2159 (2006)
4. Spectrum Efficiency Working Group. Report of the Spectrum Efficiency Working Group. Technical Report, FCC, Washington, DC (November 2002)
5. Berger, R.J.: Open Spectrum: A Path to Ubiquitous Connectivity. FCC ACM Queue 1(3) (May 2003)
6. Peha, J.M.: Approaches to a Spectrum Sharing. IEEE Communication Magazine 43(2), 10–12 (2005)
7. Peng, C., Zheng, H., Zhao, B.Y.: Utilization and Fairness in Spectrum Assignment for Opportunistic Spectrum Access. In: ACM Mobile Networks and Applications (MONET) (May 2006)
8. Cao, L., Zheng, H.: Distributed Spectrum Allocation via Local Bargaining. In: Proc. IEEE/DySPAN (2005)
9. Etkin, R., Parekh, A., Tse, D.: Spectrum Sharing for Unlicensed Bands. In: Proc. IEEE DySPAN (2005)
10. Fu, F., van der Schaar, M.: Learning to Compete for Resource in Wireless Stochastic Game. IEEE Trans. on Vehicular Technology 58(4), 1904–1919 (2009)

11. Vickrey, W.: Counterspeculation, Auctions, and Competitive Scaled Tenders. The Journal of Finance 16(1), 8–37 (1961)
12. Kloeck, C., Jaekel, H., Jondral, F.K.: Dynamic and Local Combined Pricing, Allocation and Billing Systems with Cognitive Radios. In: Proc. 1st IEEE Int. Symp. DySPAN, vol. 2(1), pp. 73–81 (November 2005)
13. Clarke, E.: Multipart Pricing of Public Goods. Public Choice 11(1), 17–33 (1971)
14. Groves, T.: Incentives in Teams. Econometrica 41(4), 617–631 (1973)
15. Wysocki, T., Jamalipour, A.: A Spectrum Management in Cognitive Radio: Applications of Portfolio Theory in Wireless Communications. Wireless Communications, IEEE Comm. Soc. 18(4), 52–60 (2011)
16. Cremene, L.C., Dumitrescu, D.: Analysis of Cognitive Radio Scenes Based on Noncooperative Game Theoretical Modelling. IET Communications 6(13), 1876–1883 (2012)
17. Chun, S.H.: Secondary Spectrum Trading – Auction-based Framework for Spectrum Allocation and Profit Sharing. IEEE/ACM Trans. on Networking 21(1), 176–189 (2013)
18. Harsanyi, J.C.: Games with Incomplete Information Played by Bayesian Players, I–III. Management Science 50(12), 1804–1817 (2004)
19. Mas-Colell, A., Whinston, M., Green, J.: Microeconomic Theory. Oxford Univ. Press, London (1995)
20. Nisan, N., Ronen, A.: Computationally Feasible VCG Mechanisms. In: Proc. ACM Conf. Electron., pp. 242–252 (2000)
21. Tennenholtz, M.: Some Tractable Combinatorial Auctions. In: Proc. AAAI, pp. 98–103 (2000)

Performance Modeling of Opportunistic Networks

Jerzy Martyna

Institute of Computer Science, Jagiellonian University,
ul. Prof. S. Lojasiewicza 6, 30-348 Cracow, Poland

Abstract. The influence of node mobility in Mobile Ad hoc NETworks (MANETs) has significant implications for system performance. A class of MANETs characterised by a sparse density of nodes coupled with a relatively short range of radio communication, results in a network topology that is disconnected most of the time. This wireless mobile ad hoc network is called an "opportunistic network" or "delay-tolerant" network. This paper presents some models in which the rate of information propagation within the opportunistic network is considered. Furthermore, a characterisation of the multicast time in the opportunistic network is developed.

Keywords: opportunistic networks, mobility models, performance evaluation.

1 Introduction

Opportunistic networking is a new paradigm in wireless mobile ad hoc networks. These networks are an evolution of MANETs, where mobile nodes communicate with each other even if a route connecting them does not exist. The nodes that are spread access the environment form the actual network, and they are not required to possess information about the network topology. The routes between nodes are dynamically created, and any nodes can be opportunistically used as a next hop if they can bring the message closer to the destination [1–3]. The opportunistic networks are often considered as a class of Delay-Tolerant Networks (DTNs), where communication opportunities are intermittent and an end-to-end path between the source and the destination may never exist [4].

The opportunistic networks are typically used in an environment that is tolerant of long delays and high error rates. These networks naturally arise in many contexts, including battlefield communication networks, animal tag based sensor networks [5], emergency response networks. Examples of such networks also include deep-space interplanetary networks [6], networks of mobile robots [7] and vehicular ad hoc networks [8].

There are many works devoted to the subject of mobility modeling. Some popular mobility models belong the Random Waypoint [9], and the Random Direction model [10]. However, two important parameters, namely hitting and meeting times, under such models is not available. These parameters have been analysed for Random Walk mobility model [11]. Recently, these models have been extensively analysed and simulated [12, 13].

A. Kwiecień, P. Gaj, and P. Stera (Eds.): CN 2013, CCIS 370, pp. 240–251, 2013.
© Springer-Verlag Berlin Heidelberg 2013

Several opportunistic network performance analysis have been recently introduced in the literature. For instance, in by R. Groenevelt et al. [14], the message delay in mobile ad hoc networks is studied. A framework for routing performance analysis in DTN with application to non-cooperative networks has been presented by G. Resta and P. Santi [15]. Performance modelling of epidemic routing has been carried out by X. Zhang et al. [16]. In these papers, the authors have assumed the inter-contact times between any pair of nodes in the opportunistic network. For instance, in network performance models in [17–19], the simple geometric, random-walk mobility models have been used. On the other hand, these models have been derived based on much simpler mobility models. Recently, the impact of mobility in information spreading has been studied by A. Clementi et al. [20]. Theoretical performance bounds on broadcast time in opportunistic networks as the size n of the network grows are derived by L. Becchetti et al. [21].

In this paper, we present for the first time new results aimed at characterising opportunistic network performance. More specifically, we present two possible network models based on epidemic routing and two-hop multicopy routing. Both models concern pairwise inter-contact between nodes in the network. Additionally, we present a realistic model with inter-contact time between nodes, namely the Home-MEG model. Next, we introduce an upper-bound on the multicast time for the network with n nodes and the the k-ary multicast tree. As our results below demonstrate, our bound suggests a different performance measure than that predicted by the previous studies based on on various mobility models.

The paper is organized as follows. First, we describe the pairwise inter-contact patterns between nodes models of opportunistic networks in Sect. 2. In Section 3, we present the Home-MEG model of opportunistic network. In Section 4, we introduce new performance measures for the opportunistic networks such as the upper multicast time and the lower multicast time. Section 5 concludes the paper.

2   Opportunistic Network with Pairwise Inter-contact Patterns between Nodes

In this section, we present a performance analysis of the opportunistic network with pairwise inter-contact patterns between nodes.

Let be an opportunistic network composed of n nodes. We assume that a randomly selected source, node S, wants to deliver a message, M, to a randomly selected destination node $D(D \neq S)$. In order to characterize performance in this opportunistic network, the following assumption is made: each message is delivered to its destination within a certain time TTL (Time to Live) since its generation. Three different routing protocols are used in the analysis: (1) epidemic routing, and (2) two-hops routing.

Ad (1). In the epidemic routing, described as a.o. by A. Vahdat et al. [22], source node S delivers a copy of message M to all nodes it encounters.

If the two nodes meet, they exchange message copies with each other until the message is delivered to the destination node.

Ad (2). Two-hops routing was introduced by M. Grossglauser et al. [23] and further studied in several articles (see: [12, 14]). In this routing scheme, the source node generates up to L copies of the message and delivers $L - 1$ (encountered new nodes). Any node holding a message or a copy of the message can send it only to the destination node. These nodes cannot send the message further.

Next, the following assumptions are made:

(a) Pairwise contacts are instantaneous, but sufficiently long as to entirely transfer the message M between nodes.
(b) The nodes move according to an arbitrary mobility model with exponentially distributed meeting time, with rate $\frac{1}{\mu}$ between arbitrary node pairs. The meeting time of a mobility model is defined as the time that has elapsed between a random time and the first "meeting" of an arbitrary node pair. It was formally proved for some mobility models, i.e. random walks [24]; and such mobility models as a random waypoint, random direction, etc. [12].
(c) The transmission range is equal to r. This means that any communicating pair of nodes communicating at the same time can do so without interference.

2.1   Epidemic Routing

In the following pare, we consider the performance of epidemic routing. Our focus will be on the *expected delivery delay*, which is formally defined by Groenevelt et al. [14].

Definition 1. *Let T_D be the random variable corresponding to the time elapsing between the time instant at which message M is generated at node S and the time at which the message M is first delivered to node D. The expected value of random variable T_D and is denoted $E[T_D]$.*

Under the assumption of exponential pairwise distribution of inter-contact times in the epidemic routing, the expected delivery delay with epidemic routing is as follows:

Theorem 1. *[14] The expected delivery delay with epidemic routing and exponentially distributed pairwise inter-contact times with parameter λ is given by:*

$$E_{\exp}\left[T_D^{\mathrm{epi}}\right] = \frac{1}{\lambda(n-1)}\left(\log n + 0.57721 + O\left(\frac{1}{n}\right)\right) \qquad (1)$$

where n is size of network.

The asymptotic behaviour of $E_{\exp}[T_D^{\mathrm{epi}}]$ requires the investigation of the influence of n. Thus, for large n is satisfied:

$$\lim_{n\to} E_{\exp}\left[T_D^{\text{epi}}\right] = \lim_{n\to\infty} \frac{\log n}{n} = 0 \ . \tag{2}$$

When the bundle size is smaller than the link size, bundles may perform up to $\lfloor\frac{1}{\alpha}\rfloor$ hops during one time step. Thus, the expected delivery time is as follows:

Theorem 2. *The expected delivery delay with epidemic routing and inter-contact times obeying the parameter $\alpha(0 < \alpha < 1)$ grows unboundedly with n, that is:*

$$\lim_{n\to\infty} E_{\exp}[T_D^{\text{epi}}] = +\infty \ . \tag{3}$$

In other hand, the expected delivery time of epidemic routing in network with a boundle size smaller then the link size grows unboundedly.

2.2 Two-Hops Routing

Here, we give an overview of two main theories on two-hops routing. The first was derived by Groenevelt et al. [14] and states that an expected delivery delay with two-hops routing under the Assumption is distributed in inter-contact time.

Theorem 3. *The expected delivery delay with two-hops routing and exponentially distributed pairwise inter-contact times with parameter λ is equal to*

$$E_{\exp}\left[T_D^{2h}\right] = \frac{1}{\lambda}\left(\sqrt{\frac{\pi}{2(n-1)}} + O\left(\frac{1}{n}\right)\right) \ . \tag{4}$$

The above theorem implies the following asymptotic trend for $E_{\exp}\left[T_D^{2h}\right]$, that is

$$\lim_{n\to\infty} E_{\exp}\left[T_D^{2h}\right] = \lim_{n\to\infty} \frac{1}{\sqrt{n}} - 0 \ . \tag{5}$$

In particular, when the source node S delivering the copy of the message to the first node meets relay node R, and then node R is in charge of delivering message M to the destination node D, then multi-copy version of two-hops routing is faster than its single-copy counterpart in delivering a message to its destination. The expected routing delay for two-hops routing can be considered as an upper bound to the expected routing delay under the multicopy twohops routing presented in the following theorem.

Theorem 4. *The expected delivery delay with single-copy two-hops routing and inter-contact times with truncated Pareto distribution of parameters γ, 1, b, with $0 < \gamma \leqslant 2$, $1 < b$, is equal to*

$$E_{\text{Pareto}}\left[T_D^{s2h}\right] = \frac{1}{E_{\text{Pareto}}[T_{u,v}](1-b^{-\gamma})}\left(\frac{b^{2-\gamma}-1}{2-\gamma} - \frac{b^2-1}{2b^\gamma}\right) \tag{6}$$

where $E_{\text{Pareto}}[T_{u,v}]$ is the expected inter-contact time with truncated Pareto distribution and is defined as follows:

$$E_{\text{Pareto}}[T_{u,v}] = \begin{cases} \left[\ln \frac{b}{1-b^{-1}} \right] & \text{if } \gamma = 1 \\ \frac{\gamma}{\gamma-1} \left(\frac{1-b^{1-\gamma}}{1-b^{-\gamma}} \right) & \text{otherwise} \end{cases} \quad . \tag{7}$$

This theorem implies the assumption

$$\lim_{n\to\infty} E_{\text{Pareto}}\left[T_D^{s2h}\right] \leq \lim_{n\to\infty} E_{\text{Pareto}}\left[T_D^{s2h}\right] = c \tag{8}$$

where $c > 0$.

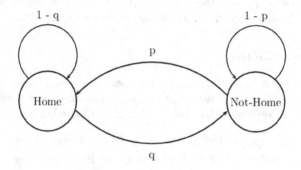

Fig. 1. The Home-MEG model

3   Opportunistic Network with the Inter-contact Time between Nodes

In this section the contacts between nodes in the opportunistic network is modeled. This approach allows us to produce simpler models of contact patterns amenable for use in performance analysis between network nodes.

3.1   The Home-MEG Model

A simple pairwise contact model for opportunistic networks, called the Home-MEG (Markovian Evolving Graph), has been introduced by L. Becchetti et al. [25]. This model is used to model the occurrence and/or disappearance of wireless links between pairs of nodes. Thus, Home-MEG is a discrete-time model, according to which the existence of the link between u and v can change state at time $t, t+1$, etc. The state transition diagram of the Home-MEG model is given in Fig. 1. We can thus interpret the state transition in Fig. 1 diagram as follows: the pair of arbitrary nodes, u and v, can be in one of two states: the *Home* state, corresponding to the situation in which both nodes u and v are at one of their home locations; and the *Non-Home* state, corresponding to the complementary situation in which one of the nodes (or both) are not in a home location. The *Home* state is based on the well-known observation made by T. Karagiannis et al. [26]

that pairs of individuals in the network tend to repeatly meet in a few locations, known as "home" (or "meeting") location.

In accordance with the observation made by T. Karagiannis et al., in the Home-MEG model the probability of establishing an instaneous communication opportunity (e.g., a contract between u and v) at time t depends on the state of the pair: it is α with $0 < \alpha < 1$ where the pair is in state *Home*, and it is β with $0 < \beta \leq \alpha$ when the pair is in the *Non-Home* state. Two further parameters of the model are the probability q of making a transition from *Home* to *Non-Home* state, and the probability p of making a transition from *Non-Home* to *Home* state. Finally, the Home-MEG model is fully characterised by four parameters: the state transition probabilities p and q, and the contract opportunity probabilities α and β.

The Home-MEG model was introduced to model *communication opportunities or contacts u and v*, where a communication opportunity is intended as an instantaneous event during which an arbitrary large of message can be exchanged. The stationary probabilities can be derived as follows. Let IC be the random variable corresponding to the number of time steps elapsing between two consecutive contacts. The stationary probabilities p_H and p_{NH} of finding the pair of nodes in the *Home* state and the *Non-Home* state respectively are given by:

$$p_H = \frac{p}{p+q} \quad \text{and} \quad p_{NH} = 1 - p_H \ . \tag{9}$$

The probability of finding the prior state *Home* (*Non-Home*), conditioned on the event that a contact occurs can be computed by the use of the Bayes' theorem

$$P(H \mid Contact) = \frac{P(Contact \mid H) \cdot P(H)}{P(Contact)} = \frac{\alpha \cdot p}{p \cdot \alpha + q \cdot \beta} \tag{10}$$

$$P(NH \mid Contact) = \frac{\beta \cdot q}{p \cdot \alpha + q \cdot \beta} \tag{11}$$

For any $k \geq 1$ the probability that $P(IC = k)$ is recursively defined as follows:

$$P(IC = k)P(H \mid Contact)P_{kH} + P(NH \mid Contact)P_{kN} \tag{12}$$

where

$$P_{1H} = (1 - q)\alpha + \beta, \quad P_{1N} = (1 - p)\beta + p\alpha \tag{13}$$

$$P_{iH} = q(1 - \beta)P_{(i-1)N} + (1 - q)(1 - \alpha)P_{(i-1)H} \tag{14}$$

$$P_{iN} = (1 - p)(1 - \beta)P_{(i-1)N} + p(1 - \alpha)P_{(i-1)H}, \quad i = 2, \ldots, k \ . \tag{15}$$

Various aspects of the Home-MEG model validation have been studied in the past few years.

The obtained results [25] proved that the Home-MEG model is able to accurately reproduce pairwise contact patterns observed in real-world traces. The Home-MEG model can be modeled by use of a four-states Markov chain. In the states HC and NC (where -C is modeling the existence), we say the edge exists and the nodes u and v are in *Home* and *Non-Home* states respectively. In the HD and ND (where -D is modeling non-existence) states, we say that the edge does not exist. The Markov chain over such state space is given in Table 1.

Table 1. The transition matrix of the Markov chain used to model existence/non-existence of a pairwise wireless link in the Home-MEG model

	HC	HD	NC	ND
HC	$(1-q)\alpha$	$(1-q)(1-\alpha)$	$q\beta$	$q(1-\beta)$
HD	$(1-q)\alpha$	$(1-q)(1-\alpha)$	$q\beta$	$q(1-\beta)$
NC	$p\alpha$	$p(1-\alpha)$	$(1-p)\beta$	$(1-p)(1-\beta)$
ND	$p\alpha$	$p(1-\alpha)$	$(1-p)\beta$	$(1-p)(1-\beta)$

The Markov chain used to model existence/non-existence of a pairwise link in the Home-MEG model is ergodic (irreducible and aperiodic). To reproduce the pairwise contact patterns in a network with n nodes is sufficient to generate $n(n-1)/2$ identical and independent copies of the above Markov chain (one for each possible paiwise link) in the network.

3.2   The Broadcast Time in the Opportunistic Networks with the Home-MEG Model

An upper bound on the broadcasting time for the network with n nodes defined by dynamic graphs obtained when for any pair of nodes $u, v \in V$ is determined by the above Markov chain is proved by Becchetti [25] in the following theorem.

Theorem 5. *Assume the dynamic graph $G = (V, E_T)$ is defined according to the Home-MEG model of parameters (q, p, α, β) as described above, that the initial edge set E_0 is randomly chosen according to the stationary distribution of the underlying Markov chain, and that*

$$\left\lceil \frac{\Lambda}{n} \right\rceil \le \min \left\{ \frac{1}{\alpha}, \frac{1}{4 \cdot p} \right\} \tag{16}$$

where Λ is defined as follows

$$\Lambda = \frac{4(q+p)}{q \cdot \alpha} . \tag{17}$$

Then the upper bound on the broadcasting is given by

$$B(G) = O\left(\frac{\log n}{\log\left(1 + \frac{1}{\Lambda}\right)} \right) . \tag{18}$$

It is easy to see that for the values $q \ll p$, $\alpha \gg \beta$, and $q + p \ll 1$ is satisfied a special case of the above theorem.

Theorem 6. *Under the same assumption for Theorem 3 let p, q, α, β be defined as follows:*

$$q = \frac{1}{n^{1+\epsilon}}, \quad p = \frac{1}{n}, \quad \alpha = \frac{1}{n^{1-\epsilon}}, \quad \beta = \frac{1}{n^2} \tag{19}$$

the broadcasting time in the dynamic graph is given by $O(\log n)$ with high probability.

4   Multicast Time in Opportunistic Networks under Exponential Inter-contact Times

In order to evaluate the inter-contact time of multicast, we concentrate on a one-to-many communication where a source node sends a different message to m, uniformly distributed destinations from the source to each destination. The multicast economises on the number of links travelled: the message is only copied at each branch point of the multicast tree to m destinations. Let us note that $H_n(m)$ is the number of links in the shortest path tree (SPT) to m uniformly chosen nodes. We define the multicast gain $g_n(m) = E[H_n(m)]$ as the average number of hops in the SPT rooted at a source to m randomly chosen distinct destinations. Thus, $g_n(m) \leq f_n(m)$, where $f_n(m) = m \cdot E[H_n]$ is the average number of hops to a uniform location in the graph with n nodes.

We now assume that the multicast delivers packets along the shortest path from a source to each of the m destination. We also assumed that the m multicast group member nodes are uniformly chosen out of the total number of nodes n. Next, we will give a following theorem.

Theorem 7. *For any connected graph with n nodes the number of m, the destinations are given by*

$$m \leq g_n(m) \leq \frac{n \cdot m}{m+1}. \tag{20}$$

For the multicast tree, we can consider the k-ary tree of depth D with the source at the root of the tree and m receivers at randomly chosen nodes. We recall that the depth D is equal to the number of hops from the root to a node at the leaves. In a k-ary tree the total number of nodes is given by

$$n = 1 + k + k^2 + \cdots + k^D = \frac{k^{D+1} - 1}{k - 1}. \tag{21}$$

Thus, we can formulate the following theorem.

Theorem 8. *For the k-ary multicast tree the gain is given by*

$$g_{n,k(m)} = N - 1 - \sum_{j=0}^{D-1} k^{D-j} \frac{\left(N - 1 - \frac{k^{j+1}-1}{k-1} \atop m \right)}{\left(N-1 \atop m \right)} \tag{22}$$

which indicates the average number of hops in this multicast tree.

Thus, we can give the boundaries on the multicast time for the network with n nodes and the k-ary multicast tree of depth D. The following theorem quantifies the multicast time.

Theorem 9. *An upper bound on the multicastt time for the opportunistic network with n nodes and the k-ary multicast tree of depth D is given by*

$$B_M^{(u)}(G) = g_{n,k(m)} \frac{\log n}{\log \left(1 + \frac{n}{\Lambda} \right)} \tag{23}$$

where $g_{n,k(m)}$ is given by Equation (22), Λ is approximately by Equation (17).

The lower bound is attained in a star topology, where $k = n - 1$, $D = 1$, and the average hopcount is $E[H_n] = \frac{n}{2}$. Thus, in a k-ary tree the average hopcount, $E[H_n] = g_n(1)$, and is given by

$$E[H_n] = N - 1 - \sum_{j=0}^{D-1} k^{D-j} \frac{N - 1 - \frac{k^{j+1}-1}{k-1}}{N-1} . \tag{24}$$

For large n, we can obtain

$$E[H_n] = \log_k n + \log_k \left(1 - \frac{1}{k} \right) - \frac{1}{k-1} + \frac{\log_k(n)}{n} . \tag{25}$$

Theorem 10. *A lower bound on the multicast time for the opportunistic network with n nodes and the k-ary multicast tree of depth D is as follows:*

$$B_M^{(l)}(G) = E[H_n] \frac{\log n}{\log \left(1 + \frac{n}{\Lambda} \right)} \tag{26}$$

where $E[H_n]$ is given by Equation (25), Λ is approximated by Equation (17).

5   Simulation Results

In this section we compare the analytical bounds with the simulation results.

We have performed a set of simulations, in which n ($1 \leq n \leq 2000$) nodes are initially distributed uniformly at random in a square area of 5 km side. Nodes have a transmission range of 250 m, and move according to the random waypoint (RWP) mobility model with no pause time and fixed speed v. A random selected node generates the messages directed towards a randomly selected destinations. We performed a large set of such experiments for each parameter setting.

The results of this experimental estimation of the average number of hops in shortest path tree (SPT) to m uniformly chosen nodes as the destination nodes is depicted in Fig. 2. As shown in Fig. 2, $g_n(m)$ is monotonously decreasing in k. In other words, we observe that the deeper D (or the smaller value of k), the more overlap is possible.

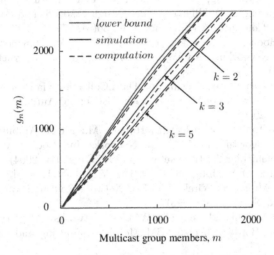

Fig. 2. The multicast gain $g_n(m)$ obtained for the k-arry tree with three values of k

6   Conclusions

In this paper, we have analysed the encounter statistics for modelling contacts in opportunistic networks. Instead of employing an unrealistic mobility model or mobility models that are derived from the user's mobility traces: temporal connectivity models that are developed from stochastic behaviour on human seems to be promising. We have explained the definitions and specific performance indices of interest to opportunistic networks, such as the delivery delay or the multicast time in these networks. Our delay analysis has many applications. For example, it can be used to predict the storage requirement at any nodes that prevents data loss due to buffer overflow.

References

1. Chaintreau, A., Hui, P., Crowcroft, J., Diot, R.G.C., Scott, J.: Pocket Switched Networks: Real-world Mobility and Its Consequences for Opportunistic Forwarding. Thech. Report, UCAM-CL-TR-617, Univ. of Cambridge, Computer Lab (February 2005)

2. Kathiravelu, T., Perars, A.: What and When: Distributing Content in Opportunistic Networks. In: Proc. of Mobile of the Int. Conf. on Wireless and Mobile Computing (ICWMC 2006), Buucharest, Romania (July 2006)

3. Su, J., Chin, A., Popivanova, A., Goel, A., de Lara, E.: User Mobility for Opportunistic Ad Hoc Networking. In: 6th IEEE Workshop on Mobile Computing Systems and Applications (WMCSA 2004) (2004)

4. Huang, C.-M., Lan, K.-C., Tsai, C.-Z.: A Survey of Opportunistic Networks. In: 22nd Int. Conf. on Advanced Information Networking and Applications, pp. 1672–1677 (2008)

5. Juang, P., Oki, H., Wang, Y., Martonosi, M., Peh, L., Rubenstein, D.: Energy-efficient Computing for Wildlife Tracking: Design Tradeoffs and Early Experiences with Zebranet. Computer Architecture News 30(5), 97–107 (2002)

6. Burleigh, S., Hooke, A., Torgerson, L., Fall, K., Cerf, V., Dungerst, B., Scott, K.: Delay-tolerant Networking: An Approach to Interplanetary Internet. IEEE Comm. Mag. 41 (2003)

7. Winfield, A.F.: Distributed Sensing and Data Collection via Broken Ad Hoc Wireless Connected Networks of Mobile Robots. Distributed Autonomous Robotic Systems, 273–282 (2000)

8. Wu, H., Fujimoto, R., Guensler, R., Hunter, M.: Mddv: Mobility-centric Data Dissemination Algorithm for Vehicular Networks. In: Proc. of ACM SIGCOMM Workshop on Vehicular Ad Hoc Networks (VANET 2004) (2004)

9. Broch, J., Moltz, D.A., Johnson, D.B., Hu, Y.-C., Jetcheva, J.: A Performance Comparison of Multi-hop Wireless Ad Hoc Network Routing Protocols. In: Mobile Computing and Networking (1998)

10. Bettsteter, C.: Mobility Modeling in Wireless Networks: Categorization, Smooth Movement, and Border Effects. ACM Mobile Computing and Communications Review (2001)

11. Shah, R.C., Roy, S., Jain, S., Brunnette, W.: Data Mules: Modeling and Analysis of a Three-tier Architecture for spare Sensor Networks. Elsevier Ad Hoc Networks Journal (2003)

12. Spyropoulos, T., Psounis, K., Raghavendra, C.S.: Performance Analysis of Mobility-assisted Routing. In: MobiHoc 2006, Florence (2006)

13. Martyna, J.: Simulation Study of the Mobility Models for the Wireless Mobile Ad Hoc and Sensor Networks. In: Kwiecień, A., Gaj, P., Stera, P. (eds.) CN 2012. CCIS, vol. 291, pp. 324–333. Springer, Heidelberg (2012)

14. Groenevelt, R., Nain, P., Koole, G.: The Message Delay in Mobile Ad Hoc Networks. Performance Evaluation 62(1-4), 210–228 (2005)

15. Resta, G., Santi, P.: A Framework for Routing Performance Analysis in Delay Tolerant Networks with Application to Non Cooperative Networks. IEEE Trans. on Parallel and Distributed Systems 23(1), 2–10 (2012)

16. Zhang, X., Neglia, G., Kurose, J., Towsley, D.: Performance Modeling of Epidemic Routing. Computer Networks 51, 2867–2891 (2007)

17. Jacquet, P., Mans, B., Rodolakis, G.: Information Propagation Speed in Mobile and Delay Tolerant Networks. IEEE Trans. on Information Theory 56, 5001–5015 (2010)

18. Peres, Y., Sinclair, A., Sousi, P., Stauffer, A.: Mobile Geometric Graphs: Detection, Coverage and Percolation. In: Proc. 22nd ACM-SIAM SODA, pp. 412–428 (2011)

19. Pettarin, A., Pietracaprina, A., Puci, G., Upfal, E.: Tight Bounds on Information Dissemination in Sparse Mobile Networks. In: Proc. 30th ACM PODC, pp. 355–362. ACM (2011)

20. Clementi, A., Monti, A., Pasquale, F., Silvestri, R.: Information Spreading in Stationary Markovian Evolving Graphs. IEEE Trans. on Parallel and Distributed Systems 22(9), 1425–1432 (2011)
21. Becchetti, L., Clementi, A., Pasquale, F., Resta, G., Santi, P., Silvestri, R.: Flooding Time in Opportunistic Networks Under Power Law and Exponential Inter-Contact Times (2012) Internet Draft available at,
    http://arxiv.org/abs/1107.5241v3
22. Vahdat, A., Becker, D.: Epidemic Routing for Partially Connected Ad Hoc Networks. Tech. Rep. CS-200006, Duke Univ. (April 2000)
23. Grossglauser, M., Tse, D.N.C.: Mobility Increase the Capacity of Ad Hoc Wireless Networks. In: Proc. IEEE INFOCOM, pp. 1360–1369 (2001)
24. Aldous, D., Fill, J.: Reversible Markov Chains and Random Walks on Graphs,
    http://www.stat.berkeley.edu/verb/~aldous/RWG/
25. Becchetti, L., Clementi, A., Pasquale, F., Resta, G., Santi, P., Silvestri, R.: Information Spreading in Opportunistic Networks is Fast (2011) Internet Draft available at, http://arxiv.org/abs/1107.5241
26. Karagiannis, T., Le Boudec, J.Y., Vojnovic, M.: Power Law and Exponential Decay of Inter Contact Times Between Mobile Devices. In: Pin Proc. of ACM MobiCom, pp. 183–194 (2007)
27. Van Miehem, P., Janic, M.: Stability of a Multicast Tree. In: IEEE INFOCOM, vol. 2, pp. 1099–1108 (2002)

Differential Two-Pulses Position Modulation for Synchronized Wireless Optical Communications

Mehdi Rouissat and Riad Ahmed Borsali

Telecommunications Laboratory of Tlemcen (TLT),
Dept Electronic, Faculty of Technology
Abou Bekr Belkaid University, PB 119
Tlemcen, Algeria

Abstract. In Wireless Optical Systems (WOS), Multi-pulse Pulse Position Modulation and Differential Pulse Position Modulation are two famous modulation schemes compatible with free space optical communication system. In this paper we present a modified form of the existing DPPM based on MPPM coding, called Differential Two-Pulses Position Modulation (D2PPM). In this new form, every symbol ends with a pulse, thus it doesn't require any synchronization, and moreover it shows an improvement in terms of spectral efficiency and data rate. We present the expressions of spectral efficiency, power requirements, and the data rate improvement normalized to PPM, and we present also comparison results to MPPM modulation scheme.

Keywords: MPPM, DPPM, D2PPM, optical wireless communications, performance analysis.

1 Introduction

In the past decade, an unprecedented demand for wireless communication technologies has been taking place. Radio and WOS also called FSO (Free Space Optic) technologies are the main parts of the electromagnetic spectrum used to transmit data wirelessly. Despite the fact that the most commonly used medium for wireless communications thus far is radio, FSO is becoming more popular every day; and it is being preferred (due to its inherent advantages) over its radio counterpart for a number of applications [1].

A free space optical communication system is a system that uses laser or visible light as a carrier to transmit information between two ends, and the free space (the atmosphere) as transmission medium, it can function over several kilometers as long as there is a clear line of sight (LOS) between the transmitter and the receiver [2]. It's based on connectivity between optical wireless units, each consisting of an optical transceiver with a transmitter and a receiver to provide full-duplex (bi-directional) capability (Fig. 1) [3].

In free space optical communication systems each modulation scheme has its own features than make it desirable and a strong candidate for those systems that require that parameter of strength. In this paper we present a modified form

A. Kwiecień, P. Gaj, and P. Stera (Eds.): CN 2013, CCIS 370, pp. 252–257, 2013.
© Springer-Verlag Berlin Heidelberg 2013

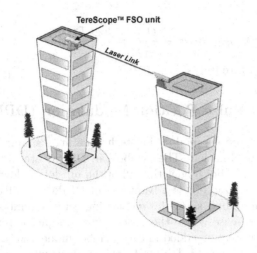

TereScope™ FSO unit

Laser Link

Fig. 1. FSO transceivers mounted on the top of two buildings in point to point connection

of the existing DPPM (Differential Pulse Position Modulation) [4] and based on MPPM coding (Multi-pulse Pulse Position Modulation) [5], called Differential Two-Pulses Position Modulation (D2PPM). In this new form of modulation every symbol ends with a pulse, thus it doesn't require any synchronization, moreover it shows an improvement in term of spectral efficiency and also in term of data rate.

2   Multi-Pulse Pulse Position Modulation (MPPM)

MPPM is also called combinatorial PPM [6], is proposed as a modulation method to improve the spectral efficiency and the data rate in optical pulse position modulation scheme (PPM). In MPPM, each sequence of b bits is mapped into one of $L = 2^b$ symbols and transmitted to the channel, each symbol interval of duration T is partitioned into M slots, each of duration T/M, and the transmitter sends w optical pulses every symbol duration, the number of possible symbols is:

$$L_{\text{MPPM}} = \frac{M!}{w!(M - w!)} \; . \tag{1}$$

The average power requirement by the MPPM normalized to the average power required by OOK (On off Keying) is given in [7]:

$$\frac{P_{\text{MPPM}}}{P_{\text{OOK}}} = \frac{2w}{\sqrt{M \cdot d \cdot \log_2(L_{\text{MPPM}})}} \tag{2}$$

where d is the Hamming Distance.

The bandwidth B is roughly M/T the inverse of the slot duration. This paper defines the band utilization efficiency η as the ratio D to B, i.e.:

$$\eta_{MPPM} = \frac{D}{B_{MPPM}} = \frac{\log_2 L_{MPPM}}{M} \tag{3}$$

where D is the data rate [bit/s].

3 Differential Pulse Position Modulation (DPPM)

DPPM is an anisochronous modulation technique; it is simple modification of the existing PPM modulation. The symbols in DPPM are obtained from the corresponding PPM symbol by deleting all of the *off* slots following the *on* slots [8], as shows Table 1. One of the advantages of DPPM over PPM is the symbol synchronization ability, since the *on* slot indicates a symbol boundary [9]. In order to avoid the case where the time between adjacent pulses is zero, an additional guard slot may be added in each symbol immediately after the pulse. Thus, a symbol which encodes b bits of data is represented by a pulse in one slot, after k slots of zero power, where $1 \leq k \leq M$, $(M = 2^b)$. The minimum and the maximum symbol lengths are $2\,Ts$ and $(M+l)\,Ts$ respectively, Table 1. So the mean symbol length is $(M+3)\,Ts/2$.

Table 1. Mapping of source bits to transmitted slots for 2MPPM, DPPM and D2PPM

PCM	2MPPM	DPPM	D2PPM
000	11000	10	110
001	10100	01	01010
010	10010	0010	10010
011	10001	00010	100010
100	01100	000010	0110
101	01010	0000010	01010
110	01001	00000010	010010
111	00110	000000010	00110

The averages power requirements by the DPPM and PPM schemes normalized to OOK modulation (On off Keying) are given respectively by [8] and [9]:

$$\frac{P_{DPPM}}{P_{OOK}} = \frac{2M}{M+3}\sqrt{\frac{2}{M\log_2 M}} \tag{4}$$

$$\frac{P_{PPM}}{P_{OOK}} = \sqrt{\frac{2}{M\log_2 M}} \cdot \tag{5}$$

The bandwidth required to support communication at a bit rate based on the average symbol duration relative to OOK, is given in [8].

$$B_{DPPM} = \frac{(M+3)D}{2\log_2 M} \cdot \tag{6}$$

The band utilization efficiency is given in [7]:

$$\eta_{\text{DPPM}} = \frac{2 \log_2 M}{M + 3} . \tag{7}$$

4   Differential Two-Pulses Position Modulation

The D2PPM is a modified form of DPPM and based on MPPM coding, where two pulses are used for coding the bit stream. Instead of coding the data sequence according to the positions of one pulse, in this modified DPPM two pulses are used for coding the bit stream, which is coded by the time interval between the two pulses and the previous pulse of the previous symbol, and all the slots following the last pulse are deleted from within each symbol. Table 1 presents a mapping example of source bits to transmitted slots for the modulations 2MPPM, DPPM and D2PPM.

The D2PPM as DPPM doesn't require any synchronization, as every symbol ends with a pulse. In order to avoid symbols in which the time between adjacent pulses of two symbols is zero, an additional guard slot may be added to each symbol immediately following the second pulse. The minimum and the maximum symbol lengths are $3\,Ts$ and $(M+1)\,Ts$ respectively, so the mean symbol length is $(M + 4)\,Ts/2$.

The number of possible symbols in D2PPM is the same as MPPM and it's given by (1).

The normalized average power required by the D2PPM is the same as in MPPM, but based on mean symbol length, and it is given by:

$$\frac{P_{\text{D2PPM}}}{P_{\text{OOK}}} = \frac{2W}{\sqrt{\frac{M+4}{2} \cdot d \cdot \log_2(L_{\text{MPPM}})}} . \tag{8}$$

Also based on average symbol length, the band utilization efficiency of D2PPM is given by:

$$\eta_{\text{D2PPM}} = \frac{D}{B_{\text{D2PPM}}} = \frac{2 \log_2 L_{\text{MPPM}}}{M + 4} . \tag{9}$$

Figure 2 shows the power requirement based on the spectral efficiency (bit/s/Hz) of 2MPPM and D2PPM for different values of M from 6 to 48. The figure shows clearly that D2PPM is more efficient in term of bandwidth for all the values of M. On the other hand, D2PPM shows degradation in term of power efficiency, since two pulses are used in shorter time compared to MPPM. In this paper, we use the parameter R [2] to show the improvement in data rate, which presents the ratio in term of data rate of any modulation to that of PPM modulation.

$$R = \frac{D_M}{D_{\text{PPM}}} = \frac{\log 2 L_M}{\log_2 M} \tag{10}$$

$D_M = D_{\text{MPPM}}$ or D_{D2PPM} and $L_M = L_{\text{MPPM}}$ or L_{D2PPM}.

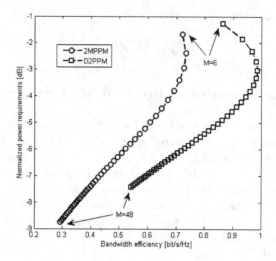

Fig. 2. Normalized power requirement based on bandwidth efficiency for 2MPPM and D2PPM

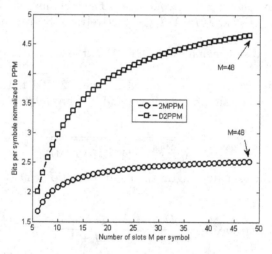

Fig. 3. Normalized data rate based on number of slots per symbol for 2MPPM and D2PPM

The throughput that can be achieved with D2PPM based on mean symbol length is:

$$D_{\text{D2PPM}} = \frac{\log_2 L}{T_{\text{mean}}} \ . \tag{11}$$

The ratio R in the average data rate based on average symbol length of D2PPM modulation scheme to that of PPM is:

$$R = \frac{D_{\text{D2PPM}}}{D_{\text{PPM}}} = \frac{(2M)\log_2 L}{(M+4)\log_2 M} \ . \tag{12}$$

Figure 3 shows the ratio R of 2MPPM and D2PPM for different values of M from 6 to 48. The figure shows the outperformance in term of data rate of the new form over MPPM for all the values of M, where for $M = 48$ the data rate archived with D2PPM is almost 1.9 times that of MPPM.

5   Conclusion

In this paper, we have presented and investigated a new modified form of modulation compatible with wireless optical communication systems called Differential Two-Pulses Position Modulation (D2PPM). It is a modified form of the existing DPIM based on MPPM coding. In D2PPM every symbol ends with a pulse, thus it doesn't require any synchronization, moreover it shows an improvement in terms of spectral efficiency and data rate. The proposed modulation form may be a good addition in the modulations field, and a solution for those wireless optical communication systems that require synchronization and good data rate.

References

1. Ramirez-Iniguez, R., Idrus, S.M., Sun, Z.: Optical Wireless Communications, IR for Wireless Connectivity. Taylor and Francis Group, LLC (2008)
2. Rouissat, M., Borsali, A.R., Chikh-Bled, M.E.: Dual Amplitude-Width Pulse Interval Modulation for Optical Wireless Communications. IJCSI International Journal of Computer Science Issues 9(3) (1), 187–191 (2012)
3. http://www.pulsewan.com
4. Shiu, D., Kahn, J.M.: Differential pulse-position modulation forpower-efficient optical communication. IEEE Trans. Commun. 47, 1201–1209 (1999)
5. Sugiyama, H., Nosu, K.: MPPM: a method for improving the bandutilization efficiency in optical PPM. J. Lightwave Technol. 7, 465–472 (1989)
6. Budinger, J.M., Vanderaar, M., Wagner, P., Bibyk, S.: Combinatorial pulse position modulation for power-effcient free-space laser communications. In: SPIE Proc., vol. 1866 (January 1993)
7. Park, H., Barry, J.R.: Modulation Analysis for Wireless Infrared Communications. In: IEEE International Conference on Communications, ICC 1995, Seattle, pp. 1182–1186 (1995)
8. Chadha, D., Rathore, P.K.: Performance of pulse modulation schemes for infrared wireless communications. In: Microwave Conference, Asia-Pacific, Sydney, NSW, pp. 946–949 (2000)
9. Sethakaset, U., Gulliver, T.A.: Performance of Differential Pulse-Position Modulation (DPPM) with Concatenated Coding over Indoor Wireless Infrared Communications. In: IEEE 63rd Vehicular Technology Conference, pp. 1792–1796. Vic., Melbourne (2006)

Extending the TLS Protocol by EAP Handshake to Build a Security Architecture for Heterogenous Wireless Network

Krzysztof Grochla* and Piotr Stolarz

Proximetry Poland Sp. z o.o.,
Al. Rozdzienskiego 91, 40-203 Katowice, Poland
{kgrochla,pstolarz}@proximetry.pl
http://www.proximetry.com

Abstract. The Extensible Authentication Protocol, or EAP, is an authentication framework used frequently in wireless networks and point-to-point connections. The Transport Layer Security (TLS) provides a secure communication layer, using asymmetric cryptography for key exchange, symmetric encryption for privacy and message authentication codes for message integrity. In this work we propose to replace the TLS handshake mechanism with the EAP authentication, which allows authentication to be easily integrated into multiple wireless network technologies using EAP-TLS.

Keywords: wireless networks, wireless security, EAP-TLS, handshake.

1 Introduction

The Transport Layer Security (TLS) protocol is designed to enable communication in a computer network in a manner which prevents eavesdropping and tampering [1]. It is based on client-server connections. The TLS extends the Secure Socket Layer (SSL) [2] protocol developed by Netscape. It guarantees confidence and integrity of transmitted data by creating a socket connection, which virtually any transmission can be carried on. The TLS also allows the server and the client sides to be authenticated on the basis of X.509 certificates and asymmetric cryptography [3].

Communication over the TLS channel starts by negotiation of a stateful connection with use of a handshake procedure, as described in [1]. During the handshake, the client and server agree on parameters required to establish the connection's security. The handshake procedure of the TLS/SSL protocol is also responsible for establishing or resuming sessions. The main goals of this layer are to negotiate cipher suites and compression algorithms, authenticate the server

---

* This work was partially supported by the "Grid AirSync – The Network Management System for Intelligent Grid" project subsidized by Polish Agency of Enterprise Development by the grant no POIG.01.04.00-24-022/11.

A. Kwiecień, P. Gaj, and P. Stera (Eds.): CN 2013, CCIS 370, pp. 258–267, 2013.
© Springer-Verlag Berlin Heidelberg 2013

to the client, and optionally the client to the server, and to exchange a common secret in a secure way. Two types of the TLS handshake procedure can be distinguished: simple (only the server is authenticated using a certificate) and client-authenticated (both connection end-points are authenticated).

The Extensible Authentication Protocol (EAP) defined by [4] allows different methods of client and server authentication to be implemented in a unified and hardware independent way. It is based on client / server architecture, where the server is the authenticator and the client is the supplicant. A large number of authentication methods for EAP makes it very flexible. Some of the EAP authentication methods allow mutual authentication of peers with a secure key exchange mechanism to be performed without a need to re-design the underlying protocol. It is widely used, particularly in wireless networks, e.g. in IEEE 802.11 WPA and WPA2, and IEEE 802.16.

Currently, many enterprises use the EAP authentication in their wireless network deployments. The same enterprises provide intranet access or other network services to employees, which are authenticated separately. Application of a single authentication mechanism is more convienient for the user and decreases the number of possible security threats. Typically, the infrastructures for authentication of layer 2 network access and layer 4 communication are separated. This solution requires deployment of different authentication mechanisms, thus increasing the cost of the network infrastructure.

It is possible to use a single authentication server, instead of two, due to flexibility of the EAP authentication which can be combined with the TLS protocol. Nir et al in [5] proposed to use TLS with EAP Extension (TEE). In this approach, the EAP based authentication, which is designed to improve security in HTTP connections and performed during the TLS handshake process, occurs within the TLS protected tunnel. We propose a slightly different solution where the original TLS handshake, as specified for the TLS protocol, is replaced with the EAP authentication method which offers similar functionality as the TLS protocol in terms of the mechanism for mutual authentication and key exchange. Examples of such EAP methods include EAP-TLS [6], EAP-TTLS [7] or PEAP [8], among others. The above-mentioned approach allows us to leverage a wide range of mutual authentication methods, such as server to client authentication using an X.509 certificate and client to server authentication using TLS tunnelled PAP offering a very high level of security [9]. Furthermore, since the whole authentication process is based on EAP methods, we are able to maintain a single point of authentication in the system – the AAA server. Therefore, AAA can be used seamlessly for both L2 wired/wireless and application layer authentications. The proposed protocol allows for protection of datagram and stream transported application data (UDP/TCP). Later in this paper, we refer to the proposed TLS protocol with EAP handshake as TLSE.

2   Architecture of the Proposed System

We analyze a system with the architecture presented in Fig. 1. The user authentication, authorization and accounting is carried out by an AAA server composed of two elements: a RADIUS server and a Credentials Database. Wireless access points and base stations serve as authenticators for wireless clients. Mobile or fixed clients are authenticated to access the network and to maintain their connections using the TLSE protocol with the EAP based authentication mechanism.

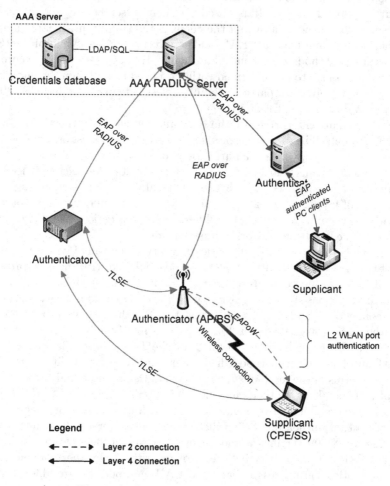

Fig. 1. Architecture of the proposed system

The proposed architecture allows to use a single authentication point for both authentication of TLS connections and wireless and wired clients. The AAA server is queried using EAP methods for authentication of the clients connected to wireless access points and base stations. The same AAA server is also used for the EAP handshake while establishing TLS connections. Authenticators can be placed in remote locations which provide network services over the TLSE protocol to the clients and communicate with AAA using the RADIUS protocol.

3   Changes to TLS Handshake Mechanism

3.1   Handshake Messages Format

We propose the following extension to the TLS handshaking protocol introduced for transport of EAP authentication messages as well as for negotiation of cipher suites used for secure and tamper protected transportation of application content. The format of a handshake message is defined below:

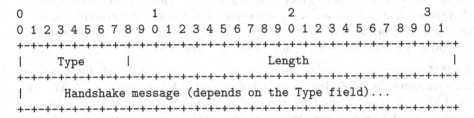

Type – specifies the type of the handshake message. TLSE supports the following messages:

- HelloRequest – a request from the server side for re-handshake,
- ClientHello – a client's hello message,
- ServerHello – a server's hello message,
- Finished – end of a handshaking marker,
- EAPMessage – a placeholder for transportation of EAP messages.

The *Length* field specifies the length of the entire handshake message and is followed by the proper handshaking message. The format of TSLE handshaking protocol messages is specified below:

HelloRequest: the message may be sent by a server at any time to notify a client about a need for new re-handshaking. It may be ignored by the client if the handshaking process is already pending. If the server sent a *HelloRequest* message but did not receive *ClientHello* for the new handshaking, it may close the connection. After sending the message, the server should not resend it until the handshaking process is finished.

The *HelloRequest* message has no data frame associated with it – it is just an empty frame recognized by the client on the basis of the Type field in the handshake message header.

ClientHello: when a client connects to a server, it is required to send *Client-Hello* as its first message. The client can also send *ClientHello* in response to *HelloRequest* or on its own initiative in order to renegotiate the security parameters in the existing connection. The *ClientHello* message contains the same fields as the standard TLSE message: random bytes used for session keys derivation, session ID determining the secure session which the client requests to re-establish and the used cipher suites. The session ID may point to a session of a closed or existing connection if the client wants to establish several secure connections

with the same master secret key. In case a new connection is requested, the
TLSE Session ID field must be empty.

ServerHello: a message transmitted in response to *ClientHello* when the server
is able to find an acceptable cipher algorithm. Otherwise it closes the connection.

Finished: the last handshake message which is sent just before establishing
a secure connection. The message contains signatures of all the handshaking
messages exchanged for this session, thus protecting the handshaking protocol
against man-in-the-middle attacks. Recipients of finished messages must verify
that the contents are correct. Once one of the sides has sent the *Finished* message
and then received and validated the *Finished* message from its peer, it may begin
to send and receive application data over the connection.

EAPMessage: a container for EAP message transportation. Single *EAPMessage*
must contain only one EAP message delivered to TLSE from the higher EAP protocol layer. It is up to the EAP protocol to divide long, logical EAP messages
(such as EAP-TLS messages transporting a chain of certificates) into several,
physical EAP messages which are next transported over TLS in the *EAPMessage* frames (refer to [4] for more information about the EAP fragmentation
process).

The actual length of a transported EAP message is deduced from the header
of the handshake message. Since the maximum length of a handshake message
is 2^{24} and the maximum length of an EAP message is 2^{16}, there is no need for
fragmentation on the TSL handshaking layer. In practice, the maximum length of
an EAP message is further constrained, usually by a configuration parameter in
the Supplicant and AAA Radius servers. Because of Radius and UDP protocols
limits, it is set to $1\,024$ bytes in most cases.

3.2   Sequence of Messages in the Proposed Authentication

The sequence of messages exchanged between the Supplicant, Authenticator and
AAA server is presented in Fig. 2. The following message exchange is required:

1. The TLSE Client (Supplicant) requesting establishment of a new TLSE session, begins the handshaking process by sending two consecutive handshake
   messages:
   - ClientHello – in this case, it contains a request for a new session estab-
     lishment along with client preferred cipher suites,
   - EAPMessage – it contains the *EAP/Identity message* identifying the
     connecting peer.
2. The TLSE Server (Authenticator) forwards the *EAP/Identity* message to
   the AAA Radius server (via the Radius protocol as the EAP transport).
   EAP/Identity is the first message starting a new EAP authentication conversation for the AAA server.
3. The AAA server responds with the EAP-TLS/EAP-TTLS Start message
   which prompts the connecting peer to proceed according to the specific authentication method.

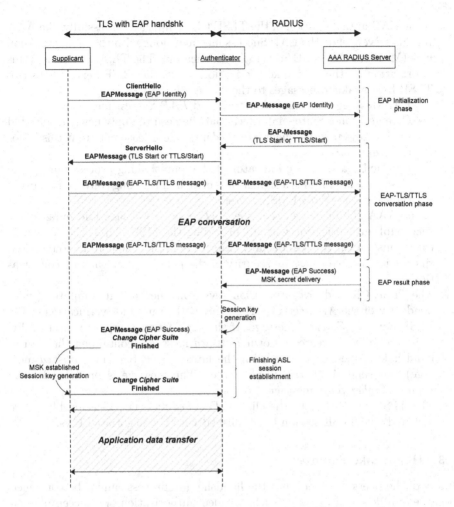

Fig. 2. Sequence of messages in the proposed handshake procedure

4. The TLSE server forwards the EAP-TLS/EAP-TTLS Start message to the TLSE Client in the TLSE's *EAPMessage* container. This message is preceded by *ServerHello* informing the TLSE client about a new TLSE session identifier associated with this conversation as well as the cipher suite selected by the TLSE Server from the client's list.

5. Next, the EAP conversation occurs. Each EAP message is carried by TLSE via the *EAPMessage* container. The TLSE server (Authenticator) serves as a relay forwarding the EAP messages between the AAA Radius server and TLSE Client (Supplicant). It packs/unpacks them into the appropriate transport protocol (TLSE and Radius). The Authenticator does not interpret the content of the forwarded EAP messages.

6. If the EAP authentication of the TLSE Client finishes successfully, the AAA Radius server sends the EAP Success message along with the Master Session Key (MSK) to the TLSE Server (Authenticator). The TLSE Server uses this information for the session key derivation. Next, the TLSE Server sends two TLSE handshaking messages to the TLSE Client:
   - EAPMessage – contains the forwarded *EAP Success* message.
   - Change Cipher Suite – informs about the need to apply newly negotiated session keys to subsequent *Finished* message. Please note that this is not the *Handshake* message.
   - Finished – acts as the indicator of the handshaking process finish. The message also contains a sign of the entire handshaking conversation using the newly negotiated MSK secret.

   If the AAA Radius server sent the *EAP Success* message informing about successful authentication without providing the MSK secret, the Authenticator must finish the session preventing the TLSE Client from connection, since it is not able to assure security of the pending handshaking conversation.

7. The TLSE Client derives its session keys from the MSK it computed (independently of the AAA server). The MSK is then used for verification of the TLSE Server's *Finished* message. If the verification completes successfully, the TLSE Client creates its own *Finished* message (containing the sign of handshaking messages along with the latest TLSE Server's *Finished* message) and sends it to the TLSE Server. The message is preceded by the *Change Cipher Suite* message.

8. The TLSE Server checks the client's *Finished* message. If no problems are detected, the TLSE session is established with the negotiated keys.

3.3   Handshake Failures

If any of the peers discovers that the handshaking process cannot be continued because of a detected problem, such as failed authorization or unacceptable cipher suites, it must close the network connection associated with the established session and release all the resources of this session. If the discovering peer is the server and the client tries to continue with the legacy session which is already discarded by the server, the server must inform the client about the need to start a new session by sending the *HelloRequest* message.

3.4   Session Resumption

The proposed TLSE protocol allows a session to be resumed or duplicated in the same way as the TLS protocol. If the client and server decide to resume a previous session or duplicate an existing session instead of negotiating new security parameters, the client sends the *ClientHello* message. If the server finds a matching session, it responds with *ServerHello*. After the cipher spec is changed, the peers may start to exchange the application data. In this way, the TLSE also allows several connections to be established with use of the same MSK secret, without additional load on the AAA server.

The sequence of events used for session resumption is presented in Fig. 3. At the beginning, the client sends the *ClientHello* message to the server with the TLSE session ID pointing to the session it wants to resume. *EAPMessage* contains EAP/Identity identifying the peer. If the server accepts the session resumption, it announces it to the client by providing the same TLSE session ID together with the cipher suite in the *ServerHello* message, as in case of the renewed session. Otherwise, the new TLSE session ID is presented in the message and further conversation proceeds as specified in Fig. 2. Finally, both sides exchange *Finished* preceded by *Change Cipher Suite* messages authenticating the conversation based on the MSK of the renewed session. After completing this exchange, the application data can be transported.

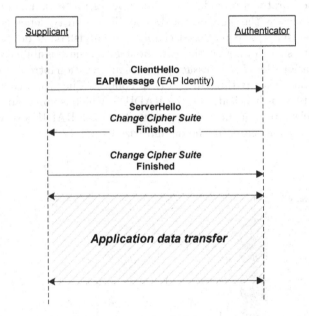

Fig. 3. Sequence of messages in the session resume procedure

The session resumption must be performed only within a context of the same user. The Authenticator must check if *EAP Identity* from the session resumption message matches the Supplicant EAP ID session parameter of the session is to be resumed. If they do not match, full authentication is required. If they match and the requested session may be resumed, the new resumed session inherits the Supplicant AAA ID session parameter from its base session.

4   Security and Performance Considerations

Integration of the TLS and EAP authentication (EAP-TLS) together with the proposed mechanism for transporting messages of these protocols, offers a possibility to use a single AAA server and single mechanism for authentication for both transport layer communication and client access to the IEEE 802.11 and 802.16 wireless networks. This approach decreases the number of potential points of attack, thus increasing the overall security of the system. The proposed TLSE protocol allows consolidating control of the user access to different network services and L2 access into one point, which simplifies the architecture of the system and lowers network management costs. High flexibility of the EAP mechanism is another advantage of bundling the EAP authentication with TLS in the proposed TLSE protocol, which offers running authentication protocols that fit a specific deployment environment.

The proposed protocol provides the same level of protection against man-in-the-middle attacks as the TLS protocol. The TLS security depends greatly on the used cipher suite. The proposed changes to handshaking do not influence the cipher suite selection, since they are neutral to the possibility to decipher the communication. The *EAPMessage* protocol information introduced for TLSE handshaking contains only the required EAP content. The proposed architecture relies on clear text authentication using RADIUS, which is vulnerable to known-plaintext attacks [10], but most deployments both the RADIUS server and the Authenticator are collocated in the demilitarized zone.

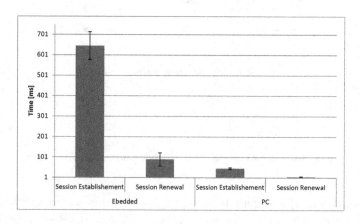

Fig. 4. Session establishement and renewal times for the TLSE protocol, with error bars showing the standard deviation

4.1   Tests and Measurements

The proposed TLSE protocol was implemented in Linux to analyze its performance. The implementation was tested on both a PC and an embedded device.

The OpenSSL library was used for key generation. The time required to perform a full handshake and a session resumption was measured in for 2 048 bits keys. Each test was executed 30 times to get the average and standard deviation. A dual core CPU without hardware encryption support running at 2.66 GHz was used on the PC and a Power PC 440EPx CPU at 667 MHz was used on the embedded platform. The results are presented in Fig. 4. On the PC platform, the session was established in approximately 50 ms, on the embedded platform in over 600 ms, which shows that the computing power heavily influences the amount of time required to set up the communication. The session resumption in both cases was much faster, since there was no need to generate the key.

5   Conclusion

The presented TLSE protocol can be an extension to the TLS security mechanism, allowing a single AAA server to be used for authentication of both TLS connections and wireless and wired devices using EAP authentication. It provides a high level of security, due to well-established and verified standards it is based on.

References

1. Dierks, T., Rescorla, E.: RFC 4346: The Transport Layer Security (TLS) Protocol. Version 1.1 (April 2006)
2. Freier, A., Kocher, P., Karlton, P.: The Secure Sockets Layer (SSL) Protocol Version 3.0, http://tools.ietf.org/html/rfc6101
3. Information technology – Open Systems Interconnection – The Directory: Public-key and attribute certificate frameworks. ISO/IEC standard 9594-8:2005
4. Aboba, B., Levkowetz, H., Vollbrecht, J.R., Blunk, L.J., Carlson, J.: Extensible Authentication Protocol (EAP), http://tools.ietf.org/html/rfc3748
5. Tschofenig, H., Sheffer, Y., Nir, Y., Gutmann, P.: A Flexible Authentication Framework for the Transport Layer Security (TLS) Protocol using the Extensible Authentication Protocol (EAP), http://tools.ietf.org/html/draft-nir-tls-eap-13
6. Aboba, B., Simon, D.: PPP EAP TLS Authentication Protocol, http://tools.ietf.org/html/rfc2716
7. Funk, P., Blake-Wilson, S.: EAP Tunneled TLS Authentication Protocol, http://tools.ietf.org/html/draft-ietf-pppext-eap-ttls-05
8. Zorn, G., Palekar, A., Simon, D., Josefsson, S.: Protected EAP Protocol (PEAP) Version 2, http://tools.ietf.org/html/draft-josefsson-pppext-eap-tls-eap-10
9. Asokan, N., Niemi, V., Nyberg, K.: Man-in-the-middle in tunnelled authentication protocols. In: Christianson, B., Crispo, B., Malcolm, J.A., Roe, M. (eds.) Security Protocols 2003. LNCS, vol. 3364, pp. 28–41. Springer, Heidelberg (2005)
10. Bidgoli, H.: Handbook of Information Security, Information Warfare, Social, Legal, and International Issues and Security Foundations, vol. 2. Wiley (2006)

Conservative Graph Coloring: A Robust Method for Automatic PCI Assignment in LTE

Lukasz Chrost and Krzysztof Grochla

{lchrost,kgrochla}@proximetry.com

Abstract. The article presents a new approach in the field of automatic Physical Cell Id (PCI) assignment in LTE-advanced Self Organized Networks. The presented algorithm allows for multi-target optimization for different network deployment scenarios in order to minimize the number of PCI conflicts and confusions. Unlike some other competitive algorithms, the presented mechanism does not require any changes in the LTE specification, as the operation is based on standard information already available in the management system. The algorithm allows for specific limitations of the physical layer, e.g. setting the maximum number of co-existing PCIs in range in order to restrain interference. The presented algorithm is compared with other solutions through numerous simulations.

Keywords: LTE, Automatic PCI assingment, SON, LTE-Advanced.

1 Introduction

The diverse, multilevel network structure proposed by the LTE standard requires a new approach to network management and configuration. Unlike the classic cellular 2G/3G networks, LTE operates utilizing smaller, heterogeneous cells. The cells can be either deployed by the network operator in a pre-planned manner (typical core macro cells), or in much less predictable approach related to smaller cell deployment and special UE handling (Home eNodeB and CSG handling).

Traditional GSM deployment incorporates the mixture of on-site maintenance and measurements combined with the centralized computer-aided management. The heterogeneous nature of the LTE network makes the classic approach to the network deployment impossible due to high operational effort and cost. In order to minimize the operating expenses and deployment time, the provision of Self Organizing Networks (SON) is one of the key objectives in LTE.

Multiple SON use cases have been specified by 3GPP for three areas: self-configuration (covering pre-operational phases of network deployment, such as planning and initial configuration), self-optimization (optimization during the operational phase) and self-healing (maintenance, recovery from faults).

The idea of the automatic Physical Cell Identification assignment has been introduced in 3GPP TS 36.300 Release 10. According to the standard, the automatic PCI assignment is part of the wider Operation and Management (OAM) functionality and can be performed in two manners: centralized and distributed.

A. Kwiecień, P. Gaj, and P. Stera (Eds.): CN 2013, CCIS 370, pp. 268–276, 2013.
© Springer-Verlag Berlin Heidelberg 2013

During the centralized operation, the OAM server provides a single PCI value to be used. This stands in contrast to the distributed approach, where the OAM provides a set of possible PCI values. Some of this values are rejected by the e-NodeB based on locally gathered information. The PCI value is then selected randomly from the remaining set.

Background and related work is presented in Sect. 2, the algorithm description is provided in Sect. 3, while Sect. 4 provides evaluation results for the presented solution. The conclusions are presented in Sect. 5.

2   Background and Related Work

Many research articles are dedicated to the field of PCI auto-assignment. Bandh et al. [1] propose a traditional graph-coloring approach for solving the discussed problem. Many other articles seem to be based or influenced by their work, especially concerning the graph-coloring approach.

Furqan et al. [2] prepared an evaluation of different, distributed graph-coloring algorithms proposed in the field. Their work is not only interesting as the algorithm evaluation, but is also a very good starting point for starting a PCI assignment research, as it references many other interesting proceedings in the field. Additionally, the authors propose and use various metrics allowing for validation and comparison of the PCI assignment algorithm. Bandh and his team continued their work on graph-coloring optimization of their previous achievements. Their research led to [3], which describes a fully developed solution in the field. Several other approaches to this subject have been described by other authors. One of the papers by Zhang et al. [4] is interesting, as it is the only paper that actually considers a multi-level, heterogeneous nature of LTE deployments. The main drawback is the use of undocumented, custom-designed simulator during the research. The other papers are hard to judge, as the authors did not include any references to the contemporary state of the art in the field, and their results are incomparable with other works. Wu et al. [5] present such an approach. The authors propose a potentially interesting algorithm, but the simulation results are compared to the worst-case random PCI selection algorithm.

3   Graph Coloring Model for PCI Assignment

As stated in the previous section, many authors consider the problem of automatic PCI assignment as classic graph coloring problem. We use the same approach, as not only it is natural for this group of problems, but also provides a numerous opportunities for future development [3]. Our algorithm relies on data available for any OAM system conforming with the 3GPP TS 36.300.

The LTE network is represented by an undirected graph G consisting of g_i; $i \in [1 \ldots n]$ vertices representing the LTE cells. The (g_i, g_j) edge represents the direct neighbor relation of the respective cells. The direct neighbor relation data is acquired from the OAM system and is based on the reports and measurements performed by the mobile equipment. The data required for this mapping does not exceed the information gathered during the standard neighbor relation table (NRT) management and is already available in the OAM system. This approach does not require changes in LTE specification.

Each PCI value is represented by a corresponding color c. $C(1 \ldots k)$ represents an ordered set of the currently used colors k, while $c(g_i)$ represents the PCI value of the particular cell. The k value is used by the algorithm to limit the number of colors required for graph coloring. The C set consists of the first k elements from $K(1 \ldots 504)$ ordered set of colors (PCI values) available in LTE. The c value of uncolored vertex is 0.

Despite the 504 possible PCI values, the LTE physical layer specification imposes some limitations. For optimal utilization of the physical uplink shared channel (PUSCH), it should be possible to assign cell-IDs so that the same sequence-group hopping pattern and sequence-group shift offset (hence the same base sequences are used in adjacent cells. The cell-id directly influences the sequence-hopping group. Since 30 sequence-groups are defined, an optimal sequence-group assignment is possible for up to 30 neighboring cells in normal LTE operation [6]. In case of collaborative multipoint transmission (CoMP) this value is limited to 6. Therefore, the K set ordering and creation should be carefully considered.

The number of direct neighbors of g_i cell is represented by $\deg(g_i)$. Unlike some other authors, we do not resolve the PCI confusion by mapping it to the basic color collision via connecting second level neighbors. We rather differentiate the confusion and collision events using a wider concept of n-level neighboring nodes.

3.1   The Conservative Algorithm Using the Centralized Approach

According to the 3GPP TS 36.300, the centralized automatic PCI assignment algorithm should select a single PCI for use by the configured cell. As stated before, our conservative algorithm requires only basic information regarding the network, which is already available in case of a centralized OAM system. The algorithm uses the concept of color rejection based on the neighbor relations. In order to simplify its description, several additional symbols are needed:

- $U(g_i, l)$ represents a set of colors already used by up to l-level neighbors of g_i. $F(g_i) = U(g_i, 2)$ is a special set representing the colors that are forbidden for a specified node, because they are either causing the PCI conflict ($l = 1$) or the PCI confusion in case of the second-level neighbors.
- $N[g_x \ldots g_y]$ represents an ordered set of g_i. By default, the ascending $\deg(g_i)$ order is used.
- $l_{\min}, l_{\max}$ are the minimum and maximum depth for color elimination respectively. Both of the parameters are input parameters for the algorithm and are constant during all operations.

The algorithm is represented by the below pseudo-code:

```
 1: k ← 2
 2: for all gᵢ ∈ N do
 3:     A = F(gᵢ)
 4:     if c(gᵢ) = 0 ∨ c(gᵢ) ∈ A then
 5:         finished ← False
 6:         a ← lₘᵢₙ
 7:         while a ≤ lₘₐₓ do
 8:             B = C(k) − U(gᵢ, a)
 9:             if |B| > 0 then
10:                 c(gᵢ) ← B[0]
11:                 finished ← True
12:             end if
13:             a ← a + 1
14:         end while
15:         M = C(k) − F(gᵢ)
16:         if not finished then
17:             if |M| > 0 then
18:                 c(gᵢ) ← M[0]
19:                 finished ← True
20:             end if
21:         end if
22:         if not finished then
23:             k ← k + 1
24:             c(vᵢ) = C[k]
25:         end if
26:     end if
27: end for
```

The algorithm operates using the color mappings provided by the previous mappings. The cells are checked for the colorization, conflicts and confusions in an ordered manner. The colors are changed only if necessary, and only for the affected nodes.

The algorithm attempts to select a color from a relative complement of k-available colors and the set of "unwanted" colors (B). The unwanted set consists of colors already used by all the l-neighboring vertices. The algorithm tries to maximize the distance for B creation from l_{min} hops to l_{max}. As $C(k)$ is an ordered set, the g_i vertex is colored with the first available color during each attempt.

If no colors are available in $l_{min} \ldots l_{max}$ range, a coloring attempt is made with the first non-conflicting and non-confusing value. In case of unsuccessful attempt, the number of available colors is increased and the vertex gets the newly available color. The algorithm is a subject of a pending patent application No. 61649912.

3.2   Algorithm Objectives

The algorithm has two objectives: minimizing the number of re-colorings and re-duction of used PCIs. The first objective is achieved by maximizing the distance between the cells with the same color. In order to reduce the number of utilized colors, the value of k is increased based on the demand.

4   Performance Evaluation

The performance evaluation incorporated comparison with the advanced Welsh-Powell based algorithm presented in [3], as both algorithms utilize the centralized approach, operate on the same input data and require no modifications in the LTE specification. Moreover, the Welsh-Powell based algorithm is very effective in terms of minimizing the number of required PCI values.

4.1   Test Environment

The algorithms were implemented using Python programming language. Net-work model is based on the NetworkX graph processing library. The graphs were generated using the Watts-Strogatz algorithm [7]. Each graph consisted of 500 cells. Figures 1 and 2 represent the distribution of number of direct neigh-bors and second-level neighbors in the network. The distributions are consistent with the networks used in [1] and [2].

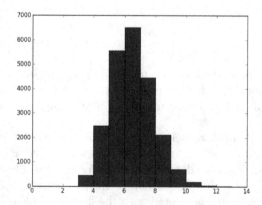

Fig. 1. Direct neighbor distribution for the evaluated networks

For the comparison we used few configurations of Welsh-Powell algorithm with different number of nodes considered as neighbours (up to 5th level neighbor relations where considered). The comparison in each point selected the variant giving best results (lowest number of reassignments).

The values of $l_{\min}$ and $l_{\max}$ are set to 4 and 5 respectively in all experiments.

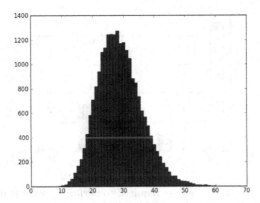

Fig. 2. 2-hop neighbor distribution for the evaluated networks

4.2   PCI-Reassignments

In this experiment, the simulation covers the situation, when all the base stations have been set up by the field operators and the whole network is turned on. The neighbor relation tables are assumed to be empty and all the cells perform the initial PCI assignment. According to the LTE specification, the NRT related data is acquired from the mobile equipment during the normal network operation. The objective is to verify network PCI convergence during the network life-time.

The described use-case is simulated by removing all of the graph edges. The edges are then added one-by-one in random order. Each time the central system performs the collision-or-confusion verification and initiates the appropriate PCI algorithm when needed.

Both algorithms were tested using the same 100 randomly generated networks. The neighbor relation information has been added using 30 different random orders, same for both algorithms.

Figure 3 depicts the distribution of number of PCI values (the size of PCI pool) used in the experiments. It can be easily observed, that our conservative algorithm is only slightly worse than the Welsh-Powell algorithm in terms of number of required colors.

Figure 4 depicts the average number of PCI reassignments remaining until full network topology is acquired from mobile equipment. Our algorithm not only requires much lower overall number of reassignments, but is also decreasing in linear manner.

The average number of reassignments in function of network age is presented in Fig. 5. While the previous statistics might not be a problem, as normally the network is allowed to run in pre-operation mode for some time, the linear nature of our conservative algorithm gives it an indisputable performance gain over the standard approach. The main reason for that is that our conservative algorithm requires only one reassignment per edge report during the whole network lifetime. This is not the case with the standard approach, as due to the

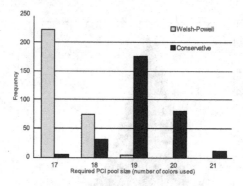

Fig. 3. Distribution of number of colors required to perform a successful mapping using the conservative and Welsh-Powell algorithms. The conservative approach typically requires one to two more colors than the classic Welsh-Powell algorithm.

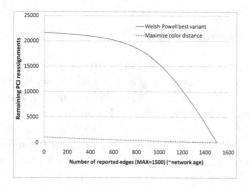

Fig. 4. Average number of remaining PCI reassignments before the full network topology information is gathered

nature of the Welsh-Powell algorithm, the addition of single edge may result in PCI reassignments in whole network.

4.3   Addition of a New Cell

During this experiment a new cell is added to the existing LTE network. The OAM system has full knowledge of the network relations before the new cell is added. After the addition, the new neighbor data is collected using standard LTE mechanism. The edges are added in random manner.

Both algorithms where tested using 100 identical, randomly generated networks. During the experiment, the topology has been generated and a single, randomly selected cell was removed, to be added later. After the removal, the whole network was colored using the appropriate algorithm. The cell was added to the network afterwards, and the new cell relations were reported using the same mechanism as in the previous experiment.

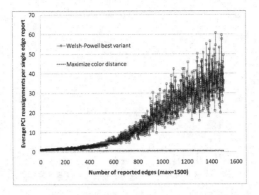

Fig. 5. Average number of PCI reassignment as a function of the number of already known edges. The reassignment is triggered by addition of a previously unknown edge.

The results are presented in Fig. 6 and 7. The advantage of our approach over the standard method is overwhelming.

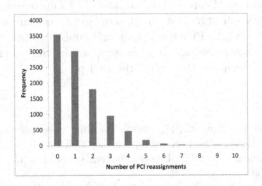

Fig. 6. Distribution of the number of PCI reassignments in case of new cell addition using conservative algorithm

4.4   Cell Prioritization

In real-world scenario, the LTE cells have different priorities in terms of PCI assignment and reassignment. The typical LTE use-case suggests that the macro-cells, covering large area, should minimize the number of PCI re-assignments in order to maintain high level of user experience. Therefore the cost of PCI re-assignment varies for different cells. It can be easily observed, that our conservative algorithm deals with this kind of problems via ordering of the $N[1 \ldots n]$. The probability of PCI reassignment is the highest for the cells with the lowest position in the set. Therefore, the PCI reassignment can be prevented by changing the way the set is ordered.

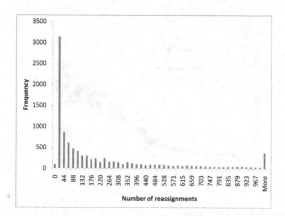

Fig. 7. Distribution of the number of PCI reassignments in case of new cell addition using Welsh-Powell algorithm

5 Conclusion

The paper presents a novel approach in the field of PCI auto-assignment, allowing robust LTE network self-organization and configuration. The number of forced PCI re-assignments due to PCI confusions and conflicts is much lower than in case of other algorithms. The algorithm is comparable with other graph-coloring based algorithms in terms of the size of the PCI pool.

References

1. Bandh, T., Carle, G., Sanneck, H.: Graph coloring based physical-cell-ID assignment for LTE networks. In: Proceedings of the 2009 International Conference on Wireless Communications and Mobile Computing: Connecting the World Wirelessly, pp. 116–120 (2009)
2. Furqan, A., Olav, T., Matti, P., Juha-Matti, K., Chia-Hao, Y., Mikko, A., et al.: Distributed graph coloring for Self-Organization in LTE networks. Journal of Electrical and Computer Engineering (2010)
3. Bandh, T., Carle, G., Sanneck, H., Schmelz, L., Romeikat, R., Bauer, B.: Optimized network configuration parameter assignment based on graph coloring. In: Network Operations and Management Symposium (NOMS), pp. 40–47. IEEE (2010)
4. Zhang, X., Zhou, D., Xiao, Z., Liu, E., Zhang, J., Glasunov, A.: Dynamic group PCI assignment scheme. In: The Seventh International Conference on Wireless and Mobile Communications, ICWMC 2011, pp. 101–106 (2011)
5. Wu, T., Rui, L., Xiong, A., Guo, S.: An automation PCI allocation method for eNodeB and home eNodeB cell. In: 2010 6th International Conference on Wireless Communications Networking and Mobile Computing (WiCOM), pp. 1–4. IEEE (September 2010)
6. Sesia, S., Toufik, I., Baker, M.: LTE-The UMTS long term evolution. From Theory to Practice 66 (2009)
7. Watts, D., Strogatz, S.: Collective dynamics of "small-world" networks. Nature 393(6684), 440–442 (1998)

Network Coding-Based QoS and Security for Dynamic Interference-Limited Networks

Amin Mohajer, Mojtaba Mazoochi, Freshteh Atri Niasar,
Ali Azami Ghadikolayi, and Mohammad Nabipour

Integrated Network Management Group Cyber Space Research Institute (CSRI),
Tehran, Iran
a.mohajer@ieee.org

Abstract. In this paper first the problem of secure minimum-cost multicast with network coding is studied, while the maximum end-to-end parameters, such as security, delay, and rate are assumed to be bounded in one multicast session. We present a decentralized algorithm that computes minimum-cost QoS flow subgraphs in network coded multicast networks. In next stage we generalize this idea to interference-limited dynamic networks where capacities are functions of the signal-to-noise-interference ratio (SINR). Since dynamic link capacities can be controlled by varying transmission powers, minimum-cost multicast must be achived by jointly optimizing network coding subgraphs with power control schemes. Simulation results shows this approach provides an efficient way for solving the optimization problem. The optimization numerical results show that using power control algorithm, higher success ratio is obtained, in comparison with previous algorithms.

Keywords: network coding, Quality of Service, network flow optimization, security, dynamic networks.

1 Introduction

Nowadays multimedia services is widely used as a part of the real-time data broadcasting (i.e. voice over IP, video over IP, IP television). These types of services have stringent QoS (Quality of Service) requirements [1] and if they are not being met by the underlying network, the user experience will be degraded significantly.

In order to provide QoS, routing algorithms must be modified to support QoS and be able to discriminate between different packet types in the process of finding the best route that satisfies QoS metrics. The goals of QoS routing are in general twofold: selecting routes that satisfy QoS requirement(s), and achieving global efficiency in resource utilization [2].

It is well-known that network performance can be significantly improved through such network coding approach [3]. As an important example, use of network coding makes the once intractable optimal multicast routing problem tractable [4]. As shown in [5–7], with network coding the achievable throughput

A. Kwiecień, P. Gaj, and P. Stera (Eds.): CN 2013, CCIS 370, pp. 277–289, 2013.
© Springer-Verlag Berlin Heidelberg 2013

of a multicast session can be acquired by running max-flow algorithm from the source to each individual receiver, and choosing the minimal value.

For a good survey on network coding see [8]. Because of these advantages, many algorithms that used routing are being modified to incorporate network coding instead [9].

All optimum sub-graph does not directly consider hard QoS although those relies on optimization schemes to determine proper flow sub-graphs to minimize given cost functions [10, 11]. In this paper we introduce a path-based mixed integer programming (MIP) model. We apply decomposition method, which provides primal solutions at all iterations and it is able to tackle the path-flow model [12]. Note that the path flow formulation of problem has a very simple constraint structure. The primal dual procedure considers only paths that do not violate the QoS constraint. We show by our computational experiment that a carefully implemented secure QoS NC approach provides an efficient way for solving the problem.

But in the case of extending to actual condition, the performance gains offered by network coding point to their promising application in dynamic networks, where multi-user interference, channel fading, energy constraints, and the lack of centralized coordination present new challenges. Initial studies on the application of network coding in dynamic networks shown in [5, 13]. In [5], the minimum-energy multicast problem is studied by exploiting the "dynamic multicast advantage". The work in [13] introduces a distributed protocol which supports multiple unicast flows efficiently by exploiting the shared nature of the wireless medium.

In this work, we extend the optimization framework and distributed algorithms in [14] to achieve minimum-cost multicast with network coding in interference-limited dynamic networks. To address this, first we determine best subgraph that provide user quality limits with the best cost in safe conditions. Second we design a set of node-based distributed gradient projection algorithms which iteratively adjust local control variables so as to converge to the optimal power control, coding subgraph configuration. We consider dynamic networks where link capacities are functions of the signal-to-interference-plus noise ratio (SINR) at the receiver.

In this context, dynamic link capacities can be controlled by varying transmission powers. To achieve minimum-cost multicast, the coding subgraphs must now be jointly optimized with power control schemes at the physical layer. Moreover, this joint optimization must be carried out in the network without excessive control overhead. To solve this problem, we design a set of node-based scaled gradient projection algorithms which iteratively adjust local control variables at network nodes so as to converge to the optimal power control, coding subgraph, and congestion control configuration. These algorithms are distributed in the sense that network nodes can separately update their control variables after obtaining a limited number of control messages from their neighboring nodes. We explicitly derive the scaling matrices required in the gradient projection algorithms for fast, guaranteed global convergence, and show how the scaling matrices can be computed in a distributed manner.

The rest of paper is organized as follows. We first develop a precise formulation of secure minimum-cost QoS multicast problem over coded packet networks in Sect. 2. In Section 3, the proposed problem is solved by our scheme. In Section 4 we will generalize this applied method to the extended situation in more practical networks. Computational results and conclusion are presented in Sect. 5 and 6.

2   Secure Network Coding-Based Quality of Service for Multicast Session

In this section, we present the network model and explain the details of our network coding-based QoS algorithm by the security viewpoint.

2.1   Network Model and Notations

The communication network is represented by a network $G = (V, E)$, where V is a set of nodes with $|V| = N$, and ε is a set of links (arcs) with $|E| = L$, consider single multicast session, a source node, $S \in V$ must transmits an integer number of R packets per unit time to every node in a set of terminals, $k \subset V$. Let Z_l denotes the rate at which coded packets are injected onto link l. linear, separable cost C_l denotes the cost per unit rate of sending coded packets over link $l \in \varepsilon$. Link weight often can be considered as delay or jitter or security in this problem. Also let the data flow toward destination T that is passing through link l is indicated by $X_l^{(t)}$.

We assign the limitations and costsonly to the links. If we also consider the requirements for node we can use the node splitting technique to transform node costs and weights into link costs and weights [15]. The source has to provide multicast receivers with their required flows and also guarantee the desired quality. There are M quality classes in the network. Each class has a minimum required rate $R^{(c)}$ and a maximum tolerable security weakness constraint, $S^{(c)}$. Meanwhile we assign link insecurity degree instead of security level in order to convenience.

The source S either refuse to send requested flows to the receivers or if it has accepted the request, it has to guarantee that the end-to-end security weakness of each flow toward destination is less than $S^{(c)}$ and the rate of flow is at leas $R^{(c)}$.

The date and flow toward the destination t that belongs to class c and is passing through link l is indicated by $X_l^{(t)(c)}, Z_l^{(t)(c)}$. In order to account for the end-to-end security experienced by network flows, we have assumed each link to have a constant security weakness degree of S_l for a given period of time. We have assumed that before running our algorithm, each node has measured the security of its outgoing links. Therefore, S_l could be viewed as the short term average security of link. Again, We emphasize that our mean the security level of links is the link weakness level against attacks. This security includes constant propagation security plus any additional security caused by MAC layer. Some other types of security, like the security caused by nodes (attacker nodes), could be handled by including them in the cost function [16].

As we use network coding instead of routing, flows of each classes are coded together. However, interclass coding is not permitted. This indicates that if flows of different classes are coded together into one flow, all classes experience the same quality level and it would not be possible to provide the required quality level for different classes. Can achieve this goal with separating the different classes quality and security levels, in data submissions, and combining each class data by the same class.

We indicate the coded flow of class c by $Z^{(c)}$. These coded flows are sent across each link independently. The total flow passing through link l in class c is indicated by $Z_l^{(c)}$. Subsequently, we denote the set of all $X_l^{(t)(c)}$ by the matrix X and the set of all $Z_l^{(c)}$ by the matrix Z.

2.2   Problem Formulation

The goal is to find a minimum cost multicast connection such that it is guaranteed that the end-to-end weight of security and delay in each flow toward destination T is less than $D^{(t)}, S^{(t)}$ respectively.

For each destination T, let $P^{(t)}$ denote the collection of all directed paths from the source node s to the destination node t in the underlying network G. In the path flow formulation, we define the variable $f(P)$ as the flow on path.

In this formulation, We would like to determine network flows such that the final solution satisfies the rate and security constraints for all classes. If there are more than one solution with these properties, the minimum cost solution will be selected. After identifying the flow subgraphs, any coding method, such as [17] can be used to determine network codes.

Let $\delta_l(p)$ be a link-path indicator variable, that is, $\delta_l(p)$ equals 1 if link l is contained in path p, and is 0 otherwise. The link flow $X_l^{(t)}$, is computed from path flow by the following relation.

$$X_l^{(t)} = \sum_{p \in P^{(t)}} \delta_l(P)g(P) \ . \tag{1}$$

Taking into account the required constraints, the minimum cost QoS multicast over coded packet networks is then given by the following optimization model.

It is shown in [4] that decoupling these two tasks, namely finding flow subgraphs and determining network codes does not change the optimality of the solution. The network cost of each link is assumed to be a convex and non-decreasing function of total flow (Z_l) over each link. The total cost is the sum of all link cost functions and total flow is:

$$Z = \min \sum_{l \in E} C_l Z_l \ . \tag{2}$$

End-to-end transmitting level insecurity will be equal to insecurity of the most insecure path in the subgraph. Therefore, we formulate the QoS network coding flow optimization problem as follows:

$$\min_{\underline{x}} \sum_{l \in E} f(Z_l) \tag{3}$$

where:

$$Z_l = \sum_{c=1}^{M} Z_l^{(c)} \tag{4}$$

$$Z_l^{(c)} = \max_{t \in T} \left\{ x_l^{(t)(c)} \right\} \tag{5}$$

$$0 \leq Z_l \leq a_l \quad \forall l \in E . \tag{6}$$

During the entire route will be:

$$\sum_{p \in P^{(t)}} \delta_l(p) f(p) \leq Z_l \quad \forall l \in E, \ t \in T \tag{7}$$

subject to:

$$r^{(c)} \leq R^{(c)} \quad c = 1, \ldots, M \tag{8}$$

$$\sum_{c=1}^{M} r^{(c)} \leq R \tag{9}$$

and

$$\sum_{l \in \varepsilon} s_e^{(t)} \delta_e(P) y_p \leq S^{(t)} \quad \forall t \in T \tag{10}$$

$$x_l^{(t)(c)} \geq 0 \quad \forall l \in E, \ t \in T, \ c = 1, \ldots, M . \tag{11}$$

And for security constraints we have too:

$$S(t,c) = \sum_{l \in E} S_l \left(x_l^{(t)(c)} \right) \leq S^{(c)} \quad \forall t \in T, \ c = 1, \ldots, M \tag{12}$$

$$S(t,c) = \max_{p \in P_l} S_p^{(t)(c)} \leq S^{(c)} \quad t \in T, \ c = 1, \ldots, M \tag{13}$$

where

$$S_l \left(x_l^{(t)(c)} \right) = \begin{cases} s_l & \text{if} \quad x_l^{(t)(c)} \, 0 \\ 0 & \text{if} \quad \text{otherwise} \end{cases} . \tag{14}$$

In above equations, $f(z_l)$, a_l, e_l and S_l are the cost function, capacity, cost coefficient and security of link l respectively. $R^{(c)}$ is the minimum required rate and $S^{(c)}$ is the maximum tolerable security of class c. R is the max-flow-min-cut rate of the network and $r^{(c)}$ refers to the actual rate of class c. Equation (6) is the capacity constraint. It guarantees that the total flow on link l is less than its capacity. Constraint (8) indicates that the rate of class c must be greater than the minimum required value determined by the user. Constraint (9) makes sure that the total rate of all classes is less than the max-flow rate of network. Equation (11) formulates the non-negative flow conservation constraint. And constraint ensured the non-negativity of X. Equation (13) indicates the security constraint, ensuring that the security weakness of class c is less than the maximum tolerable weakness of the class. By solving problem (4), we obtain flow subgraphs that satisfy the specified constraints and are also minimum cost among all feasible answers. In the next section, we provide a distributed and simple solution for this problem.

Note that in order to reduce the complexity simplify the implementation of the optimization algorithm, we slightly modify the problem definition as [18].

3   Problem Solving

To solve the problem (3), the primal dual decomposition method [18] is used, in which the problem is first broken into two subproblems using primal decomposition [12]. Then, each subproblem is solved using dual Lagrangian method [19]. This approach will result in a simple, distributed solution.

Simplify the algorithm, we have assumed that the actual rate of each class is equal to the minimum accepted rate [20] then in this scheme we utilize (13) instead of (12) and we spot constant applied users constraints for all receivers.

More specifically, we have assumed that for all classes, $r^{(c)} = R^{(c)}$ and seek the solution that satisfies this condition. Consequently, the constraint (8) is removed. Moreover, since is now fixed, the constraint (9) becomes a feasibility condition which should be checked before the algorithm is run. In other words the source node should check this condition to see if the problem is feasible. If not, the users request is withdrawn.

Ability to communicate will be checked in two stages. First with respect to the permissible rate range and security which depends on network nature, the reasonableness of the request will be evaluated. If it is reasonable, according to constraints for each class of user, we examine possiblity of sub graph existence in second stage.

Since we have assumed the cost function to be convex, problem (3) is a convex optimization problem and the duality gap of decomposition method will be zero (see Section 5.2.3 of [21]). Subsequently, in order to break the problem (3) into two subproblems, we first assume x to be constant and solve the problem over z. Then, the resulting cost function is minimized over x [19]. More specifically, the problem (3) is decomposed into the following subproblems: **Subproblem A**:

$$\min_{z} \sum_{l \in E} f(z_l) \tag{15}$$

subject to:

$$x_l^{(t)(c)} \leq z_l^{(c)} \quad \forall t \in T, \ \forall l \in E, \ c = 1, \ldots, M \tag{16}$$

and **Subproblem B** [20]:

$$\min_x \sum_{l \in E} f^*(x) \tag{17}$$

subject to constraints (8) and (9). In the above equation, f^* indicates the solution of subproblem (A). In order to solve subproblem (A), we use its Lagrangian equivalent. Define:

$$L(z, \lambda, \beta) = \sum_{l \in E} f(Z_l) + \lambda^T (Z - a) + \sum_{l \in E} \sum_{t \in T} \sum_{c=1}^{M} \beta_{ltc} \left(x_l^{(t)(c)} - z_l^{(c)} \right) \tag{18}$$

then subproblem (A) is equivalent to:

$$\min_z L(z, \lambda, \beta) \tag{19}$$

and

$$\max_{\lambda, \beta} L^*(\lambda, \beta) \tag{20}$$

where L^* is the solution of problem (19). We could decouple the above solution into L subproblems, one for each link:

$$\max_{\lambda, \beta} f(Z_l^*) + \lambda_l (Z_l^* - a_l) + \sum_{t \in T} \sum_{c=1}^{M} \beta_{ltc} \left(x_l^{(t)(c)} - z_l^{*(c)} \right) . \tag{21}$$

Subproblem (21) could be solved using subgradients method as follows [19]:

$$z_l^{(c)}(\tau + 1) = \left[z_l^{(c)}(\tau) - \alpha(\tau) \left(\nabla f_l^c - \sum_{t \in T} \beta_{ltc} \right) \right]_{z \in Z} \tag{22}$$

$$\lambda_l(\tau + 1) = [\lambda_l(\tau) + \alpha(\tau) (Z_l^* - a_l)]_+ \tag{23}$$

$$\beta_{ltc}(\tau + 1) = \left[\beta_{ltc}(\tau) + \alpha(\tau) \left(x_l^{(t)(c)} - z_l^{*(c)} \right) \right]_+ \tag{24}$$

where $[]_+$ indicates that α and β must be non-negative and $[]z \in Z$ ensures that the updated z lies in the feasible region. $\nabla f_l(c)$ represents the (l, c) member of $\nabla f_l(z)$. The parameter $\alpha(\tau)$ is the algorithm step size, chosen such that convergence of the algorithm is guaranteed. In our algorithm we have $\alpha(\tau) = 1/\tau$ which although guarantees the convergence of the subgradient method, any diminishing step-size may be used as well (see section 6.3.1 of [19]). In the same manner, subproblem (17) could be solved using subgradients method. In each iteration, x and its lagrange multiplier are updated according to (25), (26) as follows:

$$x_l^{(t)(c)}(\tau + 1) = \left[x_l^{(t)(c)}(\tau) - \alpha(\tau) \left(\nabla f_l^{*(c)} + \beta_{ltc} + \sum_{\substack{p \in P_t \\ l \in P}} \vartheta_P^{(t)(c)} s_l \right) \right]_{x \in X} \tag{25}$$

and

$$\vartheta_p^{(t)(c)}(\tau+1) = \left[\vartheta_p^{(t)(c)}(\tau) + \alpha(\tau)\left(S_p^{(t,c)} - S^{(c)}\right)\right]_+ \tag{26}$$

where "$x \in X$" means that the updated x should be projected onto feasible x region. The parameter θ is the Lagrange multipliers vector for security constraint. Each node should solve subproblem (21) for its outgoing links and derive $z_l^{(c)}$, update x and Lagrange multipliers according to (22), (23), (24) and (25), respectively. Then, it has to exchange these multipliers with its neighbors until convergence is achieved. This procedure leads to a distributed, simple solution in which each node n must solve at most $M.\ outdegree(n)$ where $outdegree(n)$ is the number of the output links of node n and M is the number of classes.

To summarize, each node has to perform the following procedure to solve the problem (3):

1. Parameters initialization.
2. Solve subproblem (21) for each outgoing link.
3. Update Lagrange multipliers via (22),(23) and (24).
4. Update x according to (25).
5. Update constraints Lagrange multipliers via (26).
6. Repeat till convergence.

Separable Cost Function. In the previous section, a distributed solution of the problem (3) was presented. So far, convexity is the only assumption that was made about the cost function. But if the cost function could be decomposed itself, each subproblem could also be decoupled into other subproblems. Some examples of these kind of cost functions presented in [20]:

$$f(Z_l) = e_l Z_l = \sum_{c=1}^{M} e_l z_l^{(c)} = \sum_{c=1}^{M} f\left(z_l^{(c)}\right) . \tag{27}$$

The cost function (27) represents energy consumption and the cost function (28) may be considered as a representative of queuing costs. In this scheme, energy or monetary costs are more important for network operators then we use cost function (27).

4   To Generalize the Algorithm to Practical Conditions

The problem of jointly optimal power control, and network coding in wireless networks with multiple multicast sessions is investigated in [22] that explicitly derive the scaling matrices required in the gradient projection algorithms for fast, guaranteed global convergence, and show how the scaling matrices can be computed in a distributed manner. In our optimization framework some consepts of this paper is used. Our scheme yields to a feasible set of transmission powers, link capacities, as well as a set of network coding subgraphs We assume that the wireless network is interference limited, so that the capacity of the link (i, j), denoted by C_{ij}, is a nonnegative function of the signal-to-interference-plus noise ratio $(SINR)$ at the receiver of the link, i.e. $C_{ij} = C(SINR_{ij})$. We further assume $C(.)$ is increasing, concave, and twice continuously differentiable. For $(i,j) \in E$ the receiver node calculates $SINR$ in terms of G using equation (28).

$$SINR_{ij}(p) = \frac{G_i P_{ij}}{G_{ij} \sum_{n \neq j} P_{in} + \sum_{m \neq i} G_{mj} \sum_n P_{mn} + N_j} . \qquad (28)$$

In above equations, G_{ij} is gain of link ij which obtain corresponding matrix G elements according to (29). N_j is the noise power at the receiver of the node j.

$$G = \lfloor g_{ij} \rfloor_{i=1,\dots,8,\ j=1,\dots,8,\ g_{ij}=g_{ji},\ g_{ii}=0} . \qquad (29)$$

Assume every node i is subject to an individual power constraint: $\sum_j P_{ij} = P_i \leq \overline{P_i}$, where P_{ij} is allocated power to link ij by node i, P_i is power of node i, and $\overline{P_i}$ is constraint on power allocated to node i. The set of all feasible nodes power allocated vectors denoted by vector P, $P = [P_1, \dots, P_N]$ where $\pi = \{P : \sum_j P_{ij} \leq \overline{P_i}, \forall i \in V, P_{ij} \geq 0, \forall i,j \in V\}$.

In the previous work [4, 22], it has shown that in wireline multicast networks with network coding, coding subgraph optimization can be achieved using a routing methodology. We now extend this consept to wireless networks, where in contrast to wireline networks, link capacities can be further controlled by varying transmission powers.

Large-scale wireless networks usually lack centralized coordination, and it is desirable to distribute the control functionalities to individual nodes. In this method we should permit each node to independently adjust the sub-session flow rates on its outgoing links [23]. But in this scheme we use random power allocation. which leads to a little more complexity.

For each node, we calculate $SINR$ value as a parameter of transmission reliability in terms of interference. Taking calculated $SINR$ value of each link, we review the possibility of a link failure occurrence network. New links capacity, are determined according to calculated $SINR$ (30).

$$C_{ij} = C(SINR_{ij}) . \qquad (30)$$

In a practical network, for accessibility to transmission reliability, receiver nodes can compare $SINR$ value of link with their sensitivity. That sensivity vector is a set of integers that are considered equal to a threshold for simplicity.

The link failure is considered if the $SINR$ value of link is less than the sensitivity. In this situation, answer subgraphs will not include failed link. It is clear that in this case, possibility of finding optimal subgraph which is restricted to users limitations decreases. It also results increase in cost. In this mode, when $SINR$ of outgoing links are less than the threshold (receiver sensivity), transmitter node attempts to maintain $SINR$ by allocating more power to transmission with increasing links power (that is equivalent to link capacity). In this situation that link exits from temporary failure. Note that total power dedicated to the collection of nodes in the network is fixed.

5   Simulation Results

5.1   Secure Network Coding-Based Quality of Service

In this section, we investigate the performance of our proposed algorithm based on several simulations. We have considered two different cases, ordinary condition and interference condition.

We evaluate the performance of the proposed algorithm. For illustrative purposes, we first simulate the suggested algorithm for some basic cases. Then, more general and complex cases will be taken into account in the simulations.

We try to provide more quality parameters in secure condition (Table 1) and compare the obtained results with QoSNC [20] and MCM [4] that these methods have no security constraints (Table 2). Also we compare proposed algorithm (SQoSNC) results in interference condition with power control algorithm and without power control algorithm (Table 3).

In order to examine the performance of our algorithm in such cases, we have considered several random networks in which there are one multicast source and two quality classes. In basic network, constraint (security) and cost of each link is determined. In simulation condition we assume that there is one multicast source, s, two multicast sinks, t_1 and t_2 and two quality class with desired rate 1 and 2 for each class and maximum tolerable security will determine by user.

As Table 1 shows with increased degree of restrictions (as security restriction) we reach to the ideal answer with far fewer iterations. also, due to more choice in determining the optimal subgraphs and the various transmitting rate in links, cost functions reduce too.

Table 1. Number of iteration and total cost in various security constraints

Class	Constraints	Output	Iteration	Cost
I	4	[2 2]	219	11.41
II	5	[3 3]		
I	6	[2 2]	53	10.65
II	8	[8 8]		
I	8	[2 4]	95	9.62
II	10	[8 8]		
I	10	[3 5]	72	2.69
II	12	[6 6]		
I	11	[6 6]	21	1.67
II	13	[7 7]		

5.2   Simulation Results in Gereralize the Algorithm to Practical Conditions

In the second section of simulation, we tries to generalize the algorithm to the dynamic wireless networks in practical conditions. As expected, previous algorithms provide poor performance in presence of interference, with respect to reliability. A power control scheme is proposed as a solution to this problem.

Table 2. Cost of various algorithms in different number of nodes

No.of Nodes	Algorithm	Total Cost
7	MCM	16.19
	QoSNC	19.31
	SQoSNC	11.15
10	MCM	18.25
	QoSNC	21.22
	SQoSNC	25.29
20	MCM	25.13
	QoSNC	27.76
	SQoSNC	31.43

The Table 3 show comparison the performance of algorithm with power control and without power control on illustrated graph in Fig. 1.

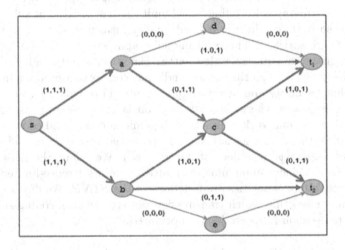

Fig. 1. One muilticast session network that we have implemented the power control on it

This results suggest that previous proposed algorithm with determine the threshold for its links $SINR$, will not accountable in higher limitation. While the same algorithm with power control will have coordination feature with network random conditions and will determine the corresponding subgraph except the limits are too high (strong limit).

As the Table 3 shows, only in the very low limitation (unrestricted), both algorithms give nearly identical answers. Subgraphs of the same cost and different iterations will be answers in this case.

Table 3. SQoS algorithm with power control and without power control

Factors Constraints	With Power Control	Without Power Control
(2,3)	-------	-------
	-------	-------
(4,6)	Total Cost: 25.29	-------
	Iteration: 1000	-------
(8,10)	Total Cost: 28.78	Total Cost: 25.57
	Iteration: 21	Iteration: 90
(10,12)	Total Cost: 8.32	Total Cost: 21.89
	Iteration: 13	Iteration: 32

6   Conclusion and Future Works

We have proposed a new decentralized algorithm that computes secure minimum cost QoS flow subgraphs in coding-based multicast networks. The subgraphs are determined so that certain user-defined quality measures, such as security and throughput, are satisfied. The resulting subgraphs are chosen to achieve minimum cost among all subgraphs that satisfy the given quality and security constraints. The proposed algorithm can handle any convex cost function in addition to user-defined security and quality requirements. This makes our algorithm an ideal choice for cases where the cost function is linear (energy consumption). We will Extending our work to practical dynamic networks and considering network dynamism such as interference, and potential link failures and reluctate interference impact is a subject of our research. We adopt the network coding approach to achieve minimum-cost multicast in interference-limited wireless networks where link capacities are functions of the $SINR$. We develop a node-based framework within which transmission powers, network coding subgraphs, and admitted session rates are jointly optimized.

Acknowledgment. We would like to thank all the advisors on the subject of decomposition methods and distributed algorithms for network utility maximization, especially, A.H. Salavati for his helpful ideas. This work was supported by Cyber Space Research Institue (CSRI).

References

1. Chen, S., Nahrstedt, K.: An Overview of Quality-of-Service Routing for the Next Generation High-Speed Networks: Problems and Solutions. IEEE Network, Special Issue on Transmission and Distribution of Digital Video, 64–79 (November/December 1998)
2. Wang, B., Hou, J.C.: Multicast Routing and Its QoS Extension: Problems, Algorithms, and Protocols. IEEE Network 14(1), 22–36 (2000)
3. Zhang, X., Li, B.: Optimized multipath network coding in lossy wireless networks. In: Proc. of ICDCS 2008 (2008)

4. Lun, D.S., Ratnakar, N., Medard, M., Koetter, R., Karger, D., Ho, T., Ahmed, E., Zhao, F.: Minimum-cost multicast over coded packet networks. IEEE Trans. Inform. Theory 52(6), 2608–2623 (2006)
5. Low, S.H., Lapsley, D.E.: Optimization flow control, I: basic algorithm and convergence. IEEE/ACM Transactions on Networking 7(6), 861–874 (1999), http://netlab.caltech.edu
6. Li, S.-Y.R., Yeung, R.W., Cai, N.: Linear network coding. IEEE Trans. Inform. Theory 49(2), 371–381 (2003)
7. Jaggi, S., Sanders, P., Chou, P.A., Effros, M., Egner, S., Jain, K., Tolhuizen, L.: Polynomial Time Algorithms for Multicast Network Code Construction. IEEE Transactions on Information Theory 51(6), 1973–1982 (2005)
8. Fragouli, C., Boudec, J.-Y.L., Widmer, J.: Network Coding: An instant primer. SIGCOMM Comput. Commun. Rev. 36, 63 (2006)
9. Chou, P.A., Wu, Y., Jain, K.: Practical network coding. In: Allerton Conference on Communication, Control, and Computing (2003)
10. Julian, D., Chiang, M., O'Neill, D., Boyd, S.: QoS and fairness constrained convex optimization of resource allocation for wireless cellular and ad hoc networks. In: Proc. IEEE INFOCOM, New York, USA, pp. 477–486 (June 2002)
11. Chen, S., Shavitt, Y.: A Scalable Distributed QoS Multicast Routing Protocol. In: Proc. IEEE International Conference on Communications (ICC), vol. 2, pp. 1161–1165 (2004)
12. Desaulniers, G., Desrosiers, J., Solomon, M.M. (eds.): Column generation. Springer, Berlin (2005)
13. Yeung, W., Yen, S., Li, R., Cai, N., Zhang, Z.: Network Coding Theory. Now Publishers Inc. (June 2006)
14. Li, Z., Li, B.: Network Coding in Undirected networks. In: Proc. 38th Annu. Conf. Information Sciences and Systems, Princeton, NJ (March 2004)
15. Salavati, A.H., Aref, M.R.: A New Framework for Solving Multi constraint Network QoS Provisioning Problems in Polynomial Time (2009)
16. Yang, H., Luo, H., Ye, F.: Security and Privacy in Sensor Networks. IEEE Wireless Communications 11, 38–47
17. Ho, T., Koetter, R., Mdard, M., Karger, D.R., Effros, M.: The Benefits of Coding Over Routing in a Randomized Setting. In: Proc. IEEE Int. Symp. Information Theory, Yokohama, Japan, p. 442 (June/July 2003)
18. Palomar, D., Chiang, M.: A Tutorial on Decomposition Method and Distributed Network Resource Allocation. IEEE J. Sel. Areas Commun. 24, 1439 (2006)
19. Bertsekas, D.P.: Nonlinear Programming, 2nd edn. Athena Scientific, Belmont (1999)
20. Salavati, A.H.: Quality of Service Network Coding. Msc thesis. Sharif university of technology (September 2008)
21. Boyd, S., Vandenberghe, L.: Convex Optimization. Cambridge Univ. Press, Cambridge (2004)
22. Xi, Y., Yeh, E.M.: Distributed algorithms for minimum cost multicast with network coding. In: Proceedings of the 43rd Allerton Annual Conference on Communication, Control, and Computing (September 2005)
23. Gallager, R.: A minimum delay routing algorithm using distributed computation. IEEE Transactions on Communications 25(1), 73–85 (1977)

Simple Communication with FPGA Device over Ethernet Interface

Marcin Kucharczyk and Grzegorz Dziwoki

Silesian University of Technology, Institute of Electronics,
ul. Akademicka 16, 44-100 Gliwice, Poland
{marcin.kucharczyk,grzegorz.dziwoki}@polsl.pl

Abstract. Individual components of the communication system can be implemented in different environments on the stage of design and prototyping. Some of them are simulated in a high-level language environment, like matlab, and some are implemented in a target device, FPGA or signal processor, using evaluation boards. The verification of hardware implementation needs the communication interface sufficiently fast to transmit high amount of data. The Ethernet interface widely available on modern boards seems to be useful. The paper presents a set of modules, compiled as applications as well as implemented in FPGA, used to send data between PC and FPGA device over Ethernet omitting the TCP/IP protocol stack. The data are send and received by direct access to the Ethernet frames. Simple transmission protocol with acknowledgment was proposed to provide reliable data transmission.

Keywords: transmission protocol, Ethernet, FPGA, LDPC codes.

1 Introduction

Any digital communication system consists of independent modules in transmitter and receiver: source encoder and decoder, channel encoder and decoder, modulator, demodulator, equalizer, etc. These individual parts of the system can be designed and tested independently, but at some stage of the project they should be connected together to verify the whole system.

All modules can be created and simulated in high-level programming environments, like matlab, and then communication between them is easy to realize. Each module receives block of data in the defined format and after processing sends it to the next one. The block of data is the variable defined in the program. If any of the modules is implemented in different environment, particularly in hardware, then the idea of communication between modules remains the same, but its realization becomes slightly more complex. It is necessary to select an interface and a communication protocol available in both environments and then the communication modules need to be implemented on both sides.

The presented work is a part of the larger project directed to design and evaluation of transmission systems using non-binary $GF(2^q)$ LDPC (Galois Field Low-Density Parity Check) codes [1]. The project includes mainly research of

A. Kwiecień, P. Gaj, and P. Stera (Eds.): CN 2013, CCIS 370, pp. 290–299, 2013.
© Springer-Verlag Berlin Heidelberg 2013

software and hardware LDPC decoding unit in the transmission system with different modulations. The different $GF(2^q)$ LDPC codes are tested for quality of error correction. One of then goals of the research is creating the VHDL description of a hardware decoder of non-binary LDPC codes [2,3]. The VHDL description is hardware independent but the FPGA device should have enough capacity to accommodate the whole hardware. The ML605 evaluation board with Virtex-6 LX240T FPGA from Xilinx is used for the designed decoder implementation. After the verification of the proper operation of the decoder in HDL simulator (the Xilinx ISE Simulator was used) it was required to process higher amount of data to compare the performance of the hardware decoder with the software one.

2   Related Work

The best performance of the multiplatform system with an FPGA device can be achieved connecting the PC with the FPGA by PCI/PCIe interface [4]. The evaluation board used in the project has the PCI Express (PCIe) connector avilable [5], but such a solution has several drawbacks. To establish communication a dedicated driver for PCIe on the PC is required, which increases programming requirements. Another drawback at the stage of the decoder verification is necessity to install the board permanently in the PC with the PCIe x8 connector available. It causes the board is powered on all the time and the simulations can be made on the only one system. Other communication interfaces available on the board: the USB UART (serial port via USB), the USB 2.0 and the Ethernet can be used with most of the modern PC systems and allows hot-plug operation with the board which is powered on only when required computation is running.

The UART is too slow for transmission of high amount of data. From the two other interfaces the Ethernet was chosen because of full-duplex operation, available higher rates and possibly easier programming because no additional driver needs to be created. The Virtex-6 device used in the project has got embedded Tri-Mode Ethernet MAC (TEMAC) [6] units inside and the implementation using one of them shouldn't consume a lot of remaining FPGA resources. But the Ethernet itself is insufficient for data transmission between devices, it includes only two bottom layers of the OSI networking model: physical and data-link [7]. To handle partitioning and combination of data the fourth layer protocol is required. If the Internet protocol suite, commonly known as TCP/IP, is used than the TCP or the UDP protocol can be used. The example of full implementation of the TCP/IP stack designed for embedded systems is named LightWeight IP (lwIP) [8]. The lwIP requires the MicroBlaze or other processor core to be included in a project. Such a solution would be efficient when the FPGA includes the processor core inside, otherwise one of the available soft cores need to be implemented [9]. It would consume FPGA resources that are required by the LDPC decoder itself. It is possible to implement the TCP/IP stack without the processor, but the resource requirements for such a solution are still high [10].

If the main goal of the project is data transmission between directly connected FPGA and PC the TCP protocol can be omitted, which results in UDP/IP stack. The commercial [11] and the open source [12] solutions can be found after some bibliographical search. These solutions are much less resource hungry than the whole TCP/IP stack but the cost of using UDP protocol only is possible data loss. The data can be checked by higher layers protocols and retransmitted if required. In the solution presented in this paper we have decided to make another simplification to the protocol stack. The IP layer was removed completely, because it is not necessary if the devices are connected directly. The transport layer protocol is still required because longer blocks of data needs to be sent in parts. Simple transmission protocol proposed for this task was described.

3   Communication Protocol

Two devices connected directly by the Ethernet cable can be identified by the MAC address. So the routing is unnecessary in direct communication and the IP layer can be omitted. The Ethernet frames can be sent directly between devices. The fields of Ethernet frame defined in the standard: source and destination addresses are enough for the sender and the recipient identification. If the size of a payload in one frame is enough to transfer data between devices then transmission can be realized by direct data encapsulation into frame. Otherwise data need to be partitioned and some kind of transmission protocol, the fourth layer of the OSI model, is required. In that case the payload of an Ethernet frame contains a datagram. The datagram includes fields that used for parts identification and the transmitted data themselves.

The proposed datagram format is defined as follows: first 10 octets are used as a header and the rest is the payload. So the maximum payload size is 1490 bytes, it is 10 less than the maximum size of payload in the Ethernet frame. The datagram header was divided into 4 fields. The first field is the datagram identifier. It is 4 octets in length and includes a specific sequence of bytes to check if the datagram contains data transmitted between devices. All datagrams with different identifier are dropped. The direction of transmission is determined by the destination and source MAC addresses included in the Ethernet frame. The block number is defined by fifth and sixth octets of the datagram. The next two octets form the datagram number in the block of data. The size of the payload is included in the last two octets of the header.

The block is a piece of data that need to be acknowledged by the receiver. If the receiver doesn't read all parts of the block then NAK answer with the appropriate block number and lost datagrams numbers in payload is sent, otherwise the answer contains ACK. Bad parts of the block need to be retransmitted. The two bytes was provided for the datagram number, but only the second of them is used. The first byte is used to identify special datagrams like ACK and NAK. The acknowledgment frequency depends on the length of the block.

There is no CRC field in the datagram. It is not necessary. The assumption was that the data link layer drops the frame if its CRC is not correct. Thus the datagram encapsulated in it is dropped too. Not all parts of the block are received and the NAK answer is sent to the transmitter.

4 Protocol Implementation

The protocol described previously need to be implemented at both ends of the transmission system. The PC and the evaluation board with the FPGA device are used. Our target project consists the main program written in matlab. The program encodes a code word with a non-binary LDPC code. The resulted bitstream is used in the modulator and than the signal is transmitted by the disturbed channel. After receiving the data the demodulator makes soft decisions on them based on the channel parameters. The soft decisions are 2^q probabilities for each codeword element. The values calculated are converted to $Qn.m$ integer format and then the transmission block is prepared. The block is sent to the decoder implemented in the Virtex 6 FPGA device [2]. The result of decoding is sent back to the PC through the same Ethernet port and than it is compared with the original data to determine error rate. The Figure 1 depicts the system corresponding to this description.

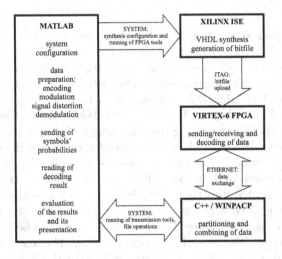

Fig. 1. The multiplatform communication system

Software Implementation. The communication module on the PC is build as a console application in the Visual Studio 2010. The code was implemented in C++ language using WinPcap library, a tool for link-layer network access in Windows environments [13]. The most important feature of the library is ability to capture and transmit network packets bypassing system's protocol stack. It gives well documented API that allows the developer to create an application that can read and/or write the data directly from/to Ethernet frames. Microsoft

Visual Studio IDE gives enough tools to create an application and thoroughly check its functioning. It seems rather easy to create the transmission and reception applications to check communication protocol assumptions and measure real data rate.

The source and destination MAC addresses and the name of the file to transmit/receive are defined as the application's runtime parameters. The use of external file is versatile method for communication between different environments. Functions for reading and writing a file are available probably in each development environment. In this model, the transmission block size is limited only by the limitations of the filesystem. Created application reads the file, divides it into blocks of defined size and then the blocks are divided into the finite number of datagrams. The datagrams are sent using WinPcap library as a payload in data-link layer frames. The reception site performs the inverse procedure and writes the received data into another disk file.

Generated frame passed to WinPcap function includes all fields except the preamble with SFD (Start Frame Delimiter) and the FCS (Frame Check Sequence). These fields are generated by the network adapter itself. Otherwise, when the FCS of received frame is broken the frame is dropped before passing to WinPcap driver, so the application assumes that if only the frame is received it is received properly.

Hardware Implementation. After the transmission protocol correctness was verified on the PC it has been implemented in the FPGA device. The Xlinx ML605 Evaluation Board has been used as a target platform [5]. It consists Virtex-6 LX240T FPGA device with embedded 10/100/1000 Mb/s Ethernet MAC unit inside. The Xilinx ISE environment includes free LogiCORE IP called: Virtex-6 FPGA Embedded Tri-Mode Ethernet MAC Wrapper [6]. The IP core utilizes internal MAC unit, leaving the remaining FPGA resources free for other projects. An internal MAC unit is responsible for control of the data-link layer of the Ethernet standard. The physical layer is controlled by Marvell 88E1111 device (an Integrated 10/100/1000 Ultra Gigabit Ethernet Transceiver). The transceiver is connected directly to the FPGA on the evaluation board.

The MAC wrapper core has two communication interfaces: the first connected to the external ports to the Marvell device and the second accessible by the target device (application). The second interface, called AXI4-Stream User Interface, enables communication similar to interface provided by the WinPcap library. The developer transmits only the content of the Ethernet frame by the AXI4-Stream interface. The preamble, the SFD and the checksum are added by the embedded MAC wrapper. When receiving the frame the wrapper removes these fields from the passed data.[1] Unlike the WinPcap, the AXI4-Stream interface sends the data as a stream of bytes, not as a buffer. The error in the transmission is indicated if required after the whole frame is received/transmitted. For example: When the frame is corrupted and the FCS doesn't correspond to data received the user

---

[1] More precisely: transfer of the FCS field to user depends on the configuration of the wrapper, but in the presented solution it has been turned off.

gets first the whole frame and then the appropriate flag signalizes error. The frame should be dropped then.

The receiver and transmitter devices of the proposed protocol are described in VHDL language and synthesized using Xilinx ISE software. The structure of the receiver and the transmitter are determined by the AXI4-Stream interface on the one hand and on the other hand by the created $GF(2^q)$ LDPC decoder. The decoder defines own communication interface for data exchange. It requires the input data to be placed in the memory of defined size. When the memory is filled the data ready flag is set and codeword decoding starts. After the decoding the decoder sets the suitable output flags. The receiving module is able to overwrite data in the input memory of the decoder and the transmitting module is allowed to transmit the decoding result. It needs to be noticed, that the decoder output memory can be filled only if the transmitter is ready to send. It means that previous result was already sent by the Ethernet port.

The structure of the whole system with the designed transmission module is depicted in Fig. 2 (the control signals were omitted to save picture readability). The Ethernet wrapper has been generated using the LogiCORE tool. Than it has been included in the project and connected to the physical interface in 10/100 Mbps mode. The wrapper passes the RX and TX clocks from Marvell's transceiver to the AXI4-Stream interface. The frequency of both clocks is about 2.5/25 MHz and it is less than the 66 MHz clock available for the user on the evaluation board. These three clocks are used in particular units of the project. The RX and TX clocks are used by the units connected directly to the Ethernet wrapper. The rest of the project uses the user clock. Synthesis results of the implemented LDPC decoder, the main part of the project resulted in maximum clock frequency about 100 MHz (it changes slightly with parameters of the decoder) so we decided to use in the project the mentioned clock available directly in the board. The maximum clock of the decoder caused also resignation of the 1000 Mbps mode which requires higher clock rate.

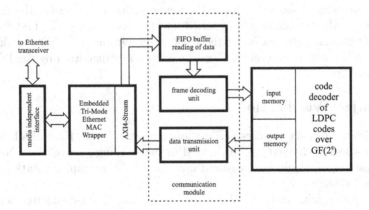

Fig. 2. The structure of hardware implementation of the communication module

The transmission module itself consists of three blocks. The first of them is the FIFO memory. It receives data from the Ethernet wrapper by the AXI4-Stream interface to subsequent pages of the memory. The size of the page is 2048 bytes. It is enough to save in the standard Ethernet frame of any size and the power of 2 value simplifies modulo operations in buffer implementation. If the jumbo frame is received on the Ethernet interface the successive bytes above the page size will overwrite the previous ones. The system doesn't accept the jumbo frames so the frame will be dropped anyway. The whole FIFO buffer has got 32 pages. Both values create a buffer long enough to save any frame received on the Ethernet interface before processing by the next unit. As it was stated before, the data are written to the FIFO with the speed of RX clock. When the whole frame is received and no error is indicated then a flag is set and the frame is analyzed by the frame decoder, the second unit of the module. It uses faster clock and the analysis time is shorter than filling of the FIFO.

The frame decoder reads the FIFO memory using different memory port and different clock speed than the memory has been written. The frame is analyzed in terms of the protocol described previously. If it is addressed to the FPGA and includes the appropriate tags defined in the protocol definition than the payload of the frame is copied from the FIFO buffer to the input memory of the decoder. The address of the data in the input memory is calculated from the part number field and the number of byte in the datagram. It is not necessary to send the parts by order. There is no explicit information about the end of the transmission block. If the size of data in received datagram is lower than the maximum defined or the number of datagram is equal to the limit than the end of the transmission block is detected. The frame decoder unit checks if all parts were received and, if so, initializes the main unit of the system: the $GF(2^q)$ LDPC decoder.

The decoder reads data from the input memory, decodes the codeword and writes the result to the output memory. Before the output memory is written the transmitter state must be checked. If the result of previous decoding was sent, the decoder is allowed to write the memory. After storing the the result in the memory, the transmitter is switched into the send mode. The first frame of the result includes also ACK for the pack of datagrams including the decoded codeword. If no ACK is received than the data is retransmitted by the PC after the defined timeout.

5   Project Verification

The presented project requires verification at several levels. First of all the proposed protocol itself needs to be checked in the working environment. Next the transmission rate would be measured and the hardware implementation would be verified then.

Two systems with installed WinPcap library have been configured with the sending application on one of them and the receiving application on the second. After proper verification of transmission of the individual frames with the Wireshark tool [14], the large file was transmitted between the systems. The about

1 GB file has been sent with average speed of 84 Mbps. It has been received completely without errors. Theoretical maximum speed of transmission the proposed protocol can achieve is about 96 Mbps on the 100 Mbps Ethernet link. It is achievable when all data are send without any retransmission and the Ethernet frame payload is full in every frame so the datagram uses maximum 1490 bytes payload. The assumption of that calculation is zero processing time of the frames before the acknowledgment would be sent. For comparison, the same file was sent between these computers using the FTP and SMB/CIFS protocols. Obtained speeds are about 90 Mbps and 82 Mbps respectively.

The protocol assumptions and its performance has been verified creating the PC to PC communication system. Afterwards the evaluation board has been connected to the PC and the protocol has been realized in the FPGA. Reading the above description of the protocol implementation in the FPGA it can be noticed, that the proposed protocol has been further simplified in the hardware. It has been made for two reasons: to reduce the consumption of resources in an FPGA device and because the communication of the PC with the hardware decoder runs in a specific way. The FPGA device is always used as a slave in the transmission system. The PC application sends data to be processed in the hardware decoder and waits for the decoding result. If timeout occurs, no result is received in the specified time, than the data are transmitted to the FPGA again. The block number field in datagram is used to link the result with the prepared input data. If some blocks are lost it is possible to retransmit particular data or intentionally remove them from the subsequent analysis. The decision of removing or retransmitting data depends on target application.

In the created non-binary LDPC code decoder one block of data corresponds to one codeword. The data transmitted to the FPGA include pack of probabilities for each symbol of the codeword and the decoded codeword itself is returned. If the result doesn't contain some of transmitted code words they wouldn't be processed by the next system units.

The resource utilization of the project presented depends on the decoder configuration. The major parameters are the codeword length, the code rate, the size of Galois Field and the distribution of non-zero symbols in the parity check matrix of the code. The minor ones are the parameters describing the precision of numbers inside the decoder. As an example the 2400 bit long LDPC codes with 0.5 code rate and different symbol size have been used. In Table 1 the results of synthesis are presented for the whole system and the system without the decoder unit.

It is hard to directly measure the data transmission speed between the PC and the FPGA. However, it can be estimated from the available measurements. The time of processing one data block containing defined number of the codewords (t_0) can be measured. That time includes: the time of running external application from the matlab (t_1), the time of data transmission between the PC and the FPGA (t_2) and the decoding time (t_3). The decoding time was estimated using simulation in the Xilinx ISE environment. It was assumed that the time of

Table 1. Synthesis results for the LDPC(600,300,2^4) and LDPC(342,171,2^7) codes

Resource name	LDPC(600,300,2^4)		LDPC(342,171,2^7)	
	The complete system	The system without decoder	The complete system	The system without decoder
Slices	15151 (10%)	412 (0%)	125397 (83%)	543 (0%)
Block RAM	43 (10%)	16 (3%)	87 (20%)	16 (3%)
DSP48E1s	32 (4%)	0 (0%)	256 (33%)	0 (0%)

running application is constant for each transmission. The times of the transmission and the decoding depends on the number of codewords in the data block N: $t_0 = t_1 + Nt_2 + Nt_3$. So changing the length of the transmitted data block the t_1 and t_2 values can be calculated. After the calculations the speed of transmission between the PC and the FPGA has been estimated to about 30 Mbps.

6 Conclusion

The idea of a simple communication over the Ethernet and its implementation in operating system and inside FPGA device was presented in the paper. The data are transmitted using direct access to the Ethernet frame on both ends of the link. The transmission protocol with datagrams numbering, acknowledgments and the ability to retransmit was proposed. The datagrams were encapsulated directly into Ethernet frames and the network layer was omitted. Reliability of the protocol was confirmed building the PC to PC transmission system. Its performance is good but slightly lower than expected. It is probably caused by using the WinPcap library, which gives the access to data-link layer with some overhead. It was observed when switching between send and receive mode.

The resource utilization in FPGA is insignificant compared to requirements of the decoder. Putting all the blocks of the system together the working multi-platform configuration was configured and successfully run. The distorted data generated in the software environment on the PC has been decoded by the external hardware device and it was possible to compare the results of fixed point hardware decoder with floating point software one. The obtained transmission speed is only about the third part of the interface capabilities but the estimation still contains excess associated with switching between the transmit and receive mode. Average transmission time can be decreased by sending longer codewords or by including more than one codeword in the transmission block. Unfortunately, the second option requires a design of additional logic in the FPGA to adjust different memory sizes of the communication module and the decoder unit.

The project has been implemented on a particular platform, the Matlab in Windows operating system and the Xilinx Virtex-6 FPGA, but it is not limited to only this one. The application for data transmission and reception can be run on any system with WinPcap library installed. Implementation in unix-like systems can be done easily, based on Windows code, using libpcap library. It can be also done by direct access to sockets at data-link layer, which is possible in Linux. Proper implementation could increase the performance of proposed protocol. The VHDL hardware description is useful for any system including MAC wrapper with the AXI4-Stream interface. On the second end the input and the output memory of defined size need to be made available.

Acknowledgement. This work was supported by the Polish National Science Centre under Grant number 4698/B/T02/2011/40.

References

1. Davey, M.C., MacKay, D.: Low-Density Parity Check Codes over GF(q). IEEE Comunications Letters 2(6), 165–167 (1998)
2. Sułek, W., Dziwoki, G., Kucharczyk, M.: GF(q) LDPC decoder design for FPGA implementation. In: 10th Annual IEEE Consumer Communications and Networking Conference, Las Vegas, pp. 445–450 (2013)
3. Sułek, W.: Pipeline processing in low-density parity-check codes hardware decoder. Bulletin of the Polish Academy of Sciences Technical Sciences 59(2), 149–155 (2011)
4. Falcao, G., et al.: Shortening design time through multiplatform simulations with a portable OpenCL golden-model: the LDPC decoder case. In: 20th IEEE International Symposium on Field-Programmable Custom Computing Machines, pp. 224–231 (2012)
5. Xilinx Inc.: ML605 Hardware User Guide. UG534 (v1.8) (October 2012), http://www.xilinx.com/support/documentation/boards_and_kits/ug534.pdf
6. Xilinx Inc.: Virtex-6 FPGA Embedded Tri-Mode Ethernet MAC – User Guide. UG368 (v1.3) (March 2011), http://www.xilinx.com/support/documentation/user_guides/ug368.pdf
7. IEEE Computer Society: IEEE Std 802.3-2008 – Part 3: Carrier sense multiple access with Collision Detection (CSMA/CD) Access Method and Physical Layer Specifications. New York, USA (2008), http://standards.ieee.org/about/get/802/802.3.html
8. Sarangi, A., MacMahon, S.: LightWeight IP (lwIP) Application Examples. XAPP1026 (v3.2) (October 2012), http://www.xilinx.com/support/documentation/ application_notes/xapp1026.pdf
9. 1-CORE Technologies: Soft CPU Cores for FPGA, http://www.1-core.com/library/digital/soft-cpu-cores/
10. Dollas, A., Ermis, I., Koidis, I., Zisis, I., Kachris, C.: An Open TCP/IP Core for Reconfigurable Logic. In: 13th Annual IEEE Symposium on Field-Programmable Custom Computing Machines, Napa, pp. 297–298 (2005)
11. Löfgren, A., Lodesten, L., Sjöholm, S., Hansson, H.: An analysis of FPGA-based UDP/IP stack parallelism for embedded Ethernet connectivity. In: 23rd IEEE NORCHIP Conference, 94–97. Oulu, Finland, pp. 94–97 (2005)
12. Alachiotis, N., Berger, S.A., Stamatakis, A.: Effcient PC-FPGA Communication over Gigabit Ethernet. In: 10th IEEE International Conference on Computer and Information Technology, Bradford, UK, pp. 1727–1734 (2010)
13. Riverbed Technology: WinPcap – The industry-standard windows packet capture library, http://www.winpcap.org/
14. Riverbed Technology: Wireshark – The world's foremost network protocol analyzer, http://www.wireshark.org/

Evaluation and Development Perspectives
of Stream Data Processing Systems

Marcin Gorawski[1,2], Anna Gorawska[2], and Krzysztof Pasterak[2]

[1] Wroclaw University of Technology, Institute of Computer Science,
Wybrzeze Wyspianskiego 27, 50-370 Wroclaw, Poland
`Marcin.Gorawski@pwr.wroc.pl`
[2] Silesian University of Technology, Institute of Computer Science,
Akademicka 16, 44-100 Gliwice Poland
`{Marcin.Gorawski,Anna.Gorawska,Krzysztof.Pasterak}@polsl.pl`

Abstract. The following paper describes some common aspects of stream data processing systems. The paper consists of two main parts – first showing the short description, tests results and conclusions of an implemented system – the AGKPStream, while the second part focuses on proposed solutions, created upon experiences gained during development of mentioned system, as well as knowledge collected during learning about some concepts of a StreamAPAS system. The first discussed issue is a tuple construction – basic data representation. It concerns tuple time model, tuple schema and a tuple decorator. Afterwards, the stream query and scheduling problems are described.

Keywords: stream data processing, tuple, tuple time model, tuple schema, joined tuples decorator, stream query, stream schedulers.

1 Introduction

Designing a system for events monitoring, which is efficient and scalable is a major interest in recent studies. Processing multiple data connected with critical events is a crucial issue. The increasing number of data generated by on-line sources created unpredictable data streams processing a challenging problem that cannot be solved using traditional data bases. Therefore it is highly probable that the paradigm of data stream processing will became an important part of managing such data volumes.

The Data Stream Processing Management System (DSMS) assumes that data sources, called data streams, produce data continuously in an unpredictable matter. Data streams are considered to be open-ended and theoretically unbounded in size. The system does not have any control over data volumes arrival order and structure (schema), since DSMS is usually connected to remote sources and sinks. Moreover, once a tuple is taken from a data stream it is processed and then archived or discarded. Unless there is a data warehouse [1–10] storing historical data it is almost impossible to retrieve the processed tuple. Data stream processing systems are considered to support rapidly changing data sources.

A. Kwiecień, P. Gaj, and P. Stera (Eds.): CN 2013, CCIS 370, pp. 300–311, 2013.
© Springer-Verlag Berlin Heidelberg 2013

Detailed description of the data stream processing paradigm can be found in [11, 12, 3, 13–17].

Main aspects of the stream data system described in this paper concern tuple time model, tuple schema and tuple decorator. Next discussed issues are stream queries and scheduling policies.

First part of the paper shows main ideas of a prototype Data Stream Management System called AGKPStream. The goal of the research is to assure continuity of reading data from the stream sources. Minimization of delay between measuring and storing the data is main criteria of evaluation. If it is possible, we want to assure successful recovery of the interrupted processing without losing any stream data. Next part is dedicated to presentation of new ideas, which are planned to be implemented in a forthcoming system [18–24].

2 Implemented Aspects of Stream Data Processing Systems

The AGKPStream system is based on the data stream processing paradigm. Therefore, all fundamental ideas and conditions consistent with this model had to be met. This part focuses on several interesting aspects of the AGKPStream system, which results from applying mentioned paradigm. Similarly to previously described system, a StreamAPAS system allows user to define temporal data analysis as well as positive/negative tuples processing. Some of further mentioned aspects are also described in [25, 26].

During all tests presented in Sect. 2, three main metrics were measured. First, a tuple response time, represents the time interval between tuple departure and arrival time. Next metric is a tuple slowdown [11], which is a ratio between tuple response time and ideal tuple response time (i.e. the time tuple would be processed by the system without waiting in queues). This metric illustrates how big delay each tuple is experiencing during processing. Algorithm used to measure ideal processing time can be found in [20]. The last tested metric was memory usage calculated by measuring both the number of tuples present in the system and total system memory consumption.

2.1 Tuple Time Model

The AGKPStream system adopts a temporal tuple time model. Each tuple contains, apart from data (attributes), two timestamps – the beginning and the end of tuples life time, determining the slice of time, where event described by tuple is valid. There were some operators, which had to store their input tuples in internal data structures. Those structures were cleaned each time new tuple was taken by those operators. During such cleaning operation the remove condition was based on the tuples life time – tuples with end timestamp value less than current system time were deleted.

The temporal tuple time model was satisfying for the AGKPStream system applications. It is worth noticing that the simplicity of the selected model distinguishes it among other time models. The biggest asset to the temporal time model is that, according to [25, 26], it reduces the amount of transmitted data doubly.

2.2 Tuple Schema

In the AGKPStream system, a tuple schema was assigned to the stream. Each tuple contained data, but proper interpretation of them could be obtained only when the tuple was stored or read from the particular stream. As a consequence, an isolated tuple could not be used (or even its existence was indefinite) without a stream. The only way to properly extract the desired attributes value was to address tuples attribute using name taken from the schema.

Tuple schemas for all streams were calculated before appropriate data processing. When the first tuple arrived to the system, all streams already knew their schemas. Due to such strategy, all query defining and system configuration errors are detected before the data processing start.

Other advantage is the construction of a projection operator. It was intended to trim tuple schemas. Since all schema calculations were performed before data processing, i.e. in the system's initialization phase, projection operator simply set trimmed input schema to its output stream before the first tuple arrived to the system. After that, its work was equal to sending input tuples to the output. They contain untrimmed data indeed, although the view of that data was trimmed by the schema located in the stream. Besides all these advantages, schema-in-stream strategy had also some disadvantages, such as complications of union operator (detailed description can be found at [11]).

2.3 Tuple Decorator

A joined tuples decorating mechanism was implemented in the AGKPStream system and was used as a part of join operators algorithm. The idea was to substitute two input tuples with a single output tuple of mentioned operators without any data copying. In the case of multiple subsequent joins, the final structure of the output tuple was formed by a number of decorators, pointing each other.

The attribute mapping in a tuples decorator was based on a simple strategy of finding first matching pattern. The method is current until the condition of unique attribute names in each decorated tuple is preserved, however, this assumption cannot be always assured. Therefore, the AGKPStream system had been provided with an attributes name prefix mechanism, which was based on the origin of the tuple (i.e. a tuples source). That strategy allowed to ensure that each attribute of the two joined tuples would be unique (unless join operation was performed over two copies the same tuple).

The aim of performed tests of join operators, was to verify the thesis which assumes that using a joined tuples decorator instead of a simple data copying approach to newly created output tuple is better in the issue of memory usage, due to limitation of total tuple number. Furthermore, the secondary aim was to show the differences between other metrics (such as tuple response time or tuple slowdown). This chart (Fig. 1) clearly shows that using the joined tuples decorating mechanism visibly reduces total memory usage of tested cross-join query. In addition, the amplitude of changes is significantly less in contrast to the copying and creating new tuple strategy. Therefore it has positive influence on the stability of the system.

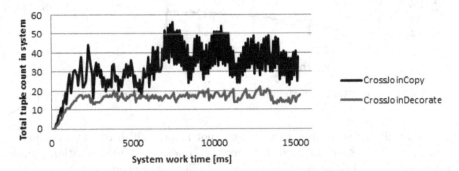

Fig. 1. Joined tuples strategies: total tuple count in system vs. system work time [ms]

The following charts (Fig. 2 and 3) reflect the results of comparison between those two joining strategies with respect to measured tuple response time and tuple slowdown.

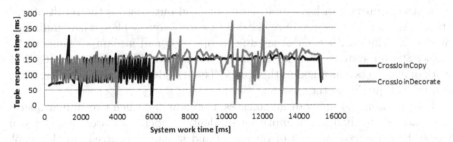

Fig. 2. Joined tuples strategies: tuple response time [ms] vs. system work time [ms]

Tuple slowdowns in both strategies are comparable, in favor of decorating strategy. However, the amplitude changes more visibly than in the copying strategy, which is quite balanced after the initial peak. Thus, it is hard to point a better strategy with respect to tuples slowdown optimization.

In all considered and tested implementations of join operator (i.e. cross-join, equi-join and theta-join), the tuple decorating strategy consumes less memory than corresponding copying technique. Therefore, in case of memory usage optimization, it is suggested to use joined tuples decoration strategy.

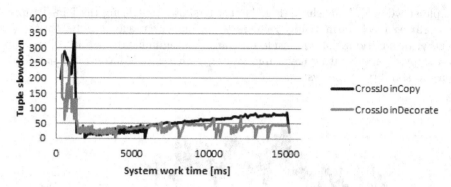

Fig. 3. Joined tuples strategies: tuple slowdown vs. system work time [ms]

Test results of two other metrics have not shown which of these compared strategies were better. In each of tested join operator implementations, tuple response time and tuple slowdown have had different dependences.

2.4   Queries

The AGKPStream system allows to define and create stream queries only before system's start. It was done during manual editing of multiple configuration files.

Each stream query is processing tuples in separation of the other queries, both with respect to scheduling algorithms and operator sharing. Furthermore, each stream query can have multiple inputs and outputs. As the result of such structure, defining query is more like defining whole query plan, where individual queries interlace with each other due to operator and stream sharing. In the AGKPStream system, each stream query has a local scheduler, which controls work of operators contained in this query. There is also a global scheduler, which controls all queries in the system.

This solution had many disadvantages. For example, when two similar queries (i.e. queries that had a nonempty set of identical operators) were registered in the system, the redundancy of operators and streams occurred (in the whole query plan, there was more than one element, which had been performing the same operations). However, there was a possibility of manual optimization of such query plan, by creating a double-output query, where data collected from each input would be identical as in two regular queries. Such joined query should not contain any redundant operators nor streams, although the end user skills need to be involved in an optimization process.

2.5   Schedulers

There were six schedulers implemented in the AGKPStream system. They could be divided into categories: simple schedulers (FIFO, Round Robin) [27, 28] and priority schedulers (Greedy, HR, HNR, MTIQ) [11, 27, 28]. Detailed information on their algorithms was described in [29]. We distinguish two levels of scheduling policies: contained operator-level and query-level scheduling. Each query had a local scheduler, while all queries were controlled by a global scheduler.

The aim of performed tests was to choose the best priority scheduler of all four tested strategies (Greedy, HR, HNR, MTIQ). All of them were based on Round Robin scheduling policy. Three metrics were measured: tuple response time, tuple slowdown and memory usage (showed as total tuple count and total memory usage). Undoubtedly, the best priority scheduler was MTIQ according to Fig. 4–7. It is probably caused by simplicity of priority counting algorithm. The three remaining priority schedulers (Greedy, HR, HNR) were determining operator priority by performing complex calculations over selected operator statistics (such as selectivity and cost), which were often derived by expensive recursive floating point operations. The MTIQ scheduling strategy used stream tuple count. This statistic's value is simply actual buffer size.

3   Proposed Solutions

In the previous part, several aspects of the data stream processing AGKPStream system was presented. The analysis presented in Sect. 2 led to conclusions and ideas presented in following paragraphs.

3.1   Tuple Time Model

Knowledge of tuples life time was required in the AGKPStream system all of the time. When a new tuple arrived to the system, its end timestamp had to be known. The impossibility of changing that value and also lacks of such information in real life causes some difficulties for a certain area of applications. E.g. real-time traffic monitoring systems, where the character of events described by tuples is not determined (especially in terms of tuple life time). The need for creating more universal system led to development of different, more flexible tuple time model.

In the following system it is proposed to implement a punctuation-negative tuple time model, which assumes an existence of two kinds of tuples: regular data tuples and special punctuation tuples. The first one contain data, determined by proper schema. The punctuation tuples inform the system about groups of data tuples, which are outdated and have to be deleted from system. This model combines the features of punctuation model with the negative tuples model (signed tuples model).

In such model, operators would perform their data structure cleaning only when a punctuation tuple arrives (in the previous system, cleaning was done before processing each tuple). Moreover, there is no need for knowing the moment of tuple deactivation, i.e. tuple end timestamp is set when a new tuple is created, because it is determined by creation of the other – punctuation tuple. In graphical representation, such tuples (data and punctuation) can be described as points. In temporal model, tuples could be represented as lines.

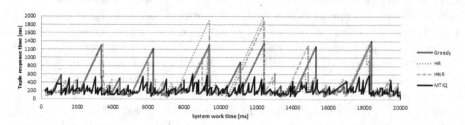

Fig. 4. Priority schedulers: tuple response time [ms] vs. system work time [ms]

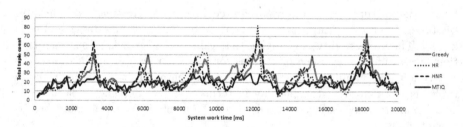

Fig. 5. Priority schedulers: tuple slowdown vs. system work time [ms]

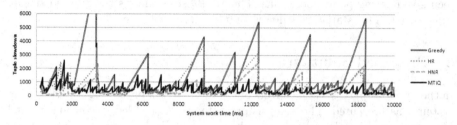

Fig. 6. Priority schedulers: total tuple count in system vs. system work time [ms]

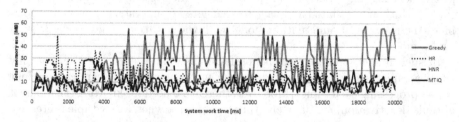

Fig. 7. Priority schedulers: total memory usage [MB] vs. system work time [ms]

3.2   Tuple Schema

In the previous section, a tuple schema realization in the AGKPStream system was described. Even though it had advantages mentioned in Sect. 2.2, such approach complicates the system. Checking validity of each stream and operator (with respect to tuple schema) with such precision led to unexpected difficulties.

The proposed solution reduces checking mechanisms for operators and streams in favor of increasing their fault tolerance. In consequence each operator should adjust to input data (with reasonable limits), e.g. when it needs to extract nonexistent attribute, it is better not to throw exception and stop the whole system, but to attempt to process that tuple as precisely as possible.

3.3   Tuple Decorator

Basing on experiences gained during creating and testing the AGKPStream system, joined tuples decorating strategy seems to be the most promising approach. Therefore, it is planned to implement this mechanism in the forthcoming stream data processing system.

In order to overcome problems connected with the proper tuple attributes addressing, it is proposed to extend each tuple with a special field, determining its source, called tuple origin (in the AGKPStream system, each attribute had special prefix). Moreover, tuple decorator ought to map tuple origin values to the decorated tuple pointers. Thus, when decorator is given a full attribute name (i.e. attribute name preceded by origin value), requested attribute value will be extracted from origin defined tuple, which is also solution for ambiguous attribute name in joined tuples problem. Furthermore, an origin based tuple mapping mechanism is flexible for multiple tuples being decorated, instead of only two tuples limit in the AGKPStream system.

3.4   Queries

In the forthcoming system, it is planned to introduce a different stream query model. First of all, proposed model ought to handle dynamic query registration and removal. Each stream query should have one output – representing user desired data. As input data for queries, it is assumed that each system's source (input stream, which is connected to the system) as well as actual running query results could be used. In terms of scheduling, it is planned to apply one global scheduler to all operators working in the system.

To solve operator and stream redundancy problem, a dynamic query merging and a splitting mechanism is intended to be used. The aim of such operation is to provide a query plan without redundant elements, but containing shared operators and streams [30]. In the AGKPStream system, similar result could be obtained only manually, but in future system it ought to be a regular procedure.

Two basic query operations should be defined. The result of merging two stream queries is a query, where all operators and streams are unique. It means that for each pair of identical elements in these two merged queries, the final query will contain only one of them. It is also important to define when two elements (operators, streams) are identical. In the AGKPStream system, operators were distinguished by name. There was no such mechanism for streams. The forthcoming system will assume that two operators are identical while performing the same operations on the same data streams. Therefore two streams are identical when they are connecting identical operators.

On the other hand, when a stream query is no longer needed it should be removed from the query plan. The result of splitting one query with a query pattern are two separate queries. When a primary query contains some shared elements, they ought to be split into two separate copies of the same element, each in one of result queries. When removing an unnecessary query from a query plan, a primary query becomes query plan, while removed query is a query splitting pattern. One of the result queries is a new structure of a query plan, while the other is equal to the splitting pattern and is no longer used.

Currently developed system will implement query merging and splitting mechanism using a special stream attribute called sharing factor. It is simply the number of queries sharing that stream. Each query merging operation increments this value (only in affected elements) and each split operation decrements it. When sharing factor is equal to 0 it means that no queries share this stream and it could be removed from the query, also with operator connected to its output. When new element is inserted to the query, its sharing factor is equal to 1.

3.5   Schedulers

In the future, it is planned to limit the number of complex priority scheduler statistics and instead of them use simpler one. Most of all it is the stream size and optionally tuple wait time (time which each tuple has to wait to be processed) that will be used. These values do not require any calculations in order to obtain expected data. It is planned to test many different priority ratios, which can be obtained from these statistics.

In the AGKPStream system, each stream query had its own local scheduler. In the forthcoming system that strategy cannot be applied because of sharing query elements – it is impossible to determine clear borders between particular queries. Thus, it is proposed to apply only one global scheduler to all operators contained in query plan (which was mentioned in the previous part).

Different scheduling policies are considered, but each is based on simple statistics. The general rule is to call operator with the highest priority. Quite similar strategy is to creating a group of operators which are called periodically (like in Round Robin strategy). Operators which belong to such group are chosen from all operators and the choice is based on priority. It is also possible to split all working operators to many groups and to call all of them, but the frequency of calls depends on group priority, which depends on included operators priority.

4 Conclusions

In this paper we introduce interesting aspects of data stream processing that are connected to the prototype AGKPStream system or the StreamAPAS system [25]. Described ideas are the result of tests and conclusions drawn from mentioned systems. Although in the AGKPStream temporal model was satisfying, in the forthcoming work we consider punctuation-negative tuple time model as more suitable and flexible. Described measurements of joined tuples decorating mechanism performance in comparison to the results of copying and creating strategy shows that using tuple decorator has positive influence on system's efficiency. Disadvantage of the AGKPStream system was that queries were allowed to be defined only in system's initialization phase. A new system ought to handle dynamic query registration and removal. Mentioned features will make continuous system more flexible and efficient. With this motivation in mind it is likely that the forthcoming system will be more advanced than the previous one, e.g. by applying the distributed model [18, 19, 31, 21, 22] instead of a single application. In further future, it is also planned to include presented ideas in development of the stream data warehouse [1, 7, 22, 8, 23, 24]. Applying data stream processing mechanism is also being considered in addition to problems which descriptions can be found in [32–36, 29, 37–39].

References

1. Abhirup, C., Ajit, S.: A Partition-based Approach to Support Streaming Updates over Persistent Data in an Active Data Warehouse. In: Proceedings of the 2009 IEEE International Symposium on Parallel & Distributed Processing, IPDPS 2009, pp. 1–11. IEEE Computer Society, Washington, DC (2009)
2. Gorawski, M.: Extended Cascaded Star Schema and ECOLAP Operations for Spatial Data Warehouse. In: Corchado, E., Yin, H. (eds.) IDEAL 2009. LNCS, vol. 5788, pp. 251–259. Springer, Heidelberg (2009)
3. Gorawski, M.: Time complexity of page filling algorithms in Materialized Aggregate List (MAL) and MAL/TRIGG materialization cost. Control and Cybernetics 38(1), 153–172 (2009)
4. Gorawski, M., Gorawski, M.: Balanced spatio-temporal data warehouse with RMVB, STCAT and BITMAP indexes. In: PARELEC 2006: International Symposium On Parallel Computing In Electrical Engineering, pp. 43–48 (2006)
5. Gorawski, M., Malczok, R.: Indexing Spatial Objects in Stream Data Warehouse. In: Nguyen, N.T., Katarzyniak, R., Chen, S.-M. (eds.) Advances in Intelligent Information and Database Systems. SCI, vol. 283, pp. 53–65. Springer, Heidelberg (2010)
6. Gorawski, M., Marks, P.: Checkpoint-based resumption in data warehouses. In: Software Engineering Techniques: Design for Quality. IFIP, vol. 227, pp. 313–323. Springer, US (2006)
7. Gorawski, M., Marks, P.: Resumption of data extraction process in parallel data warehouses. In: Wyrzykowski, R., Dongarra, J., Meyer, N., Waśniewski, J. (eds.) PPAM 2005. LNCS, vol. 3911, pp. 478–485. Springer, Heidelberg (2006)
8. Gorawski, M., Morzy, T., Wrembel, R.: Special Issue on: Techniques of Advanced Data Processing and Analysis Introduction. Control and Cybernetics 38(1) (2009)

9. Kozielski, S., Wrembel, R. (eds.): New Trends in Data Warehousing and Data Analysis. Annals of Information Systems, vol. 3. Springer, US (2009)
10. Morzy, T.: Extraction, Transformation, and Loading Processes. In: Data Warehouses and Olap: Concepts, Architectures and Solutions, pp. 88–110 (2007)
11. Brian, B., Shivnath, B., Mayur, D., Rajeev, M., Dilys, T.: Operator scheduling in data stream systems. VLDB J. 13(4), 333–353 (2004)
12. Gorawski, M.: Advanced Data Warehouses. Habilitation, Studia Informatica 30(3B). Pub. House of Silesian Univ. of Technology (2009)
13. Gorawski, M., Chrószcz, A.: Synchronization Modeling in Stream Processing. In: Morzy, T., Härder, T., Wrembel, R. (eds.) Advances in Databases and Information Systems. AISC, vol. 186, pp. 91–102. Springer, Heidelberg (2013)
14. Gorawski, M., Malczok, R.: Towards stream data parallel processing in spatial aggregating index. In: Wyrzykowski, R., Dongarra, J., Karczewski, K., Wasniewski, J. (eds.) PPAM 2007. LNCS, vol. 4967, pp. 209–218. Springer, Heidelberg (2008)
15. Gorawski, M., Malczok, R.: Answering Range-Aggregate Queries over Objects Generating Data Streams. In: Kitagawa, H., Ishikawa, Y., Li, Q., Watanabe, C. (eds.) DASFAA 2010. LNCS, vol. 5982, pp. 436–439. Springer, Heidelberg (2010)
16. Gorawski, M., Marks, P.: Distributed stream processing analysis in high availability context. In: Proceedings of the Second International Conference on Availability, Reliability and Security, ARES, pp. 61–68 (2007)
17. Roger, S.B., Jonathan, G., Mohamed, H.A., Hong, M.: Consistent Streaming Through Time: A Vision for Event Stream Processing. In: Third Biennial Conference on Innovative Data Systems Research, CIDR 2007, Asilomar, CA, USA (2007)
18. Gorawski, M.: Architecture of Parallel Spatial Data Warehouse: Balancing Algorithm and Resumption of Data Extraction. In: Proceedings of the 2005 conference on Software Engineering: Evolution and Emerging Technologies, pp. 49–59. IOS Press, Amsterdam (2005)
19. Gorawski, M., Chroszcz, A.: Optimization of operator partitions in stream data warehouse. In: Proceedings of the ACM 14th international workshop on Data Warehousing and OLAP, pp. 61–66. ACM, New York (2011)
20. Gorawski, M., Gorawski, M.: Modified R-MVB tree and BTV algorithm used in a distributed spatio-temporal data warehouse. In: Wyrzykowski, R., Dongarra, J., Karczewski, K., Wasniewski, J. (eds.) PPAM 2007. LNCS, vol. 4967, pp. 199–208. Springer, Heidelberg (2008)
21. Gorawski, M., Marks, P.: Towards reliability and fault-tolerance of distributed stream processing system. In: DEPCOS-RELCOMEX 2007 International Conference on Dependability of Computer Systems, pp. 246–253. IEEE Computer Society, Washington, DC (2007)
22. Gorawski, M., Marks, P., Gorawski, M.: Collecting data streams from a distributed radio-based measurement system. In: Haritsa, J.R., Kotagiri, R., Pudi, V. (eds.) DASFAA 2008. LNCS, vol. 4947, pp. 702–705. Springer, Heidelberg (2008)
23. Waas, F., Wrembel, R., Freudenreich, T., Theile, M., Koncilia, C., Furtado, P.: On-Demand ELT Architecture for Right-Time BI: Extending the Vision. International Journal on Data Warehousing and Mining (to appear, 2013)
24. Wrembel, R.: A Survey of Managing the Evolution of Data Warehouses. IJDWM 5(2), 24–56 (2009)
25. Gorawski, M., Chroszcz, A.: StreamAPAS: Query Language and Data Model. In: Proceedings of the Third International Conference of Complex, Intelligent and Software Intensive Systems, CISIS 2009, pp. 75–82. Springer, Heidelberg (2009)

26. Gorawski, M., Chrószcz, A.: Query Processing Using Negative and Temporal Tuples in Stream Query Engines. In: Szmuc, T., Szpyrka, M., Zendulka, J. (eds.) CEE-SET 2009. LNCS, vol. 7054, pp. 70–83. Springer, Heidelberg (2012)

27. Mohamed, A.S., Panos, K.C., Alexandros, L., Kirk, P.: Efficient scheduling of heterogeneous continuous queries. In: Proceedings of the 32nd International Conference on Very Large Data Bases, VLDB 2006, pp. 511–522. Endowment (2006)

28. Timothy, M.S., Bradford, P., Zhu, Y., Luping, D., Elke, A.R.: An Adaptive Multi-Objective Scheduling Selection Framework for Continuous Query Processing. In: Proceedings of the 9th International Database Engineering & Application Symposium, IDEAS 2005, pp. 445–454. IEEE Computer Society, Washington, DC (2005)

29. Jestratjew, A., Kwiecien, A.: Performance of HTTP Protocol in Networked Control Systems. IEEE Trans. Industrial Informatics 9(1), 271–276 (2013)

30. Patroumpas, K., Sellis, T.: Subsuming multiple sliding windows for shared stream computation. In: Eder, J., Bielikova, M., Tjoa, A.M. (eds.) ADBIS 2011. LNCS, vol. 6909, pp. 56–69. Springer, Heidelberg (2011)

31. Gorawski, M., Marks, P.: Fault-tolerant distributed stream processing system. In: International Workshop on Database and Expert Systems Applications – DEXA, pp. 395–399 (2006)

32. Gorawski, M., Malczok, R.: AEC Algorithm: A Heuristic Approach to Calculating Density-Based Clustering *Eps* Parameter. In: Yakhno, T., Neuhold, E.J. (eds.) ADVIS 2006. LNCS, vol. 4243, pp. 90–99. Springer, Heidelberg (2006)

33. Gorawski, M., Malczok, R.: Towards automatic *Eps* calculation in density-based clustering. In: Manolopoulos, Y., Pokorný, J., Sellis, T.K. (eds.) ADBIS 2006. LNCS, vol. 4152, pp. 313–328. Springer, Heidelberg (2006)

34. Gorawski, M., Marks, P.: Towards automated analysis of connections network in distributed stream processing system. In: Haritsa, J.R., Kotagiri, R., Pudi, V. (eds.) DASFAA 2008. LNCS, vol. 4947, pp. 670–677. Springer, Heidelberg (2008)

35. Gorawski, M., Lorek, M., Gorawska, A.: CUDA Powered User-Defined Types and Aggregates. In: International Workshop on Engineering Object-Oriented Parallel Software (IEEE AINA_EOOPS-2013). IEEE CS (to appear, 2013)

36. Jestratjew, A., Kwiecień, A.: Using Cloud Storage in Production Monitoring Systems. In: Kwiecień, A., Gaj, P., Stera, P. (eds.) CN 2010. CCIS, vol. 79, pp. 226–235. Springer, Heidelberg (2010)

37. Kwiecień, A., Sidzina, M.: Dual Bus as a Method for Data Interchange Transaction Acceleration in Distributed Real Time Systems. In: Kwiecień, A., Gaj, P., Stera, P. (eds.) CN 2009. CCIS, vol. 39, pp. 252–263. Springer, Heidelberg (2009)

38. Kwiecień, A., Opielka, K.: Industrial Networks in Explosive Atmospheres. In: Kwiecień, A., Gaj, P., Stera, P. (eds.) CN 2011. CCIS, vol. 160, pp. 367–378. Springer, Heidelberg (2011)

39. Skrzewski, M.: Analyzing Outbound Network Traffic. In: Kwiecień, A., Gaj, P., Stera, P. (eds.) CN 2011. CCIS, vol. 160, pp. 204–213. Springer, Heidelberg (2011)

The Use of a Cloud Computing
and the CUDA Architecture
in Zero-Latency Data Warehouses

Marcin Gorawski[1,2], Damian Lis[1,*], and Michal Gorawski[3]

[1] Silesian University of Technology, Institute of Computer Science,
ul. Akademicka 16, 44-100 Gliwice, Poland
{Marcin.Gorawski,Damian.Lis}@polsl.pl
[2] Wroclaw University of Technology, Institute of Computer Science,
Wybrzeze Wyspianskiego 27, 50-370 Wroclaw, Poland
Marcin.Gorawski@pwr.wroc.pl
[3] Institute of Theoretical and Applied Informatics, Polish Academy of Sciences,
Baltycka 5, 44-100 Gliwice, Poland
mgorawski@iitis.pl

Abstract. The growing importance of data warehousing [1–3] and the need to provide up-to-date information, changed procedures of data processing [4–8]. Classic data warehouses which are based on a traditional ETL process, proved to be ineffective and limited further development, due to the need of time-sharing of an access time between updates and analysis [9, 10]. Introduction of the zero-latency data warehouse, solved the problem of data mining time limit, however it enforces the need to use larger computing power for processing updates and queries in the ETL process. The article presents two ETL systems for zero-latency data warehouses which implement the WINE-HYBRIS algorithm. The first ETL system processes tasks in CUDA and CPU architectures, while the second uses Cloud Computing. The purpose of the article is to describe advantages and disadvantages of each solution.

Keywords: ETL, CUDA, WINE-HYBRIS, zero-latency data warehouse, cloud computing, Windows Azure.

1 Introduction

The design of a zero-latency data warehouse [11–15], based on a traditional ETL process (Extract, Transform, Load) [16–20] proved to be inefficient. The distribution of the access time between data updates and data analysis is simply ineffective. Due to the continuous flow of data, and the need to receive up-to-date information, each modification of source data or change of the zero-latency data warehouse environment must be applied automatically and instantly. Such

* Project co-financed by the European Union under the European Social Fund. Project no. UDA-POKL.04.01.01-00-106/09.

A. Kwiecień, P. Gaj, and P. Stera (Eds.): CN 2013, CCIS 370, pp. 312–322, 2013.
© Springer-Verlag Berlin Heidelberg 2013

action is not possible using classical ETL process. Therefore, the main problems are: (a) time-sharing of the access time, and (b) data loading method.

In this paper we introduce the data extraction system, created to support zero-latency data warehouse based on the WINE-HYBRIS algorithm. The first version is based on two different architectures-CUDA and CPU, the second version is based on the computing cloud. Presented systems are based on the WINE [21] and the LEMAT [22] solutions. Systems utilize the principle of an effective data loading, which involves loading only the necessary data (use of actualizations prioritization). Such approach allows to relieve the data warehouse resources and increase its availability. This increases the number of users queries processed and assures that obtained data keep the proper freshness. Both versions – the CUDA architecture as well as the Cloud Computing aim to increase the efficiency of the queries processing and data loading in the zero-latency data warehouse environment.

2 ETL Process in a Zero-Latency Data Warehouse

The standard ETL process, incorporated in classical data warehouses, consists of: (1) extracting data from underlying sources, (2) cleaning and transforming, and (3) loading data to the data warehouse [22]. However, such ETL process is not appropriate for the zero-latency data warehouse, because during the actualization time frame, data warehouse is offline [9]. Also, during data update (loading) time, all updates are initialized sequentially, regardless of the actual users demand. During queries processing time, no update is done and approved, which leads to processing out of date data and obtaining less quality information.

The introduction of the advanced ETL process, fixes issues related to the use of independent timelines for updates and analysis (Fig. 1), and introduces a common timeframe in which updates and queries can be processed. By developing this solutions, it is possible to select which updates must be completed before processing a user's query, in order to maintain the required freshness of data (respectively for the current response to a query).

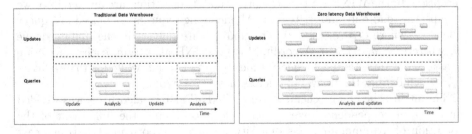

Fig. 1. The division of time for updates and queries in data warehouses

3   WINE-HYBRIS Algorithm

The WINE-HYBRIS initiated a new era in the management of ETL processes in zero-latency data warehouses. The use of Workload Balancing Unit (WBU) (Fig. 2), controls all operations related to the processing and analysis of both updates and queries in the data warehouse (DW).

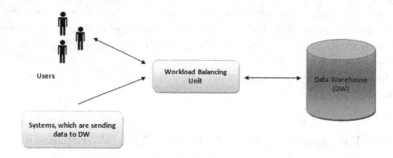

Fig. 2. The use of the Workload Balancing Unit in data warehouses

The base that was used to create the WINE-HYBRIS algorithm is a two-level scheduling WINE algorithm. Using the particular characteristics of the WINE algorithm, it was possible to incorporate user preferences, balancing and prioritization of queries and updates. Such approach enables balancing of: (a) maintaining the freshness of data at the appropriate level, and (b) answers to users' queries. The algorithm also uses the term partition that can be presented as a subset of the data warehouse's data table. When creating a WINE-HYBRIS algorithm, method, which is responsible for packet prioritization queries and updates, was enhanced with the schema of two algorithms with FIFO Group [22]. It is based on the simplest solution, resulting in minimizing the number of operations on queues during processing queries and updates. However, algorithms, FIFO-QH (First In First Out-FIFO), and High Query-UH (First In First Out-Update High) does not define levels of data quality and the speed of their reception. In the WINE-HYBRIS algorithm, all transactions are divided into two groups: the read-only and write-only. All incoming requests are placed in a query queue Q, where $Q = q_i \mid i \geq 0$. The updates are queued in the update queue U, where $U = u_j \mid j \geq 0$. According to the WINE algorithm, each transaction is assigned a timestamp (for query timestamp t_q is used, while for the update – t_u), which defines the arrival of queries or system update.

The main scheduling scheme in the WINE-HYBRIS algorithm was divided into two interdependent stages which follows only after the fulfillment of the condition of a minimum items quantity in queues – both must have at least one element. A basic, non-scheduled level (first processing phase) is followed by the allocation of resources, that have a higher priority. At the next level (the second phase of processing) queries and updates are tested and processed – QoS values-queries and QoD values – updates. Thanks to this solution, it is possible to predict a higher priority queue.

In the primary scheduling phase, the summation of the QoS and QoD values is executed for all the queries, that are currently in the query queue. When sums are calculated, in the next step, it is possible to set queues priorities. For example analyzing data from Fig. 3, totals for queries sent by users: $\sum QoS = 1.6$ and $\sum QoD = 2.4$. Since the sum of Qod is greater than the sum QoS, priority is set to the queue update. After the allocation of priorities, the second scheduling phase is executed. In this stage the transaction (query or update) is selected, which is dependent on the result obtained in the first stage of scheduling – in this example it will be the update.

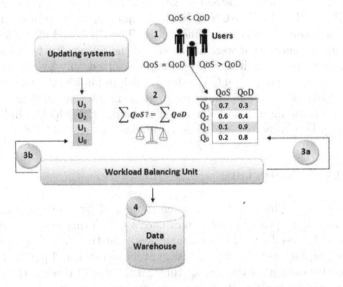

Fig. 3. The main processing scheme in the algorithm

The first phase of a scheduling, is followed by a check which queue has a higher priority. If it was a query queue (the first substage of the second scheduling phase), transactions (requests) located in the queue are scheduled descending by the QoS value and ascending by the t_{qi} timestamp. Also, due to the possibility of neglecting and lack of implementation of the queries, with a low QoS value, it is incremented by the value of delta d [21]:

$$d = \frac{1}{|Q| \cdot r_{\max}} . \tag{1}$$

The value of a Delta parameter depends on parameters, both – the queue length $|Q|$, and the $r_{\max}$ parameter.

During the second substage of the second scheduling phase (priority of update queue), the updates (selected as the most desired by user) are approved.

However, before next operation is executed, updates are scheduled, ordered by decrementing $w(u)$ value and incrementing the time of arrival to the system (timestamp) t_{ui}. The $w(u)$ value [21], is calculated before each iteration during the second sub-phase of the second scheduling phase:

$$w(u) = \sum_{\forall q_i, |P_{qi} \cap P_u| = 1} \frac{qod_{qi}}{1 + pos_{qi}} .$$

(2)

Using parameters defining the query (pos_{qi} – position of query in the queue; qod_{qi} – query QoD value) sequence of updates depends on the actual need for specific information, the timestamp assures that no information will be overwritten.

If one of the queues – queries queue or the updates queue does not have any transactions the third substage is executed. The use of the FIFO algorithm schemes FIFO-QH and FIFO-UH, reduces the number of operations and calculations by using only transactions timestamps. The FIFO-QH algorithm, allows the execution of queries that were at the end of the queue, due to the low value of the QoS parameter (high QoD). Queries are sorted only by the timestamp value, (there is no scheduling of QoS value), which minimizes the time needed to find the next transactions and allows the increase in intensity of their execution. In the third scheduling all queries have the same priority which is independent of QoS and QoD parameters. Also, empty queries queue, enables data update.

4   CUDA Architecture

The CUDA architecture [23, 19] has been developed by NVidia. It allows the use of the multicore processors graphics card's computing power. This allows the considerable speed up of the calculations. The transfer of computational tasks to the graphic card enables the usage of its memory and parallelization of calculations (for example engineering) through the use of threads (Fig. 4). The CUDA accelerates the encryption and compression, and creates special effects in computer graphics.

To develop applications using CUDA architecture in JAVA language, can be used jCUDA. It provides access to CUDA for Java programmers.

5   ETL Process on Windows Azure: Microsoft Cloud Platform

With the release by Microsoft Windows Azure libraries for Java [24], it is possible to use the Java to create workload balancing unit using WINE-HYBRIS algorithm as a servlet. The main part of the system has been placed in the cloud (Fig. 5), and is avalible through the web browser. With such approach, users do not need workstations with high processing power, because all operations are performed on the primary server (in the cloud). Through the use of Cloud Computing, at any time, it is possible to increase or decrease the power, and thereby

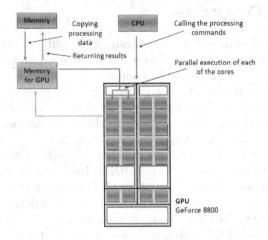

Fig. 4. Data processing based on CUDA architecture [23]

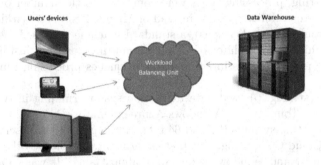

Fig. 5. The use of Cloud Computing in data warehousess

reduce the cost to a minimum. In addition, the use of Windows Azure allows to access the system from any place with active Internet connection.

Smartphone users also have the possibility to connect to the system and receive query response from the data warehouse. While using the Cloud Computing, users gain access to the data warehouse and the ability to process their queries from anywhere in the world.

Data update delivery devices, are especially designed for such data warehouse systems. Heterogeneous hardware and software environment, in combination with a long-term control system, creates problems in such systems design. However, the Cloud Computing approach, used together with modern controllers(which act as an HTTP client), allows such devices to be used at any location, regardless where da are stored. The use of Cloud Computing as well as additional Cloud storage (can be stored) reduces the infrastructure costs [25, 26].

6   Tests and Comparison

In this section we present the results of tests of the author extraction system, developed to support zero-latency data warehouses based on two architectures (1) CUDA and CPU and (2) Cloud Computing.

It was executed almost 1000 tests with using different series of queries of varying complexity. However, to show a variety of behavior for these different systems, results are presented for a single data packet, which had the most diverse input information. All experiments were performed on two different machines, in order to verify the behavior of data extraction in different hardware environments. The first machine was equipped with 3 GB of DDR2 RAM, and an Intel E8400 processor. In addition, for possible tests using CUDA architecture, the computer was equipped with a graphics card NVIDIA GeForce 9600GT (512 MB) and Windows 7, 32 bit OS. The second computer has an Intel Core i5-2410M clocked at 2.30 GHz and 4.00 GB RAM size, and equipped with graphic card GTX 260. and Windows 7 x64 OS.

During testing, processing times were compared to the number of objects in the queue. For credibility reasons, instead of Microsoft's servers (which you can use to test your own applications), a standard machine was used (as described above), on which the simulator of Cloud Computing has been installed. While using such approach for testing, differences in queues processing time could be observed.

The first test (Fig. 6) used data extraction system running in two different environments – TomCat, and Windows Azure simulator. By using two different Windows operating systems (32 and 64 bit), it was possible to observe different behavior of Cloud Computing. The first experiment, (32-bit OS) the processing times of Tomcat and Windows Azure were almost alike. It was caused by the fact that the majority of the system features in Windows Azure is unavailable, thus limiting its possibilities. Also, due to the weaker performance of the first machine (used in the sample), queue processing time increases (compared to processing time received on the second machine) with each additional query (update) incoming to the system.

For the 64-bit OS (the second machine used in the test) we observed, that the Cloud Computing simulator use significantly improved the queues processing time. In conclusion it is possible to increase the efficiency and the number of processed queries or updates when using the Cloud Computing as the basic platform. Moreover, not only the platform, but also the hardware and operating system impacts the performance.

During the second test the same machines were used, except that applications were not run on TomCat as servlets and have not been tested in Windows Azure. Both machines run standard Java applications. Analysis of Fig. 7 shows that the queues processing time, using both CPU and GPU architecture, it was possible to perceive some very interesting behavior, which did not occur during startup on Tomcat and Cloud Computing. In this case "limit value" is used [22], which allows to estimate the number of queries or updates, above which it is cost effective to use the CUDA architecture.

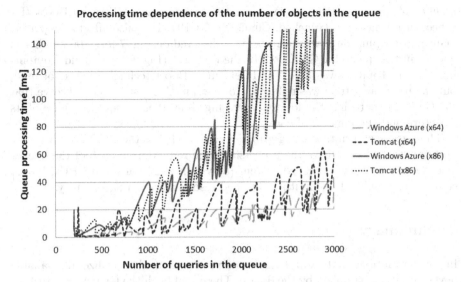

Fig. 6. Comparison of Windows Azure and Apache Tomcat

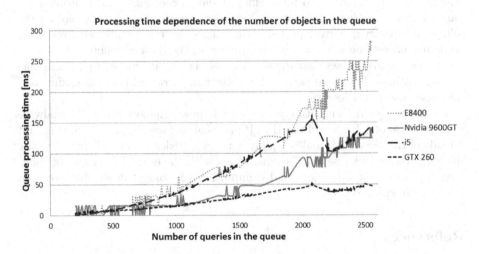

Fig. 7. Comparison of CPU and GPU architectures

For computer, which was based on i5 processor and GTX 260 graphics card, the limit is lower, which means that the CUDA architecture, is used with a smaller number of items in the queue. The second difference is the queues processing time. For the PC based on processor E8400, both the CPU processing time and GPU processing time is higher in comparison to other machine used for testing. The new technology used in the i5 processor (Intel Turbo Boost), while increasing the CPU load, it is able to obtain results comparable to the 9600GT graphics card (older generation card). This behavior can be observed on Fig. 7 for 2100 number of queries in the queue, when the technology Intel Turbo Boost (for processor i5) has been launched. Changing i5 CPU clock also affects the performance for graphic card, what can be also observed on Fig. 7 for 2100 number of queries in the queue for GTX 260. In the first test (Fig. 6) the Cloud Computing and the TomCat was used, the Intel Turbo Boost technology was not used, but the processing time was reduced. However, in this test, we can observe that the GTX 260 card has the lowest processing time. It is achieved because of its structure and the number of cores.

These tests, shown that using graphics card, and thus the CUDA architecture in data warehouses, allows the processing time reduction in each of the queues in comparison to the standard system (a system using only the CPU), but in comparison to Cloud Computing obtained results were not satisfactory.

7   Summary

In the zero-latency data warehouse systems it is crucial to utilize all available power resources, supplied by the device. The use of both the CUDA architecture as well as the Windows Azure platform (Cloud Computing) increases performance, what translates into more efficient data processing. An additional advantage of Cloud Computing, is the costs optimization, – you only pay for the allocated capacity in a given period of time. This also relieves allocated resources while not processing any queries and updates in the data warehouse.

The article also shows that the selection of inadequate runtime platform may not fully utilize the device potential, resulting in increased processing time and decreased users satisfaction.

Using the Cloud Computing technology, exposes the system to the malware risks (the use of a LAN and the HTTP protocol and keeping the balancing unit outside of the closed internal company network). Risks associated with malware infected systems, can be quite serious, along with the critical data leak or theft. The approach based on the Cloud Computing technology, should be protected by e.g. honeypot [27].

References

1. Kozielski, S., Wrembel, R. (eds.): New Trends in Data Warehousing and Data Analysis. Annals of Information Systems, vol. 3. Springer (2009)
2. Wrembel, R.: A Survey of Managing the Evolution of Data Warehouses. IJDWM 5(2), 24–56 (2009)
3. Gorawski, M., Morzy, T., Wrembel, R.: Special Issue on: Techniques of Advanced Data Processing and Analysis Introduction Control and Cybernetics 38(1), 5–8 (2009)
4. Andrzejewski, W., Wrembel, R.: GPU-WAH: Applying GPUs to Compressing Bitmap Indexes with Word Aligned Hybrid. In: Bringas, P.G., Hameurlain, A., Quirchmayr, G. (eds.) DEXA 2010, Part II. LNCS, vol. 6262, pp. 315–329. Springer, Heidelberg (2010)

5. Gorawski, M., Marks, P., Gorawski, M.: Collecting data streams from a distributed radio-based measurement system. In: Haritsa, J.R., Kotagiri, R., Pudi, V. (eds.) DASFAA 2008. LNCS, vol. 4947, pp. 702–705. Springer, Heidelberg (2008)
6. Gorawski, M., Bańkowski, S., Gorawski, M.: Selection of Structures with Grid Optimization, in Multiagent Data Warehouse. In: Fyfe, C., Tino, P., Charles, D., Garcia-Osorio, C., Yin, H. (eds.) IDEAL 2010. LNCS, vol. 6283, pp. 292–299. Springer, Heidelberg (2010)
7. Morzy, T.: OLAP with a Database Cluster. In: Wrembel, R., Koncilia, C. (eds.) Data Warehouses and OLAP: Concepts, Architectures and Solutions, pp. 230–252 (2007)
8. Cichon, P., Huzar, Z., Mazur, Z., Mrozowski, A.: Managing Adaptive Information Projects in the Context of a Software Developer Organizational Structure. In: B.L. (ed.) BIS. LNI, vol. 85 GI, pp. 242–255 (2006)
9. Gorawski, M., Gorawski, M.: Modified R-MVB tree and BTV algorithm used in a distributed spatio-temporal data warehouse. In: Wyrzykowski, R., Dongarra, J., Karczewski, K., Wasniewski, J. (eds.) PPAM 2007. LNCS, vol. 4967, pp. 199–208. Springer, Heidelberg (2008)
10. Gorawski, M., Gorawski, M.: Balanced spatio-temporal data warehouse with R-MVB, STCAT and BITMAP indexes. In: PARELEC 2006, pp. 43–48 (2006)
11. Bruckner, R., Min Tjoa, A.: Capturing Delays and Valid Times in Data Warehouses – Towards Timely Consistent Analyses. J. Intell. Inf. Syst. 19(2), 169–190 (2002)
12. Bruckner, R.M., List, B., Schiefer, J.: Striving towards Near Real-Time Data Integration for Data Warehouses. In: Kambayashi, Y., Winiwarter, W., Arikawa, M. (eds.) DaWaK 2002. LNCS, vol. 2454, pp. 317–326. Springer, Heidelberg (2002)
13. Gorawski, M., Marks, P.: Towards automated analysis of connections network in distributed stream processing system. In: Haritsa, J.R., Kotagiri, R., Pudi, V. (eds.) DASFAA 2008. LNCS, vol. 4947, pp. 670–677. Springer, Heidelberg (2008)
14. Gorawski, M., Marks, P.: Towards reliability and fault-tolerance of distributed stream processing system. In: DepCoS – RELCOMEX 2007, pp. 246–253 (2007)
15. Gorawski, M.: Architecture of Parallel Spatial Data Warehouse: Balancing Algorithm and Resumption of Data Extraction. In: Software Engineering: Evolution And Emerging Technologies, FAIA, vol. 130, pp. 49–59 (2005)
16. Gorawski, M.: Extended Cascaded Star Schema and ECOLAP Operations for Spatial Data Warehouse. In: Corchado, E., Yin, H. (eds.) IDEAL 2009. LNCS, vol. 5788, pp. 251–259. Springer, Heidelberg (2009)
17. Gorawski, M., Marks, P.: Resumption of data extraction process in parallel data warehouses. In: Wyrzykowski, R., Dongarra, J., Meyer, N., Waśniewski, J. (eds.) PPAM 2005. LNCS, vol. 3911, pp. 478–485. Springer, Heidelberg (2006)
18. Gorawski, M., Marks, P.: Checkpoint-based resumption in data warehouses. In: Sacha, K. (ed.) IFIP International Federation For Information Processing. Software Engineering Techniques: Design for Quality, vol. 227, pp. 313–323. Springer, Boston (2006)
19. Rahm, E., Hai Do, H.: Data Cleaning: Problems and Current approches. Bulletin of the Technical Committee on Data Engineering 23 (2000)
20. Waas, F., Wrembel, R., Freudenreich, T., Theile, M., Koncilia, C., Furtado, P.: On-Demand ELT Architecture for Right-Time BI: Extending the Vision. International Journal on Data Warehousing and Mining (to appear, 2013)
21. Thiele, M., Fischer, U., Lehner, W.: Partition-based Workload Scheduling in Living Data Warehouse Environments. In: DOLAP 2007. ACM, Portugal (2007)
22. Gorawski, M., Lis, D.: Architektura CUDA w bezpoznieniowych hurtowniach danych. Studia Informatica 32, 157–167 (2011)

23. CUDA in research, http://www.nvidia.pl/object/cuda_home_new_pl.html
24. Windows Azure, SDK for Java, http://www.windowsazure4j.org/
25. Jestratjew, A., Kwiecien, A.: Performance of HTTP Protocol in Networked Control Systems. IEEE Trans. Industrial Informatics 9(1), 271–276 (2013)
26. Jestratjew, A., Kwiecień, A.: Using Cloud Storage in Production Monitoring Systems. In: Kwiecień, A., Gaj, P., Stera, P. (eds.) CN 2010. CCIS, vol. 79, pp. 226–235. Springer, Heidelberg (2010)
27. Skrzewski, M.: Monitoring Malware Activity on the LAN Network. In: Kwiecień, A., Gaj, P., Stera, P. (eds.) CN 2010. CCIS, vol. 79, pp. 253–262. Springer, Heidelberg (2010)

MViewer: Visualization of Protein Molecular Structures Stored in the PDB, mmCIF and PDBML Data Formats

Dawid Stanek, Dariusz Mrozek, and Bożena Małysiak-Mrozek

Institute of Informatics, Silesian University of Technology,
Akademicka 16, 44-100 Gliwice, Poland
{dariusz.mrozek,bozena.malysiak-mrozek}@polsl.pl
http://zti.polsl.pl/dmrozek

Abstract. Molecular viewers allow visualization of spatial structures of proteins and other biological molecules. However, most of them are not able to read molecular structures stored in the PDBML format provided by the world-wide Protein Data Bank (PDB). In the paper, we present main features of the MViewer that we have designed and developed in order to support a man-protein structure interaction. The unique feature of the MViewer is the ability to open, read and visualize structures stored in all three formats (i.e. PDB, mmCIF, and PDBML) available from the PDB repository.

Keywords: protein structure, visualization, molecular viewer, Protein Data Bank, PDB, mmCIF, PDBML, data formats.

1 Introduction

Visualization of macromolecular structures of biological molecules is crucial for their detailed analysis. This takes a special place in the visualization of protein structures that are said to be molecules of life, involved in many cellular processes and molecular networks [1]. Graphical tools, called viewers, that enable protein structure visualization are parts of the standard software equipment of bioinformaticians and molecular biologists; and all these people who study complex mechanisms of how proteins interact to each other in cellular reactions and signaling pathways [2], and what parts of protein structures are directly involved in the reaction.

Development of molecular viewers requires not only skills in computer graphics and computer programming, but also knowledge on the construction of protein molecules and other biological molecules. Four different representation levels of protein structures can be distinguished: primary, secondary, tertiary and quaternary structure [3]. Each of the levels describes different aspects of protein structure and can be studied jointly or separately while conducting a research on protein functions, interactions and activities.

The main, but not the only, source of information on protein structures is the Protein Data Bank (PDB) [4] (http://www.pdb.org), the world-wide repository

A. Kwiecień, P. Gaj, and P. Stera (Eds.): CN 2013, CCIS 370, pp. 323–333, 2013.
© Springer-Verlag Berlin Heidelberg 2013

for macromolecular structures of proteins and other biological molecules. These structures are available for the scientific community for free through the Internet in the form of atomic coordinate files in three different file formats: PDB, mmCIF and PDBML.

While PDB [5] format seems to be the most popular, other two formats provide some additional benefits. PDB files were designed for storage of crystal structures and related experimental information on biological macromolecules, primarily proteins, nucleic acids, and their complexes. PDB files contain first of all the Cartesian coordinates of the atoms forming the molecule, primary and secondary structures (in the case of proteins), bibliographic citations, and crystallographic structure factors as well as X-ray diffraction or NMR (Nuclear Magnetic Resonance) experimental data [6]. The mmCIF (Macromolecular Crystallographic Information File) [7] format on the other hand covers the same data, but is specified by the CIF dictionary and the data files based upon that dictionary. This gives the benefit of validation of data and provides higher level of detail in describing various features of molecular structures. PDBML (Protein Data Bank Markup Language) [8] files are XML files. The description of this format is provided in XML schema of the PDB Exchange Data Dictionary. This schema is produced by direct translation from the mmCIF format [7]. What is important, the use of XML language in the description of molecular structures guarantees the automatic validation of the files describing the structures, what is important while exchanging macromolecular data between various software applications, systems and platforms.

Although the PDBML format has been used for several years, there are no much molecular viewers accepting this format. A short review of some of them is done in Sect. 2. In this paper, we present the MViewer – a software application that allows visualization of molecular structures of proteins and other molecules, based on the macromolecular data stored in the PDB, mmCIF and PDBML data formats.

2   Existing Solutions

In this section we provide examples of existing tools dedicated for protein visualization. The section outlines strengths and weaknesses of chosen tools comparing them with respect to their features.

One of the most popular molecular visualization applications is RasMol [9]. RasMol is a free, open source application intended for the visualization of proteins, nucleic acids and small molecules in three dimensions. RasMol can display molecules in a variety of color schemes (monochrome, CPK, shapely, group, chain, temperature, structure, model or user defined) and molecule representations (wireframe, backbone, sticks, spacefill, balls and sticks, ribbons, strands, cartoons and molecular surface). Its main advantage is the remarkable speed of visualization even on older machines without the good 3D accelerator and the ability to recognize and visualize most types of molecule file formats (including PDB and mmCIF). Chime [10] (also referred to as MDL Chime), developed by

Tim Maffett and Bryan van Vliet, is based on RasMol engine and uses many of its components as command processor and visualization component. The main difference between both applications is the execution environment. RasMol runs as a stand-alone application and Chime is designed to run inside the web browser as a plug-in. Jmol [11] was designed and implemented to replace MDL Chime. It is available in three different packages: JmolApplet, Jmol application and JmolViewer. The JmolApplet is the web browser applet that can be integrated into web pages as a plug-in. The Jmol application is a standalone application that runs on Windows, Mac OS X, and Linux/Unix systems. The JmolViewer is an SDK that can be used by developers to create other applications. Jmol supports PDB files and can download them directly from the PDB repository. QMOL [12] is another free application for viewing molecular structures in three dimensions. QMOL provides multiple tools for molecular analysis and manipulations (superimposing two molecules by rigid body rotation and translation) with arbitrary atom selection, interactively computing RMSD between two molecules (with arbitrary atom selection), aligning all structures in a trajectory against the initial structure, interactively measuring bond lengths, bond angles and torsion angles. QMOL supports PDB files and can download them directly from the PDB repository. PyMOL [13] offers very rich functionality in comparison to RasMol, Chime, Jmol and Qmol. Application supports both command line and GUI controls. PyMOL allows users to open multiple file formats and display a molecule in many different color schemes and representations, select and modify components of the molecule, perform measurements, create movies and so forth. Cn3D [14] is dedicated to simultaneously display biomolecular structures, sequences, and sequence alignments from NCBI's Entrez database. Cn3D can be run inside WWW browser or can also be used as a standalone application. The main feature that distinguish the Cn3D apart from other molecular visualization applications is the ability to correlate structure and sequence information. Finally, Protein Molecular Viewer (PMV) [15] is a free and open-source program, which visualizes protein molecular structures. It can be used by anyone who wants to see molecular structures represented in the PDBML format from the well-known Protein Data Bank (PDB). The PMV is available as a stand-alone application or web browser applet. Although of limited capabilities in terms of visualization, the possibility of loading and presenting protein structures from the PDBML files makes the PMV one of a few tools in the world having this unique function.

3   Protein Visualization with MViewer

MViewer provides reach capabilities in terms of protein structure visualization, including various representations of protein structure, coloring modes, selection modes, and structure transformation (rotation, translation and zooming). In this section we show different visualization features of the MViewer for sample atomic coordinate files. If possible test results were compared with the RasMol's results obtained for the same atomic coordinate files.

3.1   MViewer Working Environment

MViewer supports five different input file types: atomic coordinate files script files, txt/html files, graph/report files and animation files. Obviously the most important for the application are atomic coordinate files. Three common file formats are supported: PDB, mmCIF and PDBML. MViewer allows users to open the file, parse it and translate it into collection of displayable objects such as models, chains, atoms and bondings (Fig. 1). The atomic coordinate file can be opened from the local or network disk or it can be directly downloaded from PDB repository or any other repository through the application's FTP downloader. The application uses suitable parsers to open the atomic coordinate files and to translate them into collections of objects that are used in the visualization. These objects are *models* that are collections of *chain* objects; each *chain* object is a collection of *atom* objects; and each *atom* object contains collection of *bonding* objects that are references to other *atom* objects.

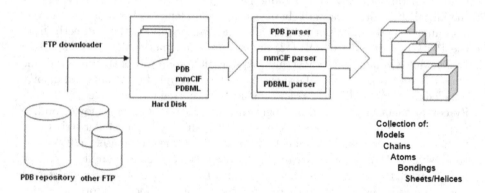

Fig. 1. Downloading, opening, parsing and translating of atomic coordinate files

3.2   Representations

Representation modes allow users to focus on concrete visual details of the chosen molecular structure. For example, some users can be interested just in the secondary structure elements that are present in the 3D structure and other users want to see only bonds linking atoms of the structure. In all these cases they have to switch to particular representation mode. MViewer supports five representations of molecular structures: (1) sticks and balls showing atoms and bonds, (2) sticks showing only bonds between atoms, (3) wireframe bonds (backbone) focusing on the reduced representation of the structure, (4) space-filling (CPK) spheres presenting the external shape of the molecule and atoms on the surface, and (5) macromolecular ribbons revealing secondary structure elements (α-helices, β-sheets, and turns). In Figure 2 we present different representations for Human Estrogen (PDB ID: 3ERT) with comparison to RasMol's results.

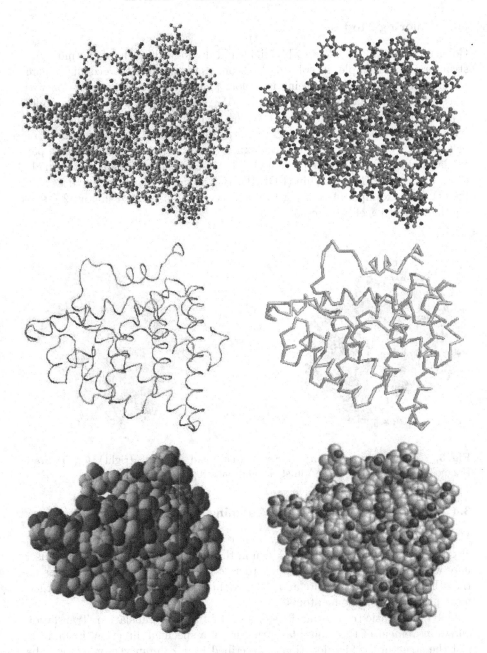

Fig. 2. Comparison of representation modes in MViewer (left) and RasMol (right) for the molecular structure of Human Estrogen (PDB ID: 3ERT): sticks and balls representation (top), wireframe representation (middle), spacefill representation (bottom)

3.3 Coloring Modes

Coloring modes allow users to highlight and distinguish parts of the molecular structure. For example, in coloring by atom users are able to recognize chemical element of an atom by seeing its color. This facilitates the interaction with the structure and navigation through it. MViewer supports five coloring modes: (1) coloring the molecule by atom type, (2) chain (to distinguish chains in quaternary structure), (3) group (e.g. to distinguish domains) or (4) monochrome or (5) to one of custom color modes defined by the user (e.g. to distinguish particular atoms). The example presented in Fig. 3 shows comparison of coloring by chain for T State Haemoglobin (PDB ID: 1GZX) in MViewer (left) and RasMol (right). And example of coloring by atom type was presented in Fig. 2 (sticks and balls and spacefill representations).

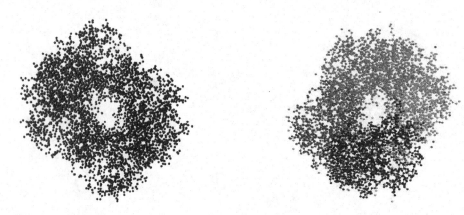

Fig. 3. Coloring by chain in MViewer (left) and RasMol (right) for T State Haemoglobin (PDB ID: 1GZX) in sticks and balls representation

3.4 Rotating, Translating and Zooming

MViewer supports interactive translation, rotation, and zooming in. These operations allow users to see what is hidden in the original view. For example, a user may wish to rotate the structure in order to see the arrangement of atoms on the other side of the structure and then, zoom in the molecule to see the larger, more precise view of some atoms.

One of the basic transformations of a set of atoms is translation. Translation allows movement of the molecular structure in a specified direction. Each atom i of the structure, which location is described by a column vector $p^{(i)}$, can be translated by a vector v to the location described by a column vector $q^{(i)}$ by applying the following Equation (1):

$$q^{(i)} = p^{(i)} + v \ .$$

(1)

Translation is also often used when we want to move the entire set of atoms so that their centroid is at the origin of the coordinate system. New coordinates for the set of atoms $\{x^{(i)}|i = 1, 2, \ldots, N\}$ can be calculated according to the Equation (2):

$$x^{(i)} = p^{(i)} - p^{(c)} \tag{2}$$

where: N is a number of atoms in the protein structure, and $p^{(c)}$ is a column vector describing the protein centroid calculated according to the Formula (3):

$$p^{(c)} = \frac{1}{N} \sum_{i=1}^{N} p^{(i)} . \tag{3}$$

Rotation is another tranformation, which can be used in order to change the current view of molecular structure. In $\Re^3$, the rotation transformation of the each atom $p^{(i)}$ belonging to molecular structure can be accomplished by premultiplication of p by a special 3×3 matrix R (Equation (4)):

$$q^{(i)} = Rp^{(i)} . \tag{4}$$

According to the Euler's rotation theorem any rotation can be given as a composition of rotations about three axes. Therefore, the rotation matrix R can be a product of three elementary matrices $R_z(\gamma), R_y(\beta), R_x(\alpha)$ (Equation (5)).

$$R = R_z(\gamma)R_y(\beta)R_x(\alpha), \text{ where: } R_x = \begin{bmatrix} 1 & 0 & 0 \\ 0 & \cos\alpha & -\sin\alpha \\ 0 & \sin\alpha & \cos\alpha \end{bmatrix},$$

$$R_y = \begin{bmatrix} \cos\beta & 0 & \sin\beta \\ 0 & 1 & 0 \\ -\sin\beta & 0 & \cos\beta \end{bmatrix}, \quad R_\gamma = \begin{bmatrix} \cos\gamma & -\sin\gamma & 0 \\ \sin\gamma & \cos\gamma & 0 \\ 0 & 0 & 1 \end{bmatrix} . \tag{5}$$

Scaling allows zooming in and out a set of atoms and can be accomplished by multiplication of coordinates of each atom $p^{(i)}$ of the structure by a scaling matrix S_v (Equation (6)).

$$q^{(i)} = S_v p^{(i)}, \text{ where } S_v = \begin{bmatrix} v_x & 0 & 0 \\ 0 & v_y & 0 \\ 0 & 0 & v_z \end{bmatrix}, \tag{6}$$

and v_x, v_y, v_z are elements of the scaling vector v.

We performed several tests with satisfactory results for different molecules in order to check implemented transformations. In Figure 4 we show results for zooming the structure of Lysozyme C molecule (PDB ID: 2VB1) with comparison to the original view.

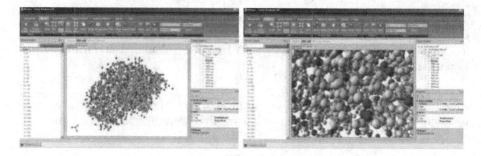

Fig. 4. Zooming the structure of the Lysozyme C (PDB ID: 2VB1) in MViewer: original view (left) and zoomed view (right) in sticks and balls representation

3.5   Selection Modes and Tools

MViewer provides three selection modes: chain, group or atom mode, and two selection tools: selection by pick and selection by box. Selection by pick allows to choose and highlight the item (atom, chain, or group) that was indicated by the mouse click. Selection by box allows to choose and highlight items contained in the area selected by a mouse cursor. Several tests for different molecules were performed with satisfactory results. In Figure 5 we show results for selecting by pick in atom mode (top) and selecting by box in chain mode (bottom) for the Cyclin Dependent Kinase 5 (PDB ID: 1UNL) molecule.

3.6   Additional Tools and Features of MViewer

MViewer provides a variety of tools that support and complement the visualization of molecular structures. These tools make it feasible to perform the following tasks:

- Creating documentation. Users can create plain text and HTML files in documentation editors that can be used to document solution, project or project item. HTML documentation files allow using different text formatting, nesting images, hyperlinks and so forth.
- Creating reports and graphs. The application supports generation of simple reports and graphs for the given atomic coordinate file, e.g. atom structure grouped by chains or atom distribution in the molecule.
- Creating and executing scripts. The application provides its own script language that can automate some routines and speed up common operations. It supports most of the application's visualization features (e.g. representation modes, coloring modes, transformations) and is accessible by both script editor and command line.
  Example of a MViewer Script

```
\\Title: Sample script
\\Description: Script showing sample commands of MViewer
\\Date: 2013-02-20 02:01:33
```

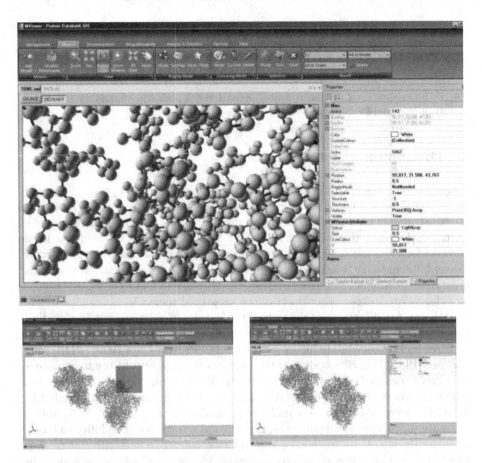

Fig. 5. Selecting by pick in atom mode (top) and selecting by box (bottom) in MViewer for Cyclin Dependent Kinase 5 (PDB ID: 1UNL)

```
\\Author: D. Stanek, B. Malysiak-Mrozek and D. Mrozek
Commands.ZoomFit();
Commands.SetColouring(ByAtomType);
Commands.SetDisplayMode(SpaceFill);
Commands.ClearSelection();
Commands.SelectChain(A);
Commands.CameraRotate(10,0,0,10,10,30);
```

– Creating and viewing animations. Users are able to create simple animations as compilations of rotation, translation and zoom transformations (Fig. 6).
– Customizing user interface. Users can customize the graphical interface of the MViewer. All windows inside the application can be shown or hidden, docked or undocked and moved to any position. Additionally, the applications supports skins changing its appearance.

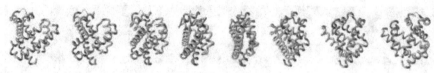

Fig. 6. Successive steps of the rotation around the Y axis for Myoglobin (PDB ID: 3HC9) in the sample animation with MViewer

4   Discussion and Conclusions

Visualization tools for molecular structures of DNA/RNA and proteins are very important if we want to get appropriate knowledge of how particular molecules are constructed in more details. This is especially important for structural bioinformatics [1] and clinical bioinformatics [16]. Various representations of a molecule allow to focus on the specific type of information (e.g. only secondary structure elements) concealing what is not required in order not to obscure the viewing area. Additionally, selection and coloring modes allow to highlight what is needed or differentiate it from other parts of the visualized molecule.

MViewer provides most of the important features of existing protein visualization applications. It also introduces some additional features, such as support for all three atomic coordinate file formats from the Protein Data Bank (PDB, mm-CIF and PDBML), rich customizable user interface, ability to organize atomic coordinate files and other files into projects and solutions through Solution Explorer, ability to browse and search components of the molecule through Elements Explorer, ability to view multiple visualizations at the same time, ability to create supplementary documents, reports and graphs and possibility to create and execute scripts in application's script language. The possibility of reading PDBML files is still unique among molecular viewers available world-wide. The main disadvantage of the MViewer in comparison to some of the existing protein visualization applications is the fact the application is not platform independent and can not be embedded within a web browser. Also the performance of the application can be an issue in case of large molecules. Despite both drawbacks are mainly due to chosen development framework (.NET Framework and EyeShot library), the decision to use it was reasonable.

Future works will cover further development of the viewer focused on the analysis and statistics module (currently the application supports only some sample reports and graphs). We also want to improve the quality and performance of the visualization and provide additional possibility of downloading molecular structures from the PDB repository by appropriate invocation of available Web services [17]. And perhaps, we will also add features to not only visualize existing molecular models without changing them, but also to create the model or modify it by adding or removing components, changing bonds, bond angles, conformations or non-covalent interactions. This, however, can move the application from visualization to modeling domain [18].

Acknowledgments. This work was supported by the European Union from the European Social Fund (grant agreement number: UDA-POKL.04.01.01-00-106/09).

References

1. Burkowski, F.: Structural Bioinformatics: An Algorithmic Approach, 1st edn. Chapman and Hall/CRC (2008)
2. Malysiak, B., Mrozek, D., Kozielski, S., Znamirowski, L.: Signal transduction simulation in nanoprocesses using distributed database environment. In: Modelling, Simulation, and Optimization: Fifth IASTED International Conference Proceedings, pp. 17–22 (2005)
3. Branden, C., Tooze, J.: Introduction to Protein Structure, 2nd edn. Garland Science (1999)
4. Berman, H., et al.: The Protein Data Bank. Nucleic Acids Res. 28, 235–242 (2000)
5. Westbrook, J., Fitzgerald, P.: The PDB format, mmCIF, and other data formats. Methods Biochem. Anal. 44, 161–179 (2003)
6. Gasteiger, J., Engel, T.: Chemoinformatics: a textbook. Wiley-VCH (2003)
7. Bourne, P., Berman, H., Watenpaugh, K., et al.: The macromolecular crystallographic information file (mmCIF). Methods Enzymol. 277, 571–590 (1997)
8. Wesbrook, J., Ito, N., Nakamura, H., Henrick, K., Berman, H.: PDBML: the representation of archival macromolecular structure data in XML. Bioinformatics 21(7), 988–992 (2005)
9. Sayle, R.: RasMol, Molecular Graphics Visualization Tool. Biomolecular Structures Group, Glaxo Welcome Research & Development, Stevenage, Hartfordshire (1998), http://www.umass.edu/microbio/rasmol/ (February 5, 2013)
10. Chime and Jmol Homepage: Molecular Visualization Resources, http://www.umass.edu/microbio/chime/
11. Jmol Homepage: Jmol: an open-source Java viewer for chemical structures in 3D, http://www.jmol.org
12. Gans, J., Shalloway, D.: Qmol: a program for molecular visualization on Windows-based PCs. J. Mol. Graph. Model. 19(6), 557–559 (2001)
13. The PyMOL molecular graphics system, ver. 1.3r1. Schrödinger, LLC (2010), http://www.pymol.org
14. Hogue, C.: Cn3D: a new generation of three-dimensional molecular structure viewer. Trends Biochem. Sci. 8, 314–316 (1997)
15. Mrozek, D., Mastej, A., Malysiak, B.: Protein Molecular Viewer for visualizing structures stored in the PDBML format. In: Pietka, E., Kawa, J. (eds.) Information Technologies in Biomedicine. AISC, vol. 47, pp. 377–386. Springer, Heidelberg (2008)
16. Bellazzi, R., Masseroli, M., Murphy, S., Shabo, A., Romano, P.: Clinical Bioinformatics: challenges and opportunities. BMC Bioinformatics 13(suppl. 14), S1 (2012)
17. Mrozek, D., Małysiak-Mrozek, B., Siaznik, A.: search GenBank: interactive orchestration and ad-hoc choreography of Web services in the exploration of the biomedical resources of the National Center For Biotechnology Information. BMC Bioinformatics 14, 73 (2013)
18. Burczyński, T., Poteralski, A., Szczepanik, M.: Topological evolutionary computing in the optimal design of 2D and 3D structures. Engineering Optimization 39(7), 811–830 (2007)

CASSERT: A Two-Phase Alignment Algorithm for Matching 3D Structures of Proteins

Dariusz Mrozek and Bożena Małysiak-Mrozek

Institute of Informatics, Silesian University of Technology,
Akademicka 16, 44-100 Gliwice, Poland
{dariusz.mrozek,bozena.malysiak-mrozek}@polsl.pl
http://zti.polsl.pl/dmrozek

Abstract. Protein structure alignment allows assessment of protein similarities and leads to the knowledge of the nature of proteins themselves. In this paper, we present a new version of the two-phase alignment algorithm for matching protein structures, called CASSERT. The algorithm can be used in scanning databases of protein structures while searching protein similarities. Effectiveness of the CASSERT was studied comparing its results to those returned by DALI algorithm. Performed tests confirm that the CASSERT algorithm exhibits high effectiveness in protein structure similarity searching and can be a useful tool in the identification of proteins and their functions.

Keywords: structural bioinformatics, alignment, protein structure, similarity, structure matching.

1 Introduction

Protein structure similarity searching is one of the key, but most difficult tasks of modern structural bioinformatics [1]. While searching similarities based on amino acid sequence (primary structure [2]) is often carried out by using operations on strings, protein structure comparison is more problematic due to the complex construction of proteins on a molecular level. If we assume that an average size protein is made up of several hundreds of amino acids, and each amino acid is made up of several atoms, then a comparison of only one pair of protein structures is a challenge. If we also want to compare the structure of the protein with the entire database of proteins, for example, to compare mutant structures to each other, then taking into account the increasing number of protein structures in databases, such as the Protein Data Bank (PDB) [3], this task becomes even more complicated.

However, protein structure similarity searching is a very important task for the modern structural bioinformatics. Based on the information about similar protein structures we can conclude about common ancestry of organisms and thus, we can study the evolution of organisms over millions of years. The analysis of protein structures by their comparison allows us to search for substitutes

A. Kwiecień, P. Gaj, and P. Stera (Eds.): CN 2013, CCIS 370, pp. 334–343, 2013.
© Springer-Verlag Berlin Heidelberg 2013

for biological molecules critical for certain cellular processes, whose lack or inadequate design can cause dysfunction of the body or serious diseases.

In this paper we present a new version of the two-phase alignment algorithm [4] for matching protein 3D structures, called CASSERT. The name of the algorithm is an acronym from the words defining the representative features of protein structures taken into account in the comparison process – C_α atom (C), angle defined by vectors between successive C_α atoms (A), Secondary Structure Element (SSE), Residue Type (RT). Presented algorithm can be used in protein structure similarity searching. In the paper, we also present tests that we have performed in order to examine the effectiveness and efficiency of the algorithm.

2   Related Works

Several algorithms for protein structure similarity searching have been developed in the last two decades, including VAST [5], DALI [6,7], LOCK2 [8], FatCat [9], ClusCo [10], CTSS [11], CE [12], DEDAL [13], RAPIDO [14], FAST [15], MICAN [16] and others [17,18]. Taking into account the complexity of protein structures, existing algorithms use various representations of these structures in the similarity searching process.

For example, the CTSS [11] algorithm includes local geometric features and selected biological characteristics. For each residue in a protein structure the algorithm calculates shape signatures based on C_α atom positions, torsional angles, and type of the secondary structure. DALI algorithm [6,7] makes use of distance matrices in the comparison process. These matrices are built for each of compared proteins. The distance between the C_α atoms in amino acids i and j in the protein is stored in each cell of the distance matrix. Distance matrices are then decomposed to so-called contact patterns, which are fragments of 6×6 elements of the matrix, and compared to find the best match. On the other hand, the VAST algorithm [5] uses secondary structure elements (SSEs) forming the cores of compared proteins (α-helices and β-sheets). SSEs are then mapped to the representative vectors, which simplifies the analysis process. During the comparison, the algorithm attempts to match a set of vectors for pairs of protein structures. The SSE representation of protein structures is also used in the comparison method applied in the LOCK2 [8]. The CE [12] algorithm uses the combinatorial extension of alignment path formed by aligned fragment pairs (AFPs). AFPs are fragments of both structures indicating a clear structural similarity and are described by local geometrical features, including positions of C_α atoms. The idea of AFPs is also used in the FATCAT [9].

3   Two-Phase Alignment Algorithm for Matching Protein 3D Structures

Protein similarity searching is typically performed by comparing the query protein (Q) specified by the user with successive proteins (D) from the database of protein structures. In this section we present our newly developed algorithm for fast and accurate comparison of two protein structures that can be used in scanning databases of proteins in order to find similar biological molecules.

3.1   First Phase – Comparison of Secondary Structures

In the first phase of the algorithm, protein structures Q and D are compared by aligning their reduced chains of secondary structures formed by secondary structure elements SE_i:

$$Q = (SE_1^Q, SE_2^Q, \ldots, SE_n^Q) \text{ and } D = (SE_1^D, SE_2^D, \ldots, SE_m^D) , \qquad (1)$$

where: n is a number of secondary structures in the chain of the query protein Q, m is a number of secondary structures in the chain of the database protein D. Elements SE_i^Q and SE_j^D, hereinafter referred to as SE regions or SE fragments, are built from groups of adjacent amino acids forming the same type of secondary structure (e.g. α-helix or β-strand, Fig. 1).

Fig. 1. Secondary structure elements: (left) four α-helices in sample structure [PDBID: 1CE9], (right) two β-strands joined by a loop in sample structure [PDB ID: 1E0Q]; visualized by MViewer [19]

We call this phase of the algorithm as the *low resolution alignment*. Each element SE_i, which is a chain part isolated on the basis of its secondary structure, is characterized by two values:

$$SE_i = [SSE_i, L_i] , \qquad (2)$$

where: SSE_i describes the type of secondary structure, L_i is the length of the i-th element SE_i (measured in residues). In the presented method, we distinguish three basic types of secondary structures:

- α-helix,
- β-sheet or β-strand,
- undetermined structure, which represents loops, turns or coils.

In order to match the structures Q and D we use the modified version of the Smith-Waterman alignment algorithm [20]. In the course of the algorithm, we build the similarity matrix SSE of the size $n \times m$, where n and m describe the number of secondary structures in compared chains of proteins Q and D, i.e. the number of fragments of Q and D chains of recognized secondary structure. Successive cells of the SSE matrix are filled according to the following rules: for $0 \leq i \leq n$ and $0 \leq j \leq m$:

$$SSE_{i,0} = SSE_{0,j} = 0 \ , \tag{3}$$

$$SSE_{i,j}^{(1)} = SSE_{i-1,j-1} + \delta_{ij} \ , \tag{4}$$

$$SSE_{i,j}^{(2)} = \max_{1 \leq k \leq n} \{SSE_{i-k,j} - \omega_k\} \ , \tag{5}$$

$$SSE_{i,j}^{(3)} = \max_{1 \leq l \leq m} \{SSE_{i,j-l} - \omega_l\} \ , \tag{6}$$

$$SSE_{i,j}^{(4)} = 0 \ , \tag{7}$$

$$SSE_{i,j} = \max_{v=1...4} \{SSE_{i,j}^{(v)}\} \ . \tag{8}$$

where: δ_{ij} is a similarity reward, determining the similarity degree between two components SE_i^Q and SE_j^D of proteins Q and D, ω_k, ω_l are possible, horizontal and vertical penalties for inserting a gap of the length k and l. The similarity reward δ_{ij} takes values from the interval $\langle 0; 1 \rangle$, where 0 means no similarity, while 1 means the highest possible similarity. The degree of similarity is calculated using the formula:

$$\delta_{ij} = \sigma_{ij} - \left(\sigma_{ij} * \frac{|L_j^D - L_i^Q|}{(L_j^D + L_i^Q)} \right) \ , \tag{9}$$

where: L_i^Q, L_j^D are lengths of compared regions SE_i^Q and SE_j^D, while σ_{ij} describes the similarity degree of secondary structures building i-th and j-th SE regions of compared proteins Q and D. This parameter can take three possible values according to the following rules:

1. $\sigma_{ij} = 1$, when both SE regions have the same secondary structure of α-helix or β-strand;
2. $\sigma_{ij} = 0.5$, when at least one of the regions has undefined secondary structure;
3. $\sigma_{ij} = 0$, when one of the regions has the construction of α-helix and the second the construction of β-strand.

3.2   Second Phase – Alignment of Structural Signatures

Molecules that passed the first phase of the aligment (based on the user-defined cut off value) are being further aligned in the second phase. A pair of aligned molecules Q and D is now represented by structural signatures:

$$Q = (s_1^Q, s_2^Q, \ldots, s_q^Q) \text{ and } D = (s_1^D, s_2^D, \ldots, s_d^D) \ , \tag{10}$$

where: q is a length of the query protein Q (i.e. a number of its amino acids), d is a length of the database protein D, and each s_i corresponds to the i-th amino acid in the chain of the protein Q or D and is defined by the following vector of features:

$$s_i = \langle |C_i|, \gamma_i, SSE_i, r_i \rangle \ , \tag{11}$$

where: $|C_i|$ is a length of vector between C_α atoms of the i-th and $(i+1)$-th amino acid in a protein chain, γ_i is an angle between successive vectors C_i and C_{i+1}, SSE_i is a type of the secondary structure, which is formed by the i-th residue, r_i is a type of amino acid (Fig. 2).

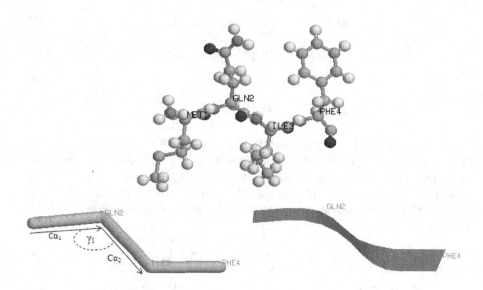

Fig. 2. Structural features included in structural signatures: (top) atomic representation with four residues visible (Met, Gln, Ile, Phe), (bottom left) vectors between C_α atoms, and γ angle, (bottom right) secondary structure element for particular residues (β-strand in the presented case)

Alignment of structural signatures is performed similarly as in the first phase. However, it takes into account the result of the alignment of secondary structure elements SE_i (SE regions) matched in the first phase. Results of the *low resolution alignment* are projected onto the new similarity matrix. For each aligned pair of regions SE_i^Q and SE_j^D from the previous phase, we calculate a new similarity matrix of the size $L_i^Q \times L_j^D$ and we align structural signatures that are inside the regions SE_i^Q and SE_j^D. This matching process is called as a *high resolution alignment*, which refines the results of the *low resolution alignment*, since it processes more structural features and in higher resolution.

The course of the *high resolution alignment* itself is analogical to the *low resolution alignment*. The main difference is the way how CASSERT calculates the similarity reward for two compared elements, which in this case are two structural signatures s_i and s_j. While calculating the similarity of two structural signatures the algorithm takes into account primary, secondary and tertiary structures of each single protein element (corresponding to one amino acid). The similarity reward is calculated according to the following formula:

$$ss_{ij} = w_C * \sigma_{ij}^C + w_\gamma * \sigma_{ij}^\gamma + w_{SSE} * \sigma_{ij}^{SSE} + w_r * \sigma_{ij}^r \; , \tag{12}$$

where: σ_{ij}^C is a similarity degree of vectors C_i^Q and C_j^D describing the location of C_α carbon atoms of residues i and j in proteins Q and D, σ_{ij}^γ is a similarity of angles γ_i^Q and γ_j^D in proteins Q and D, σ_{ij}^{SSE} is a similarity degree of secondary structures of residues i and j (calculated according to the rules 1–3, as in the first phase), σ_{ij}^r is a similarity degree of residues defined by means of the BLOSUM62 substitution matrix normalized to range of $\langle 0; 1 \rangle$, $w_C, w_\gamma, w_{SSE}, w_r$ are weights for all of the components (with default value of 1).

Similarity of vectors C_i^Q and C_j^D is defined according to the formula:

$$\sigma_{ij}^C = \exp\left(-\left(|C_i^Q| - |C_j^D|\right)^2\right) \; , \tag{13}$$

and similarity of angles γ_i^Q and γ_j^D is defined as follows:

$$\sigma_{ij}^\gamma = \exp\left(-\left(|\gamma_i^Q| - |\gamma_j^D|\right)^2\right) \; . \tag{14}$$

The value of the similarity degree of structural signatures ss_{ij} (Equation (12)) substitutes the similarity reward δ_{ij} (Equation (4)) in the *high resolution alignment*.

3.3   Assessment of Protein Structure Similarity

In order to assess the similarity between two chains of structural signatures, we use the *Score* measure. This measure is obtained for the optimal alignment path in the similarity matrix (labeled in the second phase of the algorithm as matrix S). The value always accumulates all the possible rewards for a match, mismatch penalties, and penalties for inserting gaps in the alignment (in accordance with Equations (3)–(8)) and is equal to the highest value in the similarity matrix S:

$$Score = \max\{S_{ij}\} \; , \tag{15}$$

where $i = 1, \ldots, q$, $j = 1, \ldots, d$, q is the length of the query protein Q, and d is the length of the database protein D.

The participation of each component in the similarity searching (Equation (12)) can be controlled by means of participation weights, which are set by a user. For example, researchers looking only for surprising structural similarities, but indicating no sequence similarity at the same time, can disable the component of the primary structure by setting the value of 0 for this particular component.

4   Results and Discussion

The effectiveness of the CASSERT algorithm was examined during various tests. These tests were performed with the use of DALI database [6] storing 47 697 molecular structures and 106 858 chains. The database was intalled locally on the MS SQL Server 2008R2 database management system working under controll of the Windows XP operating system. The size of the database was 12 GB. In the first part, results of the CASSERT algorithm were compared to results returned by DALI algorithm. In the second part, we compared alignments generated by CASSERT and DALI algorithms. In the third part, we compared alignment times for CASSERT, DALI, CE, and FATCAT algorithms.

In the first series of tests, we compared lists of one hundred most similar protein structures that were identified by both algorithms: CASSERT and DALI. For this purpose, we have arbitrary chosen a set of query proteins (Q). This set contained query molecules with different lengths and representing different structural classes according to the SCOP classification [21]: all α, all β, $\alpha\&\beta$, $\alpha+\beta$. In this way, we could verify the efficacy of the *low resolution alignment* phase. Query protein structures differed in size (length). We identified three groups of protein structures – short-chain proteins (up to 100 amino acids), medium-sized (up to 500 amino acids), and long chains (over 500 amino acids). Each group of protein structures was tested using each of the two algorithms. Convergence of results was observed at the level of 99.8 % for short-chain proteins, 94.2 % for medium-sized proteins, and 90.6 % for long-chain proteins.

In the second series of experiments, we verified structural alignments that were generated by both algorithms. Also in this case we observed a large convergence of results. However, analyzing the results, we also found some cases, where the alignments were slightly different for the CASSERT and DALI algorithms.

One of the cases is presented in Fig. 3. It shows structural alignment for a pair of sample structures [PDBID: 1KDD, chain A] [22] and [PDB ID: 1CE9, chain B] [23] from the DALI database. In the secondary structure of both molecules exist several α-helices. Alignments generated by the two algorithms are slightly different. In the alignment performed by DALI algorithm we can see the structural similarity of only two amino acid residues (marked by a vertical line |), while in the structural alignment performed by the CASSERT algorithm structural convergence of residues is much higher (marked by ':' symbol). Since with the structural convergence, we can also observe the convergence of amino acids at the same time (Fig. 3, right), it allows us to think that the structural alignment of these positions are correct. The similarity or even identity of residues in the compared chains, especially those observed on several successive elements, very often involves similar formation of the spatial structures of protein molecules. On this basis we conclude that including the sequence similarity into the structural alignment has a positive effect on the final result of the alignment. This was also observed in [24].

In the third series of experiments, we tested the performance of selected alignment methods. We examined CASSERT and three popular algorithms: DALI, CE and FATCAT, measuring the time of the alignment performed for pairs of

```
                                      CASSERT Structure 1: 1KDDA, Structure 2: 1CE9B
                                      Score: 76, RMSD: 2,34845524855252
DALI Query=1kddA Sbjc=1ce9B rmsd=2.4  :       ....*....|....*....|....*....|....*....|...
DSSP  LHHHHHHHHHHHHHHHHHHHHHHHHHHHHHHHHHHHHl    SSE1: LHHHHHHHHHHHHHHHHHHHHHHHHHHHHHHHHHHHH
Query EVKQLEAEVEELESEIWHLENEVARLEKENAECea    35  O001: EVKQLEAEVEELESEIWHLENEVARLEKENAEA 0033
ident       |              |                   :       .::.::..::::...::::::!:::::..:..:~
Sbjct MSVKELEDKVEELLSKNYHLENEVARLKKLVGE-r    34  O002: SVKELEDKVEELLSKNYHLENEVARLKKLVGER 0034
DSSP  LLHHHHHHHHHHHHHHHHHHHHHHHHHHHHHHHHLL-1       SSE2: LHHHHHHHHHHHHHHHHHHHHHHHHHHHHHHHHHLLL
```

Fig. 3. Structural alignment generated by DALI algorithm (left) and CASSERT algorithm (right) for sample structures [PDBID: 1KDD, chain A] and [PDB ID: 1CE9, chain B]

molecules from the three groups of protein structures. Taking into account that protein structures are very complex and the search space is huge, the alignment algorithms are usually time consuming. All algorithms complete the alignment process within several to tens of seconds. Alignment time highly depends on sizes (lengths) of compared structures. In Figure 4 we show alignment time for examined algorithms for a pair of sample molecules from the group of medium-sized proteins. Both compared structures had the length of 170 amino acids and represented different conformations of the same protein – human RAB5A and human RAB5A with a single mutation at Ala^{30}. Tests were performed on the PC workstation with the Intel Xeon CPU and 2 GB RAM. In Figure 4 we can observe that among all compared algorithms CASSERT has the lowest alignment time. This tendency was observed for all tested cases.

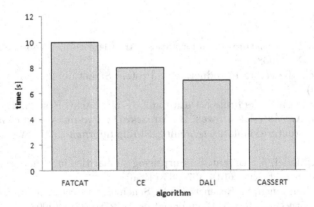

Fig. 4. Processing time while aligning a pair of molecular structures [PDB ID: 1N6H] vs. [PDB ID: 1N6N] for four tested algorithms: FATCAT, CE, DALI, and CASSERT

Having the processing time for just a single alignment, we can now easily imagine how much time will take the process of finding similar protein structures in a database storing 106 858 chains. This could take several days without any additional acceleration or filtering. Our research presented in [25,26] confirmed that for the FATCAT algorithm this takes 25 hours by implementing the process on 20 alignment agents working in parallel. CASSERT divides the time at least by two, which we consider as a good result.

5   Concluding Remarks

The CASSERT algorithm manifests high effectiveness in protein structure similarity searching. It is also characterized by a good precision, which was achieved
by including in the comparison process a set of various features regarding protein construction. The CASSERT algorithm, with the implemented method of
two-phase alignment of protein structures, returns very good results. The result
sets returned by the CASSERT algorithm are comparable to that, which were
returned by the DALI algorithm. It has been shown in our tests. The alignments
generated by both algorithms are similar, while in the course of the research we
have found cases, in which the CASSERT algorithm gave better alignment paths
than popular DALI. Moreover, the computational complexity of the CASSERT
algorithm is lower than competitive DALI, CE and FATCAT and through the
use of the phase of *low resolution alignment*, our algorithm requires fewer iterative alignments for successive structures from the database. This makes the
CASSERT a very useful tool in the identification of proteins on the basis of their
structures and in the identification of potential functions of these proteins.

Acknowledgments. This work was supported by the European Union from
the European Social Fund (grant agreement number: UDA-POKL.04.01.01-00-
106/09).

References

1. Burkowski, F.: Structural Bioinformatics: An Algorithmic Approach. Chapman
   and Hall/CRC (2008)
2. Branden, C., Tooze, J.: Introduction to Protein Structure, 2nd edn. Garland Science (1999)
3. Berman, H., et al.: The Protein Data Bank. Nucleic Acids Res. 28, 235–242 (2000)
4. Krygowski, A., Małysiak-Mrozek, B., Mrozek, D.: Two-phase alignment algorithm
   for protein structure similarity searching. Studia Informatica 33(2A)(105), 525–541
   (2012)
5. Gibrat, J., Madej, T., Bryant, S.: Surprising similarities in structure comparison.
   Curr. Opin. Struct. Biol. 6(3), 377–385 (1996)
6. Holm, L., Kaariainen, S., Rosenstrom, P., Schenkel, A.: Searching protein structure
   databases with DaliLite v.3. Bioinformatics 24, 2780–2781 (2008)
7. Holm, L., Sander, C.: Protein structure comparison by alignment of distance matrices. J. Mol. Biol. 233(1), 123–138 (1993)
8. Shapiro, J., Brutlag, D.: FoldMiner and LOCK2: protein structure comparison and
   motif discovery on the web. Nucleic Acids Res. 32, 536–541 (2004)
9. Ye, Y., Godzik, A.: Flexible structure alignment by chaining aligned fragment pairs
   allowing twists. Bioinformatics 19(2), 246–255 (2003)
10. Jamroz, M., Kolinski, A.: ClusCo: clustering and comparison of protein models.
    BMC Bioinformatics 14, 62 (2013)
11. Can, T., Wang, Y.: CTSS: A robust and efficient method for protein structure
    alignment based on local geometrical and biological features. In: Proceedings of
    the 2003 IEEE Bioinformatics Conference (CSB 2003), pp. 169–179 (2003)

12. Shindyalov, I., Bourne, P.: Protein structure alignment by incremental combinatorial extension (CE) of the optimal path. Protein Engineering 11(9), 739–747 (1998)
13. Daniluk, P., Lesyng, B.: A novel method to compare protein structures using local descriptors. BMC Bioinformatics 12, 344 (2011)
14. Mosca, R., Brannetti, B., Schneider, T.R.: Alignment of protein structures in the presence of domain motions. BMC Bioinformatics 9, 352 (2008)
15. Zhu, J., Weng, Z.: FAST: A novel protein structure algorithm. Proteins 58, 618–627 (2005)
16. Minami, S., Sawada, K., Chikenji, G.: MICAN: a protein structure alignment algorithm that can handle Multiple-chains, Inverse alignments, Ca only models, Alternative alignments, and Non-sequential alignments. BMC Bioinformatics 14, 24 (2013)
17. Sam, V., Tai, C.H., Garnier, J., Gibrat, J.F., Lee, B., Munson, P.J.: Towards an automatic classification of protein structural domains based on structural similarity. BMC Bioinformatics 9, 74 (2008)
18. Yuan, C., Chen, H., Kihara, D.: Effective inter-residue contact definitions for accurate protein fold recognition. BMC Bioinformatics 13, 292 (2012)
19. Stanek, D., Mrozek, D., Małysiak-Mrozek, B.: MViewer: Visualization of protein molecular structures stored in the PDB, mmCIF and PDBML data formats. In: Kwiecień, A., Gaj, P., Stera, P. (eds.) CN 2013. CCIS, vol. 370, pp. 323–333. Springer, Heidelberg (2013)
20. Smith, T., Waterman, M.: Identification of common molecular subsequences. J. Mol. Biol. 147, 195–197 (1981)
21. Murzin, A., Brenner, S., Hubbard, T., Chothia, C.: SCOP: A structural classification of proteins database for the investigation of sequences and structures. J. Mol. Biol. 247, 536–540 (1995)
22. Keating, A., Malashkevich, V., Tidor, B., Kim, P.: Side-chain repacking calculations for predicting structures and stabilities of heterodimeric coiled coils. Proc. Natl. Acad. Sci. USA 98(26), 14825–14830 (2001)
23. Lu, M., Shu, W., Ji, H., Spek, E., Wang, L., Kallenbach, N.: Helix capping in the GCN4 leucine zipper. J. Mol. Biol. 288(4), 743–752 (1999)
24. Daniels, N.M., Nadimpalli, S., Cowen, L.J.: Formatt: Correcting Protein Multiple Structural Alignments by Incorporating Sequence Alignment. BMC Bioinformatics 13, 259 (2012)
25. Momot, A., Małysiak-Mrozek, B., Kozielski, S., Mrozek, D., Hera, Ł., Górczyńska-Kosiorz, S., Momot, M.: Improving Performance of Protein Structure Similarity Searching by Distributing Computations in Hierarchical Multi-Agent System. In: Pan, J.-S., Chen, S.-M., Nguyen, N.T. (eds.) ICCCI 2010, Part I. LNCS, vol. 6421, pp. 320–329. Springer, Heidelberg (2010)
26. Małysiak-Mrozek, B.z., Momot, A., Mrozek, D., Hera, Ł., Kozielski, S., Momot, M.: Scalable System for Protein Structure Similarity Searching. In: Jędrzejowicz, P., Nguyen, N.T., Hoang, K. (eds.) ICCCI 2011, Part II. LNCS, vol. 6923, pp. 271–280. Springer, Heidelberg (2011)

Transfers of Entangled Qudit States in Quantum Networks

Marek Sawerwain[1] and Joanna Wiśniewska[2]

[1] Institute of Control & Computation Engineering
University of Zielona Góra, ul. Licealna 9, Zielona Góra 65-417, Poland
M.Sawerwain@issi.uz.zgora.pl
[2] Institute of Information Systems, Faculty of Cybernetics,
Military University of Technology, ul. Kaliskiego 2, 00-908 Warsaw, Poland
jwisniewska@wat.edu.pl

Abstract. The issue of quantum states' transfer – in particular, for so-called Perfect State Transfer (PST) – in the networks represented by the spin chains seems to be one of the major concerns in quantum computing. Especially, in the context of future communication methods that can be used in broadly defined computer science. The paper presents a definition of Hamiltonian describing the dynamics of quantum data transfer in one-dimensional spin chain, which is able to transfer the state of unknown qudits. The main part of the paper is the discussion about possibility of entangled states' perfect transfer, in particular, for the generalized Bell states for qudits. One of the sections also contains the results of numerical experiments for the transmission of quantum entangled state in a noisy quantum channel.

Keywords: quantum information transfer, qudits chains, entangled quantum states, numerical simulations.

1   Introduction

The problem of Perfect State Transfer (PST), raised i.a. in [1], generates a very important area of quantum computing, which is naturally combined with the context of information transfer in quantum channels. More information on PST can be found in [2] and [3].

A quantum state's transfer from the specified start position to another position is called a perfect state transfer, if the Fidelity value of the initial and the final quantum state is equal to one. The possibility of perfect transmission not only in XX-like and XY-like spin chains is shown in [4–6] and in previously cited [2, 3]. The individual chain's elements are inhomogeneously coupled. The spin chains with homogeneous coupling provide the transfer only for short chains – containing three or four elements. It is important that the transfer does not require any additional intervention except the influence of dynamics described by the corresponding Hamiltonian. The techniques of spin construction for PST are also significant – examples in [7, 8].

A. Kwiecień, P. Gaj, and P. Stera (Eds.): CN 2013, CCIS 370, pp. 344–353, 2013.
© Springer-Verlag Berlin Heidelberg 2013

Naturally, the main task of perfect state transfer protocols is addressed to transfer quantum state in qubits/qudits 1D chains, and also in more complex quantum networks, e.g. 2D grid or graphs. However, the use of chain to transfer entangled states [9] is also possible, but at least two chains must be used – see paper [10].

It should be also mentioned, that the perfect state transfer systems are mainly used to construct other primitive protocols e.g. GHZ and W states preparation, general entanglement generation, initialisation of system's state, signal amplification in measurements, realisations of quantum walks and universal quantum computation.

The main objective of this paper is to verify whether the entangled state can be also transferred to a higher-dimensional space with use of spin chains with suitably chosen dynamics.

The content of this paper is as follows: in Sect. 2 the definition of Lie algebra's generator is presented and it is combined with the Hamiltonian definition, which is responsible for sending a quantum state by the spin chain. The spin chain described in introduction of publication [11] is briefly characterised in Sect. 3. The definition of transferred entangled states and the construction of a spin chain is presented in Sect. 4. In Section 5 are shown: numerical simulations of entangled states' transfer and exemplary simulations of transfer process in an environment where noise is present. The summary and short term objectives are outlined in Sect. 6. The paper's last section consists of acknowledgments and a list of cited literature.

2   Generators of Lie Algebra

In proposed definition of the XY-like Hamiltonian for qudits' chain (given in the Sect. 4 of this paper), Lie algebra's generators for a group $SU(d)$, where $d \geq 2$, will be used to define a suitable operator which is responsible for transfer's dynamics. For clarity, the following well known set construction procedure of $SU(d)$ generators will be recalled: in the first step, a set of projectors is defined

$$(P^{k,j})_{v,\mu} = |k\rangle\langle j| = \delta_{v,j}\delta_{\mu,k}, \quad 1 \leq v, \ \mu \leq d \ . \tag{1}$$

The first suite of $d(d-1)$ operators from the group $SU(d)$ is specified as

$$\Theta^{k,j} = P^{k,j} + P^{j,k}, \quad \beta^{k,j} = -i(P^{k,j} - P^{j,k}) \ , \tag{2}$$

and $1 \leq k < j \leq d$.

The remaining $(d-1)$ generators are defined in the following way

$$\eta^{r,r} = \sqrt{\frac{2}{r(r+1)}} \left[\left(\sum_{j=1}^{r} P^{j,j} \right) - rP^{r+1,r+1} \right] \ , \tag{3}$$

and $1 \leq r \leq (d-1)$. Finally, the $d^2 - 1$ operators belonging to the $SU(d)$ group can be obtained.

Remark 1. For $d = 2$ obtained suite of $SU(d)$ operators is the set of Pauli operators:

$$\sigma_x = X = \begin{pmatrix} 0 & 1 \\ 1 & 0 \end{pmatrix}, \quad \sigma_y = Y = \begin{pmatrix} 0 & -i \\ i & 0 \end{pmatrix}, \quad \sigma_z = Z = \begin{pmatrix} 1 & 0 \\ 0 & -1 \end{pmatrix}, \qquad (4)$$

while for $d = 3$ the set of Gell-Mann operators λ_i will be obtained:

$$\lambda_1 = \Theta^{1,2} = \begin{pmatrix} 0 & 1 & 0 \\ 1 & 0 & 0 \\ 0 & 0 & 0 \end{pmatrix}, \quad \lambda_2 = \beta^{1,2} = \begin{pmatrix} 0 & -i & 0 \\ i & 0 & 0 \\ 0 & 0 & 0 \end{pmatrix}, \quad \lambda_3 = \eta^{1,1} = \begin{pmatrix} 1 & 0 & 0 \\ 0 & -1 & 0 \\ 0 & 0 & 0 \end{pmatrix},$$

$$\lambda_4 = \Theta^{1,3} = \begin{pmatrix} 0 & 0 & 1 \\ 0 & 0 & 0 \\ 1 & 0 & 0 \end{pmatrix}, \quad \lambda_5 = \beta^{1,3} = \begin{pmatrix} 0 & 0 & -i \\ 0 & 0 & 0 \\ i & 0 & 0 \end{pmatrix}, \quad \lambda_6 = \Theta^{2,3} = \begin{pmatrix} 0 & 0 & 0 \\ 0 & 0 & 1 \\ 0 & 1 & 0 \end{pmatrix},$$

$$\lambda_7 = \beta^{2,3} = \begin{pmatrix} 0 & 0 & 0 \\ 0 & 0 & -i \\ 0 & i & 0 \end{pmatrix}, \quad \lambda_8 = \eta^{2,2} = \frac{1}{\sqrt{3}} \begin{pmatrix} 1 & 0 & 0 \\ 0 & 1 & 0 \\ 0 & 0 & -2 \end{pmatrix}.$$

$$(5)$$

To define the XY-like dynamics all listed above operators are not necessary – only operators $\Theta^{k,j}$ and $\beta^{k,j}$ according to Equation (2) are used.

3   Definition of Hamiltonian for Qudits

In this paper the following Hamiltonian H^{XY_d} (firstly presented in [11]) is used to realise the perfect transfer of quantum information in qudits chains. It is also claimed that each qudit has the same level and $d \geq 2$:

$$H^{XY_d} = \sum_{(i,i+1) \in \mathcal{L}(G)} \frac{J_i}{2} \left(\Theta_{(i)}^{k,j} \Theta_{(i+1)}^{k,j} + \beta_{(i)}^{k,j} \beta_{(i+1)}^{k,j} \right), \qquad (6)$$

where J_i is defined as follows: $J_i = \frac{\sqrt{i(N-i)}}{2}$ for $1 \leq k < j < d$ and $\Theta_{(i)}^{k,j}$, $\beta_{(i)}^{k,j}$ are $SU(d)$ group operators defined by (2) applied to the (i)-th and $(i+1)$-th qudit. The Hamiltonian (6) will be also called the transfer Hamiltonian.

It is not hard to show that

$$\left[H^{XY_d}, \sum_{i=1}^{N} \eta_{(i)}^{r,r} \right] = 0 \qquad (7)$$

for $1 \leq r \leq (d-1)$, as in the definition of the XY-like Hamiltonian for qubits.

4   Transfer's Example for Entangled States

The entangled states – so-called Bell states – for two qubits are:

$$|\psi^{\pm}\rangle = \frac{1}{\sqrt{2}} \left(|00\rangle \pm |11\rangle \right), \quad |\phi^{\pm}\rangle = \frac{1}{\sqrt{2}} \left(|01\rangle \pm |10\rangle \right). \qquad (8)$$

It means there are four Bell states for two-qubit system (respecting the sign). For qudits with d levels, more complex units than qubits, the equivalent of state $|\psi^{\pm}\rangle$ is

$$|\psi^{\pm}\rangle = \frac{1}{\sqrt{d}}\left(|00\rangle \pm |nn\rangle\right) , \qquad (9)$$

where $n = d - 1$.

The definition of generalised Bell states for qudits with freedom level d is as follows

$$|\psi_{pq}\rangle = \frac{1}{\sqrt{d}}\sum_{j=0}^{d-1} e^{2\pi ijp/d}|j\rangle|(j+q)\bmod d\rangle \quad 0 \le p, q \le d-1 , \qquad (10)$$

where i represents an imaginary unit. Generally exist d^2 Bell states for two d-level qudits.

It is convenient to express the last equation as a circuit of quantum gates for generalised EPR pair.

$$|\psi_{pq}^d\rangle = (I_d \otimes X_d)^q \cdot (H_d \otimes I_d) \cdot (Z_d \otimes I_d)^p \cdot \text{CNOT}_d \cdot |00\rangle \qquad (11)$$

Remark 2. The above form of EPR pair is also used in a process of transferred state's correction – for the transfer of maximally entangled qudits, the transferred qudit is still maximally entangled, but the values of amplitudes usually represent other maximally entangled state. Other examples of the use of entangled states in transfer procotol can be found i.e. in [12].

For unknown pure state of one qudit

$$|\psi\rangle = \alpha_0|0\rangle + \alpha_1|1\rangle + \ldots + \alpha_{d-1}|d-1\rangle \quad \text{and} \quad \sum_{i=0}^{d-1}|\alpha_i|^2 = 1, \text{ where } \alpha_i, \in \mathbb{C}, \quad (12)$$

the transfer process (or transfer protocol) in a one-dimensional chain of n qudits is expressed as a transformation of the state $|\Psi_{\text{in}}\rangle$ into the state $|\Psi_{\text{out}}\rangle$:

$$|\Psi_{\text{in}}\rangle = |\psi\rangle|\underbrace{000\ldots0}_{n-1}\rangle \implies |\Psi_{\text{out}}\rangle = |\underbrace{000\ldots0}_{n-1}\rangle|\psi\rangle . \qquad (13)$$

If the transfer is performed on entangled state of two qudits, the transfer protocol realises the transmission of both qudits' quantum state. Naturally, the entanglement – as the state's feature – have to be also transferred. The state $|\Psi_{\text{out}}\rangle$ corresponds to $|\Psi_{\text{in}}\rangle$ according to selected value of Fidelity:

$$|\Psi_{\text{in}}\rangle = |\psi_{pq}\rangle|\underbrace{000\ldots0}_{n-2}\rangle \implies |\Psi_{\text{out}}\rangle = |\underbrace{000\ldots0}_{n-2}\rangle|\psi_{pq}\rangle . \qquad (14)$$

where $|\psi_{pq}\rangle$ is a two-qudit system. The spin chains for transferring qudits' states are briefly shown at the Fig. 1.

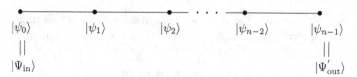

(a) one-dimensional chain to transfer state of one qudit

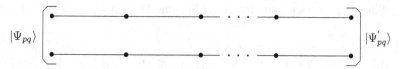

(b) two one-dimensional chains to transfer of entangled state of two qudits

Fig. 1. The realisation of information transfer in qudits chains for single (a) and entangled states (b), the interactions between qudits are performed only between adjacent qudits

An exemplary transfer of Bell state $|\psi^+\rangle$ with use of Hamiltonian (6) for four-qubit spin chain is a perfect state transfer. It means there is no additional correction needed. After the transfer process the obtained state is $|\psi^+\rangle$ Bell state. Generally, during the transfer the chain's state is:

$$|\Psi_t\rangle = \alpha_0|0000\rangle + \alpha_1|0011\rangle + \alpha_2|0110\rangle + \alpha_3|1001\rangle + \alpha_4|1100\rangle , \qquad (15)$$

and the final chain's state may be expressed as:

$$|\Psi\rangle = \alpha_0|0000\rangle + \alpha_1|0011\rangle . \qquad (16)$$

For Bell state $|\phi^+\rangle$ its transfer is a perfect state transfer in terms of Fidelity value. However, it has to be taken into account that the amplitudes' values are multiplied by imaginary unit i.

Remark 3. Naturally, the Hamiltonian (6) for qubits is XY-like Hamiltonian [1, 10].

It is easy to calculate the index of each node for a n-node spin chain. Of course, these values depend on transferred state. For the transmission of Bell state $|\psi^+\rangle$ – using binary codding and realising bit shift operations – the states are used where bit value $(11)_2$ is shifted through all chain's nodes:

$$\begin{array}{c} |11000\ldots000\rangle \\ |01100\ldots000\rangle \\ |00110\ldots000\rangle \\ |00011\ldots000\rangle \\ |00000\ldots110\rangle \\ |00000\ldots011\rangle \end{array} \qquad (17)$$

the similar action for $(1001)_2$:

$$
\begin{aligned}
&|1001000\ldots00000\rangle \\
&|0100100\ldots00000\rangle \\
&|0010010\ldots00000\rangle \\
&|0001001\ldots00000\rangle \\
&|0000000\ldots10010\rangle \\
&|0000000\ldots01001\rangle
\end{aligned}
\tag{18}
$$

where the complement states are:

$$|1000\ldots001\rangle \quad \text{and} \quad |00000\ldots000\rangle \ .$$

The entangled states may also be changed (according to the length of spin chain), if the qudits are transferred with use of the Hamiltonian (6). Mentioned change respects to the phase form and to the entangled state itself. However, the obtained state is still maximally entangled.

Just like for $|\psi^+\rangle$ the transfer of state:

$$|\psi\rangle = \frac{1}{\sqrt{3}} \left(|00\rangle + |11\rangle + |22\rangle\right) \tag{19}$$

is perfect (PST) and no additional conversion is needed. Just like before, adding the fifth qudit to the spin chain will change the state after the transfer as follows:

$$|\psi\rangle = \frac{1}{\sqrt{3}} \left(|00\rangle - |11\rangle - |22\rangle\right) \ . \tag{20}$$

For the transfer of state:

$$|\psi\rangle = \frac{1}{\sqrt{3}} \left(|01\rangle + |12\rangle + |20\rangle\right) \tag{21}$$

the state after transfer is:

$$|\psi'\rangle = \frac{1}{\sqrt{3}} \left(|02\rangle + |10\rangle - |12\rangle\right) \ . \tag{22}$$

The double use of gate X is needed on the second qudit of state $|\psi'\rangle$ to obtain the entangled state. Of course, this local operation will not change the level of entanglement.

Generalising, the length of spin chain for transferring entangled states affects on amplitudes' phase shifts (it is important if the number of nodes is odd or even). The transmission of state $|\psi^+\rangle$ in spin chain with five nodes causes that the state $|\psi^-\rangle$ will be obtained. Naturally, the state will be still maximally entangled.

Remark 4. The issue of phase shift or amplitudes' permutation in maximally entangled states causes no problems, because there are deterministic procedures of Bell states detection both for qubits [13] and qudits [14] in generalised Bell states. Using the circuits described in the cited papers it is possible to undoubtedly identify, without any damage on EPR pair, the quantum state after the process of transfer.

The package QCS (Quantum Computing Simulator) – developed at the University of Zielona Góra – was used in the experiment for the numerical simulation of spin chain's behaviour. The program's main loop for transfer's simulation is briefly presented at Fig. 2.

```
import qcs                          # simulation loop
                                    s=0
# five qudits with d = 3           while i < 8:
q = qcs.QuantumReg(5,3)                ... other operations
q.Reset()                              q.ApplyOperator( op )
                                       ... other operation
# transfer operator                    ... e.g. noise introduction
op = q.XYTranHamiltonian(
    _fromqudit=0,
    _toqudit=4,
    _step = 8)                      # display state of
                                    # quantum register
                                    q.Pr()
# create generalised Bell state
# e.g. |00> + |11> + |22>
# at qudits zero and one
q.SetGBellState(0,1,0,0)
```

Fig. 2. A Python script using QCS package to simulate entangled states' transfer in qudit chain. The process of transfer is realised by eight simulation steps.

5   Transfers of Entangled States in Noisy Channels

The research in the field of quantum circuits, where noise is present, is a very important issue of quantum computing because of the decoherence phenomenon – e.g. the paper [15] shows the impact of noise in quantum channels on: amplitude-damping, phase-damping and bit-flip in Grover's algorithm for database search. In this paper only the phase-dumping influence on qudit chain for entangled state transfer will be presented. However, in the case of qudit the phase-damping operation does not have a unique representation. The model discussed in the work [16] is an example of the phase-damping operation and it will be used in this paper:

$$\mathcal{E}(\rho) = \sum_{i=0}^{d-1} E_i \rho E_i^\dagger, \quad E_i = \sqrt{\binom{d-1}{i}\left(\frac{1-p}{2}\right)^i \left(\frac{1+p}{2}\right)^{d-1-i}} Z^i , \quad (23)$$

where $0 \leq p \leq 1$.

Remark 5. It should be noted that expression (23) can be regarded as a special case of Weyl's channel [16]:

$$\mathcal{E}(\rho) = \sum_{m,n=0}^{d-1} \pi_{m,n} (Z^n X^m) \rho (X^m Z^n)^\dagger , \quad (24)$$

where elements of the matrix π satisfy the following conditions: $0 \le \pi_{m,n} \le 1$ and $\sum_{m,n=0}^{d-1} \pi_{m,n} = 1$. The operators Z and X are generalised Pauli matrices for the sign changing and negation operations on qudits.

At the Fig. 3 the process of transfer with noise-adding operation is shown. The noise type is phase-damping – presented in Equation (23). The diagram of Fidelity value for entangled qudit state transfer $|\psi\rangle = 1/\sqrt{3}\,(|00\rangle + |11\rangle + |22\rangle)$, with $d = 3$, is shown at the Fig. 4. Increasing the value of p parameter means the reduction of phase-damping impact on transfer process and the value of Fidelity will be raising in accordance with next stages of the process. It should be pointed that the value of Fidelity is insensitive for the change of the sign or probability amplitudes' phase-shift. However, in the case analysed in this paper, the distortion of transfer process does not change the transferred state, but only distorts the level of entanglement. Of course, the transferred state is not a maximally entangled state. It means that the small noise level, for example $p = 0.05$, is acceptable and still provides the high value of Fidelity.

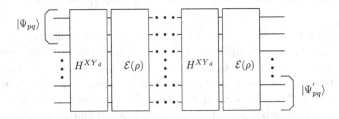

Fig. 3. The transfer of entangled state in a noisy channel implemented as a quantum circuit. The first part of the transfer is the operation X^{XY_d}. The second part of the circuit represents the noise. The transfer process consists of discrete stages, so the stages of operation X^{XY_d} and noise operation must be repeated.

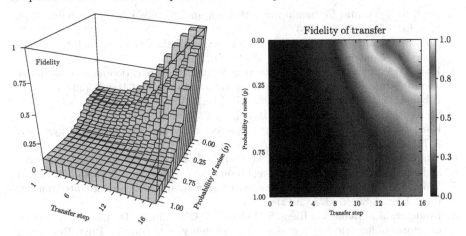

Fig. 4. The diagram of Fidelity value for maximally entangled state transfer in five qutrits' chain. The p parameter describes the phase distortion for each qudit. The transfer process consists of sixteen discrete steps.

6 Conclusions

Naturally, for entangled states transfer the next step is multi-qudit-state transmission. It is expected that this type of operation is possible, although the number of additional local corrections performed on transferred state will be greater than the number of these operations for EPR pairs.

The analytical solutions for spin chains are direct consequence of the correctness of the Hamiltonian (6). However, their form and the correctness of Hamiltonian's construction is analysed in presently prepared work [17]. It should be also mentioned that the analytical form of spin chain, with strictly specified number of nodes, makes the effective numerical calculations possible, because the number of nodes in qudit chain is much lower than the number of states in quantum register.

Another very important issue is the entangled states engineering, e.g. [18-20], which may be implemented by modified Hamiltonian describing transfer's dynamics in spin chains. The described problem is worth of further investigation in other areas of quantum computer science e.g. in [21] and [22].

Acknowledgments. We would like to thank for useful discussions with the *Q-INFO* group at the Institute of Control and Computation Engineering (ISSI) of the University of Zielona Góra, Poland. We would like also to thank to anonymous referees for useful comments on the preliminary version of this paper. The numerical results were done using the hardware and software available at the "GPU μ-Lab" located at the Institute of Control and Computation Engineering of the University of Zielona Góra, Poland.

References

1. Bose, S.: Quantum Communication through an Unmodulated Spin Chain. Phys. Rev. Lett. 91, 207901 (2003)
2. Bose, S.: Quantum Communication through Spin Chain Dynamics: an Introductory Overview. Contemporary Physics 48, 13–30 (2007)
3. Kay, A.: A Review of Perfect State Transfer and its Application as a Constructive Tool. International Journal of Quantum Information 8(4), 641–676 (2012)
4. Albanese, C., Christandl, M., Datta, N., Ekert, A.: Mirror Inversion of Quantum States in Linear Registers. Phys. Rev. Lett. 93, 230502 (2004)
5. Karbach, P., Stolz, J.: Spin chains as perfect quantum state mirrors. Phys. Rev. A 72, 030301(R) (2005)
6. Wu, L.-A., Miranowicz, A., Wang, X., Liu, Y.-X., Nori, F.: Perfect function transfer and interference effects in interacting boson lattices. Phys. Rev. A 80, 012332 (2009)
7. Vinet, L., Zhedanov, A.: How to construct spin chains with perfect state transfer. Phys. Rev. A 85, 012323 (2012)
8. Bruderer, M., Franke, K., Ragg, S., Belzig, W., Obreschkow, D.: Exploiting boundary states of imperfect spin chains for high-fidelity state transfer. Phys. Rev. A 85, 022312 (2012)
9. Horodecki, R., Horodecki, P., Horodecki, M., Horodecki, K.: Quantum Entanglement. Rev. Mod. Phys. 81, 865–942 (2009)

10. Christandl, M., Datta, N., Dorlas, T.C., Ekert, A., Kay, A., Landahl, A.J.: Perfect transfer of arbitrary states in quantum spin networks. Phys. Rev. A 71, 032312 (2005)
11. Sawerwain, M., Gielerak, R.: Transfer of quantum continuous variable and qudit states in quantum networks. In: Kwiecień, A., Gaj, P., Stera, P. (eds.) CN 2012. CCIS, vol. 291, pp. 63–72. Springer, Heidelberg (2012)
12. Ming-Ming, W., Xiu-Bo, C., Shou-Shan, L., Yi-Xian, Y.: Efficient entanglement channel construction schemes for a theoretical quantum network model with d-level system. Quantum Information Processing 11(6), 1715–1739 (2012)
13. Gupta, G., Panigrahi, P.: Deterministic Bell State Discrimination. arXiv:quant-ph/0504183 (2005)
14. Gupta, M., Pathak, P., Srikanth, S., Panigrahi, P.: Non-destructive Orthonormal State Discrimination. arXiv:quant-ph/0507096 (2005)
15. Gawron, P., Klamka, J., Winiarczyk, R.: Noise Effects in the Quantum Search Algorithm from the Viewpoint of Computational Complexity. Int. J. Appl. Math. Comput. Sci. 22(2), 493–499 (2012)
16. Fukuda, F., Holevo, H.: On Weyl-covariant channels. arXiv:quant-ph/0510148 (2005)
17. Sawerwain, M.: Perfect states transfers for qudits chains (in preparation)
18. Kowalewska-Kudłaszyk, A., Leoński, W., Peřina Jr., J.: Photon-number entangled states generated in Kerr media with optical parametric pumping. Phys. Rev. A 83, 052326 (2011)
19. Leoński, W., Kowalewska-Kudłaszyk, A.: Quantum scissors – finite-dimensional states engineering. In: Progress in Optics, vol. 56, pp. 131–185 (2011)
20. Kowalewska-Kudłaszyk, A., Leoński, W., Peřina Jr., J.: Generalized Bell states generation in a parametrically excited nonlinear couple. Physica Scripta 147, 014016 (2012)
21. Klamka, J., Węgrzyn, S., Znamirowski, L., Winiarczyk, R., Nowak, S.: Nano and quantum systems of informatics. Nano and quantum systems of informatics. Bulletin of the Polish Academy of Sciences. Technical Sciences 52(1), 1–10 (2004)
22. Klamka, J., Gawron, P., Miszczak, J., Winiarczyk, R.: Structural programming in quantum octave. Bulletin of the Polish Academy of Sciences. Technical Sciences 58(1), 77–88 (2010)

An Analysis of the Ping-Pong Protocol Operation in a Noisy Quantum Channel

Piotr Zawadzki

Institute of Electronics,
Silesian University of Technology,
Akademicka 16, 44-100 Gliwice, Poland
Piotr.Zawadzki@polsl.pl

Abstract. A generalized approach to Ping-Pong protocol analysis is introduced. The method is based on investigation of the density operator describing joint systems of communicating parties and an eavesdropper. The method is more versatile than approaches used so far as it permits on incorporation of different noise models in a unified way and make use of well grounded theory of quantum discrimination in estimation of eavesdropper's information gain. As the proof of the method usefulness an example of its application to the analysis of the protocol execution over depolarizing and dephasing channels is given.

Keywords: quantum direct communication, quantum cryptography.

1 Introduction

In the last two decades, we have witnessed several scientific discoveries which permitted to utilize quantum mechanical principles to enhance our abilities to compute and communicate [1]. Exploiting quantum nature of composite systems is of particular relevance in developing quantum technology for efficient computation [2–4] and secure communication [5] exceeding classical limits [6]. Non-locality and entanglement are the most prominent signatures of non-classicality [7]. In particular, entanglement of shared quantum states is the vital element for the success of Quantum Key Distribution (QKD) and Quantum Direct Communication (QDC) protocols, including the quantum dense information coding [8, 9] and quantum teleportation of states [10]. QKD schemes provide cryptographically secure keys [7] which are subsequently used to protect classical telecommunication links with methods known from classic cryptography [11]. In contrary, QDC protocols do not require prior key agreement and their security results from the laws of quantum mechanics [12].

The so called Ping-Pong protocol has attracted a lot of attention as it is asymptotically secure in lossless channels [13]. The theoretical success of the protocol has been closely followed by the experimental implementation and the proof on concept installation has been realized in the laboratory [14]. It has been also shown that protocol variants based on higher dimensional systems and

A. Kwiecień, P. Gaj, and P. Stera (Eds.): CN 2013, CCIS 370, pp. 354–362, 2013.
© Springer-Verlag Berlin Heidelberg 2013

exploiting dense information coding also share features of the seminal version when some improvements are introduced [15, 16].

The Ping-Pong protocol, similarly to other QDC protocols, operates in two modes: a message mode is designed for information transfer and a control mode is used for eavesdropping detection. Although the Ping-Pong protocol is asymptotically secure in perfect quantum channels, the situation looks worse in noisy environments when legitimate users tolerate some level of transmission errors and/or losses. If that level is too high compared to the quality of the channel, then an eavesdropper can peek some fraction of signal particles hiding himself behind accepted Quantum Bit Error Rate (QBER) threshold [17, 18]. But the possibility to intercept some part of the message without being detected renders the protocol insecurity. To cope with this problem an additional purely classical layer has been proposed [19]. However, estimation of security improvement offered by that layer heavily depends on observed QBER. Unfortunately the used so far methods of the protocol analysis do not offer mathematical apparatus capable to estimate QBER in noisy channels. The purpose of the following text is to fill in this gap.

The proposed method is based on the investigation of the properties of the density matrix describing the joint system of the communicating parties. This is in contrast with previous approaches in which probability distribution observed by the eavesdropper has been derived by manipulations on state vectors. The introduced approach is more general as it permits on easy incorporation of different models of noise in a unified way and make use of quantum states discrimination theory achievements [20–22] in estimation of eavesdropper's information gain and calculation of QBER observed by the receiver. As the proof of concept the example of method application to the analysis of the protocol operation over depolarizing channel is given.

2   Ping-Pong Protocol New Description

Let us consider the seminal version of the Ping-Pong protocol [13] in which the message and control mode are executed only in computational basis. The communication process is started by Bob, the recipient of information, who prepares an EPR pair

$$|\phi^+\rangle = (|0_B\rangle|0_A\rangle + |1_B\rangle|1_A\rangle)/\sqrt{2} \ . \tag{1}$$

At the same time eavesdropping Eve controls her own system, which is initially described by state $|\chi_E\rangle$. As the states of Bob and Eve are separated, the density matrix of the whole system reads

$$\rho_{BAE}^{(0)} = \rho_{BA}^{(0)} \otimes \rho_E^{(0)} = |\phi^+\rangle\langle\phi^+| \otimes |\chi_E\rangle\langle\chi_E| \ . \tag{2}$$

Next Bob sends a signal qubit A to Alice. This qubit on its way can be influenced by two factors: quantum noise because of channel imperfection and malicious activities of Eve who may entangle it with the system controlled by herself. Let us assume, that Eve is positioned close to Alice, so her action takes place on the qubit modified by the noise. The density matrix of the system just before signal qubit enters environment controlled by Alice reads

$$\rho_{\text{BAE}}^{(1)} = (\mathcal{N}_{\text{BA}} \otimes \mathcal{I}_{\text{E}}) \left(\rho_{\text{BA}}^{(0)} \otimes \rho_{\text{E}}^{(0)} \right) = \rho_{\text{BA}}^{(1)} \otimes \rho_{\text{E}}^{(0)} \,, \tag{3}$$

where it was explicitly highlighted that noise operator $\mathcal{N}$ acts only on the EPR pair ($\mathcal{I}$ denotes identity operation). Before signal qubit enters Alice's environment, Eve can entangle it with her own system

$$\rho_{\text{BAE}}^{(2)} = (\mathcal{I}_{\text{B}} \otimes \mathcal{E}_{\text{AE}}) \, \rho_{\text{BAE}}^{(1)} \,, \tag{4}$$

where entangling operator $\mathcal{E}_{\text{AE}}$ acts only on qubit A of the EPR pair and system possessed by Eve. At that point of protocol execution Alice can select a control mode or continue in information mode.

In the former case she measures received qubit in computational basis, i.e. performs von Neumann measurement using projectors $M_{x,A} = \mathcal{I}_{\text{B}} \otimes |x_A\rangle\langle x_A| \otimes \mathcal{I}_{\text{E}}$, $x = 0,1$. Probability that she finds qubit under investigation in state $|x\rangle$ (measures ± 1) is given by

$$p_A(x) = \text{Tr} \left(\rho_{\text{BAE}}^{(2)} M_{x,A} \right) \,. \tag{5}$$

After measurement the state of the whole system collapses to

$$\sigma_{x\,\text{BAE}}^{(2)} = \frac{M_{x,A} \rho_{\text{BAE}}^{(2)} M_{x,A}}{\text{Tr} \left(\rho_{\text{BAE}}^{(2)} M_{x,A} \right)} \,. \tag{6}$$

Subsequently Bob measures his qubit in computational basis using projectors $M_{y,B} = |y_B\rangle\langle y_B| \otimes \mathcal{I}_A \otimes \mathcal{I}_E$, $y = 0,1$. Probability that Bob finds his qubit in state $|y\rangle$ provided that Alice has found his qubit in state $|x\rangle$ is given by

$$p_{B|A}(y|x) = \text{Tr} \left(\sigma_{x\,\text{BAE}}^{(2)} M_{y,B} \right) \,. \tag{7}$$

It follows that errors in control mode appear with probability

$$P_{EC} = p_{B|A}(1|0)\, p_A(0) + p_{B|A}(0|1)\, p_A(1) \,. \tag{8}$$

In information mode, Alice encodes classic bit μ applying ($\mu = 1$) or not ($\mu = 0$) operator Z_A to the possessed qubit. The system state after encoding is given by

$$\rho_{\mu\,\text{BAE}}^{(3)} = (\mathcal{I}_{\text{B}} \otimes Z_A^\mu \otimes \mathcal{I}_{\text{E}}) \, \rho_{\text{BAE}}^{(2)} \left(\mathcal{I}_{\text{B}} \otimes (Z_A^\mu)^\dagger \otimes \mathcal{I}_{\text{E}} \right) \,. \tag{9}$$

The qubit A is sent back to Bob after encoding operation. Eve's task is to discriminate between states $\rho_{\mu\,\text{AE}}^{(3)} = \text{Tr}_B \left(\rho_{\mu\,\text{BAE}}^{(3)} \right)$ with maximal confidence. The system states after reception by Bob of a qubit A travelling back from Alice and in the absence of Eve measurements are given by

$$\rho_{\mu\,\text{BAE}}^{(4)} = (\mathcal{N}_{\text{BA}} \otimes \mathcal{I}_{\text{E}}) \, \rho_{\mu\,\text{BAE}}^{(3)} \,, \tag{10}$$

so Bob has to distinguish the states

$$\rho_{\mu BA}^{(4)} = \text{Tr}_E \left(\rho_{\mu BAE}^{(4)} \right) . \tag{11}$$

When Eve performs measurements, the same quantum discrimination strategy is used but Bob is unconscious that measured states are of the form

$$\tau_{\mu,\alpha BA}^{(4)} = \text{Tr}_E \left((\mathcal{N}_{BA} \otimes \mathcal{I}_E) \frac{M_{\alpha,E} \rho_{\mu BAE}^{(3)} M_{\alpha,E}}{\text{Tr}\left(\rho_{\mu BAE}^{(3)} M_{\alpha,E} \right)} \right) . \tag{12}$$

The analysis of the protocol should determine Eve's information gain I_E and probability of erroneous Bob's decoding $QBER$ as a functions of probability of error observed in control mode P_{EC} and, optionally, parameters describing noise operator $\mathcal{N}$.

3   Active Eavesdropping in the Noiseless Case

Ping-pong protocol active eavesdropping in perfect quantum channels has been analysed many times and protocol properties for this scenario are well known. The aim of this section is to show, that generalized approach presented in the previous section gives the same results. In the considered case noise operator is reduced to identity ($\mathcal{N} = \mathcal{I}$) and the most general entangling operation can be described as [13]

$$\mathcal{E}_{AE}|0_A\rangle|\chi_E\rangle \rightarrow a|0_A\rangle|0_E\rangle + b|1_A\rangle|1_E\rangle \tag{13}$$
$$\mathcal{E}_{AE}|1_A\rangle|\chi_E\rangle \rightarrow c|0_A\rangle|2_E\rangle + d|1_A\rangle|3_E\rangle \tag{14}$$

where map's coefficient are not independent: $|a|^2 + |b|^2 = 1$, $|c|^2 + |d|^2 = 1$, $|a| = |d|$, $|b| = |c|$. After some tedious calculations one gets $P_{EC} = |b|^2$ and states accessible to Eve take form

$$\begin{aligned}
\rho_{\mu AE}^{(3)} = \frac{1}{2} [\ &|a|^2 |0_A\rangle|0_E\rangle\langle 0_A|\langle 0_E| + (-1)^\mu a^* b |1_A\rangle|1_E\rangle\langle 0_A|\langle 0_E| + \\
&+ (-1)^\mu a b^* |0_A\rangle|0_E\rangle\langle 1_A|\langle 1_E| + |b|^2 |1_A\rangle|1_E\rangle\langle 1_A|\langle 1_E| + \\
&+ |c|^2 |0_A\rangle|2_E\rangle\langle 0_A|\langle 2_E| + (-1)^\mu d c^* |1_A\rangle|3_E\rangle\langle 0_A|\langle 2_E| + \\
&+ (-1)^\mu c d^* |0_A\rangle|2_E\rangle\langle 1_A|\langle 3_E| + |d|^2 |1_A\rangle|3_E\rangle\langle 1_A|\langle 3_E|] .
\end{aligned} \tag{15}$$

At this point Holevo bound is usually used to estimate Eve's information gain

$$I_E^\mathcal{H} = S\left(\frac{1}{2}\rho_{0\,AE}^{(3)} + \frac{1}{2}\rho_{1\,AE}^{(3)} \right) - \frac{1}{2}S\left(\rho_{0\,AE}^{(3)} \right) I_E^\mathcal{H} - \frac{1}{2}S\left(\rho_{1\,AE}^{(3)} \right) = H\left(P_{EC} \right) \tag{16}$$

where $S\left(\cdot\right)$ denotes von Neumann entropy and $H\left(\cdot\right)$ – entropy of a binary source. This result is in perfect agreement with data presented in literature [13, 15, 16]. However, the above estimate is an overkill in the considered scenario as it implicitly assumes that Eve has infinite number of $\rho_{\mu AE}^{(3)}$ states and she can perform a series of collective measurements. In practice, Eve can only mount an individual attack in which she has single copy of $\rho_{\mu\,AE}^{(3)}$ for the given value of μ and she can perform only one measurement. It seems that unambiguous discrimination [20] is the most reasonable approach in such situation, although some other strategies are possible [21]. Eve's information gain is then equal to

$$I_E = P_s \log_2 N \ , \tag{17}$$

where P_s is a probability of successful measurement and N denotes the number of discriminated states. In the considered protocol version $N = 2$ and the upper bound on P_s is given by [22]

$$P_s^{\max} = 1 - F(\rho_0, \rho_1) \tag{18}$$

where $F(\rho_0, \rho_1) = \mathrm{Tr}\left|\sqrt{\rho_0}\sqrt{\rho_1}\right|$ denotes fidelity, $\mathrm{Tr}\,|\mathcal{A}| = \sum_k |\lambda_k|$ where λ_k are eigenvalues of $\mathcal{A}$ and it was assumed that states ρ_μ are equally probable. Using (18) to states $\rho_{\mu\,AE}^{(3)}$ specified in (15) one gets

$$F(\rho_0, \rho_1) = (1 - 2P_{EC})^2 \tag{19}$$

what leads to the following expression for an upper bound on Eve's information gain in individual attacks

$$I_E = 4P_{EC}(1 - P_{EC}) \ . \tag{20}$$

The comparison of bounds (20) and (16) is shown on Fig. 1. It follows that in both cases undetectable attack ($P_{EC} = 0$) provides Eve no information, and an attack providing maximal information ($I = 1\,\mathrm{bit}$) is detectable by control mode with probability $1/2$. It is also visible that collective attacks provide only slight advantage compared to the individual ones.

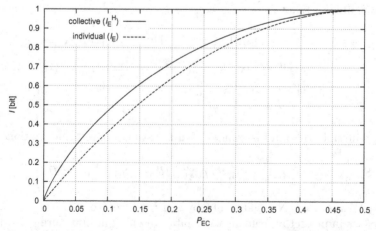

Fig. 1. Comparison of the eavesdropper's information upper bounds in collective ($I_E^{\mathcal{H}}$) and individual (I_E) attacks

4 Passive Eavesdropping in a Noisy Channel

Let us consider situation in which Eve does not entangle with a signal qubit i.e. $\mathcal{E}_{AE} = \mathcal{I}_A \otimes \mathcal{I}_E$. Such assumption results in separation of the system controlled by Eve, so it is not taken into account in further expressions.

Any interaction with the environment observed from the perspective of the principal system only can be given in operator sum (Kraus) representation [1] as

$$\mathcal{N}\rho \to \sum_k E_k \rho E_k^\dagger \qquad (21)$$

provided that $\sum_k E_k E_k^\dagger = \mathcal{I}$. Such an approach hides details of the interaction of the system under investigation with the environment. However, these details are not of immediate relevance in analysis of many quantum information processing related tasks. In such situations Kraus representation proved to be useful because it provides a unified description of many, seemingly different, physical processes.

A depolarizing channel used to model white noise [1]. is described by the following operators in the single qubit case

$$E_0 = \frac{\sqrt{1+3r}}{2}\mathcal{I} \ , \quad E_k = \frac{\sqrt{1-r}}{2}\sigma_k \ , \qquad (22)$$

where $k = 1,2,3$, σ_k are Pauli matrices and r denotes channel reliability. As the noise affects only travelling qubit, the Kraus operators for the investigated system can be obtained by simple extension [23]

$$E_{BAk} = \mathcal{I}_B \otimes E_k \ . \qquad (23)$$

With the help Equation (23) the map describing noise operator $\mathcal{N}_{BA}$ can be constructed and the quantities given by (8) and (11) are easy to find numerically. If Bob uses unambiguous discrimination, the bits are lost (measurement fails) with a probability [22]

$$QLOSS = 1 - P_s^{\max} = F\left(\rho_{0BA}^{(4)}, \rho_{1BA}^{(4)}\right) \ . \qquad (24)$$

On the other hand, if Bob uses minimum error discrimination the observed bit error rate is equal to [24]

$$QBER = \frac{1}{2}\left(1 - \frac{1}{2}\mathrm{Tr}\left(\left|\rho_{0BA}^{(4)} - \rho_{1BA}^{(4)}\right|\right)\right) \ . \qquad (25)$$

Quantities $QBER$ and $QLOSS$ as a function of control mode failure probability (8), which is a parameter directly accessible to communicating parties, are shown on Fig. 2. Both $QBER$ and $QLOSS$ do not scale linearly with P_{EC}. Moreover, the functional form of the the obtained scaling heavily depends on parameters of the noise model used, thus the correct modelling of noise is of prime importance in the estimation of the protocol operation over non-perfect quantum channels.

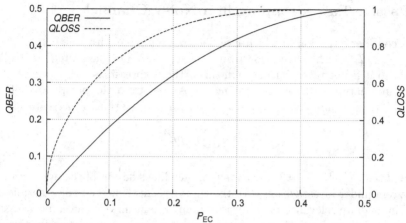

Fig. 2. Probabilities of a particle loss ($QLOSS$) or an erroneous decoding ($QBER$) as a function of control mode failure probability (P_{EC}) in protocol operation over depolarizing channel

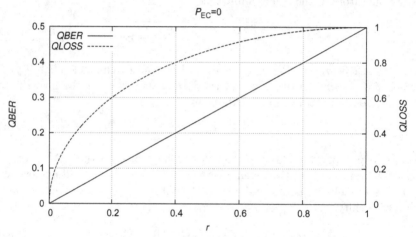

Fig. 3. Probabilities of a particle loss ($QLOSS$) or an erroneous decoding ($QBER$) as a function of dephasizig channel reliability r (see (26) for explanation)

To emphasize the above stated thesis, let us consider protocol operation over dephasing channel, which is described by the following Kraus operators [1]

$$E_0 = \begin{bmatrix} 1 & 0 \\ 0 & \sqrt{1-r} \end{bmatrix} \quad , \quad E_1 = \begin{bmatrix} 0 & 0 \\ 0 & \sqrt{r} \end{bmatrix} \tag{26}$$

where parameter $r \to 0$ for weak coupling and short interaction time and $r \to 1$ for strong coupling and/or long interaction time. Using definition (26) one can find that probability of error occurrence in control mode given by expression (8)

$$P_{EC} \equiv 0 \tag{27}$$

independent of r. On the other hand probabilities of bit loss (24) or its detection error (25) are non-zero and vary, depending on r value, in a range similar to the observed one for the depolarizing channel (see Fig. 3). It follows that some kinds of channel imperfections are not well detected by control mode, what implies, that it cannot universally (i.e. independent of occurring noise model) be used for estimation of QBER and/or QLOSS observed in information mode.

5 Conclusion

The usefulness of the general method based on density operator analysis for Ping-Pong protocol operation has been presented. As the proof of concept the example of its application to the analysis of the protocol execution over depolarizing and dephasing channels has been given. The analysis of a more complicated case of an active eavesdropping is left for future research. Although the method is more cumbersome than approach used so far, it is more versatile as it permits on incorporation of different models of noise in a unified way and make use of a well grounded theory of quantum discrimination in estimation of eavesdropper's information gain.

References

1. Nielsen, M.A., Chuang, I.L.: Quantum Computation and Quantum Information. Cambridge University Press (2000)
2. Shor, P.W.: Polynomial-time algorithms for prime factorization and discrete logarithms on a quantum computer. SIAM J. Comput. 26(5), 1484–1509 (1997)
3. Zawadzki, P.: A numerical simulation of quantum factorization success probability. In: Tkacz, E., Kapczyński, A. (eds.) Internet – Technical Developments and Applications. AISC, vol. 64, pp. 223–231. Springer, Heidelberg (2009)
4. Zawadzki, P.: A fine estimate of quantum factorization success probability. Int. J. Quant. Inf. 8(8), 1233–1238 (2010)
5. Bennett, C.H., Brassard, G.: Quantum cryptography: Public key distribution and coin tossing. In: Proceedings of International Conference on Computers, Systems and Signal Processing, New York, pp. 175–179 (1984)
6. Izydorczyk, J., Izydorczyk, M.: Microprocessor scaling: What limits will hold? IEEE Computer 43(8), 20–26 (2010)
7. Gisin, N., Ribordy, G., Tittel, W., Zbinden, H.: Quantum cryptography. Rev. Mod. Phys. 74, 145–195 (2002)
8. Bennett, C.H., Wiesner, S.J.: Communication via one- and two-particle operators on Einstein-Podolsky-Rosen states. Phys. Rev. Lett. 69, 2881–2884 (1992)
9. Wang, C., Deng, F.G., Li, Y.S., Liu, X.S., Long, G.L.: Quantum secure direct communication with high-dimension quantum superdense coding. Phys. Rev. A 71(4), 044305 (2005)
10. Bennett, C.H., Brassard, G., Crépeau, C., Jozsa, R., Peres, A., Wootters, W.K.: Teleporting an unknown quantum state via dual classical and Einstein-Podolsky-Rosen channels. Phys. Rev. Lett. 70, 1895–1899 (1993)
11. Stinson, D.R.: Cryptography: Theory and Practice, 2nd edn. Chapman & Hall/CRC (2002)

12. Long, G.L., Deng, F.G., Wang, C., Li, X.H., Wen, K., Wang, W.Y.: Quantum secure direct communication and deterministic secure quantum communication. Front. Phys. China 2(3), 251–272 (2007)

13. Boström, K., Felbinger, T.: Deterministic secure direct communication using entanglement. Phys. Rev. Lett. 89(18), 187902 (2002)

14. Ostermeyer, M., Walenta, N.: On the implementation of a deterministic secure coding protocol using polarization entangled photons. Opt. Commun. 281(17), 4540–4544 (2008)

15. Vasiliu, E.V.: Non-coherent attack on the ping-pong protocol with completely entangled pairs of qutrits. Quantum Inf. Process. 10, 189–202 (2011)

16. Zawadzki, P.: Security of ping-pong protocol based on pairs of completely entangled qudits. Quantum Inf. Process. 11(6), 1419–1430 (2012)

17. Wójcik, A.: Eavesdropping on the ping-pong quantum communication protocol. Phys. Rev. Lett. 90(15), 157901 (2003)

18. Zhang, Z., Man, Z., Li, Y.: Improving Wójcik's eavesdropping attack on the ping-pong protocol. Phys. Lett. A 333, 46–50 (2004)

19. Zawadzki, P.: The Ping-Pong protocol with a prior privacy amplification. Int. J. Quant. Inf. 10(3), 1250032 (2012)

20. Peres, A.: How to differentiate between non-orthogonal states. Phys. Lett. A 128(1-2), 19 (1988)

21. Herzog, U.: Optimal state discrimination with a fixed rate of inconclusive results: analytical solutions and relation to state discrimination with a fixed error rate. Phys. Rev. A 86, 032314 (2012)

22. Herzog, U., Bergou, J.A.: Distinguishing mixed quantum states: Minimum-error discrimination versus optimum unambiguous discrimination. Phys. Rev. A 70, 022302 (2004)

23. Miszczak, J.A.: Singular value decomposition and matrix reorderings in quantum information theory. Int. J. Mod. Phys. C 22(9), 897–918 (2011)

24. Fuchs, C.A., van de Graaf, J.: Cryptographic distinguishability measures for quantum-mechanical states. IEEE Trans. Inform. Theor. (4), 1216–1227 (1999)

Comparison of CHOKe and gCHOKe Active Queues Management Algorithms with the Use of Fluid Flow Approximation

Adam Domański[1], Joanna Domańska[2], and Tadeusz Czachórski[2]

[1] Institute of Informatics, Silesian Technical University
Akademicka 16, 44-100 Gliwice, Poland
adamd@polsl.pl
[2] Institute of Theoretical and Applied Informatics, Polish Academy of Sciences
Baltycka 5, 44-100 Gliwice, Poland
{joanna,tadek}@iitis.gliwice.pl

Abstract. In the article we examine a model of TCP connection with Active Queue Management in an intermediate IP router. We model a system where CHOKe or gCHOKe are the AQM policy. We use the fluid flow approximation technique to model the interactions between the set of TCP/UDP flows and two variants of the CHOKe algoithms. The obtained results confirm the superiority of these algorithms over a standard RED algorithm.

Keywords: active queue management, TCP flow control, RED.

1 Introduction

Congestion control mechanisms in TCP/IP networks are one of the most important topics in the field of today Internet and their modeling remains a vital problem. The development of new AQM (Active Queue Management) routers allow to improve the performance of Internet applications.

In recent years a number of analytical models of AQM in IP routers was presented, e.g. [1–5] in open-loop scenario, because of the difficulty in analyzing AQM mathematically inside the whole closed loop of TCP congestion control. This paper extends a nonlinear dynamic model of TCP proposed earlier [6, 7] to analyze the AQM systems with RED. Here, we use a similar model to investigate the performance of CHOKe or gCHOKe mechanisms.

The models based on fluid flow approximation, e.g. [8], are able to capture the dynamics of TCP flows [9] and allow to analyze networks with a large number of flows. The article describes the use of this method to compare routers having different active queue management principles (classical RED, CHOKe and gCHOKe) and transmitting TCP/UDP flows. The model allows to study not only the steady-state behavior of the network, but also the transient one when a set of TCP flows start or finish transmission. We focus on transient average router queue length for different AQM strategies.

A. Kwiecień, P. Gaj, and P. Stera (Eds.): CN 2013, CCIS 370, pp. 363–371, 2013.
© Springer-Verlag Berlin Heidelberg 2013

The rest of this article is organized as follows. The Section 2 introduces two variants of CHOKe algorithm. The Section 3 describes the fluid flow model of AQM router supporting TCP/UDP flows. The Section 4 presents the obtained results. The conclusions are presented in Sect. 5.

2   The CHOKe Algorithm and Its Variant – gCHOKe

The CHOKe (CHOose and Keep for responsive flows, CHOose and Kill for unresponsive flows), [10] is a stateless AQM algorithm slightly similar to RED (Random Early Drop), proposed not only to control TCP packets but also to prevent uncontrollable UDP connections to monopolize the links [11]. It uses incoming packages to punish streams with the highest demand for bandwidth.

Similarly to the RED mechanism, there are two threshold values: Min_{th} and Max_{th}. At the arrival of a new package, the new walking average queue length is calculated. If the average queue length is less than Min_{th}, the packet is placed in the buffer. When the average queue length is greater than Min_{th}, CHOKe pulls randomly one packet from the FIFO buffer a ("CHOKe victim") and verifies whether it belongs to the same stream as an incoming packet.

If the both packets belong to the same stream, they are removed (this situation is called "CHOKe hit"). Otherwise, the randomly selected packet is returned to the buffer and the arrived packet is placed into the queue with probability P. This probability is calculated in the same manner as in the case of RED algorithm. The event is called "CHOKe miss".

The geometric CHOKe algorithm (gCHOKe) [12] is a modification of the CHOKe having an additional, configurable parameter $maxcomp \in [1, \ldots, \infty)$. This parameter determines the maximum number of successful comparisons. As previously, the algorithm compares the incoming packet with a random packet drawn from the queue. The comparison is successful when both packages are from the same stream. The comparison ends when: the comparison is unsuccesfull or the number of comparisons exceeds $maxcomp$. In this case all matching packets (plus the incoming ones) are removed from the queue. If the first match is not successful, the random packet ("CHOKE victim") comes back to the queue. The arriving packet is placed into the queue with probability of P (see Fig. 1). Hence, CHOKe is a special case of gCHOKe algorithm ($maxcomp = 1$).

3   The Fluid-Flow Model of TCP/UDP Streams

This section presents a fluid flow model of a TCP connection having a bottleneck router with AQM policy. The router transmits also UDP packets, [8]. This model ignores the TCP timeout mechanisms and allows to obtain the average value of key network variables. This model is based on the following nonlinear differential equations:

$$\frac{dW(t)}{dt} = \frac{1}{R(t)} - \frac{W(t)W(t - R(t))}{2R(t - R(t))}p(t - R(t)) \tag{1}$$

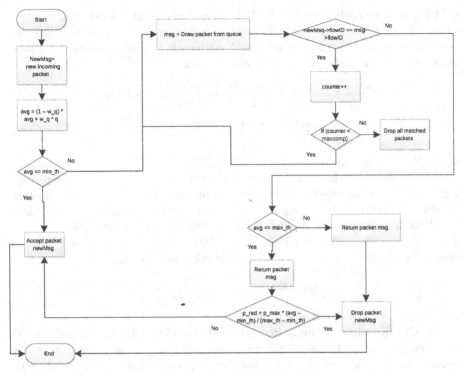

Fig. 1. Diagram of the gCHOKe algorithm

$$\frac{dq(t)}{dt} = \frac{W(t)}{R(t)}N(t) - C \qquad (2)$$

where:

W – expected TCP congestion window size (packets),
q – expected queue length (packets),
R – round-trip time $q/C + T_p$ [s],
C – link capacity [packets/s],
T_p – propagation delay [s],
N – number of TCP sessions at the router,
p – packet drop probability.

The maximum values of q and W depend on the buffer capacity and maximum window size. The dropping probability p depends on the AQM queue algorithm. The first term on the right side of the Equation (1) represents the rate of increase of congestion window due to incoming acknowledgements, the second represents the rage with which the congestion window decreases due to packet losses. The Equation (2) gives the speed of router queue changes due to incoming and leaving flows of packets.

The traffic composed of TCP and UDP streams has been considered in [13]. For this model a single router supports N TCP sessions. Each TCP stream is a TCP-Reno connection and each UDP sender is a CBR (Constant Bit Rate) source. The total rate of UDP sessions is denoted by λ, it is assumed that the rate λ is associated at equal intensity λ/N with each TCP connection. Fluid-flow equations of TCP and UDP mixed traffic become:

$$\frac{dW(t)}{dt} = \frac{1}{R'(t)} - \frac{W(t)W(t - R'(t))}{2R'(t - R'(t))}p(t - R'(t)) \tag{3}$$

$$\frac{dq(t)}{dt} = \frac{W(t)}{R'(t)}N(t) - (C - \lambda) \tag{4}$$

where $R' = $ round-trip time $ = q/(C - \lambda) + T_p$ [s].

In RED AQM mechanism, at arrival of each packet, the average queue size x is calculated as an exponentially weighted moving average using the following formula: $x_i = (1 - \alpha)x_{i-1} + \alpha q_{inst}$ where q_{inst} is the current queue length. Then the RED drop function is applied: there are two thresholds Min_{th} and Max_{th}; if $x < Min_{th}$ the packet is admitted, for $Min_{th} < x < Max_{th}$ the packet is dropped with probability p_{RED} growing linearly from 0 to p_{max}

$$p_{RED} = p_{max}\frac{x - Min_{th}}{Max_{th} - Min_{th}} \tag{5}$$

and if $x > Max_{th}$ the packet is dropped.

The CHOKe algoritm pulls from the FIFO buffer "CHOKe victim" and if the package is from the same stream as an incoming packet, we drop it. So prabability p_{CHOKe} depends on the number of packets of a stream i relative to the total buffer occupancy, $p_{CHOKe} = q_i/q$. Probability p_{CHOKE} depends also on the buffer occupancy and the number of streams. For simplicity, the model assumes that the number of packets belonging to a single stream in the queue is the same for all streams, $q_i = q/N$. It follows that the probability of packet loss is inversely proportional to the number of the streams, $p_{CHOKe} = 1/N$.

Geometric CHOKE repeats the selection of "CHOKe victim" until the packet has been drown from another stream or the number of draws exceeds $maxcomp$. For each incoming packet we can drop $1, 2, 3, \ldots, maxcomp$ packets. If we assume that the probability of selecting a package of the same stream in one selection is $\frac{1}{N}$ then

$$p_{gCHOKe} = \frac{1}{N} + \frac{1}{N^2} + \cdots + \frac{1}{N^n} = \sum_{k=1}^{n}\frac{1}{N^k} \tag{6}$$

where $n = maxcomp$. Using the sum of a geometric sequence:

$$S_n = a + aq + \cdots + aq^{n-1} = a\frac{1 - q^n}{1 - q} \tag{7}$$

we can write:

$$p_{gCHOKe} = S_n - a + q^k . \tag{8}$$

In our case $a = 1$ and $q = \frac{1}{N}$, hence

$$p_{gCHOKe} = \frac{1 - \frac{1}{n^k}}{1 - \frac{1}{n}} - 1 + \frac{1}{n^k} = \frac{n^k + n^{k-1} - n^{-1} - 1}{n^k - N^{k-1}} . \tag{9}$$

4   Numerical Results

Computations were made with the use of PyLab (Python numeric computation environment) [14] which is a combination of Python, NumPy, SciPy, Matplotlib, and IPython. The graphs shown below present transient system behavior, the time axis is drawn in seconds.

We assume the following parameters of the AQM buffer: $Min_{th} = 10$, $Max_{th} = 15$, buffer size (measured in packets) $= 20$, weight parameter $\alpha = 0.007$, and the parameters of TCP connection:

- transmission capacity of AQM router: $C = 0.075$,
- propagation delay for i-th flow: $T_{p_i} = 2$,
- initial congestion window size for i-th flow (measured in packets): $W_i = 1$.

The obtained mean queue lengths for TCP connections and various AQM policies are presented in Table 1.

Table 1. The obtained mean AQM queue lengths Q

Algorithm	Nb of streams	Nb of packets
CHOKE	1	7.98664081264
CHOKE	2	8.63146812018
CHOKE	5	10.0998514529
CHOKE	10	11.0167546717
CHOKE	11	11.7731309893
RED	1	8.57089136683
RED	2	9.05376778822
RED	5	10.3805817389
RED	10	11.1549893996
RED	11	11.7731309893
gCHOKe ($maxcomp = 2$)	1	7.88714079539
gCHOKe ($maxcomp = 5$)	1	7.81053339265
gCHOKe ($maxcomp = 10$)	1	7.78926315534
gCHOKe ($maxcomp = 2$)	2	8.61223411769
gCHOKe ($maxcomp = 5$)	2	8.60737680727
gCHOKe ($maxcomp = 10$)	2	8.61023081937
gCHOKe ($maxcomp = 2$)	10	11.0059435834
gCHOKe ($maxcomp = 5$)	10	11.0108325703
gCHOKe ($maxcomp = 10$)	10	11.0108321896

Figures 2, 3, 4, present the queue behavior in the case of two flows and respectively RED, CHOKe and gCHOKe queues. The size of congestion window increases until the buffer reaches the Min_{th} value. Algorithm drows "CHOKe victim" and the probability of removing the package is equal to $\frac{1}{2}$ (probability of removing the packet by RED mechanism is much smaller). Packets are dropped and the size of congestion window decreases causing a slow decrease of the queue length – this pattern is repeated periodically.

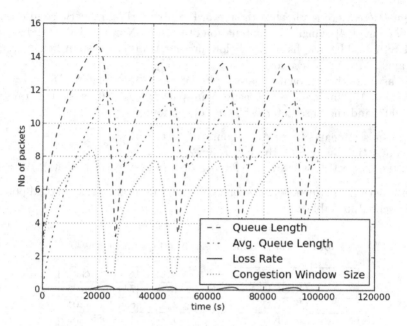

Fig. 2. RED queue, 2 TCP/UDP flows

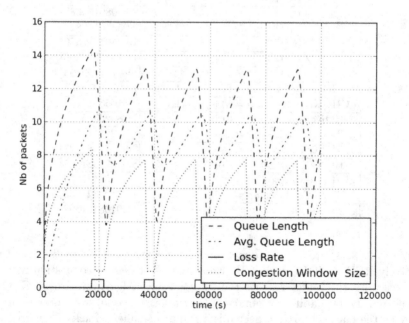

Fig. 3. CHOKe queue, 2 TCP/UDP flows

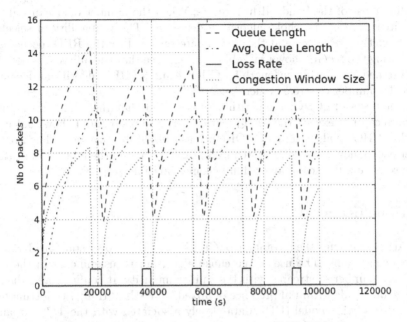

Fig. 4. gCHOKe queue ($maxcomp = 10$), 2 TCP/UDP flows

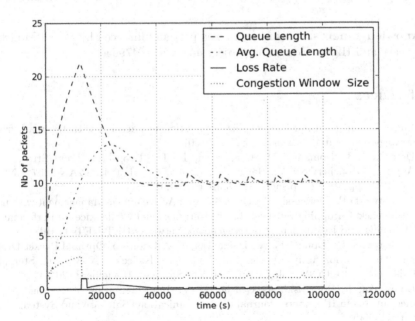

Fig. 5. gCHOKe queue ($maxcomp = 10$), 10 TCP/UDP flows

Comparing the behavior of the CHOKe algorithm with the RED algorithm one can see that the CHOKe algorithm works better in the case of aggressive (stealing most of the bandwidth) streams. When the number of streams grows, the importance of the choke algorithm decreases. The probability of selecting a good victim decreases and the packets are removed by the RED mechanism. Comparing the results shown in Table 1 one can see that the differences between the obtained average queue length for CHOKe and for RED algorithms decreases when the number of streams increases.

In our tests of gCHOKe algorithm we assumed that the maximum number of draws can not exceed the Min_{th} parameter. Figures 3 and 4 show that compared to CHOKe algorithm, gCHOKe brings a slight improvement. The average buffer occupancy is also slightly reduced (Table 1). When the number of streams increases, the influence of algorithm becomes invisible (Fig. 5).

5   Conclusions

This article confirms the advantage of CHOKe algorithm over standard-RED for aggressive streams. The use of the choke algorithm is insignificant in the case of a large number of streams with the similar intensity. It confirms also the advantage of the algorithm in presence of mixed TCP/UDP traffic. Unfortunately, weaknesses of the model (UDP data closely associated with the TCP streams) could not allow us to show the advantages of the gCHOKe algorithm in shaping the intensified traffic of UDP datagrams. Our future work will concern this issue.

Acknowledgements. This research was partially financed by Polish Ministry of Science and Higher Education project no. N N516479640.

References

1. Liu, C., Jain, R.: Improving explicit congestion notification with the mark-front strategy. Computer Networks 35(2-3) (2000)
2. Domańska, J., Domański, A., Czachórski, T.: The Drop-From-Front Strategy in AQM. In: Koucheryavy, Y., Harju, J., Sayenko, A. (eds.) NEW2AN 2007. LNCS, vol. 4712, pp. 61–72. Springer, Heidelberg (2007)
3. Augustyn, D.R., Domański, A., Domańska, J.: Active Queue Management with non linear packet dropping function. In: 6th International Conference on Performance Modelling and Evaluation of Heterogeneous Networks HET-NETs (2010)
4. Augustyn, D.R., Domański, A., Domańska, J.: A Choice of Optimal Packet Dropping Function for Active Queue Management. In: Kwiecień, A., Gaj, P., Stera, P. (eds.) CN 2010. CCIS, vol. 79, pp. 199–206. Springer, Heidelberg (2010)
5. Domańska, J., Domański, A., Czachórski, T.: Implementation of modified AQM mechanisms in IP routers. Journal of Communications Software and Systems 4(1) (March 2008)
6. Hollot, C.V., Misra, V., Towsley, D., Gong, W.-B.: On Designing Improved Controllers for AQM Routers Supporting TCP Flows. In: IEEE INFOCOM (2002)

7. Rahme, S., Labit, Y., Gouaisbaut, F.: An unknown input sliding observer for anomaly detection in TCP/IP networks. In: Ultra Modern Telecommunications & Workshops (2009)
8. Misra, V., Gong, W.-B., Towsley, D.: Fluid-based Analysis of a Network of AQM Routers Supporting TCP Flows with an Application to RED. In: ACM SIGCOMM (2000)
9. Yung, T.K., Martin, J., Takai, M., Bagrodia, R.: Integration of fluid-based analytical model with Packet-Level Simulation for Analysis of Computer Networks. In: SPIE (2001)
10. Pan, R., Prabhakar, B., Psounis, K.: CHOKe, A stateless AQM scheme for approximating fair bandwidth allocation. IEEE INFOCOM, 942–952 (2000)
11. Hollot, C.V., Misra, V., Towsley, D.: A control theoretic analysis of RED. IEEE/INFOCOM (2001)
12. Eshete, A., Jiang, Y.: Generalizing the CHOKe flow protection. Computer Network Journal (2012)
13. Wang, L., Li, Z., Chen, Y.-P., Xue, K.: Fluid-based stability analysis of mixed TCP and UDP traffic under RED. In: 10th IEEE International Conference on Engineering of Complex Computer Systems (2005)
14. www.scipy.org

Modeling Data Stream Intensity in Distributed Stream Processing System

Marcin Gorawski[1,2], Pawel Marks[1], and Michal Gorawski[3]

[1] Silesian University of Technology, Institute of Computer Science,
Akademicka 16, 44-100 Gliwice, Poland
{Marcin.Gorawski,Pawel.Marks}@polsl.pl
[2] Wroclaw University of Technology, Institute of Computer Science,
Wybrzeze Wyspianskiego 27, 50-370 Wroclaw, Poland
Marcin.Gorawski@pwr.wroc.pl
[3] Institute of Theoretical and Applied Informatics, Polish Academy of Sciences,
Baltycka 5, 44-100 Gliwice, Poland
mgorawski@iitis.pl

Abstract. In recent years energy market has changed. Consumers in many countries are free to buy energy from any of the available providers. This requires continuous reading from a huge number of energy meters to evaluate the amount of energy being bought from a particular provider. In this paper we present a fault-tolerant distributed stream processing system for continuous meter readings. The main goal of the system is to store the readings in a stream data warehouse for further analysis. We focus on modeling of the data stream intensity in order to estimate the size of buffers in a network of components composing the system. We present both the mathematical model of the intensity and the simulation results to prove the correctness of the theoretical analysis.

Keywords: stream processing, modeling, distributed system.

1 Introduction

These days it becomes more common to process continuous data streams [1]. It may have application in many domains of our life such as: computer networks (e.g. intrusion detection), financial services, medical information systems (e.g. patient monitoring), civil engineering (e.g. highway monitoring) and more.

Thousands or even millions of energy meters located in households or factories can be sources of meter-reading streams. Continuous analysis of power consumption may be crucial to efficient electricity production. Unlike other media such as water or gas, electricity is hard to store for further use. That is why a prediction of energy consumption may be very important. Real-time analysis of the media meter readings may help manage the process of energy production in the most efficient way.

There are many systems for processing continuous data streams and they are still being developed [2–6]. Various system processing stream data can also be

A. Kwiecień, P. Gaj, and P. Stera (Eds.): CN 2013, CCIS 370, pp. 372–383, 2013.
© Springer-Verlag Berlin Heidelberg 2013

found in [7–9]. In [10] the fault tolerant Borealis system is presented. This is a dedicated solution for applications where a low latency criterion is essential. Another system facing infinite data streams is described in [11]. Authors of the work deal with sensors producing data continuously, transferring the measured data asynchronously without pooling. They proposed a *Framework in Java for Operators on Remote Data Streams* (Fjords).

In our research we have focused on processing data originating from a radio-based measurement system [12, 13]. We carried research on efficient recovery of interrupted ETL jobs and proposed a few approaches [14, 15] based on the Design-Resume algorithm [16].

Based on the previous experience, we have focused on fault-tolerance and high availability in a distributed stream processing environment. In [17] we proposed a new set of modules increasing the probability that a failure of one or more modules will not interrupt the processing of endless data streams. Then we prepared a simple model [18–21] of data sources to estimate the amounts of data to be processed, useful in the configuration of the environment. In this paper we want to present a more advanced analysis of the intensity of the stream readings to be able to configure buffers of the network components properly.

In Section 2 we define the problem we want to solve. Sections 2.2, 2.3 and 2.4 contain a detailed description of the analysis we propose with the verification of the proposed model. In the last section we summarize the paper.

2   The Problem

Our research is based on a telemetric network designed for remote and automatic reading of media consumption meters. Meters of energy, water and gas transmit data to collecting nodes. In most cases wireless media is used; however, other experimental approaches are also tested (modulated transmission on AC power lines). Data from collecting nodes is sent to a local telemetric server. The data gathered in the telemetric server can be processed further, e.g. to predict media consumption based on historical data. To make such prediction possible, it is necessary to transfer the data from all of the telemetric servers (also called *data stream sources*) into the stream data warehouse. However, a data source can be also a single collecting node (not only a server). The difference in this case is that a collecting node is too simple device to buffer large amounts of data. A collecting node can be compared to a LAN switch, which only transfers data from one point to another. The data sources (collecting nodes, servers) are distributed geographically, what increases the probability of failures caused by external factors (e.g. local black-outs). Moreover, the data transmission process becomes a continuous ETL process.

The goal of the research is to assure continuity of the ETL process reading data from stream sources with the shortest possible delay between measuring and storing the data in the warehouse. If it is possible, we want to assure successful recovery of the interrupted processing without any stream data loss.

The system presented in Fig. 1 consists of : stream data sources (e.g. telemetric server or collecting node), remote stream buffers (RBF), remote persistent stream integrators (RIF), ETL modules, a module for error detection and stream integration (FTI). Our research is based on the following configuration: 4 independent stream sources, 6 RBF modules, 4 RIF modules, 3 ETL process replicas, an error detection module and a data warehouse server. The sources transmit data to RBF buffering modules, which communicate with RIF integrating modules offering persistent buffering. At this stage a replicated extraction[1] process appears. Outputs of the extraction process are connected to an FTI detecting module. The FTI is responsible for not loading of the improperly processed data (malformed during processing or transmitting). At the end of the modules chain there is a stream data warehouse and the systems using it. There are multiple connections between the modules. They are intended to provide redundant processing of all the data streams.

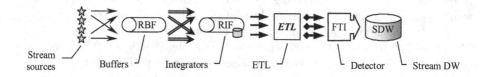

Fig. 1. Layered structure of the distributed system

2.1   Data Source Characteristics

A data source in our system is a single collecting node or optionally a local telemetric server. We assume that the source transmits data from associated media consumption meters and it has no ability to buffer any historical data. It means that there is no possibility to re-read already retrieved tuples from such a source. As a result, any interruption of the transmission from such a source leads to loss of a part of the data stream without any chance of recovering it. We assume that each source stamps the tuples keeping ascending stamp order. Consistency of stamping among all the sources in the system is not required; however, it is desirable for data analysis.

When a data source is a complex module such as a server, it can backup data received from meters on a disk to avoid data loss in case of transmission failures. Unfortunately, such a case is quite rare. To reduce system load simple collecting nodes are used. Then we have to assure that there is always a connection between the source and any receiver module, which is always ready to receive data incoming from the source. In our system RBF modules are used as receiver modules, and their buffers configuration is discussed in the following section.

---

[1] In this case extraction means a complete ETL process: Extraction + Transformation + Loading.

To configure the target RBF layer properly we have to prepare a mathematical description of such a source.

The analysed data source gathers measurements from particular meters (e.g. meters in blocks of flats, a large factory). The data can be trasmitted in one of the two modes: on-demand or asynchronous. In on-demand mode, the transmission is initiated by the collecting node and then selected meters sent out current values. This mode requires bidirectional communication between meters and a collecting node. It makes the communication easier to handle but unfortunately increases costs of such device. In asynchronous mode transmission goes in only one direction and meters can transmit data at any time. The collecting node (the source we analyse) must always be ready to handle incoming data stream.

2.2 Model Definition

Assume that there are N meters in the distributed system, and each meter transmits a reading approximately every T seconds. In other words it means that in each generation period (Fig. 2) each meter will transmit one reading. The probability that a meter transmits a reading in a particular time slot equals $p = t/T$. The total number of slots is $n_s = T/t$. The question is: how many readings will be transmitted in any number of time slots of size t?

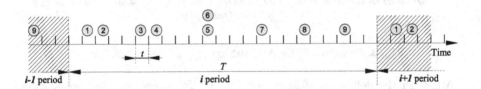

Fig. 2. Distribution of measures in a data source. The period T repeats.

Each meter is independent, so the reading events are independent also. Continuing the analysis presented in [18] the probability of k readings in any single slot t is $\binom{N}{k} p^k (1-p)^{N-k}$. Based on it we can define two recursive functions:

$$
P_=(N, p, s_s, k) =
\begin{cases}
\binom{N}{k} p^k (1-p)^{N-k} & \text{for } s_s = 1 \\[2ex]
\displaystyle\sum_{i=0}^{k} P_=(N, p, 1, i) \cdot P_=(N, p, s_s - 1, k - i) & \text{for } s_s > 1
\end{cases}
\tag{1}
$$

$$
P_\leq (N, p, s_s, k) = \begin{cases} \displaystyle\sum_{i=0}^{k} P_= (N, p, 1, i) & \text{for } s_s = 1 \\[2em] \displaystyle\sum_{i=0}^{k} P_= (N, p, 1, i) \cdot P_\leq (N, p, s_s - 1, k - i) & \text{for } s_s > 1 \end{cases} \tag{2}
$$

Both $P_=$ and $P_\leq$ functions define the probability that having N meters, in a sequence of s_s time slots we will observe exactly ($P_=$) or no more than ($P_\leq$) k readings. The results obtained for both functions are presented in Figs. 3 and 4.

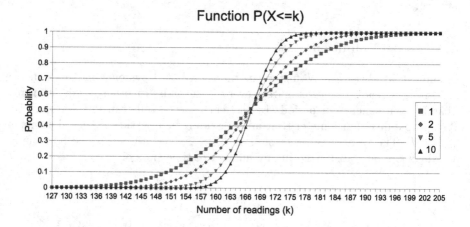

Fig. 3. Function $P_\leq$ for $N = 10^5$, $p = \frac{1}{600}$ and $s_s = 1, 2, 5, 10$

Although the definition of functions $P_=$ and $P_\leq$ is correct, their evaluation in the abovementioned recursive form is extremely complicated. That is why we propose another approach. We can make use of the hypergeometric distribution. The easiest way to understand this distribution is in terms of urn models. Suppose you are to draw n marbles without replacement from an urn containing N marbles in total, m of which are white. The hypergeometric distribution describes the distribution of the number of white marbles drawn from the urn. It is defined as follows:

$$
P_H(X = k) = f_H(k; N, m, n) = \frac{\binom{m}{k}\binom{N-m}{n-k}}{\binom{N}{n}} . \tag{3}
$$

Our model requires to be modified a little. Unchanged remain: generation period T_g, number of time slots n_s and the number of meters N. In a single generation period we still can observe up to N readings, but it may happen that all readings

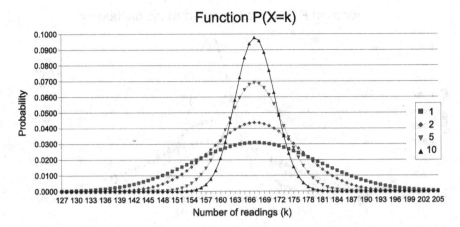

Fig. 4. Function $P_=$ for $N = 10^5$, $p = \frac{1}{600}$ and $s_s = 1, 2, 5, 10$

occur in a single time slot, and in the remaining $n_s - 1$ slots there are no readings at all. Assume that in a single generation period (urn) we have $N \cdot n_s$ events (marbles). We have N reading events (white marbles) and $N \cdot (n_s - 1)$ empty events (black marbles). We always draw $N \cdot s_s$ marbles. The probability of getting exactly k readings in a sequence of s_s time slots equals:

$$P_H(X = k) = f_H(k; N \cdot n_s, N, N \cdot s_s) \tag{4}$$

where $k \leq N$ and $0 < s_s \leq n_s$.

In Figure 5 the results obtained for P_H probability function are compared to the results for recursive $P_=$ function. The figure proves that both functions can be used interchangeably if needed. Moreover to simplify computations the hypergeometric distribution can be approximated using other distributions. Assume that $X \sim H(m, N, n)^2$ and $p = m/N$. Then:

1. If $n = 1$, then X is a Bernoulli distribution with parameter p.
2. If N and m are much greater than n, and p is not too close to 0 or 1, then $P(X \leq x) \approx P(Y \leq x)$, where Y has binomial distribution with parameters n and p.
3. If n is big, N and m are much greater then n, and p is not too close to 0 or 1, then

$$P(X \leq x) = \Phi\left(\frac{x - np}{\sqrt{np(1-p)}}\right) \tag{5}$$

where Φ is a cumulative distribution function of the standard normal distribution with parameters $\mu = np$ and $\sigma^2 = np(1-p)$.

---

[2] $H(m, N, n)$ defines hypergeometric distribution having probability function defined as $f_H(k; N, m, n)$.

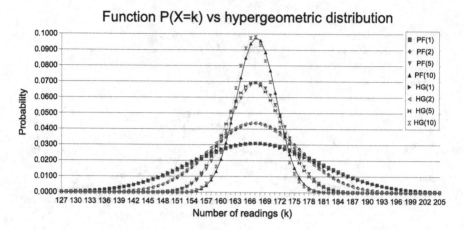

Fig. 5. Comparison of P_H and $P_=$ functions for $N = 10^5$, $p = \frac{1}{600}$ and $s_s = 1, 2, 5, 10$

Presented computations make it possible to evaluate the number of readings we can expect in a particular number of time slots if we have N meters, generation period T_g and a single time slot is t_s long. But how to compute the number of readings in case of various meter types and various generation periods?

To answer the question we need to divide all the meters into groups having the same generation period (Table 1). This way we obtain a set of meter groups Ψ, which elements are pairs $\psi_i = (N_i, T_{gi})$ for each $0 < i \leq |\Psi|$.

Table 1. Example of meter groups

Group	Meter type	Cardinality N	Generation period T_g	Std. dev. T_g
ψ_1	electricity	100 000	10 min	1 s
ψ_2	water	50 000	60 min	2 s
ψ_3	gas	50 000	24 h	5 s

For each ψ_i group we need to evaluate parameters of the distribution as described earlier in this section. This way each group will be described by the distribution of the random variable $X_i \sim N(\mu_i, \sigma_i^2)$. Knowing that the sum of any number of random variables having normal distribution still has a normal distribution we obtain a final distribution for all the meters:

$$X = \sum_{i=1}^{|\Psi|} X_i \sim N\left(\sum_{i=1}^{|\Psi|} \mu_i, \sum_{i=1}^{|\Psi|} \sigma_i^2 \right). \tag{6}$$

2.3   Model Verification

To verify our model we conducted a few experiments. Firstly, we evaluated a theoretical distribution of the the groups of meters mentioned in Sect. 2.2. Secondly, we simulated the behaviour of our system and compared it with the theoretical results.

According to Equation (5) a computation of the distribution for each group of meters goes as follows:

$$\mu_i(s_\mathrm{s}) = np = n \cdot \frac{m}{N_H} = s_\mathrm{s} \cdot N_i \cdot \frac{N_i}{N_i \cdot n_\mathrm{s}} = s_\mathrm{s} \cdot N_i \cdot \frac{N_i}{N_i \cdot \frac{T_{gi}}{t_\mathrm{s}}} = s_\mathrm{s} \cdot \frac{N_i \cdot t_\mathrm{s}}{T_{gi}} \quad (7)$$

$$\sigma_i^2(s_\mathrm{s}) = np(1-p) = n \cdot \frac{m}{N_H} \cdot \left(1 - \frac{m}{N_H}\right) = s_\mathrm{s} \cdot \frac{N_i \cdot t_\mathrm{s}}{T_{gi}} \cdot \left(1 - \frac{t_\mathrm{s}}{T_{gi}}\right) \quad . \quad (8)$$

After substitution we obtain three groups with the following distribution parameters:

$X_1 \sim N(166.67; 166.39)$ ($\sigma = 12.9$) for ψ_1 group,
$X_2 \sim N(13.89; 13.89)$ ($\sigma = 3.73$) for ψ_2 group,
$X_3 \sim N(0.58; 0.58)$ ($\sigma = 0.76$) for ψ_3 group,

The sum of the X_1, X_2, X_3 random variables gives in a result a random variable $X \sim N(181.13; 180.85)$ with $\sigma = 13.45$. The random variable X expresses the number of readings in a single time slot in a stream being the sum of three streams described in Table 1.

For the meter groups defined above we ran a simulation process. It started in the worst possible case in which all 200 000 meters generated the first measurement in the first second of work. Each meter works with a generation period T_g drawn according to the parameters of the group distribution. This way the moment of generation spreads in time as the simulation runs.

During the simulation we registered a histogram of the number of readings in a single slot. Based on it we prepared a plot of the probability function. In the first stage of the simulation histograms were collected for every 10 000 time slots. After exceeding 1 million of time slots histograms covered 100 000 slots (10 times more). Figure 6 presents the obtained simulation results compared with the theoretical curve. Simulation results where caught in the following points of time: 100 000 time slots, 200 000, 500 000, 1 million, 2 millions, 5 millions. As you can see for the first stage (100 000 slots) the probability function has quite irregular shape with at least three local peaks. At the beginning of the simulation all of the meters transmit the first reading in the first time slot. Further in the simulation the readings spread slowly according to their factory parameters distribution. We did not analyse the cause of the local peaks in details, but it results from the distribution of the factory parameters of the simulated meters. Going further the shape changes and after 1 million of slots it starts to look like a typical bell curve. Since then the simulation results are more and more similar to the shape of theoretical curve. The peak is the same as the computed above random variable $X \sim N(181.13; 180.85)$.

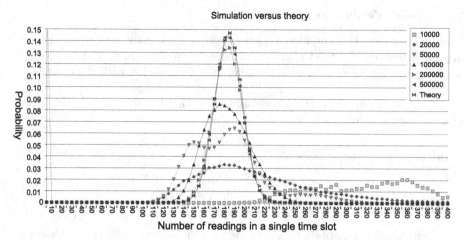

Fig. 6. Comparison of the theoretical readings distribution with the simulation results

We performed a similar calculation not for a single slot, but for a sequence of consecutive 30 times slots ($s_s = 30$). Then we rescaled the results for a single slot and the obtain results we gathered in Fig. 7. As can be seen the results for the sequence of slots are similar. The only difference is that after rescaling to a single slot, the expected number of meters in a single slot is more precise. The expected value remains unchanged and is still $\mu \approx 181.13$. But the standard deviation σ decreased from 13.45 to only 2.46. The final distribution for a sequence of 30 time slots can be computed from Equations (7) and (8). In this case it is: $\mu = 5434.03$, $\sigma^2 = 5425.58$ and $\sigma = 73.66$.

2.4 Edge Condition

Having the verified data distribution model we can try to answer the question stated at the beginning of Sect. 2.2: how many readings will be transmitted in any number of time slots t? We know the parameters of the random variable distribution. We know it is the normal distribution with parameters μ and σ. For normal distribution the density function is:

$$f(x) = \frac{1}{\sigma\sqrt{2\pi}} e^{-\frac{(x-\mu)^2}{2\sigma^2}} \ . \tag{9}$$

As the input parametr we take p_k denoting the possibility, that more than k readings will be received in the analysed period of time (number of time slots). We want to compute the smallest value of k for which the probability of receiving more than k readings is less than p_k. It leads to the following inequality:

$$p_k \geq \frac{1}{\sigma\sqrt{2\pi}} \int\limits_{k+1}^{+\infty} e^{-\frac{(x-\mu)^2}{2\sigma^2}} dx \tag{10}$$

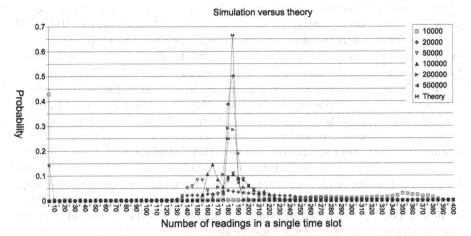

Fig. 7. Comparison of the theoretical readings distribution with the simulation results. Computation for 30 time slots scaled for a single slot.

The computed $k(p_k)$ is a base for setting the size of the buffers in RBF modules. Unfortunately there is no analytical solution for the Equation (10) due to the integration it includes. However, the result can be evaluated using numerical methods.

We can use the cumulative distribution function for the normal distribution based on the error function, after converting the distribution $N(\mu, \sigma^2)$ to the standard normal distribution $N(0, 1)$. Then we get:

$$F(x) = \frac{1}{2}\left(1 + \mathrm{erf}\left(\frac{x - \mu}{\sigma\sqrt{2}}\right)\right) \tag{11}$$

where $\mathrm{erf}(x)$ is the error function. The error function can be approximated using Taylor series:

$$\mathrm{erf}(x) = \frac{2}{\sqrt{\pi}} \sum_{n=0}^{\infty} \frac{(-1)^n x^{2n+1}}{(2n+1)n!} =$$

$$= \frac{2}{\sqrt{\pi}}\left(x - \frac{x^3}{3} + \frac{x^5}{10} - \frac{x^7}{42} + \frac{x^9}{216} - \cdots\right) . \tag{12}$$

In this case the solution is such a k, for which the following inequality is satisfied:

$$1 - p_k \geq \frac{1}{2}\left(1 + \mathrm{erf}\left(\frac{k - \mu}{\sigma\sqrt{2}}\right)\right) . \tag{13}$$

3 Summary and Conclusions

In this paper we presented briefly the distributed stream processing system we work on. Our goal was to focus on the problem of modeling the intensity of streams being processed in order to be able to configure system component properly. Firstly, we described data sources using mathematical rules. Basics of this description are included in [18–21]. Secondly, we tried to prepare a mathematical model which could be used to estimate the distribution of the number readings for a group of meters. Then the model was extended for various groups.

Obtained equations needed verification. We have built a simulation environment in which we conducted experiments. We measured how the number of readings in time periods changes. Based on the gathered data we were able to compare the simulation results with our theory. As described in the paper simulation results are very similar to the theoretical calculations.

Based on this we are able to estimate a safe buffer size, knowing the highest possible number of readings that can be received in a given time. Knowing the parameters of the distribution (σ and μ) and assuming the value of probability p_k small enough, we can find using the Equation (13) the number of readings that with probability p_k will not be exceeded.

This work lets us configure the second layer of the components network (Fig. 1) of our system. Further research needs to be done, to analyse behaviour of the other parts of the network. This is going to be the next stage of our research.

References

1. Wrembel, R.: On Handling the Evolution of External Data Sources in a Data Warehouse Architecture: Integrations of Data Warehousing. In: Taniar, D., Chen, L. (eds.) Integrations of Data Warehousing, Data Mining and Database Technologies, pp. 106–147 (2011)
2. Arasu, A., Babcock, B., Babu, S., Datar, M., Ito, K., Motwani, R., Nishizawa, I., Srivastava, U., Thomas, D., Varma, R., Widom, J.: Stream: The stanford stream data manager. IEEE Data Eng. Bull. 26(1), 19–26 (2003)
3. Gorawski, M., Chrószcz, A.: Synchronization Modeling in Stream Processing. In: Morzy, T., Härder, T., Wrembel, R. (eds.) Advances in Databases and Information Systems. AISC, vol. 186, pp. 91–102. Springer, Heidelberg (2013)
4. Gorawski, M., Chroszcz, A.: Optimization of operator partitions in stream data warehouse. In: Song, L.-Y., Cuzzocrea, A., Davis, K.C. (eds.) DOLAP 2011, pp. 61–66. ACM (2011)
5. Gorawski, M., Malczok, R.: Indexing Spatial Objects in Stream Data Warehouse. In: Nguyen, N.T., Katarzyniak, R., Chen, S.-M. (eds.) Advances in Intelligent Information and Database Systems. SCI, vol. 283, pp. 53–65. Springer, Heidelberg (2010)
6. Gorawski, M., Malczok, R.: Answering Range-Aggregate Queries over Objects Generating Data Streams. In: Kitagawa, H., Ishikawa, Y., Li, Q., Watanabe, C. (eds.) DASFAA 2010. LNCS, vol. 5982, pp. 436–439. Springer, Heidelberg (2010)
7. Gorawski, M.: Multiversion Spatio-temporal Telemetric Data Warehouse. In: Grundspenkis, J., Kirikova, M., Manolopoulos, Y., Novickis, L. (eds.) ADBIS 2009. LNCS, vol. 5968, pp. 63–70. Springer, Heidelberg (2010)
8. Kwiecień, A., Opielka, K.: Industrial Networks in Explosive Atmospheres. In: Kwiecień, A., Gaj, P., Stera, P. (eds.) CN 2011. CCIS, vol. 160, pp. 367–378. Springer, Heidelberg (2011)

9. Kwiecień, A., Stój, J.: Genius Network Communication Process Registration and Analysis. In: Kwiecień, A., Gaj, P., Stera, P. (eds.) CN 2011. CCIS, vol. 160, pp. 314–321. Springer, Heidelberg (2011)

10. Balazinska, M., Balakrishnan, H., Madden, S., Stonebraker, M.: Fault-Tolerance in the Borealis Distributed Stream Processing System. In: ACM SIGMOD Conf., Baltimore, MD (2005)

11. Madden, S., Franklin, M.J.: Fjording the stream: An architecture for queries over streaming sensor data: ICDE. In: Proceedings of the 18th International Conference on Data Engineering, pp. 555–566. IEEE Computer Society (2002)

12. Gorawski, M., Malczok, R.: Distributed spatial data warehouse indexed with virtual memory aggregation tree. In: Sander, J., Nascimento, M.A. (eds.) STDBM, pp. 25–32 (2004)

13. Gorawski, M., Bańkowski, S., Gorawski, M.: Selection of Structures with Grid Optimization in Multiagent Data Warehouse. In: Fyfe, C., Tino, P., Charles, D., Garcia-Osorio, C., Yin, H. (eds.) IDEAL 2010. LNCS, vol. 6283, pp. 292–299. Springer, Heidelberg (2010)

14. Gorawski, M., Marks, P.: High efficiency of hybrid resumption in distributed data warehouses. In: DEXA Workshops, pp. 323–327. IEEE Computer Society (2005)

15. Gorawski, M., Marks, P.: Checkpoint-based resumption in data warehouses. In: Socha, K. (ed.) IFIP International Federation for Information Processing. Software Engineering Techniques: Design for Quality, vol. 227, pp. 313–323. Springer, Boston (2006)

16. Labio, W., Wiener, J.L., Garcia-Molina, H., Gorelik, V.: Efficient resumption of interrupted warehouse loads. In: Chen, W., Naughton, J.F., Bernstein, P.A. (eds.) SIGMOD Conference, pp. 46–57. ACM (2000)

17. Gorawski, M., Marks, P.: Fault-tolerant distributed stream processing system. In: DEXA Workshops, pp. 395–399. IEEE Computer Society (2006)

18. Gorawski, M., Marks, P.: Towards reliability and fault-tolerance of distributed stream processing system. In: DepCoS-RELCOMEX, pp. 246–253. IEEE Computer Society (2007)

19. Gorawski, M., Marks, P.: Distributed stream processing analysis in high availability context. In: ARES 2007: Proceedings of the The Second International Conference on Availability, Reliability and Security, pp. 61–68. IEEE Computer Society, Washington, DC (2007)

20. Gorawski, M., Marks, P.: Towards automated analysis of connections network in distributed stream processing system. In: Haritsa, J.R., Kotagiri, R., Pudi, V. (eds.) DASFAA 2008. LNCS, vol. 4947, pp. 670–677. Springer, Heidelberg (2008)

21. Gorawski, M., Marks, P., Gorawski, M.: Collecting data streams from a distributed radio-based measurement system. In: Haritsa, J.R., Kotagiri, R., Pudi, V. (eds.) DASFAA 2008. LNCS, vol. 4947, pp. 702–705. Springer, Heidelberg (2008)

Modeling Operation of Web Service

Krzysztof Zatwarnicki and Anna Zatwarnicka

Department of Electrical, Control and Computer Engineering,
Opole University of Technology, Opole, Poland
{k.zatwarnicki,anna.zatwarnicka}@gmail.com

Abstract. This paper presents studies of modeling operation of Web service containing Web and database servers. The first part of the paper describes the method to obtain parameters of real servers. Additionally, detailed description of experiments and the results are presented. In the second part of the paper the Web and database servers models are introduced. In both of the presented models it is taken into account that the server processors are multi-core. Presented models can be used to evaluate proposed Web systems via simulation experiments.

Keywords: simulation model, web server modeling, database server modeling, queuing networks.

1 Introduction

Over the last few years the Internet has become the most innovative source of information and data. It has evolved from a medium for only privileged users into the medium we can't imagine living without. The rapid development of systems using WWW technology gives rise to the need for research on the effectiveness of the whole system delivering the required contents to the user.

The main semantic components forming the service of the Web system are: Hyper Text Markup Language (HTML), Hypertext Transfer Protocol (HTTP) and the Uniform Resource Identyfier (URI). The sources of HTML pages (Web pages) downloaded by clients (Web browsers) are Web servers. Web pages are transferred from Web servers to clients with use of the HTTP protocol. In the interaction between the client and the Web service the client sends the HTTP request to the Web server and the server sends in the response HTTP object [1].

There are a few ways of efficiency evaluation of a prospective service planned in use. One of them is the method of theoretical analysis using a proper mathematical modeling. This method is popular, however it requires simplifications of modeled system [2]. The second method enabling evaluation of proposed solutions is the simulation approach, which allows modeling of complex web services based on analysis of actual systems [3, 4].Queuing Networks usually are used in computer system simulations programs, but other approaches like Petri Nets [5] or propagation models [6] are also considered. The third method consists in building a prototype service containing real web servers [7]. The major disadvantage of this method is a high cost of carrying out experiments compared to the two previous methods.

A. Kwiecień, P. Gaj, and P. Stera (Eds.): CN 2013, CCIS 370, pp. 384–393, 2013.
© Springer-Verlag Berlin Heidelberg 2013

In this paper, methods of building a simulation model of web service are discussed. Firstly, technique enabling obtaining parameters of real Web and database servers is presented. Then the way of conducting experiments and the results of experiments are discussed. At the end, simulation models based on results of experiments are introduced.

The issue of creating the simulation model of the Web service is the topic of several publications. It is worth to start with WWW client's behavior models. The early works in this matter cover the issue of the HTTP object size modeling [8, 9]. Further, the complex client's behavior model [7] is presented as well as the client's behavior model in business service [10]. Other works concern the modeling of whole WWW services that cover the part related to client's modeling and also WWW servers, database servers, web dispatchers and wide area network [11–15]. However, there are not many works related to detailed WWW server simulation modeling [16, 17]. Authors of the paper already proposed such a model [18], however this time the presented model is designed for modern Web and database servers, the model contains much more detail and takes into account the fact that server processors are multi-core.

The paper is divided into five sections. Section 2 presents the method proposed to obtain Web and database servers parameters. Section 3 describes the conducted experiments and the results. Section 4 contains a description of simulation models. The final Section 5 presents concluding remarks.

2   Method to Obtain Parameters of Simulation Model of Web and Database Server

When creating a model of a Web system, it is important to take into account only elements of the system which have a significant impact on the request response time or can become potential bottlenecks [19]. Also, the way of operation of the system is important.

When the HTTP request is serviced, several actions have to be proceeded. At first the client opens the TCP connection with the web server. Next, it sends the HTTP request though the opened TCP connection. After that the web server prepares the response, than it looks for the requested file (in case of request for static object) or it precedes program/script preparing data for the client (in case of request for dynamic object). In the next step, the web server starts to send the response. After finishing service of the request the web server is ready to receive a subsequent HTTP request within the same TCP connection. In the end, after servicing numerous of HTTP requests send by the same client, the web server closes the TCP connection. The web server is able to service numerous HTTP requests concurrently, with the use of many different TCP connections.

The main elements of the web server which take part in the process of servicing HTTP request and have a significant impact on the response time are: processor, hard drive and the main memory, which acts as the cache memory of the file system. In the Fig. 1 the process of service of the request is presented.

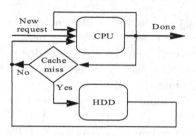

Fig. 1. Schema of process of service of HTTP request

The request service time on the processor is connected with the time to make the TCP connection $D^{\mathrm{CPU,TCP}}$, prepare the data to be sent and transfer the data through the computer network $D^{\mathrm{CPU,P+T}}$. The request service time on the hard drive involves a seek time (time to move the head of the hard drive), a rotational time (time to rotate the plate) and a transfer time, all together denoted as D^{HDD}. There are several methods to obtain service times for different elements of computer system. The first method is to modify the web server software to make the measurement of service time during the system operations [17, 13]. However, this method has some drawbacks. First of all, the access to software source codes of the Web server is required. Moreover, the changes of the software can change the way the examined system operates. These sorts of modifications can increase the load of the processor and prolong the request service time. The other method proposed by authors and presented previously in [18] is to make experiments during which the load and the throughput of the tested element of system are measured. The service demand time is then calculated on the base of service Demand Law [20]. This law specifies the dependence of service demand time D^{Resource} from the resource utilization U^{Resource} and the throughput X of the entire system according to Formula 1:

$$D^{\mathrm{Resource}} = \frac{U^{\mathrm{Resource}}}{X} . \tag{1}$$

In order to calculate service demand times several experiments and measurements had to be performed. During the experiments we intended to measure service times for processor and hard drive of web server and separately for processor and hard drive of database server supporting service of dynamic HTTP requests. We decided to obtain following values describing operations of web and database servers:

1. $D^{\mathrm{WWW,CPU,TCP}}$, the web server processor service demand time to open TCP connection;
2. $D_{\mathrm{S}}^{\mathrm{WWW,CPU,P+T}}(z)$, web server processor service demand time to prepare the static data (files of the file system) and transfer them through the computer network for different sizes z of web objects;
3. $D_{\mathrm{D}}^{\mathrm{WWW,CPU,P+T}}(z)$, service demand time of the web server processor to prepare the dynamic data and transferred them through the computer network for different sizes z of web objects;

4. $D_S^{WWW,HDD}(z)$, web server hard drive service demand time to prepare and transfer data of different sizes z to the web server processor;
5. $D_{Q1}^{DB,CPU}$, $D_{Q2}^{DB,CPU}$, $D_{Q3}^{DB,CPU}$, $D_{Q1}^{DB,HDD}$, $D_{Q2}^{DB,HDD}$, $D_{Q3}^{DB,HDD}$, data base server processor and hard drive service demand times to prepare the requested data for three different SQL queries;
6. $D_{JM}^{WWW,CPU}$, $D_{JR}^{WWW,CPU}$, $D_{JM}^{WWW,HDD}$, $D_{JR}^{WWW,HDD}$, $D_{JM}^{DB,CPU}$, $D_{JR}^{DB,CPU}$, $D_{JM}^{DB,HDD}$, $D_{JR}^{DB,HDD}$, web and data base server processor and hard drive service demand times to prepare data from a Joomla service main page (denoted as JM) and random page (denoted as JM).

3   Experiments Description and Results

A testbed containing three similar computers acting as the web server, the database server and a client were used in all described further experiments. The computers were connected to Gigabit Ethernet Repotec RP-G32224E switch (Fig. 2a).

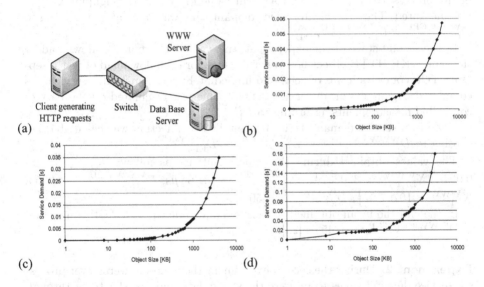

Fig. 2. Experiments and results: (a) testbed, (b) Web server processor service demand time to prepare and transfer static data in HTTP object size function, (c) Web server processor service demand time to prepare and transfer dynamic data in HTTP object size function, (d) Web server hard drive service demand time in HTTP object size function

Computers used in experiments had the following configuration: Intel Core2 Quad CPU Q6700 2.66 GHz processor, 2 x 2 GB PC2-6400 RAM memory, Barracuda ST3250410AS 250 GB 7200 RPM hard drive, Gigabyte GA-G33-DS3R FSB 1333 motherboard. Each of the computer was operating under Linux Debian 6.0.1a operating system. A computer acting as web server had installed

Apache 2.2.16 with PHP 5.3.3 while a computer acting as a database server had installed MySQL 5.6. On the third computer application acting as HTTP client was installed. The application was prepared by authors with use of curl libtaries [21], and enabled to strictly control the process of servicing HTTP request with use of several concurrent TCP connections.

Sar tool well known from unix systems enable us to observe the load of the web server resources. In order to obtain earlier mentioned servers parameters, six groups of experiments had been conducted.

Experiment 1. In the first experiment, the value of the web server processor service demand time to open TCP connection $D^{\mathrm{WWW,CPU,TCP}}$ was obtained. The experiment was divided into many separate stages. During the first stage, an HTTP object of size of $1\,\mathrm{KB}$ was downloaded with use of persistent TCP connections. The web client opened several concurrent TCP connections with the web server, and with the use of each the connection was receiving the same requested object many times. During the experiment a mean value of utilization of the processor $U_{\mathrm{E1.1}}^{\mathrm{WWW,CPU}}$ and a mean value of system throughput $X_{\mathrm{E1.1}}^{\mathrm{WWW}}$ was measured. Then the mean service demant time was calculated according to $D_{\mathrm{E1.1}}^{\mathrm{WWW,CPU}} = U_{\mathrm{E1.1}}^{\mathrm{WWW,CPU}}/X_{\mathrm{E1.1}}^{\mathrm{WWW}}$.

In the second stage of the experiment, the web client again was downloading the same $1\,\mathrm{KB}$ HTTP object however this time, for each download object a separate TCP connection was opened and immediately closed just after service was completed. Again the mean service demant time was calculated on the base of conducted measurements according to $D_{\mathrm{E1.2}}^{\mathrm{WWW,CPU}} = U_{\mathrm{E1.2}}^{\mathrm{WWW,CPU}}/X_{\mathrm{E1.2}}^{\mathrm{WWW}}$.

The mean service demand time to open TCP connection was calculated according to $D^{\mathrm{WWW,CPU,TCP}} = D_{\mathrm{E1.2}}^{\mathrm{WWW,CPU}} - D_{\mathrm{E1.1}}^{\mathrm{WWW,CPU}}$.

The values obtained during the experiment are as follow: $U_{\mathrm{E1.1}}^{\mathrm{WWW,CPU}} = 0.9946$, $X_{\mathrm{E1.1}}^{\mathrm{WWW}} = 320301\frac{1}{\mathrm{s}}$, $D_{\mathrm{E1.1}}^{\mathrm{WWW,CPU}} = 31.05\,\mu s$, $U_{\mathrm{E1.2}}^{\mathrm{WWW,CPU}} = 0.9953$, $X_{\mathrm{E1.2}}^{\mathrm{WWW}} = 19707.62\frac{1}{\mathrm{s}}$, $D_{\mathrm{E1.2}}^{\mathrm{WWW,CPU}} = 50.5\,\mu s$.

In the end the mean obtained service demand time to open TCP connection is $D^{\mathrm{WWW,CPU,TCP}} = 19.45\,\mu s$ [22].

Experiment 2. During the second experiment the values of web server processor service demand times to prepare the static data and transfer them through the computer network for different sizes of web objects $D_{\mathrm{S}}^{\mathrm{WWW,CPU,P+T}}(z)$ were obtained. The second experiment was divided into several stages, each stage was dedicated to different size of downloaded web object. During each stage the same HTTP object with use of several persistent TCP connections was downloaded millions times. The sizes of downloaded objects were: 1, 8, 16, ..., 128, 256, 384, ..., 1024, 1536, 2048, ..., 4096 [KB].

The values of the web server processor service demand were calculated according to: $D_{\mathrm{S}}^{\mathrm{WWW,CPU,P+T}}(z) = U_{\mathrm{E2}}^{\mathrm{WWW,CPU}}(z)/X_{\mathrm{E2}}^{\mathrm{WWW}}(z)$, where z is the size in KB of downloaded HTTP object.

Results of conducted experiments are presented on Fig. 2b as a diagram of mean values of web server processor service demand time in the HTTP object size function.

A linear regression algorithm was used to prepare formula letting estimate the service demand time. The formula is as follow:

$$D_S^{\text{WWW,CPU,P+T}}(z) = \begin{cases} z \cdot 268.18 \cdot 10^{-6} + 6.0043 \cdot 10^{-5}, & z \in \langle 1, 128 \rangle \\ z \cdot 1.707 \cdot 10^{-6} + 2.1128 \cdot 10^{-4}, & z \in \langle 192, 1024 \rangle \\ z \cdot 1.119 \cdot 10^{-6} + 8.2385 \cdot 10^{-4}, & z \in \langle 1536, 4096 \rangle \end{cases} \quad (2)$$

where $D_S^{\text{WWW,CPU,P+T}}(z)$ is in seconds. Obtained model well fitt to results achieved in experiments and the coefficient of determination is $R^2 = 0.9989$ [22].

Experiment 3. The third experiment was similar to the second one. The service demand times of the web server processor to prepare the dynamic data and transfer them through the computer network $D_D^{\text{WWW,CPU,P+T}}(z)$ for different sizes z of web objects were obtained. Each stage of the experiment was conducted for objects of the same size. The objects were created dynamically with use of PHP script after HTTP requests arrival. The PHP script generated random text of size entered as a parameter of the script. The sizes of downloaded objects were the same as in Experiment 2. Again the service demand times were calculated on the base of following formula: $D_D^{\text{WWW,CPU,P+T}}(z) = U_{E3}^{\text{WWW,CPU}}(z)/X_{E3}^{\text{WWW}}(z)$, where $U_{E3}^{\text{WWW,CPU}}(z)$ was the utilization of the web server processor during the experiment, and $X_{E3}^{\text{WWW}}(z)$ was the throughput of the web server.

Figure 2c presents results as a diagram of mean values of web server processor service demand time for dynamic objects in the HTTP object size function. The relation between service demand time and the HTTP object size can be also modeled in following way:

$$D_D^{\text{WWW,CPU,P+T}}(z) = \begin{cases} z \cdot 7.8986 \cdot 10^{-6} + 9.4416 \cdot 10^{-5}, & z \in \langle 1, 128 \rangle \\ z \cdot 9.1403 \cdot 10^{-6} + 2.0679 \cdot 10^{-4}, & z \in \langle 192, 1024 \rangle \\ z \cdot 8.2572 \cdot 10^{-6} + 6.9411 \cdot 10^{-4}, & z \in \langle 1536, 4096 \rangle \end{cases} \quad (3)$$

where $D_D^{\text{WWW,CPU,P+T}}(z)$ is in seconds, and the coefficient of determination is $R^2 = 0.9998$ [22].

Experiment 4. In the experiment the web server hard drive service demand times to prepare and transfer data of different sizes $D_S^{\text{WWW,HDD}}(z)$, were obtained. During the experiment the hard drive of the web server and its utilization $U_{E4}^{\text{WWW,HDD}}(z)$ was observed. Similarly to two of the previous experiments, the tests had been divided in to many stages. During each stage different files of the same size were downloaded by the web client. The total size of files of the same stage was larger then 2 GB, thanks to this the requested objects were always fetched from the hard drive, and not from the cache memory. The sizes of the files were the same as in Experiment 2. The values of the web server hard drive service demand times were calculated similarly to the way presented in Experiment 2 and 1 according to: $D_{\text{S}}^{\text{WWW,HDD}}(z) = U_{\text{E4}}^{\text{WWW,HDD}}(z) / X_{\text{E4}}^{\text{WWW}}(z)$, where $X_{\text{E4}}^{\text{WWW}}(z)$ is the throughput of the web server.

Figure 2d presents diagram of mean values of web server hard drive service demand time for dynamic objects in the HTTP object size function. A chosen function to model the hard drive service demand time is following:

$$D_{\text{S}}^{\text{WWW,HDD}}(z) = \begin{cases} z \cdot 8.559 \cdot 10^{-4} - 3.0556 \cdot 10^{-4}, & z \in \langle 1, 8 \rangle \\ z \cdot 5.4206 \cdot 10^{-5} + 1.3261 \cdot 10^{-2}, & z \in \langle 16, 128 \rangle \\ z \cdot 6.1816 \cdot 10^{-5} + 1.09 \cdot 10^{-2}, & z \in \langle 192, 1024 \rangle \\ z \cdot 6.1482 \cdot 10^{-5} + 1.473 \cdot 10^{-2}, & z \in \langle 1536, 3076 \rangle \end{cases} \tag{4}$$

where $D_{\text{S}}^{\text{WWW,HDD}}(z)$ is in seconds, and the coefficient of determination is $R^2 = 0.9955$ [22].

Experiment 5. During the fifth experiment data base server processor and hard drive service demand times $D_{\text{Q1}}^{\text{DB,CPU}}$, $D_{\text{Q2}}^{\text{DB,CPU}}$, $D_{\text{Q3}}^{\text{DB,CPU}}$, $D_{\text{Q1}}^{\text{DB,HDD}}$, $D_{\text{Q2}}^{\text{DB,HDD}}$, $D_{\text{Q3}}^{\text{DB,HDD}}$, to prepare the requested data for three different SQL queries were obtained. In order to conduct required tests a database containing tree related tables had been created on the server with MySQL database. Each of the tables contained 5 millions rows consisting of mainly text data. On the web server there had been tree PHP scripts containing SQL queries of different complexity. The first script (denoted as Q1) enables to download 10 randomly chosen rows from one table, the second one (Q2) was fetching 20 randomly chosen rows from two related tables, and in the end third script (Q3) allow to download 100 randomly chosen related rows from three tables. During each experiment client opened many concurrent connection to download data and the mean value of the load of the processor $U_{\text{Qn}}^{\text{DB,CPU}}$, hard drive $U_{\text{Qn}}^{\text{DB,HDD}}$, and the throughput $X_{\text{Qn}}^{\text{DB}}$ had been measured. The service demand times were calculated on the base of Formula 1. The results are as follow: $D_{\text{Q1}}^{\text{DB,CPU}} = 4.49\,\text{ms}$, $D_{\text{Q1}}^{\text{DB,HDD}} = 4.995\,\text{ms}$, $D_{\text{Q2}}^{\text{DB,CPU}} = 55.53\,\text{ms}$, $D_{\text{Q2}}^{\text{DB,HDD}} = 72.68\,\text{ms}$, $D_{\text{Q3}}^{\text{DB,CPU}} = 187.91\,\text{ms}$, $D_{\text{Q3}}^{\text{DB,HDD}} = 250.8\,\text{ms}$ [22].

Experiment 6. In the end, during the last experiment web and data base server processor and hard drive service demand times to prepare data from a CMS Joomla service main page and random pages were obtained. In order to conduct this experiment, very popular CMS Joomla 2.56 system was installed on web and database servers. About 50 articles containing 4 KB of HTML text were added to the CMS system. During tests client opened many concurrent TCP connection to download data. Also the load of the processors of the web and database servers and their hard drives were observed.

The mean web server processor service demand time obtained in the experiment is $D_{\text{JM}}^{\text{WWW,CPU}} = 25.464\,\text{ms}$, while the mean service demand time for the hard drive was negligibly small. The mean database server processor and hard drive demand time is $D_{JM}^{DB,CPU} = 5.043\,ms$ and $D_{JM}^{DB,HDD} = 1.333\,ms$. Results obtained for random aces pages are as follow: $D_{JR}^{WWW,CPU} = 28.215\,ms$, $D_{JR}^{DB,CPU} = 5.539\,ms$ and $D_{JR}^{DB,HDD} = 6.851\,ms$ [22].

In almost all the of conducted experiments the load of the web and database processors was high, and the load ranged from 80 to 99 %. Also the number of concurrently serviced request ranged from 50 to 100 requests. This fact is important due to the way of constructing simulation model of processors.

4   Simulation Model Construction

The most common way to build simulation model of computer system is to use queuing network model [23]. Each resource is then modeled as a separate queue containing a waiting queue and a resource or multiple resources for the same queue.

The examined computer system presented in the paper can be modeled in two ways as is shown in the Fig. 3a and 3b. In the Fig. 3a the processor is presented as a load dependent resource where the service time depends on the number of requests being serviced. The differences in the service time are due to the fact that each of the processors contains four cores acting as separate services. Therefore the service time can be calculated as follow:

$$D^{CPU}(n) = \begin{cases} D_{measured}^{CPU} \cdot (4-n) & \text{for } n < 4 \\ D_{measured}^{CPU} & \text{for } n \geqslant 4 \end{cases}, \qquad (5)$$

where $D_{measured}^{CPU}$ is processor demand time obtained in experiments, and n is number of request being serviced, and waiting in a queue to the processor.

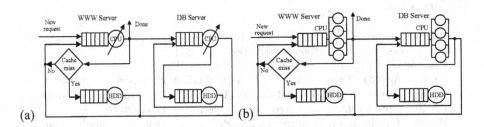

(a)   (b)

Fig. 3. Simulation models of web systems (a) processors modeled as load dependent resources, (b) processors modeled as load independent resources

In the model presented on Fig. 3b the processor is modeled as load independent resource therefore in this case the processor service time should be calculated in following way $D^{CPU}(n) = D_{measured}^{CPU} \cdot 4$.

The demand times for the hard driver resources in the simulation model can have the same values as obtained during experiments due to the fact that hard drives have single services.

The module of cache memory in the web server simulation model determine if given HTTP object is located in cache memory or should be retrieved from the hard drive. The size of cache memory in simulation can depend on requirements. The algorithm responsible for swapping HTTP objects in the memory can be the Last Recently Used policy according to which the object not used for the longest time first leaves the memory in case of cache memory overflow [16]. The service time to retrieve data from the cache memory and transmit it to the processor is included in processor demand times.

Both of the presented simulation models should make it possible to obtain similar mean request response times, however if not only the mean values are important than the second (Fig. 3b), more precise model, should be used.

The simulation program implementing proposed models can be written with the use of appropriate simulation package enabling management of discrete-event simulations. Example of such environments are CSIM 20 [24] or OMNET++ [25].

5 Summary

In the article, the way of building the simulation model of Web and database servers was presented. A method to obtain parameters of simulation model of Web and database server was introduced also the way of conducting experiments was described. With the use of service demand low the required HTTP request service demand times for the processor and the hard drive ware determined.

On the base of obtained experiments results the Web and database servers models were proposed. In both of presented models it was taken into account that the server processors are multi-core. Presented models can be used in simulation based experiments evaluating propositions of designed Web systems.

References

1. Krishnamurthy, B., Rexford, J.: HTTP/1.1, networking protocols, caching, and traffic measurement. Addison-Wesley, Boston (2001)
2. Czachórski, T.: A diffusion approximation model of web servers. In: Cellary, W., Iyengar, A. (eds.) Proceedings of IFIP TC6/WG6. 4 Workshop on Internet Technologies, Applications and Social Impact (WITASI 2002), Wroclaw, Poland, October 10-11, pp. 83–92. Kluwer Academic Publishers, Boston (2002)
3. Barford, P., Misra, V.: Measurement, Modeling and Analysis of the Internet. In: IMA Workshop on Internet Modeling and Analysis, Minneapolis, MN (2004)
4. Zatwarnicki, K.: Providing Web service of established quality with the use of HTTP requests scheduling methods. In: Jędrzejowicz, P., Nguyen, N.T., Howlet, R.J., Jain, L.C. (eds.) KES-AMSTA 2010, Part I. LNCS (LNAI), vol. 6070, pp. 142–151. Springer, Heidelberg (2010)
5. Rzońca, D., Stec, A., Trybus, B.: Data Acquisition Server for Mini Distributed Control System. In: Kwiecień, A., Gaj, P., Stera, P. (eds.) CN 2011. CCIS, vol. 160, pp. 398–406. Springer, Heidelberg (2011)
6. Olejnik, R.: A Floor Description Language as a Tool in the Process of Wireless Network Design. In: Kwiecień, A., Gaj, P., Stera, P. (eds.) CN 2009. CCIS, vol. 39, pp. 135–142. Springer, Heidelberg (2009)

7. Borzemski, L., Zatwarnicki, K.: A Fuzzy Adaptive Request Distribution Algorithm for Cluster-Based Web Systems. In: Proceedings of 11th PDP Conference, pp. 119–126. IEEE Press, Los Alamitos (2003)
8. Arlitt, M., Friedrich, R., Jin, T.: Workload characterization of a Web proxy in a cable modem environment. ACM Performance Eval. Review 27(2), 25–36 (1999)
9. Williams, A., Arlitt, M., Williamson, C., Barker, K.: Web workload characterization: ten years later. In: Tang, X., Xu, J., Chanson, S.T. (eds.) Publish Info. Web Content, pp. 3–22. Springer, New York (2005)
10. Cardellini, V., Colajanni, M., Yu, P.S.: Impact of workload models in evaluating the performance of distributed Web-server systems. In: Gelenbe, E. (ed.) System Performance Evaluation: Methodologies and Applications, pp. 397–417. CRC Press (2000)
11. Borzemski, L., Zatwarnicka, A., Zatwarnicki, K.: Global Adaptive Request Distribution with Broker. In: Apolloni, B., Howlett, R.J., Jain, L. (eds.) KES 2007, Part II. LNCS (LNAI), vol. 4693, pp. 271–278. Springer, Heidelberg (2007)
12. Borzemski, L., Suchacka, G.: Business-oriented admission control and request scheduling for e-commerce websites. Cybernetics and Systems 41(8), 592–609 (2010)
13. Riska, A., Riedel, E.: Disk Drive Level workload Characterization. In: Proceedings of the USENIX Annual Technical Conference, Boston, pp. 97–103 (2006)
14. Zatwarnicki, K.: Neuro-Fuzzy Models in Global HTTP Request Distribution. In: Pan, J.-S., Chen, S.-M., Nguyen, N.T. (eds.) ICCCI 2010, Part I. LNCS, vol. 6421, pp. 1–10. Springer, Heidelberg (2010)
15. Zhang, Q., Riska, A., Riedel, E., Mi, M., Smarni, E.: Evaluating performability of systems with background jobs. In: Proceedings of The Symposium on Dependable Systems and Networks (DSN), Philadelphia, pp. 495–505 (2006)
16. Aron, M., Druschel, P., Zwaenepoel, W.: Efficient Support for P-http in Cluster-based Web Servers. In: Proceedings of the USENIX 1999 Annual Technical Conference, Monterrey, CA (1999)
17. Cherkasova, L., Gardner, R.: Measuring CPU Overhead for I/O Processing in the Xen Virtual Machine Monitor. In: USENIX Association, Berkeley, CA, Anaheim, USA, pp. 24–27 (2005)
18. Zatwarnicki, K.: Identification of Web server. In: Kwiecień, A., Gaj, P., Stera, P. (eds.) CN 2011. CCIS, vol. 160, pp. 45–54. Springer, Heidelberg (2011)
19. Fortier, P.J., Howard, M.E.: Computer Systems Performance Evaluation and Prediction. Digital Press, USA (2003)
20. Denning, P., Buzen, J.: The Operational Analysis of Queueing Network Models. ACM Computer Survey 10(3), 225–261 (1978)
21. CURL library documentation (2012), http://curl.haxx.se/
22. Krawczyk, J.: Determination of parameters of the web and database servers. M.S. thesis, Department of Electroengineering, Automatic Control and Computer Science, Opole University of Technology, Opole, Poland (2012)
23. Menascé, D., Bennani, M.: Analytic performance models for single class and multiple class multithreaded software servers. In: Int. CMG Conference 2006, pp. 475–482 (2006)
24. CSIM20. CSIM20, Mesquite Software, 2012 Development toolkit for simulation and modeling (2012), http://www.mesquite.com
25. OMNET++, Development toolkit for simulation and modeling (2012), http://www.omnetpp.org/

Total Volume Distribution
for Multiserver Queueing Systems
with Random Capacity Demands

Oleg Tikhonenko[1] and Magdalena Kawecka[2]

[1] Institute of Theoretical and Applied Informatics, Polish Academy of Sciences
Baltycka 5, 44-100 Gliwice, Poland
oleg.tikhonenko@gmail.com
[2] Czestochowa University of Technology, Institute of Mathematics
ul. Dabrowskiego 69, 42-201 Czestochowa, Poland
kawecka_magdalena@wp.pl

Abstract. We investigate multiserver queueing systems with Poisson arrivals, identical servers and random capacity demands under assumption that demand's service time is proportional to its capacity having an exponential distribution. For such systems statistical characteristics of the total demands' volume are determined.

Keywords: queueing system, demand's capacity, total demands' volume, Laplace-Stieltjes transform.

1 Introduction

We consider multiserver queue $M/M/n/m \leq \infty$, where n is the number of identical servers, m is the number of waiting positions in the common queue. Let a be the rate of the arrival flow. Suppose that 1) each demand is characterized by some non-negative random capacity ζ which is independent of capacities of other demands; 2) service time ξ of the demand depends on its capacity only.

Such systems have been used to model and solve the various practical problems occurring in the design of computer and communicating systems [1,2].

The common distribution of ζ and ξ random variables is determined generally by the following joint distribution function:

$$F(x,t) = P\{\zeta < x, \xi < t\} \ .$$

Let $L(x) = F(x,\infty)$ be the distribution function of the demand's capacity and $B(t) = F(\infty,t)$ be the distribution function of its service time. Let $\sigma(t)$ be the total demands' volume at time instant t (the sum of capacities of all demands present in the system at this moment).

We can't determine analytically the total volume distribution for the systems of $M/G/n/m$ and $M/G/n/\infty$ type in general case. But we can do it when service time has an exponential distribution (i.e. in the case of $M/M/n/m$ and $M/M/n/\infty$ systems). So, for the further analysis, we have take into account

A. Kwiecień, P. Gaj, and P. Stera (Eds.): CN 2013, CCIS 370, pp. 394–405, 2013.
© Springer-Verlag Berlin Heidelberg 2013

two possibilities: 1) demand's capacity has an arbitrary distribution and service time is independent of it; 2) demand's capacity is distributed exponentially and service time is proportional to the demand's capacity. The first case is trivial (see ex. [1]). Hence, we shall investigate the second case.

Suppose that f is the parameter of demand's capacity (i.e. $L(x) = 1 - e^{-fx}$, $f > 0$). If ζ is the demand's capacity, then the demand's service time ξ is equal to $c\zeta$, where c is a coefficient of proportionality $(c > 0)$. Then we have $B(t) = 1 - e^{-\mu t}$, where $\mu = f/c$.

The main purpose of the paper is to determine the total demands's volume distribution in stationary mode, existing when $\rho = a/(n\mu) < \infty$, for the system $M/M/n/m < \infty$, and $\rho = a/(n\mu) < 1$, for the system $M/M/n/\infty$ [3].

2   Notation and Auxiliary Statements

Let $\eta(t)$ be the number of demands present in the system $M/M/n/m < \infty$ at time instant t. Denote as $P_i(t) = \mathbf{P}\{\eta(t) = i\}$, $i = \overline{0, n+m}$. Assume that the stationary mode exists $(\rho < \infty)$. Let η and σ be the number of demands present in the system and total demands' volume in stationary mode consequently $(\eta(t) \Rightarrow \eta$ and $\sigma(t) \Rightarrow \sigma$ in the sense of a weak convergence). Let $D(x) = \mathbf{P}\{\sigma < x\}$ be the distribution function of the random variable σ and $\delta(s) = \int_0^\infty e^{-sx} \, dD(x)$ be the Laplace-Stieltjes transform (LST) of the distribution function $D(x)$.

Suppose that there are i demands in the system at time instant t, $i = \overline{1, n+m}$. Then $k = \min(i, n)$ demands are served at this moment. Denote as $\chi_j(t)$ the duration of service from the beginning to time instant t of j-th demand being served at time instant t, $j = \overline{1, k}$. In stationary mode we have $\chi_j(t) \Rightarrow \chi_j$ in the sense of a weak convergence. Let $R_i(t, x_1, \dots, x_k) dx_1 \dots dx_k$ be the probability that there are i demands in the system at time instant t and the duration of service down to the moment t of k demands being served are equal to $\chi_1(t) \in [x_1; x_1 + dx_1), \dots, \chi_k(t) \in [x_k; x_k + dx_k)$:

$$R_i(t, x_1, \dots, x_k) dx_1 \dots dx_k =$$

$$= \mathbf{P}\{\eta(t) = i, \chi_1(t) \in [x_1; x_1 + dx_1), \dots, \chi_k(t) \in [x_k; x_k + dx_k)\} \ ,$$

$$i = \overline{1, n+m}, k = \min(i, n) \ .$$

If stationary mode exists, the following limits exist too:

$$p_i = \lim_{t \to \infty} P_i(t), \ i = \overline{0, n+m} \ ;$$

$$r_i(x_1, \dots, x_k) = \lim_{t \to \infty} R_i(t, x_1, \dots, x_k), \ i = \overline{1, n+m}, \ k = \min(i, n) \ .$$

Note that (see ex. [4])

$$p_0 = \left[\sum_{k=0}^{n} \frac{(n\rho)^k}{k!} + \frac{n^n \rho^{n+1}(1 - \rho^m)}{n!(1 - \rho)} \right]^{-1}, \ \rho \neq 1 \ ; \tag{1}$$

$$p_0 = \left[\sum_{k=0}^{n} \frac{n^k}{k!} + \frac{mn^n}{n!} \right]^{-1}, \ \rho = 1 \ ; \tag{2}$$

$$p_k = \frac{(n\rho)^k}{k!} p_0, \ k = \overline{1, n} \ ; \tag{3}$$

$$p_k = \frac{n^n \rho^k}{n!} p_0, \ k = \overline{n+1, n+m} \ . \tag{4}$$

Let us assume that at an arbitrary time instant (or in stationary mode) demands being served are numerated randomly, that is, if at the given instant there are k such demands in the system, any of $k!$ possible ways of numeration can be used with probability $1/k!$. Then the functions $R_i(t, x_1, \ldots, x_k)$ and $r_i(x_1, \ldots, x_k)$ are symmetrical about permutation of their arguments.

It's easy to show (taking into account the aforementioned symmetry) that the following equations (including boundary conditions) take place for the functions $r_i(x_1, \ldots, x_k)$:

$$\sum_{j=1}^{i} \frac{\partial r_i(x_1, \ldots, x_i)}{\partial x_j} = -\mu(i + n\rho) r_i(x_1, \ldots, x_i) +$$

$$+ (i+1)\mu \int_0^\infty r_{i+1}(x_1, \ldots, x_i, y) dy, \ i = \overline{1, n-1} \ ; \tag{5}$$

$$\sum_{j=1}^{n} \frac{\partial r_n(x_1, \ldots, x_n)}{\partial x_j} = -n\mu(1 + \rho) r_n(x_1, \ldots, x_n) \ ; \tag{6}$$

$$\sum_{j=1}^{n} \frac{\partial r_i(x_1, \ldots, x_n)}{\partial x_j} = -n\mu(1 + \rho) r_i(x_1, \ldots, x_n) + n\mu\rho r_{i-1}(x_1, \ldots, x_n) \ ,$$

$$i = \overline{n+1, n+m-1} \ ; \tag{7}$$

$$\sum_{j=1}^{n} \frac{\partial r_{n+m}(x_1, \ldots, x_n)}{\partial x_j} = -n\mu r_{n+m}(x_1, \ldots, x_n) + n\mu\rho r_{n+m-1}(x_1, \ldots, x_n) \ ; \tag{8}$$

$$r_1(0) = n\mu\rho p_0 \ ; \tag{9}$$

$$i r_i(x_1, \ldots, x_{i-1}, 0) = n\mu\rho r_{i-1}(x_1, \ldots, x_{i-1}), \ i = \overline{2, n-1} \ ; \tag{10}$$

$$r_n(x_1, \ldots, x_{n-1}, 0) = \mu\rho r_{n-1}(x_1, \ldots, x_{n-1}) + \mu \int_0^\infty r_{n+1}(x_1, \ldots, x_{n-1}, y) dy \ ; \tag{11}$$

$$r_i(x_1, \ldots, x_{n-1}, 0) = \mu \int_0^\infty r_{i+1}(x_1, \ldots, x_{n-1}, y) dy, \ i = \overline{n+1, n+m-1} \ ; \tag{12}$$

$$r_{n+m}(x_1, \ldots, x_{n-1}, 0) = 0 \ . \tag{13}$$

Lemma 1. *The following relations take place:*

$$\int_0^\infty r_{i+1}(x_1,\ldots,x_i,y)dy = \frac{n\rho}{i+1}r_i(x_1,\ldots,x_i), \; i = \overline{1, n-1} \; ;$$

$$\int_0^\infty r_{i+1}(x_1,\ldots,x_{n-1},y)dy = \rho\int_0^\infty r_i(x_1,\ldots,x_{n-1},y)dy, \; i = \overline{n, n+m-1} \; .$$

Proof. For $i < n$ we evidently have taking into account the conditional probability formula:

$$\int_{y=0}^\infty r_{i+1}(x_1,\ldots,x_i,y)dy\,dx_1,\ldots,dx_i =$$

$$= \mathbf{P}\{\eta = i+1, \chi_1 \in [x_1; x_1 + dx_1),\ldots,\chi_i \in [x_i + dx_i)\} =$$

$$= \mathbf{P}\{\chi_1 \in [x_1; x_1 + dx_1),\ldots,\chi_i \in [x_i + dx_i)|\eta = i+1\}p_{i+1} \; ,$$

where $p_i = \mathbf{P}\{\eta = i\}$.

It's clear that

$$\mathbf{P}\{\chi_1 \in [x_1; x_1 + dx_1),\ldots,\chi_i \in [x_i + dx_i)|\eta = i+1\} =$$

$$= \mathbf{P}\{\chi_1 \in [x_1; x_1 + dx_1),\ldots,\chi_i \in [x_i + dx_i)|\eta = i\} \; ,$$

whens

$$\int_0^\infty r_{i+1}(x_1,\ldots,x_i,y)dy = r_i(x_1,\ldots,x_i)p_{i+1}/p_i \; . \tag{14}$$

Then, from the relations (3) and (14) we obtain the first statement of the lemma.

For such i that $n \le i < n+m$ we have:

$$\int_{y=0}^\infty r_{i+1}(x_1,\ldots,x_{n-1},y)dy\,dx_1,\ldots,dx_{n-1} =$$

$$= \mathbf{P}\{\eta = i+1, \chi_1 \in [x_1; x_1 + dx_1),\ldots,\chi_{n-1} \in [x_{n-1} + dx_{n-1})\} =$$

$$= \mathbf{P}\{\chi_1 \in [x_1; x_1 + dx_1),\ldots,\chi_i \in [x_{n-1} + dx_{n-1})|\eta = i+1\}p_{i+1} \; ,$$

whens

$$\int_0^\infty r_{i+1}(x_1,\ldots,x_{n-1},y)dy = \frac{p_{i+1}}{p_i}\int_0^\infty r_i(x_1,\ldots,x_{n-1},y)dy \; . \tag{15}$$

From the relations (4) and (15) we obtain the second statement of the lemma. The lemma is proved.

3   Determination of LST of the Total Volume's Distribution Function

Let κ be the total service time of demands present in the system in stationary mode. Then $\kappa = c\sigma$. Denote as $M(t) = \mathbf{P}\{\kappa < t\}$ the distribution function of the random variable κ. Then we have $D(x) = M(cx)$. Let $\gamma(s)$ be LST of the function $M(t)$. Then we have $\delta(s) = \gamma(s/c)$. First, we shall determine $\gamma(s)$.

Theorem 1. *The function $\gamma(s)$ is determined by the relation:*

$$\gamma(s) = p_0 \left\{ \sum_{i=0}^{n-1} \frac{(n\rho)^i \mu^{2i}}{i!(s+\mu)^{2i}} + \frac{(n\rho)^n \mu^{2n}(s+\mu+\rho s)}{n!(s+\mu)^{2n-2}(s+\mu-\mu\rho)[(s+\mu)^2+s\mu\rho]} - \right.$$

$$\left. - \frac{n^n \mu^{2n+m} \rho^{n+m+1}}{n!(s+\mu)^{2n+m}(s+\mu\rho)} \left[\frac{s(s+\mu)+\mu^2\rho}{s+\mu-\mu\rho} - \frac{\mu^{m+1}s^2}{(s+\mu+\mu\rho)^m((s+\mu)^2+s\mu\rho)} \right] \right\} .$$

Proof. Denote as $h(s, x_1, \ldots, x_k)$ LST of conditional distribution function of total service time of k demands being served, under condition that jth demand was served during x_j time units from the beginning, $j = \overline{1,k}$, $k = \overline{1,n}$. Obviously, we have:

$$\gamma(s) = p_0 + \sum_{i=0}^{n-1} \int_0^\infty \ldots \int_0^\infty h(s, x_1, \ldots, x_i) r_i(x_1, \ldots, x_i) \mathrm{d}x_1 \ldots \mathrm{d}x_i +$$

$$+ \sum_{i=n}^{n+m} (\beta(s))^{i-n} \int_0^\infty \ldots \int_0^\infty h(s, x_1, \ldots, x_n) r_i(x_1, \ldots, x_n) \mathrm{d}x_1 \ldots \mathrm{d}x_n ,$$

where $\beta(s) = \mu/(s+\mu)$ is LST of the distribution function $B(t)$. It's clear that

$$h(s, x_1, \ldots, x_k) = \frac{\mu^k}{(s+\mu)^k} \prod_{j=1}^k e^{-sx_j} ,$$

whence

$$\gamma(s) = p_0 + \sum_{i=1}^{n-1} \frac{\mu^i}{(s+\mu)^i} \int_0^\infty \ldots \int_0^\infty \prod_{j=1}^i e^{-sx_j} r_i(x_1, \ldots, x_i) \mathrm{d}x_1 \ldots \mathrm{d}x_i +$$

$$+ \sum_{i=n}^{n+m} \frac{\mu^i}{(s+\mu)^i} \int_0^\infty \ldots \int_0^\infty \prod_{j=1}^n e^{-sx_j} r_i(x_1, \ldots, x_n) \mathrm{d}x_1 \ldots \mathrm{d}x_n . \tag{16}$$

Let us consider the integral

$$I(i,k) = \int_0^\infty \ldots \int_0^\infty \prod_{j=1}^k e^{-sx_j} r_i(x_1, \ldots, x_k) \mathrm{d}x_1 \ldots \mathrm{d}x_k \tag{17}$$

for different values of i and k.

1) $i = k = 1$.

In this case the integral (17) has the following form:

$$I(1,1) = \int_0^\infty e^{-sx} r_1(x) dx \ .\tag{18}$$

For $i = 1$, taking into account Equation (10), we have from Equation (5): $r_1(x) = -\frac{1}{\mu}\frac{\partial r_1(x)}{\partial x}$. If we substitute this result to (17), we obtain $I(1,1) = -\mu^{-1}\int_0^\infty e^{-sx}\frac{\partial r_1(x)}{\partial x} dx$. Using integration by parts and taking into account Equation (9), we obtain:

$$I(1,1) = \frac{r_1(0)}{s+\mu} = \frac{n\mu\rho p_0}{s+\mu} \ .\tag{19}$$

2) $i = k = \overline{2, n-1}$.

It follows from Equation (5) that

$$r_i(x_1,\ldots,x_i) = -(\mu(i+n\rho))^{-1}\sum_{j=1}^i \frac{\partial r_i(x_1,\ldots,x_i)}{\partial x_j} +$$

$$+ \frac{i+1}{i+n\rho}\int_0^\infty r_{i+1}(x_1,\ldots,x_i,y) dy \ ,$$

whence, taking into account the first statement of Lemma 1, we obtain:

$$r_i(x_1,\ldots,x_i) = -(i\mu)^{-1}\sum_{j=1}^i \frac{\partial r_i(x_1,\ldots,x_i)}{\partial x_j} \ .$$

If we substitute obtained function $r_i(x_1,\ldots,x_i)$ to the relation (17), implement integration by parts and take into account Equation (10), we obtain the following recursive relation:

$$I(i,i) = \frac{a}{i(s+\mu)}\int_0^\infty \cdots \int_0^\infty \prod_{j=1}^k e^{-sx_j} r_{i-1}(x_1,\ldots,x_{i-1}) dx_1 \ldots dx_{i-1} =$$

$$= \frac{a}{i(s+\mu)} I(i-1,i-1) \ .$$

From the relation (19) we finally obtain:

$$I(i,i) = \frac{a^i p_0}{i!(s+\mu)^i} \ .\tag{20}$$

3) $i = k = n$.

Determining the function $r_n(x_1,\ldots,x_n)$ from Equation (6), substituting it to the relation (17), using integration by parts and taking into account Lemma 1 and Equation (11), we analogously obtain:

$$I(n,n) =$$

$$= \frac{\rho(1+\rho)}{n\mu(s+\mu+\mu\rho)} \int_0^\infty \cdots \int_0^\infty \prod_{j=1}^{n-1} e^{-sx_j} r_{n-1}(x_1,\ldots,x_{n-1}) dx_1 \ldots dx_{n-1} =$$

$$= \frac{\rho(1+\rho)}{n\mu(s+\mu+\mu\rho)} I(n-1,n-1) = \frac{(n\mu\rho)^n(1+\rho)p_0}{n!(s+\mu)^{n-1}(s+\mu+\mu\rho)} . \qquad (21)$$

4) $i = n+1, n+m-1$; $k = n$.

If we determine the function $r_i(x_1,\ldots,x_n)$ from Equation (7), substitute it to (17), implement integration by parts and take into account Equation (12) and Lemma 1, we obtain the following recursive relation:

$$I(i,n) = \frac{(n\mu)^n \rho^{i+1}}{n!(s+\mu)^{n-1}(s+\mu+\mu\rho)} +$$

$$+\frac{\mu\rho}{s+\mu+\mu\rho} \int_0^\infty \cdots \int_0^\infty \prod_{j=1}^{n} e^{-sx_j} r_{i-1}(x_1,\ldots,x_n) dx_1 \ldots dx_n .$$

Finally, we have:

$$I(i,n) = \frac{(n\mu)^n \rho^{i+1}[(s+\mu+\mu\rho)^{i-n} - \mu^{i-n}]p_0}{n!(s+\mu)^{n-1}(s+\mu\rho)(s+\mu+\mu\rho)^{i-n}} + \frac{n^n(\mu\rho)^i(1+\rho)p_0}{n!(s+\mu)^{n-1}(s+\mu+\mu\rho)} . \qquad (22)$$

5) $i = n+m$; $k = n$.

If we determine the function $r_{n+m}(x_1,\ldots,x_n)$ from Equation (8), substitute it to (17), implement integration by parts and take into account Equation (13) and Lemma 1, we obtain the following relations:

$$I(n+m,n) = \frac{\mu\rho}{s+\mu} \int_0^\infty \cdots \int_0^\infty \prod_{j=1}^{n} e^{-sx_j} r_{n+m-1}(x_1,\ldots,x_{n-1}) dx_1 \ldots dx_n =$$

$$= \frac{\mu\rho}{s+\mu} I(n+m-1,n) = \frac{n^{n-2}\mu^{n-1}\rho^{n+m-1}[(s+\mu-\mu\rho)^{m-1} - \mu^{m-1}]p_0}{n!(s+\mu)^n(s+\mu\rho)(s+\mu+\mu\rho)^{m-1}} +$$

$$+\frac{n^n(\mu\rho)^{n+m}(1+\rho)p_0}{n!(s+\mu)^n(s+\mu+\mu\rho)^m} . \qquad (23)$$

If we substitute the relations (20)–(23) to the formula (16), we obtain the statement of the theorem after some calculations. The theorem is proved.

Corollary 1. *The function $\delta(s)$ is determined by the relation:*

$$\delta(s) = p_0 \left\{ \sum_{i=0}^{n-1} \frac{(n\rho)^i f^{2i}}{i!(s+f)^{2i}} + \frac{(n\rho)^n f^{2n}(s+f+\rho s)}{n!(s+f)^{2n-2}(s+f-f\rho)[(s+f)^2 + sf\rho]} - \right.$$

$$\left. -\frac{n^n f^{2n+m}\rho^{n+m+1}}{n!(s+f)^{2n+m}(s+f\rho)} \left[\frac{s(s+f) + f^2\rho}{s+f-f\rho} - \frac{f^{m+1}s^2}{(s+f+f\rho)^m((s+f)^2 + sf\rho)} \right] \right\} . \qquad (24)$$

Corollary 2. *The stationary first (δ_1) and second (δ_2) moments of the total demands' volume in the system under consideration are determined by the following relations:*

$$\delta_1 = \frac{p_0}{f} \left\{ 2 \sum_{i=1}^{n-1} \frac{(n\rho)^i}{(i-1)!} + \frac{(n\rho)^n}{n!(1-\rho)} \left[\frac{\rho(1-\rho^m)}{1-\rho} + 2n(1-\rho^{m+1}) - m\rho^{m+1} \right] \right\},$$

$$\delta_2 = \frac{2p_0}{f^2} \left\{ \frac{n^n \rho^{n+m}}{n!(1+\rho)^m} + \sum_{i=1}^{n-1} \frac{(2i+1)(n\rho)^i}{(i-1)!} + \frac{(n\rho)^n}{n!(1-\rho)} \left[\frac{1-\rho^{m+1}}{(1-\rho)^2} + \right. \right.$$

$$\left. \left. + \frac{2n-1-\rho^{m+1}(2n+m)}{1-\rho} + \rho - \rho^m + n(2n-1) - \frac{\rho^{m+1}(2n+m)(2n+m+1)}{2} \right] \right\},$$

for $\rho \neq 1$, where p_0 is calculated by the relation (1);

$$\delta_1 = \frac{p_0}{f} \left[2 \sum_{i=1}^{n-1} \frac{n^i}{(i-1)!} + \frac{n^n(m+1)(4n+m)}{2n!} \right],$$

$$\delta_2 = \frac{p_0}{f^2} \left\{ 2 \sum_{i=1}^{n-1} \frac{(2i+1)n^i}{(i-1)!} + \frac{n^n}{n!} \left[\frac{1}{3}(m-1)(m^2+m+6) + \right. \right.$$

$$\left. \left. + (m+1)(2n+1)(2n+m) \right] + \frac{n^n}{2^{m-1}n!} \right\},$$

for $\rho = 1$, where p_0 is calculated by the relation (2).

4 Formulas for Distribution Functions of the Total Demands' Volume

To calculate the distribution function $D(x)$ using its LST (26) we have consider three special cases:

1) $\rho \neq 0.5$ and $\rho \neq 1$. In this case we obtain:

$$D(x) = 1 + p_0 \left\{ A(b_1)e^{-b_1 f x} + A(b_2)e^{-b_2 f x} + \right.$$

$$+ \frac{(-1)^m n^n}{n!(2n+m-1)!} \sum_{k=0}^{m-1} \frac{(fx)^{m-k-1}}{(m-k-1)!} \left[\frac{S_1(k)}{\rho^{n+k-1}} + \frac{1}{\rho^{n-2}} \sum_{l=0}^{k} \frac{S_1(l)}{\rho^l} \right] e^{-(1+\rho)fx} +$$

$$+ \left[\frac{n^n \rho^{n+1}}{n!} \sum_{k=0}^{2n+m-1} \frac{(fx)^{2n+m-k-1}}{(2n+m-k-1)!} \left(\frac{(-1)^k S_2(k)}{\rho^k(m-1)!} + \right. \right.$$

$$+ \frac{\rho}{(m-1)!(1-\rho)^{k+1}} \sum_{l=0}^{k} (-1)^l \left(\frac{1-\rho}{\rho} \right)^l S_2(l) + \frac{1}{1-\rho} + \frac{1}{\rho^{k-1}(1-\rho)(1-2\rho)} -$$

$$-\frac{\rho}{(1-\rho)^{k+1}(1-2\rho)}\Bigg) - \sum_{k=1}^{n-1}\frac{(n\rho)^k}{k!}\sum_{l=0}^{2k-1}\frac{(fx)^l}{l!}+$$

$$+\frac{n^n\rho^{n-2}}{n!}\sum_{k=0}^{2n-3}\frac{(fx)^{2n-k-3}}{(2n-k-3)!}\left(\frac{1+\rho}{\rho^k}S_3(k)+\sum_{l=0}^{k}\frac{S_3(l)}{\rho^l}\right)\Bigg]\Bigg\}\,,$$

where

$$b_1 = \frac{2+\rho-\sqrt{\rho(4+\rho)}}{2},\ b_2 = \frac{2+\rho+\sqrt{\rho(4+\rho)}}{2}\,,$$

$$A(b) = \frac{(n\rho)^n}{n!(1-b)^{2n-2}(2+\rho-2b)}\left[\frac{b(1+\rho)-1}{b(1-b-\rho)}+\right.$$

$$\left.+\frac{2^{m-1}b\rho^{m+1}}{(b-\rho)(1-b)^{m+2}(1+\rho-b)^m}\right]\,,$$

$$S_1(k) = \sum_{i=0}^{k}(-1)^i\frac{(2n+m+k-i-1)!\rho^i}{(k-i)!}\sum_{j=0}^{i}\frac{(1-b_1+\rho)^{j-i-1}}{(1-b_2+\rho)^{j+1}}\,,$$

$$S_2(k) = \sum_{i=0}^{k}(-1)^i\frac{(m+k-i-1)!\rho^i}{(k-i)!}\sum_{j=0}^{i}\frac{(1-b_1)^{j-i-1}}{(1-b_2)^{j+1}}\,,$$

$$S_3(k) = \sum_{i=0}^{k}\left(\frac{\rho}{1-b_2}\right)^i\sum_{j=0}^{i}\left(\frac{1-b_2}{1-b_1}\right)^j\,,$$

p_0 is determined from the relation (1).

2) $\rho = 0.5$. In this case we obtain:

$$D(x) = 1 - p_0\Bigg\{\left[\frac{n^n}{n!2^{n+m+1}}\sum_{k=0}^{2n+m-1}\frac{(fx)^{2n+m-k-1}}{(2n+m-k-1)!}\left(2^{m+2}\sum_{l=0}^{k}(-1)^{k-l}2^{l+1}\times\right.\right.$$

$$\times\sum_{i=0}^{l}\frac{(-1)^i(l-i+1)(m+i-1)!}{i!(m-1)!} - 2^{k+m+2}\sum_{l=0}^{k}\frac{(-1)^l(k-l+1)(m+l-1)!}{l!(m-1)!}-$$

$$-\sum_{l=0}^{k}(l+1)2^l-1\Big) + \sum_{i=1}^{n-1}\frac{n^i}{i!2^i}\sum_{k=0}^{2i-1}\frac{(fx)^k}{k!} - \frac{n^n}{9n!2^{n-2}}\left(\sum_{i=0}^{2n-3}\frac{(8-3i)2^i(fx)^{2n-3-i}}{(2n-3-i)!}-\right.$$

$$-\frac{9}{2}\sum_{i=0}^{2n-3}\frac{(fx)^{2n-3-i}}{(2n-3-i)!} + \sum_{i=0}^{2n-3}\frac{(-1)^i(fx)^{2n-3-i}}{(2n-3-i)!}\right)\Bigg]e^{-fx}-$$

$$-\frac{(-1)^m2^{n-1}n^n}{n!(2n+m-1)!}\sum_{k=0}^{m-1}\frac{(fx)^{m-k-1}}{(m-k-1)!}\left(\sum_{l=0}^{k}\frac{2^l(k-l-1)(2n+m+l-1)!}{l!}-\right.$$

$$-\sum_{l=0}^{k}(-1)^{k+l}2^{k-l+2}\sum_{i=0}^{l}\frac{2^i(l-i+1)(2n+m+i-1)!}{l!}\right)e^{-3fx/2}\Bigg\}\,,$$

where $p_0 = \left[\sum_{i=0}^{n}\frac{n^i}{2^i i!} + \frac{n^n(2^m-1)}{2^{m+n}n!}\right]^{-1}$.

3) $\rho = 1$. In this case we obtain:

$$D(x) = 1 + p_0 \left\{ \frac{n^n}{n!} \left[\frac{1}{2b_1 - 3} \left(\frac{2b_1 - 1}{b_1^2(b_1 - 1)^{2n-2}} - \frac{b_1}{(b_1 - 1)^{2n+m+1}(b_1 - 2)^m} \right) e^{-b_1 f x} + \right. \right.$$

$$+ \frac{1}{2b_2 - 3} \left(\frac{2b_2 - 1}{b_2^2(b_2 - 1)^{2n-2}} - \frac{b_2}{(b_2 - 1)^{2n+m+1}(b_2 - 2)^m} \right) e^{-b_2 f x} \Bigg] +$$

$$+ \frac{(-1)^m n^n}{n!(2n + m - 1)!} \sum_{k=0}^{m-1} \frac{(fx)^{m-k-1}}{(m - k - 1)!} \left[\sum_{l=0}^{k} \frac{(-1)^l (2n + m + k - l - 1)!}{(k - l)!} \times \right.$$

$$\times \sum_{j=0}^{l} \frac{(b_1 - 2)^{j-l-1}}{(b_2 - 2)^{j+1}} + \sum_{l=0}^{k} \sum_{j=0}^{l} \frac{(-1)^j (2n + m + l - j - 1)!}{(l - j)!} \times$$

$$\times \sum_{i=0}^{j} \frac{(b_1 - 2)^{i-j-1}}{(b_2 - 2)^{i+1}} \Bigg] e^{-2fx} + \left[\frac{n^n}{n!} \sum_{k=0}^{2n+m} \frac{(fx)^{2n+m-k}}{(2n + m - k)!} \left(\frac{(-1)^k}{(m - 1)!} \times \right. \right.$$

$$\times \sum_{l=0}^{k} \frac{2^l (2l - 2k - m + 1)(m + k - l - 2)!}{(k - l)!} \sum_{j=0}^{l} \frac{(b_1 - 1)^{j-l-1}}{(b_2 - 1)^{j+1}} - k \right) -$$

$$- \sum_{k=1}^{n-1} \frac{n^k}{k!} \sum_{l=0}^{2k-1} \frac{(fx)^l}{l!} + \frac{2n^n}{n!} \sum_{k=0}^{2n-3} \frac{(fx)^{2n-k-3}}{(2n - k - 3)!} \sum_{l=0}^{k} (-1)^l \sum_{j=0}^{l} \frac{(b_2 - 1)^{j-l}}{(b_1 - 1)^j} -$$

$$- \frac{n^n}{n!} \sum_{k=0}^{2n-3} \frac{(fx)^{2n-k-3}}{(2n - k - 3)!} \sum_{l=0}^{k} \sum_{j=0}^{l} (-1)^j \sum_{i=0}^{j} \frac{(b_2 - 1)^{i-j}}{(b_1 - 1)^i} \Bigg] e^{-fx} \right\},$$

where $b_1 = (3 - \sqrt{5})/2$, $b_2 = (3 + \sqrt{5})/2$, p_0 is determined from the relation (2).

5 Formulas for the Case of Unlimited Queue

It's clear that the stationary ($\rho < 1$) queueing systems $M/M/n/\infty$ can be investigated by the same way. For this system we have [4]:

$$p_0 = \left[\sum_{k=0}^{n} \frac{(n\rho)^k}{k!} + \frac{n^n \rho^{n+1}}{n!(1 - \rho)} \right]^{-1}; \tag{25}$$

$$p_k = \frac{(n\rho)^k}{k!} p_0, \; k = \overline{1, n}; \; p_k = \frac{n^n \rho^k}{n!} p_0, \; k = n + 1, n + 2, \ldots \; .$$

For LST of total demands volume and for the first and second moments of this random value we obtain:

$$\delta(s) = p_0 \left\{ \sum_{i=0}^{n-1} \frac{(n\rho)^i f^{2i}}{i!(s + f)^{2i}} + \frac{(n\rho)^n f^{2n}(s + f + \rho s)}{n!(s + f)^{2n-2}(s + f - f\rho)[(s + f)^2 + sf\rho]} \right\};$$

$$\tag{26}$$

$$\delta_1 = \frac{p_0}{f}\left\{2\sum_{i=1}^{n-1}\frac{(n\rho)^i}{(i-1)!} + \frac{(n\rho)^n[\rho+2n(1-\rho)]}{n!(1-\rho)^2}\right\} ;$$

$$\delta_2 = \frac{2p_0}{f^2}\left\{\sum_{i=1}^{n-1}\frac{(2i+1)(n\rho)^i}{(i-1)!} + \frac{(n\rho)^n}{n!(1-\rho)}\left[\frac{1}{(1-\rho)^2} + \frac{2n-1}{1-\rho} + \rho + n(2n-1)\right]\right\} ,$$

where p_0 is calculated by the relation (25).

Explicit formulas for the distribution function $D(x)$ have the following forms.

1) $\rho \neq 0.5$.

$$D(x) = 1 + p_0\left\{\frac{n^n e^{-(1-\rho)fx}}{n!\rho^{n-3}(1-\rho)(1-2\rho)} - \frac{(n\rho)^n}{n!\sqrt{\rho(4+\rho)}}\left[\frac{e^{-b_1 fx}}{(1-b_1)^{2n-3}(1-b_1-\rho)} - \right.\right.$$

$$\left.- \frac{e^{-b_2 fx}}{(1-b_2)^{2n-3}(1-b_2-\rho)}\right] - \left[\sum_{i=1}^{n-1}\frac{(n\rho)^i}{i!}\sum_{k=0}^{2i-1}\frac{(fx)^k}{k!} - \frac{n^n\rho^{n-2}(1+\rho)}{n!}\times\right.$$

$$\times \sum_{i=0}^{2n-3}\frac{(fx)^{2n-i-3}}{(2n-i-3)!\rho^i}\sum_{j=0}^{i}\left(\frac{\rho}{1-b_2}\right)^j\sum_{k=0}^{j}\left(\frac{1-b_2}{1-b_1}\right)^k + \frac{n^n\rho^{n-2}}{n!}\times$$

$$\times \sum_{i=0}^{2n-3}\frac{(fx)^{2n-i-3}}{(2n-i-3)!}\sum_{j=0}^{i}\rho^{-j}\sum_{k=0}^{j}\left(\frac{\rho}{1-b_2}\right)^k\sum_{l=0}^{k}\left(\frac{1-b_2}{1-b_1}\right)^l\left.\right]e^{-fx}\right\} ,$$

where p_0 is calculated by the relation (25), $b_1 = \frac{2+\rho-\sqrt{\rho(4+\rho)}}{2}$, $b_2 = \frac{2+\rho+\sqrt{\rho(4+\rho)}}{2}$.

2) $\rho = 0.5$.

$$D(x) = 1 - p_0\left\{\frac{2^{n-2}n^n}{9n!}(3fx - 12n + 34)e^{-fx/2} + e^{-fx}\sum_{i=1}^{n-1}\frac{n^i}{2^i i!}\sum_{k=0}^{2i-1}\frac{(fx)^k}{k!} - \right.$$

$$- \frac{n^n e^{-fx}}{9n!2^{n-2}}\left[\sum_{i=0}^{2n-3}(8-3i)\frac{2^i(fx)^{2n-i-3}}{(2n-i-3)!} - \frac{9}{2}\sum_{i=0}^{2n-3}\frac{(fx)^{2n-i-3}}{(2n-i-3)!} + \right.$$

$$\left.+ \sum_{i=0}^{2n-3}(-1)^i\frac{(fx)^{2n-i-3}}{(2n-i-3)!}\right] - \frac{n^n e^{-2fx}}{9n!2^{n-2}}\left.\right\} .$$

Note that for $n = 1$ we obtain the known relations (see [1]). For $n = 2$ we have $p_0 = (1-\rho)/(1+\rho)$, and we obtain in this case:

$$D(x) = 1 + \frac{2\rho e^{-(1-\rho)fx}}{(1+\rho)(1-2\rho)} - \frac{2\rho^2 p_0}{\sqrt{\rho(4+\rho)}}\left[\frac{e^{-b_1 fx}}{(1-b_1)(1-b_1-\rho)} - \right.$$

$$\left.- \frac{e^{-b_2 fx}}{(1-b_2)(1-b_2-\rho)}\right] - 2\rho p_0 e^{-fx}\left(1 - \frac{1}{1-b_1} - \frac{1}{1-b_2}\right) ,$$

if $\rho \neq 0.5$, and

$$D(x) = 1 - \frac{2}{27}\left[(3fx + 10)e^{-fx/2} - e^{-2fx}\right] ,$$

if $\rho = 0.5$.

6 Conclusion

In the paper we investigated stationary queueing systems $M/M/n/m \leq \infty$ with demands of exponentially distributed random capacity when service time of the demand is proportional to its capacity. For such systems we calculated the explicit form of the total demands' volume distribution function. The obtained results can be used for estimations of loss characteristics of computer and communicated systems with the help of the known technique [1] when the total volume (capacity of the system) is limited.

Acknowledgment. This work was partially supported by grant no. 4796/B/ T02/2011/40 of Polish National Council of Science (NCN).

References

1. Tikhonenko, O.: Computer Systems Probability Analysis. Akademicka Oficyna Wydawnicza EXIT, Warsaw (2006) (in Polish)
2. Tikhonenko, O.M.: Destricted Capacity Queueing Systems: Determination of their Characteristics. Autom. Remote Control 58(6), 969–972 (1997)
3. Klimov, G.P.: Stochastic Service Systems. Nauka, Moscow (1966) (in Russian)
4. Gnedenko, B.W., Konig, D.: Handbuch der Bedienungstheorie. Akademie Verlag, Berlin (1983)

Queueing System $MAP|PH|N|R$ with Session Arrivals Operating in Random Environment

Chesoong Kim[1], Alexander Dudin[2,*], Sergey Dudin[2], and Olga Dudina[2]

[1] Sangji University, Wonju, Kangwon, 220-702, Korea
[2] Belarusian State University, 4, Nezavisimosti Ave., Minsk, 220030, Belarus
dowoo@sangji.ac.kr, dudin@bsu.by, dudin@madrid.com, dudina_olga@email.com

Abstract. We consider a multi-server queueing system with session arrivals that operates in random environment. A session consists of a random number of customers. Sessions arrive to the system in Markovian arrival flow. Inter-arrival times of customers within each session are random. The service time of a customer has a phase type distribution. A numerically stable algorithm for calculation of the stationary distribution of system states is presented. The main performance measures are calculated.

Keywords: session arrivals, Markovian arrival flow, phase type distribution, random environment.

1 Introduction

Many problems in routing, scheduling, flow control, resources allocation and capacity management in telecommunication networks can be solved with help of queueing theory. Typically, a user of a network can generate whole bunch of requests and due that batch arrivals are often assumed in the analysis of queueing systems. It is usually assumed that, at a batch arrival epoch, all requests of the batch arrive to the system simultaneously.

It was assumed in the paper [1], that requests arrive to the system in batches, but the arrival of requests of a batch is not instantaneous. Such batches are named *sessions*. The first request (customer) of a session arrives at the session arrival epoch, while the rest of requests arrive individually during random intervals. The session size is random and not known at the session arrival epoch. Such a situation is typical in IP networks, e.g., in modeling transmission of video and multimedia information, in World Wide Web with Hypertext Transfer Protocol (HTTP) where a session can be interpreted as a HTTP connection and a request as a HTTP request.

In the paper [1], the multi-server Markovian queueing model with a finite buffer and session arrivals is investigated. In the paper [2], the $MAP/PH/1/N$ queueing system with session arrivals is investigated. In [3], the mechanism of requests arrival within a session is significantly generalized comparing to the

---

* Corresponding author.

A. Kwiecień, P. Gaj, and P. Stera (Eds.): CN 2013, CCIS 370, pp. 406–415, 2013.
© Springer-Verlag Berlin Heidelberg 2013

model considered in [2] by suggesting that customers from the admitted session can arrive in groups. Session arrivals are directed by a MAP (Markovian Arrival Process) and customers' arrivals in session are directed by a $TBMAP$ (Terminating Batch Markovian Arrival Process) in [3]. In [4], the $MAP/M/N$ retrial queueing system with session arrivals was considered.

In the present paper, we analyze the multi-server queueing system with phase type (PH) service time distribution that operates in *random environment* (RE).

Models operating in RE take into account the influence of some random external factors, e.g., different level of noise in the transmission channel, change the distance of mobile user from the base station, precipitation effects, different disturbance, etc.

In consideration of a queue operating in RE, we assume that there are a queueing system and an external finite state space stochastic process called as RE. Under the fixed state of RE, the queueing system operates as a queueing system of the correspondent type. However, when RE changes its state, parameters of the queueing system (inter-arrival times distribution or arrival rate, service times distribution or service rate, etc.) can immediately change their values.

The investigation of queues operating in RE has a great importance due to their application to model the real-word communication networks.

Some information about the current state-of-the-art in the investigation of queues operating in RE is provided, e.g., in [5].

In this paper, we assume that the arrival of sessions is described by the MAP. Importance of consideration of the MAP stems from the fact the MAP catches the effect of correlation of successive inter-arrival times while information flows in modern telecommunication networks are correlated and this correlation has profound effect on system performance measures.

Also, PH service time distribution is considered. This allows to one take into account the variation of the service time. The fact that the service time distribution in modern telecommunication networks is poorly approximated by an exponential distribution, but is well described by a PH distribution is reported in [6,7].

The rest of the paper is organized as follows. In Section 2, the mathematical model is described. The steady state joint distribution of the number of customers in the system, the state of RE, number of sessions in the system, and the state of the MAP and PH underlying processes is analyzed in Sect. 3. Some performance measures of the system are presented in Sect. 4. Section 5 concludes the paper.

2   Mathematical Model

We consider a queueing system of capacity $R, 1 \leq R < \infty$. The system has $N, 1 \leq N \leq R$, servers and a buffer of capacity $R - N \geq 0$. The servers are assumed to be identical and independent of each other. The system behavior depends on the state of RE. RE is defined by means of stochastic process r_t,

$t \geq 0$, which is an irreducible regular continuous time Markov chain with state space $\{1, \ldots, L\}$ and infinitesimal generator H.

Customers arrive at the system in sessions. Sessions arrive at the system according to the MAP. It means the following. Sessions arrival in the MAP is directed by an irreducible continuous time Markov chain ν_t, $t \geq 0$, with the finite state space $\{0, 1, \ldots, W\}$. Under the fixed state r of RE, this process behaves as an irreducible continuous time Markov chain. The sojourn time of the Markov chain ν_t, $t \geq 0$, in the state ν has an exponential distribution with the parameter $\lambda_\nu^{(r)}$, $\nu = \overline{0, W}$, $r = \overline{1, L}$. Here notation such as $\nu = \overline{0, W}$ means that ν assumes values from the set $\{0, 1, \ldots, W\}$. After this sojourn time expires, with the probability $p_k^{(r)}(\nu, \nu')$ the process ν_t, $t \geq 0$, transits to the state ν' and k sessions, $k = 0, 1$, arrive at the system. The intensities of jumps from one state to another, which are accompanied by an arrival of k sessions, are combined to the square matrices $D_k^{(r)}$, $k = 0, 1$, of size $\bar{W} = W + 1$. The matrix generating function of these matrices is $D^{(r)}(z) = D_0^{(r)} + D_1^{(r)} z$, $|z| \leq 1$. The matrix $D^{(r)}(1)$ is an infinitesimal generator of the process ν_t, $t \geq 0$, for all fixed values of $r = \overline{1, L}$. The stationary distribution vector $\boldsymbol{\delta}^{(r)}$ of this process under the fixed state r of RE satisfies the equations

$$\boldsymbol{\delta}^{(r)} D^{(r)}(1) = \mathbf{0}, \ \boldsymbol{\delta}^{(r)} \mathbf{e} = 1 \ .$$

Here and in the sequel $\mathbf{0}$ is a zero row vector and $\mathbf{e}$ is a column vector of the appropriate size consisting of 1's. If the dimension of a vector is not clear from the context, it is indicated as a subscript, e.g., $\mathbf{e}_{\bar{W}}$ denotes the unit column vector of size $\bar{W}$.

The average intensity $\lambda^{(r)}$ (fundamental rate) of the MAP under the fixed value r is defined as $\lambda^{(r)} = \boldsymbol{\delta}^{(r)} D_1^{(r)} \mathbf{e}$.

The variance $v^{(r)}$ of intervals between session arrivals is calculated as $v^{(r)} = 2\lambda^{(r)-1} \boldsymbol{\delta}^{(r)} (-D_0^{(r)})^{-1} \mathbf{e} - \lambda^{(r)-2}$.

The squared coefficient c_{var} of the variation is calculated by

$$c_{\text{var}}^{(r)} = 2\lambda^{(r)} \boldsymbol{\delta}^{(r)} (-D_0^{(r)})^{-1} \mathbf{e} - 1$$

while the correlation coefficient $c_{\text{cor}}^{(r)}$ of intervals between successive arrivals is given by

$$c_{\text{cor}}^{(r)} = (\lambda^{(r)-1} \boldsymbol{\delta}^{(r)} (-D_0^{(r)})^{-1} D_1^{(r)} (-D_0^{(r)})^{-1} \mathbf{e} - \lambda^{(r)-2}) / v^{(r)} \ .$$

At the epochs of the process r_t, $t \geq 0$, transitions, the state of the process ν_t, $t \geq 0$, is not changed, but the intensities of its transitions are immediately changed.

The service time of a customer by each server has a PH distribution. This service time can be interpreted as a time until the underlying Markov process η_t, $t \geq 0$, with a finite state space $\{1, \ldots, M, M+1\}$ reaches the single absorbing state $M+1$ conditioned on the fact that under the fixed state r of RE the initial state of this process is selected among the states $\{1,\ldots,M\}$ according to the probabilistic row vector $\boldsymbol{\beta}^{(r)} = (\beta_1^{(r)},\ldots,\beta_M^{(r)})$, $r = \overline{1,L}$.

Under the fixed state r of RE transition rates of the process η_t within the set $\{1,\ldots,M\}$ are defined by the sub-generator $S^{(r)}$ and transition rates into the absorbing state (which lead to the service completion) are given by entries of the column vector $\boldsymbol{S}_0^{(r)} = -S^{(r)}\mathbf{e}$, $r = \overline{1,L}$. The mean service time is calculated by $b_1^{(r)} = \boldsymbol{\beta}^{(r)}(-S^{(r)})^{-1}\mathbf{e}$. The squared coefficient of variation is given by $c_{\text{var}}^{(r)} = b_2^{(r)}/(b_1^{(r)})^2 - 1$ where $b_2^{(r)} = 2\boldsymbol{\beta}^{(r)}(-S^{(r)})^{-2}\mathbf{e}$.

For more information about PH distribution and its usefulness for modelling in telecommunication see, e.g., [8].

We assume that admission of sessions is restricted by means of so called *tokens*. The total number of available tokens is assumed to be K, $K \geq 1$. The number K can be considered as a control parameter, and various optimization problems can be solved.

If there is no token available at a session arrival epoch, or buffer is full, the session is rejected. It leaves the system permanently. If the number of available tokens is positive and the system is not full, the session is admitted to the system and the number of available tokens decreases by one. After admission of the session, the next customer of this session can arrive at the system in an exponentially distributed time, with the parameter $\gamma^{(r)}$, $r = \overline{1,L}$.

If the system is not full at the instant of arrival of a customer from the admitted session, the customer is admitted to the system. Otherwise, the customer is rejected. Here we assume that rejection of the customer does not lead to termination of the session to which it belongs.

Under the fixed state r of RE, a new customer from the session arrives with the probability $\theta^{(r)}$, $0 < \theta^{(r)} < 1$, $r = \overline{1,L}$, and with the complementary probability $1 - \theta^{(r)}$ the arrival of session is finished.

If an exponentially distributed time with the parameter $\gamma^{(r)}$ since the arrival of the previous customer of a session expires, and a new customer does not arrive, the arrival of the session is finished. The token, which was obtained by this session upon arrival, is returned to the pool of available tokens. Customers of this session, which stay in the system during the epoch when the token is returned, must be processed by the system.

It is intuitively clear that this mechanism of arrivals restriction by means of tokens is reasonable. At the expense of rejecting some sessions, it allows to decrease the loss probability and the sojourn time and jitter for customers who belong to accepted sessions. It is important in modeling real-life systems because quality of transmission should satisfy imposed requirements of Quality of Service. Note that in concrete application of this queueing model to mobile communication networks, different interpretations of a server, a session, a customer, and a token can be suggested.

3   Joint Distribution of System States

Our goal is to calculate the main performance measures of the system under study.

To this end, we consider the following multi-dimensional process

$$\xi_t = \{i_t, r_t, k_t, \nu_t, \eta_t^{(1)}, \ldots, \eta_t^{(M)}\}, \; t \geq 0 \;,$$

where i_t, $i_t = \overline{0, R}$, denotes the total number of customers in the system; r_t, $r_t = \overline{1, L}$, is the state of RE; k_t, $k_t = \overline{0, K}$, denotes the number of sessions with a token for admission to the system; ν_t, $\nu_t = \overline{0, W}$, denotes the state of the MAP underlying process, and $\eta_t^{(m)}$ is the number of servers at the phase m of service, $m = \overline{1, M}$, $\eta_t^{(m)} = \overline{0, \min\{i_t, N\}}$, $\sum_{m=1}^{M} \eta_t^{(m)} = \min\{i_t, N\}$, at the epoch t, $t \geq 0$.

Note that the meaning of components $\eta_t^{(m)}$ is chosen according to the approach by Ramaswami and Lucantoni, see [9,10]. This approach allows significantly reduce the number of states of the underlying stochastic process comparing with the standard approach that suggests the separate account of the phase of a service in any busy server.

Let us enumerate states of the Markov chain ξ_t in the direct lexicographic order of components (k, ν) and in the reverse lexicographic order of components $(\eta^{(1)}, \ldots, \eta^{(M)})$, and refer to (i, r) as macro-state consisting of states $(i, r, k, \nu, \eta^{(1)}, \ldots, \eta^{(M)})$, $k = \overline{0, K}$, $\nu = \overline{0, W}$, $\eta^{(m)} = \overline{0, \min\{i, N\}}$, $m = \overline{1, M}$, $\sum_{m=1}^{M} \eta^{(m)} = \min\{i, N\}$.

Let Q be the generator of the Markov chain ξ_t, $t \geq 0$.

Aiming to define the generator Q, we need to introduce the following notation:

- $\gamma_-^{(r)} = \gamma^{(r)}(1 - \theta^{(r)})$, $\gamma_+^{(r)} = \gamma^{(r)}\theta^{(r)}$, $r = \overline{1, L}$;
- $K_i = \binom{i+M-1}{M-1}$, $i = \overline{0, N}$;
- $C = \text{diag}\{0, 1, \ldots, K\}$, i.e., C is the diagonal matrix with the diagonal entries $\{0, 1, \ldots, K\}$;
- I is the identity matrix, O is a zero matrix;
- $E_-^{(r)} = \begin{pmatrix} 0 & 0 & 0 & \ldots & 0 & 0 \\ \gamma_-^{(r)} & 0 & 0 & \ldots & 0 & 0 \\ 0 & 2\gamma_-^{(r)} & 0 & \ldots & 0 & 0 \\ \vdots & \vdots & \vdots & \ddots & \vdots & \vdots \\ 0 & 0 & 0 & \ldots & K\gamma_-^{(r)} & 0 \end{pmatrix}$, $r = \overline{1, L}$;
- E^+ is the square matrix of size $K + 1$ with all zero entries except the entries $(E^+)_{k,k+1} = 1$, $k = \overline{0, K-1}$;
- $\hat{E}$ is the square matrix of size $K + 1$ with all zero entries except the entry $(\hat{E})_{K,K} = 1$;
- $\otimes$ is a sign of Kronecker product of matrices;
- $\tilde{D}_l^{(i)} = \text{diag}\{I_{K+1} \otimes D_l^{(r)} \otimes I_{K_i}, r = \overline{1, L}\}$, $l = 0, 1$, $i = \overline{0, N}$;
- $\mathcal{A}_i = \text{diag}\{I_{(K+1)\bar{W}} \otimes A_i(N, S^{(r)}), r = \overline{1, L}\}$, $i = \overline{0, N}$;

- $\mathcal{D}_i = -\text{diag}\{I_{(K+1)\bar{W}} \otimes \text{diag}\{A_i(N, S^{(r)})\mathbf{e} + L_{N-i}(N, \tilde{S}^{(r)})\mathbf{e}\}, r = \overline{1,L}\}$, $i = \overline{1,N}$, $\mathcal{D}_0 = O_{\bar{W}L(K+1)}$;
- $\mathcal{L}_{N-i} = \text{diag}\{I_{(K+1)\bar{W}} \otimes L_{N-i}(N, \tilde{S}^{(r)}), r = \overline{1,L}\}$, $i = \overline{1,N}$;
- $\tilde{S}^{(r)} = \begin{pmatrix} 0 & 0 \\ S_0^{(r)} & S^{(r)} \end{pmatrix}$, $r = \overline{1,L}$;
- $\tilde{C}_i = \text{diag}\{(-\gamma^{(r)}C + E_{-}^{(r)}) \otimes I_{\bar{W}K_i}, r = \overline{1,L}\}$, $i = \overline{0,N}$;
- $\bar{C}_i = \text{diag}\{\hat{E} \otimes D_1^{(r)} \otimes I_{K_i}, r = \overline{1,L}\}$, $i = \overline{0,N}$;
- $\hat{C} = \text{diag}\{\gamma_+^{(r)}C \otimes I_{\bar{W}K_N}, r = \overline{1,L}\}$.

Here the matrices $A_i(N, S^{(r)})$ and $L_i(N, \tilde{S}^{(r)})$, $i = \overline{0,N}$, are defined in [11].

Lemma 1. *The generator Q has the following three-block-diagonal structure*

$$Q = \begin{pmatrix} Q_{0,0} & Q_{0,1} & O & \cdots & O & O \\ Q_{1,0} & Q_{1,1} & Q_{1,2} & \cdots & O & O \\ O & Q_{2,1} & Q_{2,2} & \cdots & O & O \\ \vdots & \vdots & \vdots & \ddots & \vdots & \vdots \\ O & O & O & \cdots & Q_{R-1,R-1} & Q_{R-1,R} \\ O & O & O & \cdots & Q_{R,R-1} & Q_{R,R} \end{pmatrix}$$

where the non-zero blocks $Q_{i,j}$ are computed by

$$Q_{i,i} = H \otimes I_{(K+1)\bar{W}K_i} + \tilde{D}_0^{(i)} + \tilde{C}_i + \mathcal{A}_i + \mathcal{D}_i + \bar{C}_i, 0 \leq i \leq N,$$

$$Q_{i,i} = H \otimes I_{(K+1)\bar{W}K_N} + \tilde{D}_0^{(N)} + \tilde{C}_N + \mathcal{A}_N + \mathcal{D}_N + \bar{C}_N, 0 \leq N < i < R,$$

$$Q_{R,R} = H \otimes I_{(K+1)\bar{W}K_N} + \tilde{D}_0^{(N)} + \tilde{D}_1^{(N)} + \tilde{C}_N + \mathcal{A}_N + \mathcal{D}_N + \hat{C},$$

$$Q_{i,i+1} = \text{diag}\{(E^+ \otimes D_1^{(r)} + \gamma_+^{(r)}C \otimes I_{\bar{W}}) \otimes P_i(\boldsymbol{\beta}^{(r)}), r = \overline{1,L}\}, 0 \leq i < N,$$

$$Q_{i,i+1} = \text{diag}\{(E^+ \otimes D_1^{(r)} + \gamma_+^{(r)}C \otimes I_{\bar{W}}) \otimes I_{K_N}, r = \overline{1,L}\}, N \leq i < R,$$

$$Q_{i,i-1} = \mathcal{L}_{N-i}, 1 \leq i \leq N,$$

$$Q_{i,i-1} = \text{diag}\{I_{(K+1)\bar{W}} \otimes L_0(N, \tilde{S}^{(r)})P_{N-1}(\boldsymbol{\beta}^{(r)}), r = \overline{1,L}\}, N < i \leq R.$$

Proof of Lemma 1 consists of analysis of the Markov chain $\xi_t, t \geq 0$, transitions during the infinitesimal interval of time and further combining corresponding transition intensities into matrix blocks. The value $\gamma_-^{(r)}$ is the intensity of tokens releasing due to the finish of the session arrival, $\gamma_+^{(r)}$ is the intensity of new customers in the session arrival when the state of RE is equal to r. Note that under the fixed state r of RE the matrix $P_i(\boldsymbol{\beta}^{(r)})$ defines the transition probabilities of the process $\boldsymbol{\eta}_t = \{\eta_t^{(1)}, \ldots, \eta_t^{(M)}\}, t \geq 0$, at the epoch of starting the new service, the matrix $L_{N-i}(N, \tilde{S}^{(r)})$ defines the intensities of transitions of this process at the service completion epoch, the matrix $A_i(N, S^{(r)})$ defines the intensities of transitions of the process η_t, $t \geq 0$, which do not lead to the service completion conditioned on the fact that i servers are busy. The modules of diagonal entries of the matrix $\mathrm{diag}\{A_i(N, S^{(r)})\mathbf{e} + L_{N-i}(N, \tilde{S}^{(r)})\mathbf{e}\}$ define the total intensity of leaving the corresponding states of the process η_t, $t \geq 0$, given that i servers are busy. The detailed description of matrices $P_i(\boldsymbol{\beta}^{(r)})$, $i = \overline{0, N-1}$, $A_i(N, S^{(r)})$ and $L_i(N, \tilde{S}^{(r)})$, $i = \overline{0, N}$, and algorithms for their calculation can be found in [11].

Since the Markov chain ξ_t, $t \geq 0$, is irreducible and regular, and has a finite state space, its stationary distribution always exists. Then the following limits (stationary probabilities) exist:

$$\pi(i, r, k, \nu, \eta^{(1)}, \ldots, \eta^{(M)}) =$$

$$= \lim_{t \to \infty} P\{i_t = i,\, r_t = r,\, k_t = k,\, \nu_t = \nu,\, \eta_t^{(1)} = \eta^{(1)}, \ldots, \eta_t^{(M)} = \eta^{(M)}\}\ ,$$

$$i = \overline{0, R},\, r = \overline{1, L},\, k = \overline{0, K},\, \nu = \overline{0, W}\ ,$$

$$\eta^{(m)} = \overline{0, \min\{i, N\}},\, m = \overline{1, M},\, \sum_{m=1}^{M} \eta^{(m)} = \min\{i, N\}\ .$$

Let us form the row vectors $\boldsymbol{\pi}(i, r, k)$ of probabilities $\pi(i, r, k, \nu, \eta^{(1)}, \ldots, \eta^{(M)})$, $i = \overline{0, R}$, $r = \overline{1, L}$, $k = \overline{0, K}$, enumerated as it was defined above. Then let us form the row vectors $\boldsymbol{\pi}_i$:

$$\boldsymbol{\pi}(i, r) = (\boldsymbol{\pi}(i, r, 0), \boldsymbol{\pi}(i, r, 1), \ldots, \boldsymbol{\pi}(i, r, K)),\, r = \overline{1, L}\ ,$$

$$\boldsymbol{\pi}_i = (\boldsymbol{\pi}(i, 1), \boldsymbol{\pi}(i, 2), \ldots, \boldsymbol{\pi}(i, R)),\, i = \overline{0, R}\ .$$

It is well known that the probability vectors $\boldsymbol{\pi}_i$ satisfy the following system of linear algebraic equations:

$$(\boldsymbol{\pi}_0, \boldsymbol{\pi}_1, \ldots, \boldsymbol{\pi}_R)Q = \mathbf{0},\, (\boldsymbol{\pi}_0, \boldsymbol{\pi}_1, \ldots, \boldsymbol{\pi}_R)\mathbf{e} = 1\ .$$

We can apply a modification of effective and numerically stable algorithm elaborated in [12] for calculation of such probabilities for Markov chains with a generator of tridiagonal form.

Theorem 1. *The stationary probability vectors $\boldsymbol{\pi}_i$ are calculated by*

$$\boldsymbol{\pi}_i = \boldsymbol{\pi}_0 F_i,\, i = \overline{1, R}\ ,$$

where the matrices F_i are calculated recurrently by

$$F_0 = I,\, F_i = -F_{i-1}Q_{i-1,i}(Q_{i,i} + Q_{i,i+1}G_i)^{-1},\, i = \overline{1, R-1}\ ,$$

$$F_R = -F_{R-1}Q_{R-1,R}(Q_{R,R})^{-1}\ ,$$

the matrices G_i are calculated from the backward recursion:

$$G_i = -(Q_{i+1,i+1} + Q_{i+1,i+2}G_{i+1})^{-1}Q_{i+1,i},\, i = R-2, R-3, \ldots, 0\ ,$$

under the initial condition

$$G_{R-1} = -(Q_{R,R})^{-1}Q_{R,R-1} \; ,$$

the vector π_0 is the unique solution to the following system of linear algebraic equations:

$$\pi_0(Q_{0,0} + Q_{0,1}G_0) = \mathbf{0}, \quad \pi_0 \sum_{l=0}^{R} F_l\mathbf{e} = 1 \; .$$

4   Performance Measures

As soon as the stationary distribution π_i, $i \geq 0$, has been calculated, we can calculate different performance measures of the system.

The probability distribution of the number of customer in the system is computed by

$$\lim_{t\to\infty} P\{i_t = i\} = \pi_i\mathbf{e}, \; i = \overline{0, R} \; .$$

The average number $\tilde{L}$ of customers in the system is computed by

$$\tilde{L} = \sum_{i=1}^{R} i\pi_i\mathbf{e} \; .$$

The average number B of accepted sessions (average number of busy tokens) at arbitrary moment is computed by

$$B = \sum_{i=0}^{R}\sum_{r=1}^{L}\sum_{k=1}^{K} k\pi(i, r, k)\mathbf{e} \; .$$

The average number T of customers processed by the system at unit of time (throughput of the system) is computed by

$$T = \sum_{i=1}^{R} \pi_i \mathcal{L}_{\max\{N-i,0\}}\mathbf{e} \; .$$

The probability $P_s^{(\text{loss})}$ of an arbitrary session rejection is computed by

$$P_s^{(\text{loss})} = \lambda^{-1}\sum_{r=1}^{L}\left(\sum_{i=0}^{R-1}\pi(i, r, K)(D_1^{(r)}\otimes I_{K_{\min\{i,N\}}}) + \sum_{k=0}^{K}\pi(R, r, k)(D_1^{(r)}\otimes I_{K_N})\right)\mathbf{e}$$

where the average intensity of session arrival λ is computed by

$$\lambda = \mathbf{x}\text{diag}\{D_1^{(r)}, \; r = \overline{1, L}\}\mathbf{e}$$

where $\mathbf{x}$ is the unique solution to the system

$$\mathbf{x}(H \otimes I_{\bar{W}} + \operatorname{diag}\{D_0^{(r)} + D_1^{(r)}, \; r = \overline{1, L}\}) = \mathbf{0}, \; \mathbf{x}\mathbf{e} = 1 \; .$$

The probability $P_c^{(\mathrm{loss})}$ of rejection of an arbitrary customer from an admitted session is computed by

$$P_c^{(\mathrm{loss})} = \frac{\sum\limits_{r=1}^{L} \sum\limits_{k=1}^{K} k \gamma_+^{(r)} \boldsymbol{\pi}(R, r, k)\mathbf{e}}{\sum\limits_{i=0}^{R} \sum\limits_{r=1}^{L} \sum\limits_{k=1}^{K} k \gamma_+^{(r)} \boldsymbol{\pi}(i, r, k)\mathbf{e}} \; .$$

5 Conclusion

$MAP|PH|N|R$ queueing system with session arrivals operating in random environment is investigated. Stationary distribution of system states is analyzed. Some key performance measures are calculated.

Acknowledgments. This research was supported by Basic Science Research Program through the National Research Foundation of Korea (NRF) funded by the Ministry of Education, Science and Technology (Grant No. 2011-0015214).

References

1. Lee, M.H., Dudin, S., Klimenok, V.: Queueing Model with Time-Phased Batch Arrivals. In: Mason, L.G., Drwiega, T., Yan, J. (eds.) ITC 2007. LNCS, vol. 4516, pp. 719–730. Springer, Heidelberg (2007)
2. Kim, C.S., Dudin, S.A., Klimenok, V.I.: The $MAP/PH/1/N$ queue with time phased arrivals as model for traffic control in telecommunication networks. Performance Evaluation 66, 564–579 (2009)
3. Kim, C.S., Dudin, A., Dudin, S., Klimenok, V.: A Queueing System with Batch Arrival of Customers in Sessions. Computers and Industrial Engeneering 62, 890–897 (2012)
4. Dudin, S.A.: The $MAP/M/N$ retrial queueing system with time-phased batch arrivals. Problems of Information Transmission 3, 270–281 (2009)
5. Kim, C., Dudin, A., Klimenok, V., Khramova, V.: Erlang loss queueing system with batch arrivals operating in a random environment. Computers & Operations Research 36, 674–697 (2009)
6. Pattavina, A., Parini, A.: Modelling voice call inter-arrival and holding time distributions in mobile networks. In: Proceedings of 19th International Teletraffic Congress, Performance Challenges for Efficient Next Generation Networks, pp. 729–738 (2005)
7. Riska, A., Diev, V., Smirni, E.: Efficient fitting of long-tailed data sets into hyperexponential distributions. In: Proceedings of Global Telecommunications Conference (GLOBALCOM 2002), pp. 2513–2517. IEEE (2002)

8. Neuts, M.: Matrix-geometric solutions in stochastic models – an algorithmic approach. John Hopkins University Press (1981)
9. Ramaswami, V.: Independent Markov processes in parallel. Comm. Statist. - Stochastic Models 1, 419–432 (1985)
10. Ramaswami, V., Lucantoni, D.M.: Algorithms for the multi-server queue with phase-type service. Comm. Statist. -Stochastic Models 1, 393–417 (1985)
11. Kim, C., Dudin, S., Taramin, O., Baek, J.: Queueing system $MAP|PH|N|N + R$ with impatient heterogeneous customers as a model of call center. Applied Mathematical Modelling 37, 958–976 (2013)
12. Klimenok, V.I., Kim, C.S., Orlovsky, D.S., Dudin, A.N.: Lack of invariant property of Erlang $BMAP/PH/N/0$ model. Queueing Systems 49, 187–213 (2005)

Tandem Queueing System with Correlated Input and Cross-Traffic

Valentina Klimenok[1,*], Alexander Dudin[1], and Vladimir Vishnevsky[2]

[1] Department of Applied Mathematics and Computer Science
Belarusian State University
Minsk 220030, Belarus
{klimenok,dudin}@bsu.by
[2] Closed Corporation "Information and Networking Technologies", Moscow, Russia
vishn@inbox.ru

Abstract. In this paper, we analyze a tandem queueing system consisting of R multi-server stations without buffers. The input flow at the first station is a MAP (Markovian arrival process). The customers from this flow aim to be served at all R stations of the tandem. For any r-th station, besides transit customers proceeding from the $(r-1)$-th station, an additional MAP flow of new customers arrives at the r-th station directly, not entering the previous stations of the tandem. Customers from this flow aim to be served at the r-th station and all subsequent stations of the tandem. The service time of any customer arriving at the r-th station is exponentially distributed with the service rate depending of r. We present the recursive scheme for calculating the stationary distributions and the loss probabilities associated with the tandem.

Keywords: tandem queueing system, multi-server stations, Markovian arrival process, cross-traffic, stationary state distribution, loss probabilities.

1 Introduction

Tandem queuing systems take into account that a customer may need service from several sequentially located servers. Such queues can be considered as a special case of queueing networks having a linear topology. They are good mathematical models of many real-life computer, production, transport and communication systems as well as fragments of general topology telecommunication networks. Thus, tandem queues are of great interest for researchers in telecommunications, industry and queueing theory. There are a lot of publications devoted to investigation of tandem queues, for references see [1–14] and references therein. Most of early mathematical results on tandem queueing models were obtained for exponential queues, see the survey [3]. It is clear, that the assumptions made in the study of such systems do not correspond to real nature of

---

* Corresponding author.

A. Kwiecień, P. Gaj, and P. Stera (Eds.): CN 2013, CCIS 370, pp. 416–425, 2013.
© Springer-Verlag Berlin Heidelberg 2013

traffic in modern communication networks. In particular, the adequate mathematical model of arrival process should be able to capture non-stationarity and correlation typical for real traffic in modern networks [2].

Currently, many researches consider a Markovian arrival process (MAP) as the most suitable and mathematically tractable model of real arrival process. This process includes many input flows considered previously, such as stationary Poisson, Erlangian, Hyper-Markovian, Phase-Type (PH) renewal process, Markov Modulated Poisson Process (MMPP) and their superpositions. For more detail about MAP see [15, 16]. Today, there are a number of publications devoted to tandem queues with ordinary or batch MAP, see the papers [5–14] and references therein. But all these publications deal with tandem queues consisting of two stations.

Recently, in the paper [17], the tandem system with MAP input and arbitrary finite number of stations was investigated. Shortcoming of the model discussed in the paper is that customers arrive at the tandem only entering the first station. This assumption does not allow to take into account the cross-traffic which is an integral feature of the telecommunication networks. In this study, we consider more general and realistic model. We assume that, besides transit customers, an additional MAP flows of new customers arrive at the stations of the tandem. Customer arriving at some station has to get sequentially the service at all subsequent stations of the tandem. The quality of the service in the system essentially depends on the probability of the successful service of customers arriving at the stations of the tandem. These probabilities are of great importance in performance evaluation, discovering and avoiding so called bottlenecks in the network. In our study, we solve the problem of calculating the stationary distributions and loss probabilities in the tandem using simple methods based on investigation of input and output flows at the stations.

The rest of the paper is organized as follows. In Section 2, the mathematical model of the tandem is described. In Section 3, the process of the system operation is described in terms of a multi-dimensional Markov chain. The output and input flows at the stations of the tandem are specified. The problem of calculation of the stationary distribution of the tandem and its fragments is discussed in Sect. 4. Formulas for computation of loss probability in the tandem and its fragments are presented in Sect. 5. Concluding remarks are given in Sect. 6.

2   Model Description

We consider a tandem queueing system consisting of R, $R > 1$, stations in series. The r-th station is represented by N_r queue without a buffer, $r = 1, \ldots, R$. All servers of the tandem are independent and identical. The service time of a customer at the r-th station is exponentially distributed with parameter μ_r.

Customers arrive at the first station according the Markovian arrival process. Thereafter this process will be referred as MAP_1. Arrival of customers in the MAP_1 is directed by the underlying process $\nu_t^{(1)}$, $t \geq 0$, with state space $\{0, 1, \ldots, W_1\}$.. Arrivals occur only at the epochs of the process $\nu_t^{(1)}$, $t \geq 0$, transitions. The intensity of transitions, which are accompanied by an arrival of a customer, are combined into the matrix D_1 and the intensity of transitions, which are not accompanied by an arrival, are defined by non-diagonal entries of the matrix D_0. Thus, the behavior of the MAP_1 is completely defined by the $W_1 \times W_1$ matrices D_0, D_1.

The matrix $D = D_0 + D_1$ is an infinitesimal generator of the process $\nu_t, t \geq 0$. The fundamental arrival rate λ is defined by $\lambda = \boldsymbol{\theta} D_1 \mathbf{e}$ where $\boldsymbol{\theta}$ is the row vector of stationary probabilities of the Markov chain $\nu_t, t \geq 0$. The vector $\boldsymbol{\theta}$ is the unique solution to the system $\boldsymbol{\theta} D = \mathbf{0}, \boldsymbol{\theta}\mathbf{e} = 1$. Here and in the sequel $\mathbf{e}$ is a column-vector of appropriate size consisting of 1's and $\mathbf{0}$ is a row-vector of appropriate size consisting of zeros. The coefficient of variation, c_{var}, of intervals between successive arrivals is defined by $c_{\mathrm{var}}^2 = 2\lambda\boldsymbol{\theta}(-D_0)^{-1}\mathbf{e} - 1$. The coefficient of correlation, c_{cor}, of the successive intervals between arrivals is given by $c_{\mathrm{cor}} = (\lambda\boldsymbol{\theta}(-D_0)^{-1}(D - D_0)(-D_0)^{-1}\mathbf{e} - 1)/c_{\mathrm{var}}^2$. For more information about a MAP see [15, 16].

The customers from MAP_1 aim to be served at all R stations of the tandem. For any r-th $(r = 2, \ldots, R)$ station, besides transit customers proceeding from the $(r-1)$-th station, an additional MAP flow of new customers arrives at the r-th station directly, not entering the previous stations of the tandem. We will denote this additional flow as MAP_r. MAP_r is defined by underlying process $\nu_t^{(r)}$, $t \geq 0$, with state space $\{0, 1, \ldots, W_r\}$ and the intensity transition matrices $H_0^{(r)}, H_1^{(r)}$. The customers from this flow aim to be served at the r-th, $(r+1)$-th, $\ldots$, R-th stations.

The stations of the tandem have no waiting area. So, if a customer arriving at the first station or proceeding to the r-th, $r = 2, \ldots, R$, station after service at the $(r-1)$-th station meets all servers busy he/she leaves the tandem forever (is lost).

Our aim is to calculate stationary distributions and loss probabilities associated with the tandem.

3   Process of the System States: Input and Output Flows in the Tandem

The process of the system states is described in terms of irreducible multi-dimensional continuous-time Markov chain

$$\xi_t = \{n_t^{(1)}, n_t^{(2)}, \ldots, n_t^{(R)}, \nu_t^{(1)}, \nu_t^{(2)}, \ldots, \nu_t^{(R)}, \}, \ t \geq 0 \ ,$$

where

- $n_t^{(r)}$, $n_t^{(r)} = 0, \ldots, N_r$, is the number of busy servers at the r-th station;
- $\nu_t^{(r)}$, $\nu_t^{(r)} = 0, \ldots, W_r$, $r = 1, \ldots, R$, is the state of the MAP_r underlying process at time t.

The state space of the chain is defined as

$$\{\{0, 1, \ldots, N_1\} \times \ldots \times \{0, 1, \ldots, N_R\} \times \{0, 1, \ldots, W_1\} \times \ldots \times \{0, 1, \ldots, W_R\}\} \ .$$

The row vector $\mathbf{p}$ of the steady state probabilities of the chain is calculated as the unique solution of the system of linear algebraic equations

$$\mathbf{p}Q = \mathbf{0}, \quad \mathbf{p}\mathbf{e} = 1 \tag{1}$$

where the matrix Q is an infinitesimal generator of the chain ξ_t, $t \geq 0$.

The rank of the matrix Q is equal to $\mathcal{R} = \prod_{r=1}^{R}(N_r + 1)(W_r + 1)$ and can be very large in case when one or more quantities R, N_r, W_r take a large value. As it was observed in [17], construction of this matrix by using standard technique of queueing theory is rather routine and time-consuming work, especially in the case of more or less large value of $\mathcal{R}$.

To avoid this unpleasant work we will use simple and user-friendly recursive method for construction of the generator Q and calculation of the stationary distributions associated with the tandem. The method is essentially based on the results of investigation of output flows from the stations of the tandem. These flows are specified using the following theorem.

Theorem 1. *The output flow from the r-th station, $r = 1, 2, \ldots, R$, belongs to the class of MAPs. This MAP is defined by the matrices $D_0^{(r)}$ and $D_1^{(r)}$ that are calculated using the following recurrent formulas:*

$$D_0^{(r)} = -\mu_r \mathrm{diag}\{0, 1, \ldots, N_r\} \otimes I_{K_r} + \tag{2}$$

$$\begin{pmatrix} D_0^{(r-1)} \oplus H_0^{(r)} & D_1^{(r-1)} \oplus H_1^{(r)} & 0 & \cdots & 0 & 0 \\ 0 & D_0^{(r-1)} \oplus H_0^{(r)} & D_1^{(r-1)} \oplus H_1^{(r)} & \cdots & 0 & 0 \\ 0 & 0 & D_0^{(r-1)} \oplus H_0^{(r)} & \cdots & 0 & 0 \\ \vdots & \vdots & \vdots & \ddots & \vdots & \vdots \\ 0 & 0 & 0 & \cdots & D_0^{(r-1)} \oplus H_0^{(r)} & D_1^{(r-1)} \oplus H_1^{(r)} \\ 0 & 0 & 0 & \cdots & 0 & D^{(r-1)} \oplus H^{(r)} \end{pmatrix},$$

$$D_1^{(r)} = \begin{pmatrix} 0 & 0 & \cdots & 0 & 0 & 0 \\ \mu_r & 0 & \cdots & 0 & 0 & 0 \\ 0 & 2\mu_r & \cdots & 0 & 0 & 0 \\ \vdots & \vdots & \ddots & \vdots & \vdots & \vdots \\ 0 & 0 & \cdots & (N_r - 1)\mu_{r-1} & 0 & 0 \\ 0 & 0 & \cdots & 0 & N_r\mu_r & 0 \end{pmatrix} \otimes I_{K_r}, \quad r = 1, 2, \ldots, R , \tag{3}$$

with the initial conditions

$$D_0^{(0)} = D_0, \quad D_1^{(0)} = D_1, \quad H_0^{(1)} = H_1^{(1)} = 0 . \tag{4}$$

Here $\otimes$ and $\oplus$ are the symbols of Kronecker's product and sum of matrices respectively; $D^{(r)} = D_0^{(r)} + D_1^{(r)}$, $H^{(r)} = H_0^{(r)} + H_1^{(r)}$; $\mathrm{diag}\{0, 1, \ldots, N_r\}$ denotes the diagonal matrix with the diagonal entries listed in the brackets; the value K_r is calculated by $K_r = \prod_{r'=1}^{r-1}(N_{r'} + 1) \prod_{r'=1}^{r}(W_{r'} + 1)$, $r = 2, 3, \ldots, R$; I_{K_r} stands for identity matrix of size K_r.

Proof. Let $r = 1$. It is evident that the process $\{n_t^{(1)}, \nu_t^{(1)}\}$, $t \geq 1$, describing the operation of the first station is a Markov chain. Enumerate the states of this chain in lexicographic order as follows: $(0,0), (0,1), \ldots, (0,W), (1,0), (1,1),$ $\ldots, (1,W), \ldots, (N_1,0), (N_1,1), \ldots, (N_1,W)$. Then, it is easy to see that transitions of the chain, that do not lead to service completion at the first station (and generation of arrival at the second station) are defined by the matrix $D_0^{(1)}$ calculated by Formula (2). Transitions of the chain leading to service completion at the first station (and generation of arrival at the second station) are defined by the matrix $D_1^{(1)}$. According to definition of a MAP, this means that output flow from the first station (transit flow to the second station) is a MAP defined by the matrices $D_0^{(1)}$ and $D_1^{(1)}$ that are calculated by Formulas (2)–(3).

Besides this transit flow, the MAP_2 flow of new customers arrives at the second station. This MAP_2 is defined by the matrices $H_0^{(2)}$ and $H_1^{(2)}$. Then, the total flow at the second station is the superposition of the transit flow and MAP_2 flow of new customers and is defined by the matrices $D_0^{(1)} \oplus H_0^{(2)}$ and $D_1^{(1)} \oplus H_1^{(2)}$.

Let now $r = 2$. Using precisely the same reasoning as in the case $r = 1$, we find that output flow from the second station is a MAP defined by the matrices $D_0^{(2)}$ and $D_1^{(2)}$ that are calculated by (2)–(3). The rest of the proof for $r > 2$ is implemented by induction. □

Corollary 1. *The input flow at the r-th station, $r = 2, \ldots, R$, belongs to the class of $MAPs$. This MAP is defined by the matrices*

$$\tilde{D}_0^{(r)} = D_0^{(r-1)} \oplus H_0^{(r)}, \quad \tilde{D}_1^{(r)} = D_1^{(r-1)} \oplus H_1^{(r)}, \quad r = 2, \ldots, R , \qquad (5)$$

where the matrices $D_0^{(r-1)}$, $D_1^{(r-1)}$ are calculated by Formulas (2)–(3).

Remark 1. By taking $r = 1$ in (5) and because of (4), we find the expectable relations

$$\tilde{D}_k^{(1)} = D_k, \; k = 0, 1 .$$

Remark 2. In what follows, we will denote the total input flow at the r-th station as $MAP^{(r)}$, $r = 1, 2, \ldots, R$. Note that notation $MAP^{(1)}$ as well as the above notation MAP_1 is used for input flow at the first station.

4 Calculation of the Stationary Distribution of the Tandem and Its Fragments

In this section, we present the method for calculating the collection of stationary distributions associated with the tandem based on the result of investigation of output flows given by Theorem 1.

Let $\langle r, r+1, \ldots, r' \rangle$ denote a fragment of the tandem consisting of the r-th, $(r+1)$-th,\ldots, r'-th stations, $1 \leq r \leq r' \leq R$.

Theorem 2. *The stationary distribution of the fragment $\langle r, r+1, \ldots, r' \rangle$ of the tandem is calculated as the stationary distribution of the tandem queue $MAP^{(r)}/M/N_r/N_r \to \cdot/M/N_{r+1}/N_{r+1} \to \cdots \to \cdot/M/N_{r'}/N_{r'}$ where $MAP^{(r)}$ is defined by the matrices $\tilde{D}_0^{(r)}$, $\tilde{D}_1^{(r)}$ that are calculated by Formulas (5).*

Let $\mathbf{q}_n^{(r)} = (q^{(r)}(n,1), q^{(r)}(n,2), \ldots, q^{(r)}(n, K_r))$, $n = 0, \ldots, N_r$, be the vectors whose entries define the joint stationary distribution of the number of busy servers at the r-th station and the states of the total $MAP^{(r)}$ flow at the r-th station, $r = 1, 2, \ldots, R$.

Corollary 2. *The stationary distribution $\mathbf{q}_n^{(r)}$, $n = 0, \ldots, N_r$, of the r-th station of the tandem is calculated as the stationary distribution of the queueing system $MAP^{(r)}/M/N_r/N_r$, $r = 1, 2, \ldots, R$.*

The stable algorithm for calculation of the stationary distribution of the $MAP/M/N/N$ queue was developed in [18]. For the sake of reader's convenience, we present here the main steps of this algorithm. We assume that the arrival flow is defined by the matrices D_0, D_1, and the intensity of the service is μ. Denote by $\mathbf{q}_n$ the row vector of steady state probabilities corresponding n busy servers, $n = 0, \ldots, N$.

Algorithm. *The stationary probability vectors $\mathbf{q}_n$, $n = 0, \ldots, N$, are computed by*

$$\mathbf{q}_n = \mathbf{q}_0 \Phi_n, \quad n = 1, \ldots, N ,$$

where the matrices Φ_n are computed recursively:

$$\Phi_0 = I, \quad \Phi_n = \Phi_{n-1} D_1 (n\mu I - D_0 - (1 - \delta_{n,N}) D_1 G_n)^{-1}, \quad n = 1, \ldots, N ,$$

the matrices G_n, $n = 0, 1, \ldots, N-1$, are computed from the backward recursion

$$G_n = (n+1)\mu[(n+1)\mu I - D_0 - D_1 G_{n+1}]^{-1}, \quad n = N-2, N-3, \ldots, 0 ,$$

with the terminal condition $G_{N-1} = N\mu(N\mu I - D)^{-1}$, the vector $\mathbf{q}_0$ is calculated as the unique solution to the following system of linear algebraic equations:

$$\mathbf{q}_0(D_0 + D_1 G_0) = \mathbf{0}, \quad \mathbf{q}_0 \sum_{n=0}^{N} \Phi_n \mathbf{e} = 1 .$$

Note, that the subtraction operation is avoided in this algorithm and all inverted matrices exist and are nonnegative. Thus, the algorithm is highly stable numerically.

Corollary 3. *The stationary distribution $q_n^{(r)}$, $n = 0, \ldots, N_r$, of the number of busy servers at the r-th station of the tandem is found as*

$$q_n^{(r)} = \mathbf{q}_n^{(r)} \mathbf{e}, \quad n = 0, \ldots, N_r, \, r = 1, 2, \ldots, R .$$

Theorem 3. *The joint stationary distribution* $\mathbf{p}^{(1,\ldots,r)}$ *of the first r stations of the tandem can be calculated as the stationary distribution of the output flow from the r-th station of the tandem, i.e.,*

$$\mathbf{p}^{(1,\ldots,r)} = \boldsymbol{\theta}^{(r)}$$

where the row vector $\boldsymbol{\theta}^{(r)}$ is the unique solution of the system of linear algebraic equations

$$\boldsymbol{\theta}^{(r)}(D_0^{(r)} + D_1^{(r)}) = \mathbf{0}, \ \boldsymbol{\theta}^{(r)}\mathbf{e} = 1, \ r = 1, 2, \ldots, R \ ,$$

and the matrices $D_0^{(r)}$, $D_1^{(r)}$ are calculated using recurrent Formulas (2)–(3).

It is clear that the vector $\mathbf{p}$ of the stationary distribution of the whole tandem coincides, in terms of notation of Theorem 3, with the vector $\mathbf{p}^{(1,\ldots,R)}$. Thus, the following statement is true.

Corollary 4. *The vector $\mathbf{p}$ of the stationary distribution of the tandem is calculated as the unique solution of the system of linear algebraic equations*

$$\mathbf{p}(D_0^{(R)} + D_1^{(R)}) = \mathbf{0}, \ \mathbf{pe} = 1 \ . \tag{6}$$

It follows from system (6), that the matrix $D_0^{(R)} + D_1^{(R)}$ coincides, up to a constant multiplier, with infinitesimal generator Q of the Markov chain $\xi_t = \{n_t^{(1)}, n_t^{(2)}, \ldots, n_t^{(R)}, \nu_t^{(1)}, \nu_t^{(2)}, \ldots, \nu_t^{(R)}, \}$, $t \geq 0$, describing the operation of the tandem. Moreover, it can be verified that $Q = D_0^{(R)} + D_1^{(R)}$ exactly.

Thus, in this section, we have got as a by-product, the recursive method of construction of the generator Q avoiding time-consuming work required in the direct approach to the construction of this matrix. More general conclusion is formulated as follows.

Corollary 5. *The infinitesimal generator, $Q^{(r)}$, of the Markov chain*

$$\xi_t^{(r)} = \{n_t^{(1)}, n_t^{(2)}, \ldots, n_t^{(r)}, \nu_t^{(1)}, \nu_t^{(2)}, \ldots, \nu_t^{(r)}\}, \ t \geq 0 \ ,$$

describing the operation of the first r stations of the tandem can be defined as

$$Q^{(r)} = D_0^{(r)} + D_1^{(r)}, \ r = 1, \ldots, R \ .$$

5   Loss Probabilities

Having stationary distributions associated with the tandem calculated, we can find a number of its performance measures. The most important of them are loss probabilities in the fragments of the tandem and in the whole tandem. According to ergodic theorems for Markov chains, see, e.g. [19], loss probability in a fragment of the tandem can be computed via the ratio of the rate of output flow to the fragment and input flow from this fragment. Thus, the following theorem is valid.

Theorem 4. *The loss probability in the fragment* $\langle r, r+1, \ldots, r' \rangle$ *of the tandem is calculated by*

$$P_{\text{loss}}^{(r,\ldots,r')} = \frac{\tilde{\lambda}_r - \lambda_{r'}}{\tilde{\lambda}_r}, \tag{7}$$

where $\tilde{\lambda}_r$ *and* $\lambda_{r'}$ *are the intensities of the total* $MAP^{(r)}$ *flow of customers arriving at the r-th station and MAP output flow from r'-th station respectively. These intensities are calculated as*

$$\tilde{\lambda}_r = \tilde{\boldsymbol{\theta}}^{(r)} \tilde{D}_1^{(r)} \mathbf{e}, \quad \lambda_{r'} = \boldsymbol{\theta}^{(r')} D_1^{(r')} \mathbf{e},$$

where the vectors $\tilde{\boldsymbol{\theta}}^{(r)}$, $\boldsymbol{\theta}^{(r)}$ *are calculated from the systems*

$$\tilde{\boldsymbol{\theta}}^{(r)} (\tilde{D}_0^{(r)} + \tilde{D}_1^{(r)}) = \mathbf{0}, \quad \tilde{\boldsymbol{\theta}}^{(r)} \mathbf{e} = 1,$$

$$\boldsymbol{\theta}^{(r')} (D_0^{(r')} + D_1^{(r')}) = \mathbf{0}, \quad \boldsymbol{\theta}^{(r')} \mathbf{e} = 1.$$

Theorem 5. *The loss probability of transit customers at the r-th station is calculated as*

$$P_{\text{loss/transit}}^{(r)} = \frac{\mathbf{q}_{N_r}^{(r)} (D_1^{(r-1)} \otimes I_{W_r+1}) \mathbf{e}}{\lambda_{r-1}}, \quad r = 2, \ldots, R, \tag{8}$$

The loss probability of new customers at the r-th station is calculated as

$$P_{\text{loss/new}}^{(r)} = \frac{\mathbf{q}_{N_r}^{(r)} (I_{K_{r-1}(N_{r-1}+1)} \otimes H_1^{(r)}) \mathbf{e}}{h_r}, \quad r = 2, \ldots, R. \tag{9}$$

Here λ_{r-1} *and* h_r *are the intensities of output flow from the* $(r-1)$-*th station and new flow at the r-th station, respectively.*

Proof. The numerator of the fraction in the right hand side of (8) is the intensity of the customers that arrive at the r-th station from the $(r-1)$-th station and meet all servers busy. The denominator of this fraction is the intensity of all customers arriving at the r-th station from the $(r-1)$-th station. Then the ratio of these intensities defines the probability $P_{\text{loss/transit}}^{(r)}$.

The similar reasons yield Formula (9) for the probability $P_{\text{loss/new}}^{(r)}$. □

The total loss probability $P_{\text{loss}}^{(r)}$ at the r-th station can be calculated by Formula (7) above. Using Theorem 5, it is possible to get an alternative expression for this probability. Alternative formula and some other useful formulas are given in the following statement.

Corollary 6. *The probability that an arbitrary customer arriving at station r is a transit one and he/she will be lost at this station is calculated as*

$$P^{(r)}_{\text{loss,transit}} = P^{(r)}_{\text{loss/transit}} \frac{\lambda_{r-1}}{\tilde{\lambda}_r}, \ r = 2, \dots, R \ .$$

The probability that an arbitrary customer arriving at station r is a new one and he/she will be lost at this station is calculated as

$$P^{(r)}_{\text{loss,new}} = P^{(r)}_{\text{loss/new}} \frac{h_r}{\tilde{\lambda}_r}, \ r = 2, \dots, R \ .$$

The probability that an arbitrary customer (transit or new one) arriving at station r will be lost at this station is calculated as

$$P^{(r)}_{\text{loss}} = P^{(r)}_{\text{loss,transit}} + P^{(r)}_{\text{loss,new}} \ .$$

Corollary 7. In case $R = 2$, the probability that a customer arriving at the first station will be successfully served at both two stations of the tandem is calculated by

$$P_{succ} = (1 - P^{(1)}_{\text{loss}})(1 - P^{(2)}_{\text{loss/transit}}) \ .$$

6 Conclusion

The results of the paper can be theoretically extended to the case of more general, so called PH (phase type) distribution [20] of the service times. However, such a generalization implies the essential increase of the state space of the Markov chain under study. For this reason, the computer implementation of the presented algorithms becomes much more time consuming. In this case, it is possible to approximate MAP flows at the stations by the MAPs of the smaller dimensions [21–23]. We hope to investigate this problem in the future.

Acknowledgements. This research was partially supported by the grant No. F12R-086 of the Belarusian Republican Foundation of Fundamental Research.

References

1. Balsamo, S., Persone, V.D.N., Inverardi, P.: A review on queueing network models with finite capacity queues for software architectures performance prediction. Performance Evaluation 51, 269–288 (2003)
2. Heindl, A.: Decomposition of general tandem networks with $MMPP$ input. Performance Evaluation 44, 5–23 (2001)
3. Gnedenko, B.W., Konig, D.: Handbuch der Bedienungstheorie. Akademie Verlag, Berlin (1983)
4. Perros, H.G.: A bibliography of papers on queueing networks with finite capacity queues. Performance Evaluation 10, 255–260 (1989)
5. Breuer, L., Dudin, A.N., Klimenok, V.I., Tsarenkov, G.V.: A two-phase $BMAP/G/1/N \rightarrow PH/1/M - 1$ system with blocking. Automation and Remote Control 65, 117–130 (2004)

6. Gomez-Corral, A.: A tandem queue with blocking and Markovian arrival process. Queueing Systems 41, 343–370 (2002)

7. Gomez-Corral, A.: On a tandem G-network with blocking. Advances in Applied Probability 34, 626–661 (2002)

8. Gomez-Corral, A., Martos, M.E.: Performance of two-station tandem queues with blocking: The impact of several flows of signals. Performance Evaluation 63, 910–938 (2006)

9. Klimenok, V.I., Breuer, L., Tsarenkov, G.V., Dudin, A.N.: The $BMAP/G/1/N \rightarrow \cdot PH/1/M - 1$ tandem queue with losses. Performance Evaluation 61, 17–40 (2005)

10. Klimenok, V., Kim, C.S., Tsarenkov, G.V., Breuer, L., Dudin, A.N.: The $BMAP/G/1 \rightarrow \cdot/PH/1/M$ tandem queue with feedback and losses. Performance Evaluation 64, 802–818 (2007)

11. Kim, C.S., Klimenok, V., Taramin, O.: A tandem retrial queueing system with two Markovian flows and reservation of channels. Computers and Operations Research 37, 1238–1246 (2010)

12. Kim, C.S., Park, S.H., Dudin, A., Klimenok, V., Tsarenkov, G.: Investigaton of the $BMAP/G/1 \rightarrow \cdot/PH/1/M$ tandem queue with retrials and losses. Applied Mathematical Modelling 34, 2926–2940 (2010)

13. Kim, C.S., Klimenok, V.I., Taramin, O.S., Dudin, A.N.: Tandem $MAP/G/1 \rightarrow \cdot/M/N/0$ queue with heterogeneous customers. Mathematical Problems in Engineering. Article ID 324604, 26 (2012), doi:10.1155/2012

14. Kim, C.S., Dudin, S.: Priority tandem queueing model with admission control. Computers and Industrial Engineering 60, 131–140 (2011)

15. Lucantoni, D.M.: New results on the single server queue with a batch Markovian arrival process. Communications in Statistics-Stochastic Models 7, 1–46 (1991)

16. Chakravarthy, S.R.: The batch Markovian arrival process: a review and future work. In: Krishnamoorthy, A., et al. (eds.) Advances in Probability Theory and Stochastic Processes, pp. 21–49. Notable Publications, NJ (2001)

17. Klimenok, V., Dudin, A., Vishnevsky, V.: On the stationary distribution of tandem queue consisting of a finite number of stations. In: Kwiecień, A., Gaj, P., Stera, P. (eds.) CN 2012. CCIS, vol. 291, pp. 383–392. Springer, Heidelberg (2012)

18. Klimenok, V., Kim, C.S., Orlovsky, D., Dudin, A.: Lack of invariant property of Erlang loss model in case of the MAP input. Queueing Systems 49, 187–213 (2005)

19. Skorokhod, A.: Probability theory and random processes. High School, Kiev (1980)

20. Neuts, M.: Matrix-geometric Solutions in Stochastic Models – An Algorithmic Approach. Johns Hopkins University Press, USA (1981)

21. Alfa, A.S., Diamond, J.E.: On approximating higher order MAPs with MAPs of order two. Queueing Systems 34, 269–288 (2000)

22. Heindl, A., Mitchell, K., van de Liefvoort, A.: Correlation bounds for second order MAPs with application to queueing network decomposition. Performance Evaluation 63, 553–577 (2006)

23. Heindl, A., Telek, M.: Output models of $MAP/PH/1(/K)$ queues for an efficient network decomposition. Performance Evaluation 49, 321–339 (2002)

Analytical and Numerical Means to Model Transient States in Computer Networks

Tadeusz Czachórski[1], Monika Nycz[2], Tomasz Nycz[2], and Ferhan Pekergin[3]

[1] Institute of Theoretical and Applied Informatics, Polish Academy of Sciences
Bałtycka 5, 44-100 Gliwice, Poland
tadek@iitis.gliwice.pl
[2] Institute of Informatics, Silesian University of Technology
Akademicka 16, 44-100 Gliwice, Poland
{monika,tomasz.nycz}@polsl.pl
[3] LIPN, Université Paris-Nord, 93 430 Villetaneuse, France
pekergin@lipn.univ-paris13.fr

Abstract. Transient queue analysis is needed to model the influence of time-dependent flows on congestion in computer networks. It may be applied to the networks performance evaluation and the analysis of the transmissions quality of service. However, the exact queuing theory gives us only few practically useful results, concerning mainly M/M/1 and M/M/1/N queues. The article presents potentials of three approaches: Markovian queues solved numerically, the diffusion approximation, and fluid-flow approximation. We mention briefly a software we implemented to use these methods and summarise our experience with it.

Keywords: diffusion approximation, fluid flow approximation, Markov chains, transient states.

1 Introduction

Transient queue analysis is needed to model time-dependent flows and the dynamics of changes of router queues in computer networks. It is needed in stability analysis of Internet connections, It helps to predict packet loss probability and queueing delays which are the major factors of the quality of service. Furthermore, efficient modelling tools are indispensable for coping with large network topologies typical for modern Internet.

The use of analytical models known in queueing theory is limited to M/M/1 and M/M/1/N single queues and even there the solution is complex. Transient states of these models were investigated more than half a century ago. Chapman-Kolmogorov equations (first-order linear differential equations) define state probabilities $p(n, t; n_0)$ of n customers present in the system at time t if $n = n_0$ at time $t = 0$. If we apply the Laplace transform to make these equations algebraic ones, solve them and then find the original functions of the solutions in time

A. Kwiecień, P. Gaj, and P. Stera (Eds.): CN 2013, CCIS 370, pp. 426–435, 2013.
© Springer-Verlag Berlin Heidelberg 2013

domain, we obtain [1]:

$$p(n, t; n_0 = i) = e^{(\lambda+\mu)t}\left[\varrho^{\frac{n-i}{2}} I_{n-n_0}(at) + \right.$$

$$\left. +\varrho^{\frac{n-n_0-1}{2}} I_{n+n_0+1}(at) + (1-\varrho)\varrho^n \sum_{j=n+n_0+2}^{\infty} \varrho^{\frac{-j}{2}} I_j(at)\right] \quad (1)$$

where λ is input flow intensity, μ is service intensity (i.e. $1/\mu$ is mean service time), $\varrho = \lambda/\mu$ is server utilisation factor, $a = 2\mu\sqrt{\varrho}$ and $I_k(x)$ is the modified Bessel function of the first type and order k. Similarily, transient distributions for the limited queue $M/M/1/N$ were derived [2, 3]. Some simplifications of the solution (1) were proposed, e.g. the generating function of the distribution $\bar{p}(n, s; n_0)$ may be replaced by expressions having simpler originals in time domain [4] or Bessel funcions may be replaced by easier to compute functions, [5].

These results do not fit well to the problem of modelling IP routers, where the incoming streams are not Poisson and the size of packets is not exponentially distributed. Note that the solution (1) refers to transient states but it is assumed that the model parameters, λ in particular, are constant. Hence, in case of time dependent flows we should make them piecewise constant. We need models treating constantly changing non-Poisson flows and assuming general distributions of service times. We need also the possibility to include in these models the description of self similarity of flows. The models should also be scalable to meet very large topologies characteristic to the Internet.

In the sections below we describe our experience with three approaches: Markovian queues solved numerically, diffusion approximation, and fluid-flow approximation.

2 Markov Models Solved Numerically

Markov models are essential for the evaluation of the performance of computer networks. The models support the design of new communication protocols, mechanisms for regulating the intensity of Internet transmissions and mechanisms to ensure the quality of transmission services. However, they are not scalable: the number of states is growing very rapidly with the complexity of a modelled object: each state of the Markov chain corresponds to one state of the system. It is necessary to construct and solve the system of equations defining the probability of states – the number of equations equals the number of states. The existing solvers as e.g. Markovian solver in QNAP, XMARCA, PEPS, PEPSY, PRISM consider only steady state Markov chains and solve algebraic systems of equations.

Theoretically, for any continuous time Markov chain the Chapman-Kolmogorov equations with transition matrix Q

$$\frac{d\pi(t)}{dt} = \pi(t)Q , \quad (2)$$

have the analytical transient solution:

$$\pi(t) = \pi(0)e^{Qt} \ , \tag{3}$$

where $\pi(t)$ is the probability vector and $\pi(0)$ is the initial condition. However, it is not easy to compute the expression e^{Qt} where Q is a large matrix, see e.g. [6]. It may be done by its expansion to Taylor's series

$$e^{Qt} = \sum_{k=0}^{\infty} \frac{(Qt)^k}{k!} \ , \tag{4}$$

but the task is numerically unstable, especially for large Q. Additionally, to consider $\lambda(t)$, we should make the parameters of the model piecewise constant in small intervals and apply the solution (3) at each of these intervals.

We are developping our own package Olymp. It is a library generating transition matrices of continuous time Markov chains (CTMC), solving them. Olymp uses Java language to define network nodes and the interactions between them. Due to the potentially very large sizes of the models' transition matrices, their generation is parallelized, and they can be compressed on-the-fly using a dedicated compression based on finite-state automata. Olymp has a quite different approach to represent CTMC in the comparison to typical model checkers. A move to another state involves a transfer of a token. A node that sends the token initiates the move asynchronously, in moments of time that adhere to an exponential distribution. A node that receives the token can accept it validating the move. The negotiation can be thought of as for example an agreed transfer of a packet between these two nodes or as a synchronisation on a signal, distributed to the network by a clock node. At the moment we are able to generate and solve Markov chains of the 150 million of states. The method of solution used is one of projection methods based on Krylov subspace with Arnoldi process to project the exponential of a large matrix approximately onto a small Krylov subspace, see [7]; the transition matrix is then small and the computation of the expression (3) with the use of uniformization method and Padé approximations is much easier, [8–11]. This approach is supplemented by direct numerical solution of large systems of ordinary differential equations (ODE) using uniformization, i.e. discretization of the CTMC, that is replacing the CTMC by a DTMC (a discrete-time Markov chain) and a Poisson process.

We are increasing the size of tractable Markov chains by several orders through the use of a GPU-CPU (graphical processing unit) and a better design of computational algorithms for parallel computing and optimization of memory usage, [12]. GPU capabilities go far beyond the computer graphics. It is well known that a potential computational power of GPUs is much greater than that of contemporary CPUs (in a sense of the performance measured by number of floating point operations per second). Thus, it is possible to shorten the time of computations. Due to the enormous amount of the data to be processed, methods must be developed to store vectors and matrices with intelligent management of memory.

3   Diffusion Approximation

This approach is merging states of the considered queueing system and thus needs much less computations than the Markov models. We present here the principles of the method following [13] where steady-state solution of a single G/G/1/N model was given and then extended to the network of queues in [14]. We supplemented these results with semi-analytical, semi-numerical transient state solution [15] given for constant model parameters but it could be applied also in case of time-dependent parameters if we only make them constant within small intervals, as demonstrate numerical results below.

Let $A(x)$, $B(x)$ denote the interarrival and service time distributions at a service station and $a(x)$ and $b(x)$ be their density functions. The distributions are general but not specified, the method requires only the knowledge of their two first moments. The means are denoted as $E[A] = 1/\lambda$, $E[B] = 1/\mu$ and variances are $\text{Var}[A] = \sigma_A^2$, $\text{Var}[B] = \sigma_B^2$. Denote also squared coefficients of variation $C_A^2 = \sigma_A^2\lambda^2$, $C_B^2 = \sigma_B^2\mu^2$. $N(t)$ represents the number of customers present in the system at time t.

Diffusion approximation replaces the process $N(t)$ by a continuous diffusion process $X(t)$, the incremental changes $dX(t) = X(t + dt) - X(t)$ of which are normally distributed with the mean βdt and variance αdt, where β, α are coefficients of the diffusion equation

$$\frac{\partial f(x,t;x_0)}{\partial t} = \frac{\alpha}{2}\frac{\partial^2 f(x,t;x_0)}{\partial x^2} - \beta\frac{\partial f(x,t;x_0)}{\partial x} . \tag{5}$$

This equation defines the conditional pdf of $X(t)$:

$$f(x,t;x_0)dx = P\left[x \leq X(t) < x + dx \mid X(0) = x_0\right] .$$

The both processes $X(t)$ and $N(t)$ have normally distributed changes; the choice $\beta = \lambda - \mu$, $\alpha = \sigma_A^2\lambda^3 + \sigma_B^2\mu^3 = C_A^2\lambda + C_B^2\mu$ ensures that the parameters of these distributions grow at the same rate with the length of the observation period. In the case of G/G/1/N station, the process evolves between barriers placed at $x = 0$ and $x = N$. When it comes to $x = 0$, it remains there for a time exponentially distributed with the parameter λ and then it returns to $x = 1$; when it comes to $x = N$, it remains there for a time which is exponentially distributed with the parameter μ and then to $x = N - 1$. These are not typical bordary conditions for differential equations, therefor we developed a special method to solve the Equation (5): the function $f(x,t;x_0)$ is expressed by a superposition of density functions of the diffusion process with the absorbing barriers $x = 0$ and $x = N$, [15]. The solution is obtained in terms of the Laplace transform of $f(x,t;x_0)$ and then inverted numerically.

We studied the errors of this heuristic approach with the use of a wide range numerical examples. An example is given below. Consider a G/G/1/30 queue (in fact, it is M/M/1/30 queue, as we assume $C_A^2 = C_B^2 = 1$) with the input rate $\lambda(t)$ varying in time as presented in Fig. 1. It represents a typical TCP flow with additive increases and multiplicative decreases in case of packet losses, the range of time is $[0, 100]$ time units. In computations, the values of diffusion parameters are changed each 0.5 time unit. Figs. 2–3 present results of the diffusion model compared with the simulation results (in the latter case it is the average of 500 thousands independent runs). The mean queue is presented both in linear and logarithmic scales, to see better the differences for very small values.

However, when the traffic is different from Poisson and the service times have nonexponential distributions, the errors of the approximation are growing: Fig. 4 presents some exemplary errors for $G/G/1/100$ station ($\varrho = 0.75$) as a function of C_A^2, C_B^2 (transient state, $t = 100$, at the beginning the queue was empty). The errors of the approximation increase with C_A^2, C_B^2.

In our method the density function of the diffusion process is obtained in form of its Laplace transform which is then inverted numerically with the use of Stehfest algorithm [16]. In this algorithm a function $f(t)$ is obtained from its transform $\bar{f}(s)$ for any fixed argument t as

$$f(t) = \frac{\ln 2}{2} \sum_{i=1}^{N} V_i \, \bar{f}\left(\frac{\ln 2}{t} i\right) , \tag{6}$$

where V_i are known constants and N is an even integer and depends on a computer precision; we used $N = 20$.

The algorithm brings some numerical difficulties, the inversion requires computation of hiberbolic sinus function with arguments that may exceed double-precision floating-point number exponent size which for 64-bit double permits only maximum positive value of 308. That often happens when maximum queue's buffer N is big (like 100). Possible solution would be to use 80-bit extended precision double which permits maximal positive exponent of 4932.

The method may be applied to a network of queueus of any topology [14] although in case of transient states the computations of the time-dependent flows evry small time-interval is of course time consuming. We tested it on an examplary network of 1000 stations.

4   Fluid-Flow Approximation

This approximation method assumes a much simpler flow model determining the mean values of traffic intensity and service times of network stations. In contrary to diffusion approximation, it is based only on first-order ordinary linear differential equations, so that the values are obtained in a much shorter calculation time. Therefore it is suited for modeling transient states of large TCP/IP networks, in particular the Internet. However it is quite difficult to analyze Internet-scale topology manually, thus we prepared a tool that is easily adaptable for analysis of different cases.

Mathematical model implemented in our software, that is already adapted to TCP congestion window mechanism and RED algorithm in routers, was presented in [17, 18]. A modelled network V is described by characteristic values of routers (instantaneous and average queue length and discard probability per

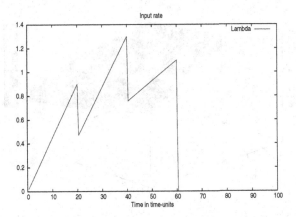

Fig. 1. Input traffic intensity $\lambda(t)$ in the studied example

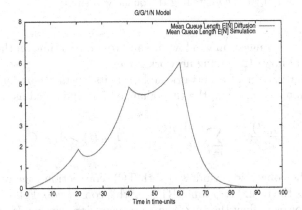

Fig. 2. Numerical example: mean number of customers (linear scale) as a function of time; diffusion approximation and simulation results

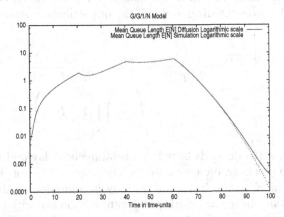

Fig. 3. Numerical example: mean number of customers (logarithmic scale) as a function of time; diffusion approximation and simulation results

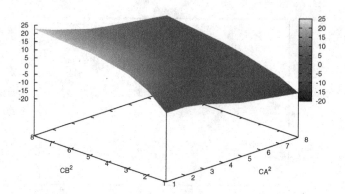

Fig. 4. G/G/1/100 station with $\varrho = 0.75$, relative error of the mean queue as a function of C_A^2, C_B^2; at $t = 100$, at the beginning the queue was empty

router) and flows (congestion window size and round trip time on the whole path of an individual flow) traversing network nodes.

The main parameter of a network node v is its mean queue length q_v; with the transmission capacity C_v, the time change of q_v is defined as

$$\frac{dq_v(t)}{dt} = \sum_{i=1}^{N} \frac{W_i(t)}{R_i(q_v(t))} - \mathbf{1}(q_v(t) > 0) \cdot C_v \tag{7}$$

where W_i is the congestion window of i-th TCP connection (one of N) crossing this node, R_i is the round trip time (RTT) of this connection. Following the TCP Reno policy, for each flow the size W of its congestion window (8) is increasing in average by 1 every RTT (the arrival of a new acknowledgement) or is divided by 2 with the intensity of losses. The loss factor is calculated on the basis of matrix **B** which stores the drop probabilities of particular routers for each TCP flows, [19].

$$\frac{dW_i(t)}{dt} = \frac{1}{R_i(\boldsymbol{q}(t))} - \frac{W_i(t)}{2} \cdot \frac{W_i(t - \tau)}{R_i(\boldsymbol{q}(t - \tau))} \cdot$$
$$\cdot \left(1 - \prod_{j \in V}(1 - B_{ij}) \right) \cdot \tag{8}$$

Round Trip Time (9) depends heavily on total queue delay and network topology, to be more precisely, on a total propagation delay (Tp_i) of the i flow route, which consists of K nodes. Total queue delay is computed as the sum of quotients of instantaneous queue length q and transmission capacity C of the routers on the route from source to destination. In our small model modification the total propagation delay is not a global flow parameter, but the sum of the links delays, where a single link is a connection between two network nodes that belongs to one flow path.

$$R_i(\mathbf{q}(t)) = \sum_{j=1}^{K} \frac{q_j(t)}{C_j} + \sum_{j=1}^{K-1} Tp_j .$$

(9)

Our software implementation allows to model large IP networks having thousands of nodes and thousands of flows, divided however into several categories, which consist of flows with identical route and starting parameters. In case of collection data of large topology we save the results in cumulative binary files, that are converted to plot files when needed. The timing results for such an exemplary topology was presented in [19].

The main drawback of our program was the necessity to manually defining the structure of nodes and flows topologies in text file. Thus, the more complex the network for analysis was, the more time was needed to create the configuration file with input parameters and flows routes. To eliminate the manual definition of flows paths, we recently implemented a converter that import the topology from network topology generator and produce a configuration file for our program. We decided to choose aSHIIP generator [20, 21] because of its ability to generate hierarchical network topologies, typical for the actual Internet networks.

Our converter parses the output file from the aSHIIP program to obtain the number of nodes, number of network layers and the directed graph that illustrates the network topology. The next step assumes to construct the flows routes based on selected options (categorization, partition method, total number of flows). Thus, we determine the border routers – the nodes that are entries points of the network – with the flag defining the node as source or destination. Then, we randomly pick out a pair of source and destination routers and perform the Dijkstra algorithm with the use of Fibonacci Heap (implementation is based on code [22] and [23]) on each source node. There are however some cases, where generated number of paths is less then required by the user. It is due to the fact that the generated hierarchical topology is a directed graph, hence it may happen that it is not possible to find a path – the existing links do not provide two-way connections between necessary nodes. The next and final step involves generation of the starting parameters values i.e. for a router: the maximum size of buffer, initial queue size, transmission capacity, weight parameter and probabilities for RED mechanism; and for flow: the initial window size, usually set to one. During the computations of the settings, the number of layers, clustering and node degree are taken into consideration. The phase ends with saving the settings to a configuration file, that is used as an input to our implementation of fluid flow approximation model.

In our next publication we will present the results of sample topology generated by aSHIIP and processed by our tool as described above.

5 Conclusions

Each of three approaches presented above has its highlights and drawbacks. Markovian models are flexible but have frequently enormous state space and therefore are time and space consuming. Fluid flow approximation is the simplest but it may be easily applied to very large configurations. Diffusion approximation is somewhere between. We believe that each of the methods has its place in the panoply of tools needed in performance evaluation and we develop appropriate software tools. We may also use simulation models. In this purpose we have developed an extension of OMNET++ (a popular simulation tool written in C++, [24]) allowing simulation of transient state models. In particular, random generators were modified to make possible the changes of their parameters as a function of time, a new software was added to collect the statistics of multiple runs and to aggregate them. We used this module to validate the diffusion approximation results. Basically, the simulation run in a transient state investigation should be repeated sufficient number of times (e.g. 500 thousands in our examples) and the results for a fixed time should be averaged. As the number of repetitions is high, the estimation of the errors is easy (confidence interval) on the basis of normal distribution. However, the number of repetitions is related to the value of the investigated probabilities and in case of rare events it should be high; this fact increases considerably the simulation time: typically in some of our examples, 5 min of computation on a standard PC station for a diffusion model is compared to 24 hours of simulations on the same machine.

Acknowledgments. This work was supported by Polish project NCN nr 4796/B/T02/2011/40 *Models for transmissions dynamics, congestion control and quality of service in Internet* and the European Union from the European Social Fund (grant agreement number: UDA-POKL.04.01.01-00-106/09) .

References

1. Champernowne, D.C.: An elementary method of solution of the queueing problem with a single server and constant parameters. J. R. Statist. Soc. B 18, 125–128 (1956)
2. Takács, L.: Introduction to the Theory of Queues. Oxford University Press (1960)
3. Tarabia, A.M.K.: Transient Analysis of M/M/1/N Queue – An Alternative Approach. Tamkang Journal of Science and Engineering 3(4), 263–266 (2000)
4. Kotiah, T.C.T.: Approximate transient analysis of some queueing systems. Operations Research 26(2), 334–346 (1978)
5. Jones, S.K., Cavin, R.K., Johnston, D.A.: An Efficient Computational Procedure for the Evaluation of the M/M/1 Transient State Occupancy Probabilities. IEEE Trans. on Comm. COM-28(12), 2019–2020 (1980)
6. Moler, C., van Loan, C.: Nineteen Dubious Ways to Compute the Exponential of a Matrix, Twenty-Five Years Later! SIAM Review 45(1), 30–49
7. Stewart, W.: Introduction to the Numerical Solution of Markov Chains. Princeton University Press, Chichester (1994)

8. Scientifique, C., Philippe, B., Sidje, R.B.: Transient Solutions of Markov Processes by Krylov Subspaces. In: 2nd International Workshop on the Numerical Solution of Markov Chains (1989)
9. Sidje, R.B., Burrage, K., McNamara, S.: Inexact Uniformization method for computing transient distributions of Markov chains. SIAM J. Sci. Comput. 29(6), 2562–2580 (2007)
10. Sidje, R.B., Stewart, W.J.: A Numerical Study of Large Sparse Matrix Exponentials Arising in Markov Chains. Computational Statistics & Data Analysis 29, 345–368 (1999)
11. Sidje, R.B.: Expokit: A Software Package for Computing Matrix Exponentials. ACM, Transactions on Mathematical Software 24(1) (1998)
12. Numerical computation for Markov chains on GPU: building chains and bounds, algorithms and applications. Project POLONIUM 2012–2013, bilateral cooperation PRISM-Université de Versailles and IITiS PAN, Polish Academy of Sciences
13. Gelenbe, E.: On Approximate Computer Systems Models. J. ACM 22(2) (1975)
14. Gelenbe, E., Pujolle, G.: The Behaviour of a Single Queue in a GeneralQueueing Network. Acta Informatica 7(fasc. 2), 123–136 (1976)
15. Czachórski, T.: A method to solve diffusion equation with instantaneous return processes acting as boundary conditions. Bulletin of Polish Academy of Sciences, Technical Sciences 41(4) (1993)
16. Stehfest, H.: Algorithm 368: Numeric inversion of Laplace transform. Comm. of ACM 13(1), 47–49 (1970)
17. Misra, V., Gong, W.-B., Towsley, D.: Fluid-based Analysis of a Network of AQM Routers Supporting TCP Flows with an Application to RED. In: ACM SIGCOMM (2000)
18. Liu, Y., Lo Presti, F., Misra, V., Gu, Y.: Fluid Models and Solutions for Large-Scale IP Networks. ACM/SigMetrics (2003)
19. Czachórski, T., Nycz, M., Nycz, T., Pekergin, F.: Transient states of flows and router queues – a discussion of modelling methods. In: Proc. of International Conference on Networking and Future Internet (ICNFI 2012), Istanbul (April 2012)
20. Weisser, M.-A., Tomasik, J.: Automatic Induction of Inter-Domain Hierarchy in Randomly Generated Network Topologies. In: 10th Communication and Networking Simulation Symposium CNS 2007 (2007)
21. Tomasik, J., Weisser, M.-A.: Internet topology on AS-level: model, generation methods and tool. In: 29th IEEE International Performance Computing and Communications Conference (IPCCC 2010) (2010)
22. Dijkstra's Algorithm for Network Optimization Using Fibonacci Heaps (September 2009),
http://www.codeproject.com/Articles/42561/
Dijkstra-s-Algorithm-for-Network-Optimization-Usin
23. Fibonacci heap implementation (March 2012),
http://code.google.com/p/gerardus/source/browse/
trunk/matlab/ThirdPartyToolbox/dijkstra.cpp
24. OMNET++ Community Site, http://www.omnetpp.org
25. Czachórski, T., Pekergin, F.: Diffusion Approximation as a Modelling Tool. In: Kouvatsos, D.D. (ed.) Next Generation Internet: Performance Evaluation and Applications. LNCS, vol. 5233, pp. 447–476. Springer, Heidelberg (2011)

The Queueing Model of a Multiservice System with Dynamic Resource Sharing for Each Class of Calls

Sławomir Hanczewski, Maciej Stasiak, and Joanna Weissenberg

Poznan University of Technology, Chair of Communication and Computer Networks,
Polanka 3, 60-965 Poznan, Poland
{shancz,stasiak}@et.put.poznan.pl, joanna@weissenberg.pl
http://nss.et.put.poznan.pl

Abstract. The paper presents a new model of a multiservice queueing system with limited queue and state-dependent dynamic resource sharing of the server between individual classes of calls. The advantage of the proposed model is the possibility to evaluate the average parameters of queues for individual classes of calls. The proposed model can be taken as a basis for further research on multiservice queueing systems.

Keywords: multiservice queueing system, resource sharing.

1 Introduction

The analysis of modern networks that have to carry integrated multiservice traffic capable of providing a sufficient quality of communication environment and quality of service of calls, clearly indicates the need for developing efficient queueing network models. Until now, however, no models that would allow researchers to evaluate the delay parameters for individual classes of calls have been developed. Only results for two boundary cases for the solution of the problem have been obtained so far. The first of them involves the application of Erlang's C formula [1–3]. Such an approach results from the assumption that all classes of calls have identical characteristics and service conditions, including the identical sustainable bit rate. [2] assumes that it is the bit rate of the traffic stream that has the highest bit rate of all streams offered to the system. The other boundary case involves the application of recursive formula for full-availability multiservice systems with elastic traffic [1, 4]. This approach, in turn, results from a simple modification to the occupancy distribution in the full-availability multiservice system with losses [5, 6]. Model [1, 4] assumes that all streams of carried traffic – in the case of the lack of free resources in the system – will be compressed, which guarantees a lossless traffic service. On the basis of [4], it is possible to adopt a certain minimum level of compression and to assume that state probabilities related to a higher compression will determine the probability of worsening the QoS parameters for all streams of offered traffic. This model leads thus only to a determination of the characteristics of the system

A. Kwiecień, P. Gaj, and P. Stera (Eds.): CN 2013, CCIS 370, pp. 436–445, 2013.
© Springer-Verlag Berlin Heidelberg 2013

for all serviced traffic streams, without a possibility of determining the same characteristics for selected traffic classes.

Both solutions indicated above assume that all traffic classes in the system are of elastic nature. Both solutions then exclude a possibility of servicing mixtures of traffic in which part of classes will not undergo the traffic compression/decompression mechanism. In real conditions, however, network systems service both elastic traffic and real-time traffic (RT traffic) with exactly determined QoS parameters [3, 7]. In such cases, the application of the boundary solutions presented above is not appropriate and may eventually lead to considerable errors. To the best knowledge of the authors, multiservice queueing systems within the context of delays related to offered classes of calls have not been hitherto addressed. In traffic theory this problem has been formulated as "an Erlang formula for the Internet" [1, 8] and, in the opinion of the present authors, is an open problem.

This paper proposes a new model of a multiservice queueing system that provides a possibility to determine average values of the lengths of queues and waiting times for individual classes of calls.

2 Multiservice System with Losses

Let us define now the multiservice system with losses $\sum M / \sum M / C$ described by the following parameters:

λ_i – intensity of a Poisson call stream of class i. The notion of the call is understood as a packet stream [3], or its part [1, 8] related to a given service,

c_i – bit-rate of a call of class i,

t_i – the number of allocation units necessary for a call of class i to be executed. The allocation unit, also known as BBU (Basic Bandwidth Unit), determines a given minimum bit rate c_{BBU} such that bit rates of all calls (packet streams) are its multiple number. Thus, [9]:

$$c_{\text{BBU}} = GCD(c_1, c_2, \ldots, c_m) \ , \tag{1}$$

$$t_i = \left\lceil \frac{c_i}{c_{\text{BBU}}} \right\rceil \ , \tag{2}$$

m – the number of classes of calls offered to the system,

C – bit rate of the system,

V – capacity of the system, expressed in BBUs:

$$V = \left\lfloor \frac{C}{c_{\text{BBU}}} \right\rfloor , \tag{3}$$

l_i – average bit length of a call of class i,

μ_i – average service intensity for a call of class i:

$$\mu_i = \frac{c_i}{l_i} \ , \tag{4}$$

a_i – average offered traffic (expressed in Erlangs) by calls of class i:

$$a_i = \frac{\lambda_i}{\mu_i} \ , \tag{5}$$

x_i – the number of serviced calls of class i.

Assume now that the value of BBU is equal to 1 bps (or 1 kbps, 1 Mbps, etc., depending on bit rate units adopted for a given network), i.e.:

$$c_{\text{BBU}} = 1 \, \text{bps} \ . \tag{6}$$

In this case, the following parameters have the identical values (though they are expressed in different units):

$$V = C \ , \tag{7}$$

$$t_i = c_i \ . \tag{8}$$

Let us adopt then – taking into account (8) – the following notation:

$$A_i = a_i t_i = \frac{a_i c_i}{1 \, \text{bps}} = a_i c_i \ . \tag{9}$$

The parameter A_i is therefore the traffic intensity of a bit stream of class i, expressed in Erlangs and not as bps, which is suggested by the product $a_i c_i$ without (6) taken into consideration.

The occupancy distribution in a full-availability multiservice system with losses [10–12] can be determined at the level of the so-called microstates. The microstate [3, 7] is defined by the number of serviced calls of particular classes in the system $X = \{x_1, x_2, \ldots, x_m\}$. The Markov process at the level of macrostates is a reversible process [5, 13]. In [14] it is proved that this process satisfies the conditions of resources sharing according to the so-called balanced fair sharing, for which the service stream between states $X = \{x_1, \ldots, x_i, \ldots, x_m\}$ and $(X - 1_i) = \{x_1, \ldots, x_i - 1, \ldots, x_m\}$ satisfies the following condition:

$$x_i c_i = \frac{\Psi(X - 1_i)}{\Psi(X)} \ , \tag{10}$$

where 1_i defines one call of class i, whereas $\Psi(X)$ is the so-called balance function [4, 14]. This function is defined as the inverse of the product of reverse streams from state $X = \{x_1, x_2, \ldots, x_m\}$ to state $0 = \{0, 0, \ldots, 0\}$. The balance function for the multiservice full-availability system with losses is defined as follows:

$$\Psi(X) = \prod_{i=1}^{m} \frac{1}{x_i! c_i^{x_i}} \ \text{for} \ 0 \leq \sum_{i=1}^{m} x_i c_i \leq C \ . \tag{11}$$

The introduction of the balance function makes it possible to write the occupancy distribution in the following way [4]:

$$[p(X)]_C = \frac{\Psi(X) \prod_{i=1}^{m} A_i^{x_i}}{\sum_{\Omega} \Psi(X) \prod_{i=1}^{m} A_i^{x_i}} \ , \tag{12}$$

where Ω is set of all possible states X.

3   Multiservice System with Elastic Traffic

Consider a full-availability multiservice system in which a number of traffic classes are elastic. This means that a lack of free resources for a new call of a given class appearing in the system is followed by a compression – a decrease in the bit rate – of currently serviced calls, which allows the new call to be serviced. The accompanying assumption is that the service time of compressed calls will be extended, which will make a lossless transfer of all data possible. The models presented in [15, 16] assume that calls of all classes are compressed to a certain pre-defined level, above which new calls will be lost. The model proposed in [3, 17] assumes that the system services both calls that undergo compression and those that do not undergo compression. Furthermore, the model assumes that calls that undergo compression can decrease the bit rate to a given pre-defined level, while after exceeding the pre-defined level the system behaves as a multiservice system with losses.

Let us consider now the model [4] in which it is assumed that all classes of calls undergo lossless compression. This means that each call can decrease – influenced by new calls that appear in the system – its bit rate indefinitely. When this is the case, the service process is a reversible process for which the balance function is determined as follows:

$$\Psi(X) = \prod_{i=1}^{m} \frac{1}{x_i! c_i^{x_i}} \qquad \text{for } 0 \leq \sum_{i=1}^{m} x_i c_i \leq C$$

$$\Psi(X) = \frac{1}{C} \sum_{i=1}^{m} \Psi(X - 1_i) \quad \text{for } \sum_{i=1}^{m} x_i c_i > C \; . \tag{13}$$

The occupancy distribution in this system can be determined on the basis of Formula (12):

$$[p(X)]_C = \frac{\Psi(X) \prod_{i=1}^{m} A_i^{x_i}}{\sum_{\Omega} \Psi(X) \prod_{i=1}^{m} A_i^{x_i}} \; . \tag{14}$$

The values of reverse streams $y_i(X)$ that correspond to the Markov process described by the distribution (14) can always be determined on the basis of the modified Formula (10):

$$y_i(X) = \frac{\Psi(X - 1_i)}{\Psi(X)} \; . \tag{15}$$

We observe that in state X such that $\sum x_i c_i > C$ the value of the service stream of class i, i.e., $y_i(X)$, is not equal to bit rate $x_i c_i$, generated by all serviced calls of class i in the condition of the lack of compression (i.e., for all states X, such that $\sum x_i c_i \leq C$).

4   Queueing Interpretation of the Model with Elastic Traffic

Consider now the queueing system $\sum M / \sum M / C / C + U / \sum sdFIFO$ with a server with the capacity C and the buffer with the capacity U. Such a system can be described by Equations (13)–(15) that enforce the way calls are serviced, because service streams in the system are described by Equation (15) and satisfy the conditions for balanced fair sharing [14]. This means that the number of serviced resources in the system in a given occupancy state depends on a division of his state between individual classes of calls, i.e., depends on the number of calls of individual classes in a given occupancy state. The considered queueing system can be called a multiservice queueing system with dynamic, state-dependent, division of the resources of the server (Fig. 1). Such a system can be also treated as the M virtual queueing systems in which the bit rates of individual servers depend on the sum of streams occurring and serviced in particular queues. Hence, the proposed name of the discipline of service: $\sum sdFIFO$ (Sum of State Dependend FIFO queues). The allocated bit rate in the server for a selected queue changes along with a change in the state of the system, i.e., the number of streams of particular classes, and can be described by Formula (15):

$$y_i(X) = \frac{\Psi(X - 1_i)}{\Psi(X)} \ , \tag{16}$$

where the balance function (16) for the considered queueing system can be rewritten in the following way:

$$\Psi(X) = \quad \prod_{i=1}^{m} \frac{1}{x_i! c_i^{x_i}} \qquad \text{for } 0 \leq \sum_{i=1}^{m} x_i c_i \leq C$$

$$\Psi(X) = \frac{1}{C} \sum_{i=1}^{m} \Psi(X - 1_i) \quad \text{for } (C + U) \geq \sum_{i=1}^{m} x_i c_i > C \ . \tag{17}$$

The occupancy distribution in the system can be expressed by Formula (14) and, with the total capacity of the system taken into consideration, takes on the following form:

$$[p(X)]_{C+U} = \frac{\Psi(X) \prod_{i=1}^{m} A_i^{x_i}}{\sum_{\Omega} \Psi(X) \prod_{i=1}^{m} A_i^{x_i}} \ . \tag{18}$$

Figure 2 shows a diagram of the Markov process for the multiservice system $\sum M / \sum M / 2 / 4 / \sum sdFIFO$ to which two traffic streams are offered with the intensity $A_1 = a_1 c_1 = 1\,\text{Erl.}$, $A_2 = a_2 c_2 = 1\,\text{Erl.}$, average bit lengths $l_1 = 1\,\text{kb}$,

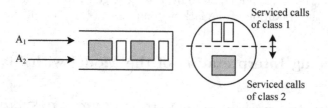

Fig. 1. Multiservice queueing system with dynamic, state-dependent, division of the resources of the server

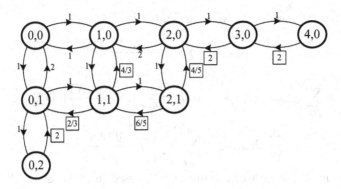

Fig. 2. Diagram of the Markov process in the system $\sum M/\sum M/2/4/\sum sdFIFO$ ($c_1 = 1$ and $c_2 = 2$, $A_1 = 1$, $A_2 = 1$)

$l_2 = 2\,\text{kb}$ and demands $c_1 = 1\,\text{kbps}$ and $c_2 = 2\,\text{kbps}$ ($c_{BBU} = 1\,\text{kbps}$). In Figure 2, the squares are showing the values of service streams – determined on the basis of Formula (17) – for those states in which some of calls are in the buffer. It is observable that the sum of service streams in these states is equal to the capacity of the server (i.e., the bit rate of the server, $C = 2\,\text{kbps}$).

5   Parameters of Queues of Individual Classes of Calls

The occupancy distribution (18) in the considered multiservice queueing system ($\sum M/\sum M/C/C+U/\sum sdFIFO$) can form a basis for a determination of the size of queues for individual classes of calls. In a given state X, the size of the queue $q_i(X)$ for calls of class i is determined by the difference between the total amount of demanded resources $x_i c_i$ and the resources allocated to service calls of class i in the server $y_i(X)$, determined by the intensity of the service stream (16). We obtained thus:

$$q_{i,\text{BBU}}(X) = x_i c_i - y_i(X) \ . \tag{19}$$

Notice that the size of the queue (19) is expressed in BBUs. The average length of the queue for calls of class i, expressed in BBU, is equal to:

$$Q_{i,\text{BBU}} = \sum_{\Omega} q_{i,\text{BBU}}(X)\,[p(X)]_{C+U} = \sum_{\Omega} [x_i c_i - y_i(X)]\,[p(X)]_{C+U} \ , \tag{20}$$

whereas the total average length of the queue in the system, expressed in BBU, is as follows:

$$Q_{\text{BBU}} = \sum_{i=1}^{m} Q_{i,\text{BBU}} = \sum_{i=1}^{m} \sum_{\Omega} [x_i c_i - y_i(X)]\,[p(X)]_{C+U} \ , \tag{21}$$

Formulae (19)–(21) express corresponding lengths of queues in BBUs. The formulas can be converted in such a way as to show the parameters of the queue in the number streams in the "waiting" state:

$$q_{i,\text{stream}}(X) = x_i - \frac{y_i(X)}{c_i} \quad , \tag{22}$$

$$Q_{i,\text{stream}} = \sum_{\Omega} q_{i,\text{stream}}(X)[p(X)]_{C+U} = \sum_{\Omega} \left[x_i - \frac{y_i(X)}{c_i} \right] [p(X)]_{C+U} \quad , \tag{23}$$

$$Q_{\text{stream}} = \sum_{i=1}^{m} Q_{i,\text{stream}} = \sum_{i=1}^{m} \sum_{\Omega} \left[x_i - \frac{y_i(X)}{c_i} \right] [p(X)]_{C+U} \quad . \tag{24}$$

By multiplying the length of the queue, expressed in streams, by the average length of the stream, expressed in bits, we get the average lengths of queues expressed in bits:

$$q_{i,\text{bit}}(X) = \left[x_i - \frac{y_i(X)}{c_i} \right] l_i \quad , \tag{25}$$

$$Q_{i,\text{bit}} = \sum_{\Omega} q_{i,\text{bit}}(X)[p(X)]_{C+U} = \sum_{\Omega} \left[x_i - \frac{y_i(X)}{c_i} \right] l_i[p(X)]_{C+U} \quad , \tag{26}$$

$$Q_{\text{bit}} = \sum_{i=1}^{m} Q_{i,\text{bit}} = \sum_{i=1}^{m} \sum_{\Omega} \left[x_i - \frac{y_i(X)}{c_i} \right] l_i[p(X)]_{C+U} \quad . \tag{27}$$

The average waiting time of a stream of class i can be determined on the basis of Little's law, e.g.:

$$T_{i,\text{bit}} = T_{i,\text{BBU}} = T_{i,\text{stream}} =$$

$$= \sum_{\Omega} q_{i,\text{stream}}(X) \frac{1}{\lambda_i} [p(X)]_{C+U} = \sum_{\Omega} \left[x_i - \frac{y_i(X)}{c_i} \right] \frac{1}{\lambda_i} [p(X)]_{C+U} \quad . \tag{28}$$

The equality of the average waiting time for streams, BBUs and bits of a given class result from the fact that the intensity of the stream is rescaled by the number of BBU units or by the average number of bits in the stream, i.e., the intensity counted in BBU units per time unit is $\lambda_i c_i$, while counted in bits per time unit is $\lambda_i l_i$.

The system $\sum M / \sum M / C / C + U / \sum sdFIFO$ is a blocking system. If the number of available BBUs in the buffer that can be allocated to a new incoming call is not sufficient enough, the call will be lost. The blocking probability is then the sum of blocking states in which the number of free BBUs in the buffer is lower than the number of BBUs required for a given call to execute a connection:

$$E(i) = \sum_{\Omega_{E(i)}} p(X)_{C+U} \quad , \tag{29}$$

where $\Omega_{E(i)}$ is the set of of blocking states for calls of class i:

$$\Omega_{E(i)} = \left\{ X : (C + U - c_i + 1) \leq \sum_{i=1}^{m} x_i c_i \leq (C + U) \right\} \quad . \tag{30}$$

6 An Example of Modelling of a Selected System

Consider the system: $\sum M / \sum M / C / C + U / \sum sdFIFO$ with the following structural parameters $C = 100\,\text{kbps}$, $U = 200\,\text{kbps}$, and the traffic parameters: $c_1 = 10\,\text{kbps}$, $l_1 = 10\,\text{kb}$, $c_2{=}20\,\text{kbps}$, $l_2 = 20\,\text{kb}$, $c_3 = 30\,\text{kbps}$, $l_3 = 30\,\text{kb}$. The system is offered traffic streams with the adopted proportion of: $A_1 : A_2 : A_3 = 3 : 2 : 1$.

The results of the analytical modeling of the system $\sum M / \sum M / 100 / 300 / \sum sdFIFO$ were compared with the results of the simulation experiments of two $\sum M / \sum M / 100 / 300 / \sum constFIFO$ systems in which the resources of the server allocated to a given class of calls had a fixed value, independent of the state of the service process in the system. The first system (SYSTEM 1) assumes that the resources allocated to each of the classes are directly proportional to offered traffic $CA_i / \sum A_j$ of this class. The other system (SYSTEM 2) assumes that each class has identical resources of the server at its disposal, equal to C/M.

Figure 3 shows the results of the analytical and the simulation modeling of the blocking probability in the three above defined systems. The solid line represents the results of the calculations for two classes of calls ($c_1 = 10\,\text{kbps}$, $c_3 = 30\,\text{kbps}$) in the system $\sum M / \sum M / 100 / 300 / \sum sdFIFO$, whereas the results of the simulation are denoted by appropriate symbols. Figure 4 presents the results of the analytical and the simulation modeling of the average length of the queue, expressed in the number of streams, for two extreme classes of calls ($c_1 = 10\,\text{kb/s}$, $c_3 = 30\,\text{kb/s}$). The solid lines represent the results of the analytical calculations in the system $\sum M / \sum M / 100 / 300 / \sum sdFIFO$. The results of the simulation experiments related to the systems $\sum M / \sum M / 100 / 300 / \sum constFIFO$ are denoted by appropriate symbols. Figures 3 and 4 do not show the values of the 95 % confidence intervals because they are lower than the values of the symbols used. The results of the analytical and the simulation modeling are expressed in relation to the average traffic offered to one BBU in the system:

$$\alpha = \frac{\sum_{i=1}^{M} A_i}{C} . \tag{31}$$

The blocking probability in a multiservice state-dependent queueing system is to be found between the values related to state-independent queueing systems (Fig. 3). This probability is the highest in the system with equal resource sharing allocation scheme in the server and the lowest in the system with proportional sharing. Such a representation of the results is the consequence of the available resources of the server. In the system $\sum M / \sum M / 100 / 300 / \sum constFIFO$ with fixed sharing, the oldest class has 1/3 of the resources at its disposal, whereas in the system with the proportional sharing this class has 1/2 of available resources. As for the blocking probability for the youngest class of calls ($c_1 = 10\,\text{kbps}$), it is the highest in the system with the proportional sharing of the resources of the server (1/6 of available resources) and the lowest in the system with equal sharing of resources (1/3 of available resources). The adopted division of resources is also decisive in the representation of the results related to the lengths of queues (Fig. 4). The highest blocking probability for the oldest class of calls in the system $\sum M / \sum M / 100 / 300 / \sum constFIFO$ with equal server resources

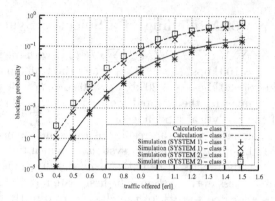

Fig. 3. The blocking probability in queueing systems

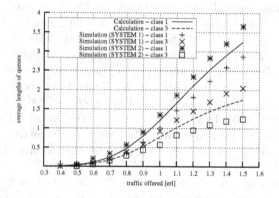

Fig. 4. Average lengths of queues in queueing systems

sharing results in a great number of rejected calls, which in consequence leads to a queue with the shortest length as compared with the queues of other analyzed systems. The identical reasoning can be applied to the youngest class of calls.

7   Conclusions

The paper presents a model of a multiservice state-dependent queueing system with limited queue and state-dependent dynamic resource sharing of the server between individual classes of calls. The advantage of the proposed model is the possibility to evaluate the average parameters of queues for individual classes of calls, which may prove to be of particular importance in engineering applications, especially in solutions concerning the analysis, dimensioning and optimization of 4G mobile networks. The proposed model can be taken as a basis for further research on multiservice queueing systems with finite and infinite buffers, different queue service disciplines, different scenarios for the division of the resources of the server and different traffic management mechanisms implemented in the functions that provide access to the system.

Acknowledgement. This work has been partially supported by a grant from Switzerland through the Swiss Contribution to the enlarged European Union (PSPB-146/2010, CARMNET).

References

1. Bonald, T., Roberts, J.: Internet and the Erlang Formula. ACM Computer Communications Review 42, 23–30 (2012)
2. de Vega, R., Pleich, R.: Validation and extension of the M/G/R Processor Sharing to dimension elastic traffic in TCP/IP networks. In: 2nd Polish-German Teletraffic Symposium, Gdansk, pp. 147–156 (2002)
3. Stasiak, M., Głąbowski, M., Wiśniewski, A., Zwierzykowski, P.: Modeling and Dimensioning of Mobile Networks. From GSM to LTE. A John Wiley and Sons, Ltd., Publication, London (2011)
4. Bonald, T., Virtamo, J.: A recursive formula for multirate systems with elastic traffic. IEEE Comm. Letters 9, 753–755 (2005)
5. Kaufman, J.: Blocking in a shared resource environment. IEEE Transactions on Communications COM-29, 1474–1481 (1981)
6. Roberts, J.: A service system with heterogeneous user requirements – application to multi-service telecommunications systems. In: Performance of Data Communications Systems and their Applications, Amsterdam, pp. 423–431 (1981)
7. Holma, H., Toskala, A.: WCDMA for UMTS: HSPA Evolution and LTE. A John Wiley and Sons, Ltd., Publication, London (2010)
8. Roberts, J.: An Erlang formula for the Internet. In: The First European Teletraffic Seminar (invited talks), Poznan (2011)
9. Roberts, J., Mocci, V., Virtamo, J. (eds.): Broadband Network Teletraffic, Report of Action COST 242. Commission of the European Communities. Springer (1996)
10. Aein, J.M.: A multi-user-class, blocked-calls-cleared, demand access model. IEEE Transactions on Communications COM-26, 378–385 (1978)
11. Inose, H.: An introduction to digital integrated communications systems. University of Tokyo Press, Tokyo (1979)
12. Gimpelson, L.A.: Analysis of mixture of wide-and narrow-band traffic. IEEE Transactions on Communications COM-13, 258–266 (1965)
13. Iversen, V.B.: Teletraffic Engineering Handbook,
    http://www.itu.int/ITU-D/study_groups/
    SGP_1998-2002/SG2/StudyQuestions/Question_16/
14. Bonald, T., Proutiere, A.: Insensitive bandwidth sharing in data networks. Queueing Systems 44, 69–100 (2003)
15. Rácz, S., Gerö, B., Fodor, G.: Low level performance analysis of a multi-service system supporting elastic and adaptive services. Journal of Performance Evaluation 49, 451–469 (2002)
16. Kallos, G., Vassilakis, V., Moscholios, I., Logothetis, M.: Performance modelling of W-CDMA networks supporting elastic and adaptive traffic. In: 4th International Working Conference on Performance Modelling and Evaluation of Heterogeneous Networks, Ilkley (2006)
17. Stasiak, M., Wiewióra, J., Zwierzykowski, P., Parniewicz, D.: Analytical Model of Traffic Compression in the UMTS network. In: Bradley, J.T. (ed.) EPEW 2009. LNCS, vol. 5652, pp. 79–93. Springer, Heidelberg (2009)

Scheduler for Virtualization of Links with Partial Performance Isolation

Tomasz Fortuna and Andrzej Chydzinski

Silesian University of Technology
Institute of Informatics
Akademicka 16, 44-100, Gliwice, Poland
{tomasz.fortuna,andrzej.chydzinski}@polsl.pl
http://inf.polsl.pl/

Abstract. In this paper we study a scheduler for creating several virtual links on one physical link. The scheduler provides partial isolation of the performance between the virtual links in a sense that the volume of traffic offered to one virtual link may influence the performance of other virtual links, but only up to some extent. Namely, the virtual links have guaranteed boundaries of the performance characteristics, no matter what happens with other links. Contrary to schedulers with full performance isolation, the studied scheduler is partially work conserving – some heavily loaded virtual links may benefit from the light load of other virtual links. We give a description of the scheduler, compute boundaries of its performance characteristics and present the actual performance in several simulated scenarios.

Keywords: virtualization of links, work-conserving scheduler, performance isolation.

1 Introduction

Network virtualization is considered as one of the important steps which can accelerate the development of the Internet. In particular, network virtualization enables a consolidation of several networks or overlays into one, based on a common hardware, physical system [1]. Each virtual network can be designed for applications of different types, with distinct QoS requirements. Using virtualized networking resources the users are able to easily create their own networks, with a desired topology, routing, QoS etc. A direct benefit of network virtualization is a reduction of hardware and maintenance costs, compared to the separate, non-virtualized systems.

Due to these reasons, one can observe an intensive world-wide research activity on network virtualization. In particular, several large scientific projects on future Internet in Europe, USA and Japan (FIA MANA, AKARI, PASSIVE FP7, GENI, IIP, to name a few) have incorporated network virtualization into their scope.

In this paper we deal with virtualization of links, which is achieved by exploiting two mechanisms. Firstly, a packet marking (e.g. using a special header)

A. Kwiecień, P. Gaj, and P. Stera (Eds.): CN 2013, CCIS 370, pp. 446–455, 2013.
© Springer-Verlag Berlin Heidelberg 2013

is required for their proper classification in nodes. Secondly, a mechanism for scheduling packet transmission from several virtual interfaces on the single physical interface is needed, so that several virtual links could share a single physical link. While the parameters of the physical link (throughput, propagation delay) cannot be modified, the parameters of the virtual links depend on the design of this scheduler. In other words, the scheduling algorithm is responsible for the performance characteristics of the virtual links.

There are several types of scheduling algorithms that can be used for this purpose. An important class o such algorithms are those with full performance isolation. We say that a scheduler provides full performance isolation if and only if the type and rate of traffic offered to one virtual link has totally no influence on the behaviour of all the remaining virtual links.

For instance, a scheduler with a full performance isolation, proposed in [2], has been used in the design of the IIP System [3], a novel Future Internet architecture, implemented using the virtualization devices described in [4]. In this scheduler the physical link attends (in a cyclic way) all the queues associated with virtual links, each queue for a constant time. The attendance time remains unaltered no matter what traffic is offered to the served link and no matter what traffic is offered to other links.

The main advantage of a scheduler with a full performance isolation is that the obtained virtual links have an unaltered, predefined performance, which can be used easily by virtual networks with QoS requirements. However, the main disadvantage connected with such approach is that the scheduler is completely non-work conserving, i.e. in the case when there is no traffic on one link and a heavy traffic on the second link, the output link is idle when the scheduler attends the first queue, even though it could potentially transmit packets from the second queue.

In this paper we will study a scheduler that is partly work-conserving. Namely, some heavily loaded virtual links may benefit from a low traffic intensity of other virtual links. Naturally, such design breaks the full performance isolation between the links. However, as will be shown, some boundaries for the performance characteristics will be preserved. Therefore, resulting virtual links can still be useful for QoS architectures.

The rest of the paper is organized as follows: In Section 2, a detailed description of the scheduler model is given. In Section 3, the analytical background is presented. In particular, in Sect. 3.1 the related work is discussed and the most relevant analytical results are quoted, while in Sect. 3.2 our analysis of the performance guarantees of the model are shown. Then, in Sect. 4, the performance of the virtual links in 15 chosen scenarios are presented. Finally, remarks concluding the paper are gathered in Sect. 5.

2   Description of the Model

In Figure 1 the model of the partially work-conserving scheduler is depicted. There are N distinct streams of packets coming from N virtual interfaces, which arrive at N separate buffers of sizes $b_1, \cdots, b_N$ packets, and form N separate queues in these buffers. It is assumed that all of the arrival streams are Poisson, with rates $\lambda_1, \ldots, \lambda_N$, respectively. If upon a packet arrival the appropriate buffer is full, the packet is dropped and lost.

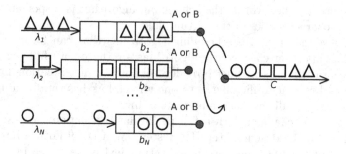

Fig. 1. The model of the scheduler

There is also a physical link of capacity C bits/s, which serves all the queues in a cyclic manner. With each queue, say j-th, a number, W_j, is associated and a type, which can be A or B. The number W_j denotes the maximum time that the physical link can spend serving the j-th queue before moving to the next queue. If a queue is of A type, the physical link serves this queue for exactly W_j time, no matter what is the buffer content (it may be empty). If a queue is of B type, the physical link attends the queue for no longer than W_j time, but, if during this W_j time the buffer becomes empty, the physical link immediately moves to the next queue. In other words, in the A-type queues the work (transmission) phase length is unaltered, while in the B-type queues the work phase length may be shortened if there are no packets for transmission.

Therefore, in order to parameterize the scheduler, we have to give all b_i's, W_i's and types of all links. For instance, we may have a scheduler for five virtual links of type ABBAB.

In both types of links, A and B, it is assumed that the packets cannot be split. This means that if the work phase W_j ends during a transmission of a packet, this packet has to remain in the buffer for the next cycle, and the whole packet will be transmitted again in the next work phase. (Alternatively, the transmission of a packet may be prevented if there is not enough time to complete it in the present phase). This assumption, necessary in packet networking, causes some loss of the throughput of the physical link – the total throughput of all physical links will be less than the physical link throughput.

As we can notice, the idea of A-type links comes from the already mentioned IIP System scheduler – a pure A-type scheduler (A...A) is the same as the one proposed in [2]. The idea of the B-type link comes from the exhaustive, time-limited disciplines studied with vacation queueing models, e.g. [5].

In this paper we will focus mainly on the mixed type, containing A and B-type links. The most suitable cases for application of the scheduler of the mixed type are when A-type links are heavily loaded while B-type links are usually lightly loaded, with occasional traffic bursts. As we may expect, the A-type links will benefit in such situations.

We will compare also the performance of the mixed type with the pure A type. We will not deal with the pure B-type scheduler. It is also a very interesting type, but some of its properties make its analysis more difficult (e.g. the cycle length may be arbitrary short in a pure B type).

Finally, note that we use Poisson arrivals herein, which can mimic neither the self-similarity nor the long-range dependence. Unfortunately, even the model without the self-similar traffic is complex and hard to analyze (see the next section). Adding the self-similarity would make it totally intractable.

3   Analytical Background

3.1   Related Work

The pure A-type schedulers have been analyzed in [6], where approximate results on the response time and buffer dimensioning can be found.

The analytical models that could be potentially useful in analysis of the mixed, AB-type scheduler fall into two categories: polling systems and vacation queues. In a polling model, the behaviour of all the queues is analyzed together, in a one model, while in the vacation models only one queue is analyzed (from the point of view of this particular queue, the time when the server is visiting other queues is a vacation period). There is a vast literature both on polling systems and vacation queues. For the bibliography on polling systems we refer the reader to [7], while for the bibliography on vacation queues to [8].

Unfortunately, according to the best of the authors' knowledge, there are no analytical results on the model of the scheduler of AB type. Some papers, which are devoted to vacation queues with time-limited service, [5, 9–11], come close to this model, but not very close. The time-limited service means that the server cannot attend the queue for longer that some constant time, called the maximum service time (MSA). Moreover, it assumed in [5, 9–11] that the service of a queue ends when the queue becomes empty. Therefore, these two assumptions are consistent with the model studied here. However, the other assumptions are different.

Firstly, the infinite buffering space is assumed in [5, 9–11]. We assume finite buffers, which is a more realistic assumption when thinking of storing packets in real networking devices. Secondly, the preemptive-resume discipline is assumed in [5, 9, 10], which means that the service of the job being served at the MSA-time-out can be preempted and resumed from the interrupted point in the next cycle. In contrast to this, we assume that service of the interrupted job must start from the beginning in the next service period.

Finally, in [11] the MSA time is an exponentially distributed random variable, which is in contrast to the constant MSA time considered herein.

The closest analyzed model, to the AB-type, is that of [9], which we will summarize now briefly. Namely, an M/G/1 queue with vacations is considered, with λ denoting the arrival rate, $f_x(t)$ denoting the density of the service time, $f_v(t)$ denoting the density of length of the vacation period and T_m denoting the MSA time. The preemptive-resume discipline is assumed. As shown in [9], this model can be solved by solving the system of equations:

$$
\begin{cases}
F_p^*(s) = V^*(\lambda - \lambda X^*(s))\, e^{(s-\lambda+\lambda X^*(s))T_m} \\
\qquad \cdot \left[F_p^*(s) - (s - \lambda + \lambda X^*(s)) \int_0^{T_m} e^{-(s-\lambda+\lambda X^*(s))y} P_0(y)\,dy \right] , \\
P_0^*(s) = F_p^*(s + \lambda - \lambda B^*(s)) , \\
B^*(s) = X^*(s + \lambda - \lambda B^*(s)) ,
\end{cases}
\tag{1}
$$

where $X^*(s)$ is the Laplace transform of the service distribution, $V^*(s)$ is the Laplace transform of the vacation period distribution, $P_0(t)$ is the probability that the queue is empty at time t (assuming that zero is an absorbing state) and $P_0^*(s)$ is the Laplace transform of $P_0(t)$. The unknown $F_p^*(s)$ denotes the Laplace transform of distribution of the amount of work (workload) at the beginning of the work phase. As system (1) is very hard to solve analytically, an approximate solution was proposed in [9]. It exploits an approximation of $P_0(t)$ by a weighed sum of Laguerre functions, i.e.: $P_0(t) \approx 1 - \sum_{n=0}^{N} a_n e^{-t/2T} L_n(t/T)$, where T is a time-scaling factor, the a_n's are unknown coefficients and L_n is the Laguerre polynomial of degree n. Using this approximation gives:

$$
P_0^*(s) \approx 1 - \sum_{n=0}^{N} a_n \frac{s(s - \frac{1}{2T})^n}{(s + \frac{1}{2T})^{n+1}} ,
\tag{2}
$$

$$
F_p^*(s) \approx 1 - \sum_{n=0}^{N} a_n \frac{\hat{s}(\hat{s} - \frac{1}{2T})^n}{(\hat{s} + \frac{1}{2T})^{n+1}}, \qquad \hat{s} = s - \lambda + \lambda X^*(s) .
\tag{3}
$$

Substituting (2) and (3) into system (1), the unknown a_n can be found and the final results can be obtained. The amount of work at the beginning of the work phase is

$$
\bar{u}_p = \sum_{n=0}^{N} (-1)^n (2T)(1 - \rho) a_n ,
\tag{4}
$$

where ρ is the offered load to the queue. The Laplace transform for the amount of work at an arbitrary time epoch is

$$
U^*(s) = \frac{(1 - \rho)s}{s - \lambda + \lambda X^*(s)} \cdot \frac{F_p^*(s)}{V^*(\lambda - \lambda X^*(s))} \cdot \frac{1 - V^*(\lambda - \lambda X^*(s))}{(\lambda - \lambda X^*(s))\bar{v}} ,
\tag{5}
$$

where $\bar{v}$ is the average vacation time, while the average amount of work at an arbitrary time is

$$
\bar{u} = \frac{\lambda \bar{x}^2}{2(1 - \rho)} + \bar{u}_p - \rho \bar{v} + \rho \frac{\bar{v}^2}{2\bar{v}} ,
\tag{6}
$$

where $\bar{v}^2$ is the second moment of vacation time, while $\bar{x}^2$ is the second moment of service time.

The presented results of [9] can be used for analysis of the mixed, AB-type scheduler in some special cases only. Firstly, they can be used for analysis of a B-type queue only, and only in the case when all the remaining queues are of A type. The A-type queue is not covered by the model at all. Secondly, the presented results were obtained under an infinite buffer assumption. In real devices we always deal with finite buffers, which may cause packet losses. A finite-buffer queue can be replaced by its infinite-buffer analog only if $\rho << 1$. This condition constitutes another limitation for the applicability of the results of [9] to the AB-type scheduler. Moreover, even in such a case the obtained results would be approximate due to the preemptive-resume discipline considered in [9].

Having said that, it becomes clear that new analytical methods have to be developed in order to carry out a full analysis of the AB type scheduler. One possible approach is a decomposition based on the renewal theory, similar to that of [12, 13]. However, for many practical purposes, it suffices to compute the boundaries for the performance characteristics of interest, instead of computing their distributions or average values. Therefore, in the next subsection the boundaries for the response time and the throughput of the virtual link will be computed.

3.2   Boundaries for the Response Time and Throughput

Recall that the response time, T, is defined as the total time that a packet spends in the scheduler. Therefore it is composed of the waiting time, in an appropriate queue, and the packet service time (transmission).

Let us consider a j-th virtual link, no matter A or B type. In order to find the upper bound for the response time, we will consider the worst case scenario. Such scenario happens when a large packet (of MTU size) arrives to the nearly full buffer, at the end of the work phase, and all the packets present in the buffer are large. In particular, there must be exactly $b_j - 1$ packets of MTU size in the buffer and the packet must arrive $MTU/C - \epsilon$ time before the end of the current work phase. Now we will calculate the response time of such packet. Here and subsequently by $\lfloor x \rfloor$ we will denote the largest integer not greater than x (floor), while by $\lceil x \rceil$ – the smallest integer not less than x (ceiling).

The number of packets of MTU size that can be transmitted during the work phase of the j-th link is

$$\left\lfloor \frac{W_j \cdot C}{MTU} \right\rfloor . \tag{7}$$

Therefore, the number of full cycles of the scheduler before the work phase, during which the worst-case packet will be served, is

$$\left\lceil b_j \Big/ \left\lfloor \frac{W_j \cdot C}{MTU} \right\rfloor \right\rceil - 1 . \tag{8}$$

Now, the number of packets left for the last work phase, during which the worst-case packet will be served, is:

$$b_j - \left(\left\lceil b_j \Big/ \left\lfloor \frac{W_j \cdot C}{MTU} \right\rfloor \right\rceil - 1 \right) \cdot \left\lfloor \frac{W_j \cdot C}{MTU} \right\rfloor . \tag{9}$$

Finally, adding the vacation time before the last work phase, $\sum_{i \neq j} W_i$, and adding the time before the first vacation phase, MTU/C, we obtain the response time for the worst-case scenario packet:

$$
T_{\max} = \frac{MTU}{C} + \left(\left\lceil b_j / \left\lfloor \frac{W_j \cdot C}{MTU} \right\rfloor \right\rceil - 1 \right) \sum_{i=1}^{N} W_i + \sum_{i \neq j} W_i
$$
$$
+ \left(b_j - \left(\left\lceil b_j / \left\lfloor \frac{W_j \cdot C}{MTU} \right\rfloor \right\rceil - 1 \right) \cdot \left\lfloor \frac{W_j \cdot C}{MTU} \right\rfloor \right) \frac{MTU}{C} . \tag{10}
$$

Naturally, formula (10) constitutes also the upper bound for the response time of j-th link. Note that if the response time is limited by (10), the variance of the response time (delay jitter) is also bounded, and the boundary can be easily computed.

In the next step we will calculate the lower bound of the throughput of the j-th virtual link under a heavy load. This bound is easy to obtain considering another worst-case scenario. It happens when a packet departure occurs $MTU/C - \epsilon$ time before the end of the work phase and there is another waiting packet of MTU size, which cannot be transmitted in the present work phase. In this way MTU/C time of each work phase can be wasted. This leads to the lower bound for the throughput of the heavily loaded j-th virtual link:

$$
\gamma_{\min} = \frac{W_j - MTU/C}{\sum_{i=1}^{N} W_i} . \tag{11}
$$

Now the partial performance isolation idea can be expressed in terms of (10) and (11): the traffic of one or more virtual links can influence the behaviour of other links, but under a heavy load the performance of none of the links can get worse than (10) and (11). Moreover, as it will be demonstrated, the A-type links can benefit from light traffic on the B-type links.

4   Results

Now we will present sample results for the schedulers creating two and three virtual links on a physical link of capacity $C = 1\,\mathrm{Gb/s}$. Three performance measures will be presented: the throughput of the virtual links, the average response time and the maximum response time. All the buffer sizes are set to 20 packets, all the work phases are set to $30\,\mu s$, all arriving packets are 500 bytes long. In each experiment, 30 s of the scheduler work is simulated using the newest OMNeT++ simulator, ver. 4.2.2 [14].

In Table 1 the obtained performance of virtual links in 15 distinct scenarios is presented. In particular, in scenarios 1–7 two virtual links were created on the physical link, while in the scenarios 8–15, three virtual links were created. For two virtual links, performance boundaries calculated by means of (10) and (11) are

$$
\gamma_{\min} = 433.33\,\mathrm{Mb/s} \ , \quad T_{\max} = 178\,\mu s \ ,
$$

per each link, while for three links are

$$\gamma_{\min} = 288.88\,\text{Mb/s} , \quad T_{\max} = 268\,\mu\text{s} .$$

As we see in Table 1, these boundaries are preserved. Now we can study the benefits from application of mixed type schedulers. They are especially visible in scenarios 5 and 12, when a heavy traffic is offered to the A-type link and a light traffic to the B-type link (links). A high throughput on the A-type link is obtained, without any influence on the B-type throughputs. Moreover, the obtained average response time is better in the mixed case than in the pure A-type case. (However, the maximum response time can be slightly higher in the mixed case – compare scenarios 2 and 5).

Table 1. The performance of the virtual links

	Link type			Input rate [Mb/s]			Output rate [Mb/s]			Avg T [μs]			Max T [μs]		
No.	1	2	3	1	2	3	1	2	3	1	2	3	1	2	3
1	A	A		500	500		463.90	463.85		117	117		176	176	
2	A	A		800	200		466.66	199.73		161	17		176	90	
3	A	B		500	500		466.27	463.93		113	117		176	176	
4	B	A		800	200		466.66	199.73		161	17		176	90	
5	A	B		800	200		740.15	199.78		73	20		176	94	
6	A	B		200	200		199.72	199.83		8	20		69	97	
7	A	B		100	100		99.79	99.96		5	19		39	63	
8	A	A	A	333	333	333	308.91	309.07	308.98	171	172	172	266	266	266
9	A	A	A	700	200	100	311.11	199.70	99.96	244	37	30	266	222	122
10	A	A	A	100	100	100	99.95	99.87	99.87	30	30	30	130	143	149
11	A	B	B	333	333	333	311.16	310.24	310.06	163	168	168	266	266	266
12	A	B	B	700	200	100	645.70	199.70	99.96	82	23	24	236	122	102
13	B	B	A	700	200	100	372.23	199.75	100.00	202	39	23	266	226	112
14	B	A	B	700	200	100	420.05	199.98	99.82	177	21	34	258	135	123
15	A	B	B	100	100	100	99.95	99.87	99.87	7	21	21	61	83	89

Another positive effect of the application of the mixed type scheduler can be observed in scenario 14. In this scenario, one B-type link benefits from a low traffic on another B type link.

Finally, a detailed comparison of the performance of the first link in schedulers of the AA type and the AB type can be observed in Figs. 2–4. They are based on the following scenarios: the arrival rate of the second link was constant and equal to 200 Mb/s, while the arrival rate of the first link varied from 10 Mb/s to 1 200 Mb/s. In Figure 2, the resulting output rate of the first link is depicted, in Fig. 3 – the maximum response time, while in Fig. 4 – the average response time of the first link. As we can see, in such scenario all presented characteristics are better in the AB-type case, in the whole range of the arrival rate.

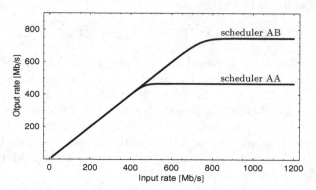

Fig. 2. The output rate of the first virtual link for schedulers AA and AB versus the input rate of this link. The input rate on the second link is of 200 Mb/s.

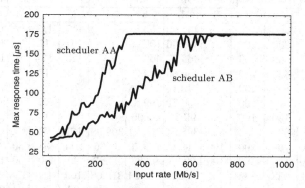

Fig. 3. The observed maximum response time of the first virtual link for schedulers AA and AB versus the input rate of this link. The input rate on the second link is of 200 Mb/s.

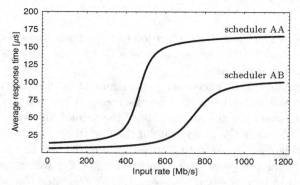

Fig. 4. The average response time of the first virtual link for schedulers AA and AB versus the input rate of this link. The input rate on the second link is of 200 Mb/s.

5   Conclusions

We studied a scheduler for virtualization of links with partial performance isolation, i.e. with guaranteed boundaries for the main performance characteristics of the virtual links. We computed these boundaries and showed the actual performance of the scheduler in several simulated scenarios. The scheduler proved to provide a benefit for some heavily loaded links when the load of the other links is low.

Acknowledgement. This work was partially supported by the Polish National Science Centre under Grant No. N N516 479240.

References

1. Tutschku, K.: Towards the Future Internet: virtual networks for convergent services. Elektrotechnik und Informationstechnik 126(7-8), 250–259 (2009)
2. Burakowski, W., et al.: Ideal device supporting virtualization of network infrastructure in System IIP. In: Proc. of KSTiT 2011, Lodz, Poland, pp. 818–823 (September 2011) (in Polish)
3. Burakowski, W., et al.: Virtualized network infrastructure supporting co-existence of Parallel Internets. In: Proc. of SNPD 2012, Kyoto, pp. 679–684 (August 2012)
4. Chydzinski, A., et al.: Virtualization Devices for Prototyping of Future Internet. In: Proc. of SNPD 2012, pp. 672–678 (August 2012)
5. Leung, K.K., Eisenberg, M.: A Single Server Queue with Vacations and Gated Time-Limited Service. IEEE Transactions on Communications 38(9), 1454–1462 (1990)
6. Sosnowski, M., Burakowski, W.: Analysis of the system with vacations under Poissonian input stream and constant service times. In: Proc. of Polish Teletraffic Symposium, Zakopane, pp. 9–13 (December 2012)
7. Boon, M., van der Mei, R., Winands, E.: Applications of polling systems. Surveys in Operations Research and Management Science 16, 67–82 (2011)
8. Tian, N., Zhang, Z.G.: Vacation Queueing Models – Theory and Applications. Springer, New York (2006)
9. Leung, K.K., Eisenberg, M.: A single-server queue with vacations and non-gated time-limited service. Performance Evaluation 12(2), 115–125 (1991)
10. Takagi, H., Leung, K.K.: Analysis of a discrete-time queueing system with time-limited service. Queueing Systems 18(1-2), 183–197 (1994)
11. Katayama, T.: Waiting time analysis for a queueing system with time-limited service and exponential timer. Naval Research Logistics 48(7), 638–651 (2001)
12. Chydzinski, A.: The M/G-G/1 oscillating queueing system. Queueing Systems 42(3), 255–268 (2002)
13. Chydzinski, A.: The oscillating queue with finite buffer. Performance Evaluation 57(3), 341–355 (2004)
14. http://www.omnetpp.org

Biometric Voice Identification Based on Fuzzy Kernel Classifier

Adam Dustor and Piotr Kłosowski

Silesian University of Technology, Institute of Electronics
Akademicka 16, Gliwice, Poland
{adam.dustor,piotr.klosowski}@polsl.pl

Abstract. This paper presents research on automatic speaker identification based on structural risk minimization and kernel functions. New approach, known as a Fuzzy Kernel Ho-Kashyap classifier FKHK, to speaker identification was applied. Instead of the most popular kernel functions like gaussian or polynomial, data dependent kernel matrix which may be interpreted in terms of linguistic values from the premises of if-then rules was applied. Classifier was tested on polish speech corpora ROBOT and obtained results were discussed.

Keywords: biometrics, security, speaker identification, voice identification.

1 Introduction

Division of Telecommunication, a part of the Institute of Electronics and Faculty of Automatic Control, Electronics and Computer Science of the Silesian University of Technology, for many years specializes in speech and speaker recognition [1–3]. This paper is devoted to speaker identification which is the process of automatically recognizing who is speaking by analysis speaker-specific information included in spoken utterances.

Speaker identification is a behavioral biometrics which means that physical and mental state of the speaking person has great impact on recognition accuracy [4]. As a result robust pattern recognition methods, which could reduce this undesired behaviour are searched for. Since speaker identification is based on similarity calculation between test utterance and reference models, the problem of a speaker model construction is crucial. Usually speaker models are divided into generative and discriminative.

Generative models are probability density estimators that attempt to capture all of the underlying fluctuations and variations of the speaker's voice. These models include Gaussian mixture models GMM [5, 6] and the broad family of nearest neighbor classifiers based on vector quantization VQ techniques such as k-means or LBG algorithm.

Discriminative models are optimized to minimize the error on a set of training samples. Discriminative approach should theoretically yield better performance than generative [7]. From this category an application in speaker recognition

A. Kwiecień, P. Gaj, and P. Stera (Eds.): CN 2013, CCIS 370, pp. 456–465, 2013.
© Springer-Verlag Berlin Heidelberg 2013

found support vector machines SVM [7], kernel Ho-Kashyap KHK [8, 2] and Fuzzy Ho-Kashyap FHK classifier [9, 10].

The paper is organized in the following way. At first fuzzy Ho-Kasyap FHK classifier based on fuzzy if-then rules and classical Ho-Kashyap procedure is discussed [3]. Next transformation to its kernel version denoted as a FKHK (fuzzy kernel Ho-Kashyap) is shown, where kernel matrix posses linguistic interpretation. At last achieved speaker identification results on Polish speech corpora ROBOT are shown.

2   FHK Classifier

Fuzzy Ho-Kashyap classifier FHK is designed on the basis of the training set, $Tr = \{(\mathbf{x}_1, y_1), \ldots, (\mathbf{x}_N, y_N)\}$ where $\mathbf{x}_i \in \mathcal{R}^t$ is a feature vector extracted from a frame of speech, N is the number of vectors and $y_i \in \{-1, +1\}$ indicates the assignment to one of two classes ω_1 or ω_2. After defining the augmented vector $\mathbf{x}_i' = [\mathbf{x}_i^T, 1]^T$ the decision function of the classifier can be defined as

$$g(\mathbf{x}_i) = \mathbf{w}^T \mathbf{x}_i' \begin{cases} \geq 0, & \mathbf{x}_i \in \omega_1 \\ < 0, & \mathbf{x}_i \in \omega_2 \end{cases} \tag{1}$$

where $\mathbf{w} = [\widetilde{\mathbf{w}}^T, w_0]^T \in \mathcal{R}^{t+1}$ is a weight vector which must be found during training of the classifier. After multiplying by -1 all patterns from ω_2 class the Equation (1) can be rewritten in the form $y_i \mathbf{w}^T \mathbf{x}_i' > 0$ for $i = 1, 2, \ldots, N$. Let $\mathbf{X}$ be the $N \times (t+1)$ matrix

$$\mathbf{X} = \begin{bmatrix} y_1 \mathbf{x}_1'^T \\ y_2 \mathbf{x}_2'^T \\ \vdots \\ y_N \mathbf{x}_N'^T \end{bmatrix}, \tag{2}$$

then (1) can be written in the matrix form $\mathbf{Xw} > 0$. To obtain solution $\mathbf{w}$ this inequality is replaced by $\mathbf{Xw} = \mathbf{b}$ where $\mathbf{b} > 0$ is an arbitrary vector called a classifier margin. If data are linearly separable then all components of error vector $\mathbf{e} = \mathbf{Xw} - \mathbf{b}$ are greater than zero. As a result the misclassification error can be approximated by

$$J(\mathbf{w}, \mathbf{b}) = \sum_{i=1}^{N} (e_i)^2 . \tag{3}$$

Vectors $\mathbf{w}$ and $\mathbf{b}$ are found by minimization the function [9]

$$J(\mathbf{w}, \mathbf{b}) = (\mathbf{Xw} - \mathbf{b})^T \mathbf{D}(\mathbf{Xw} - \mathbf{b}) + \tau \widetilde{\mathbf{w}}^T \widetilde{\mathbf{w}} , \tag{4}$$

where matrix $\mathbf{D} = \operatorname{diag}(d_1, d_2, \ldots, d_N)$ and d_i is the weight corresponding to the i-th pattern, which can be interpreted as a reliability attached to this pattern. The second term of (4) with regularization constant $\tau > 0$ is responsible for the minimization of the complexity of the classifier.

Differentiation of (4) with respect to **w** and **b** and setting the results to zero yields the conditions [9]

$$\begin{cases} \mathbf{w} = (\mathbf{X}^T \mathbf{D} \mathbf{X} + \tau \widetilde{\mathbf{I}})^{-1} \mathbf{X}^T \mathbf{D} \mathbf{b}, \\ \mathbf{e} = \mathbf{X} \mathbf{w} - \mathbf{b} = \mathbf{0}, \end{cases} \tag{5}$$

where $\widetilde{\mathbf{I}}$ is the identity matrix with the last element on the main diagonal set to zero. Solving (5) leads to the absolute error minimization procedure for classifier design which can be summarized in the following steps [9]:

1. fix $\tau \geq 0$, $0 < \eta < 1$, $\mathbf{D}^{[1]} = \mathbf{I}$, $\mathbf{b}^{[1]} > \mathbf{0}$, $k = 1$;
2. $\mathbf{w}^{[k]} = (\mathbf{X}^T \mathbf{D}^{[k]} \mathbf{X} + \tau \widetilde{\mathbf{I}})^{-1} \mathbf{X}^T \mathbf{D}^{[k]} \mathbf{b}^{[k]}$;
3. $\mathbf{e}^{[k]} = \mathbf{X} \mathbf{w}^{[k]} - \mathbf{b}^{[k]}$;
4. $d_i = 1/|e_i|$ for $i = 1, 2, \ldots, N$; $\mathbf{D}^{[k+1]} = \mathrm{diag}\,(d_1, d_2, \ldots, d_N)$;
5. $\mathbf{b}^{[k+1]} = \mathbf{b}^{[k]} + \eta \left(\mathbf{e}^{[k]} + |\mathbf{e}^{[k]}| \right)$;
6. if $||\mathbf{b}^{[k+1]} - \mathbf{b}^{[k]}||_2 > \xi$ then $k = k + 1$ and go to step 2, otherwise stop.

If step 4 is omitted then criterion (3) is minimized. Procedure is convergent to local optimum [9] and leads to the linear discriminant function $g(\mathbf{x}) = \mathbf{w}^T \mathbf{x}'$ that minimizes absolute (or squared) error.

Much better classification results should be obtained for nonlinear classifier constructed using I linear discriminant functions $g_i(\mathbf{x}) = \mathbf{w}^{(i)T} \mathbf{x}'$, $i = 1, 2, \ldots, I$. The input space is softly partitioned into I regions. Each function $g_i(\mathbf{x})$ is found by minimization the criterion [9]

$$J^{(i)} \left(\mathbf{w}^{(i)}, \mathbf{b}^{(i)} \right) = \mathbf{A}^T \mathbf{D}^{(i)} \mathbf{A} + \tau \widetilde{\mathbf{w}}^{(i)T} \widetilde{\mathbf{w}}^{(i)} , \tag{6}$$

where $\mathbf{A} = \mathbf{X} \mathbf{w}^{(i)} - \mathbf{b}^{(i)}$ and **D** equals

$$\mathbf{D}^{(i)} = \mathrm{diag} \left(\frac{F^{(i)}(\mathbf{x}_1)}{|e_1^{(i)}|}, \ldots, \frac{F^{(i)}(\mathbf{x}_N)}{|e_N^{(i)}|} \right) , \tag{7}$$

and

$$e_k^{(i)} = \mathbf{w}^{(i)T} \mathbf{x}_k' - b_k^{(i)} . \tag{8}$$

Parameters $F^{(i)}(\mathbf{x}_n)$ denote the membership function of the pattern $\mathbf{x}_n$ from the training set to the i-th region and $\mathbf{b}^{(i)}$ is the margin vector for the i-th classifier.

The patterns from each class ω_1 and ω_2 are softly partitioned into I regions by the fuzzy c-means algorithm. Each cluster is represented parametrically by the Gaussian membership function with center $\mathbf{c}^{(i)(j)}$

$$\mathbf{c}^{(i)(j)} = \frac{\displaystyle\sum_{n=1}^{N_j} u_{in}^{(j)} \mathbf{x}_n}{\displaystyle\sum_{n=1}^{N_j} u_{in}^{(j)}} , \tag{9}$$

and dispersion $s^{(i)(j)}$

$$s^{(i)(j)} = \frac{\sum\limits_{n=1}^{N_j} u_{in}^{(j)} \left[\mathbf{x}_n - \mathbf{c}^{(i)(j)}\right]^{(\bullet 2)}}{\sum\limits_{n=1}^{N_j} u_{in}^{(j)}}, \tag{10}$$

where $i = 1, 2, \ldots, I$ is a cluster index and $j \in \{1, 2\}$ is a class index, $(\bullet 2)$ denotes component by component squaring, N_j number of patterns from class ω_j and $u_{in}^{(1)}$, $u_{in}^{(2)}$ are elements of fuzzy partition matrices for class ω_1, ω_2 obtained by fuzzy c-means algorithm.

Next, I nearest pairs of clusters belonging to different classes are found. As a distance measure between centers (prototypes) the L_1 norm is used. At first, the nearest pair of clusters from ω_1 and ω_2 is found. This pair is used to construct the first linear decision function $g_1(\mathbf{x}) = \mathbf{w}^{(1)T}\mathbf{x}'$. Subsequently, after exclusion of found clusters from the set of clusters, the next nearest pair is found. Finally I nearest pairs of clusters are found. Each pair is defined by four parameters $\mathbf{c}^{(i)(1)}$, $\mathbf{s}^{(i)(1)}$, $\mathbf{c}^{(i)(2)}$, $\mathbf{s}^{(i)(2)}$.

The membership function $F^{(i)}(\mathbf{x}_n)$ is calculated using the algebraic product as the t-norm and the maximum operator as the s-norm [9]

$$F^{(i)}(\mathbf{x}) = \max \left\{ \exp\left[-\frac{1}{2}\sum_{j=1}^{t} \frac{\left(x_j - c_j^{(i)(1)}\right)^2}{s_j^{(i)(1)}}\right], \right.$$
$$\left. \exp\left[-\frac{1}{2}\sum_{j=1}^{t} \frac{\left(x_j - c_j^{(i)(2)}\right)^2}{s_j^{(i)(2)}}\right]\right\}. \tag{11}$$

The final decision of the nonlinear classifier for the pattern $\mathbf{x}$ is obtained by the weighted average

$$g(\mathbf{x}) = \frac{\sum\limits_{i=1}^{I} F^{(i)}(\mathbf{x})\mathbf{w}^{(i)T}\mathbf{x}'}{\sum\limits_{i=1}^{c} F^{(i)}(\mathbf{x})}. \tag{12}$$

This classifier can be named a mixture-of-experts classifier and works in a similar way as a Takagi-Sugeno-Kang fuzzy inference system. Unfortunately this means that none of the fuzzy consequents can be applied to decision process. This problem can be solved substituting moving singletons for fuzzy moving consequents. Position of the fuzzy sets in consequents of the If-Then rules depends on the input crisp values. As a result nonlinear discriminative function is obtained

$$g(\mathbf{x}) = \frac{\sum\limits_{i=1}^{I} \mathcal{G}\left(F^{(i)}(\mathbf{x}), w^{(i)}\right) \mathbf{w}^{(i)T}\mathbf{x}'}{\sum\limits_{i=1}^{I} \mathcal{G}\left(F^{(i)}(\mathbf{x}), w^{(i)}\right)}, \tag{13}$$

where $\mathcal{G}\left(F^{(i)}(\mathbf{x}), w^{(i)}\right)$ depends on applied fuzzy consequent [11] and for the simplicity $w^{(i)} = 1$. Each linear discriminative function $g_i(\mathbf{x})$ is found by minimization criterion (6) where $\mathbf{D}$ is given by the modified Formula (7)

$$\mathbf{D}^{(i)} = \operatorname{diag}\left(\frac{\mathcal{G}\left(F^{(i)}(\mathbf{x}_1), w^{(i)}\right)}{|e_1^{(i)}|}, \ldots, \frac{\mathcal{G}\left(F^{(i)}(\mathbf{x}_N), w^{(i)}\right)}{|e_N^{(i)}|}\right) . \tag{14}$$

Summarizing, the training procedure of this classifier denoted as a FHK (Fuzzy Ho-Kashyap) consists of the following steps:

1. fix type of fuzzy consequents – function $\mathcal{G}$ [11];
2. fix number of if-then rules I and $w^{(i)} = 1$ for $i = 1, 2, \ldots, I$;
3. fuzzy c-means clustering of data belonging to ω_1 and ω_2; compute parameters $\mathbf{c}^{(i)(1)}$, $\mathbf{s}^{(i)(1)}$, $\mathbf{c}^{(i)(2)}$, $\mathbf{s}^{(i)(2)}$ for $i = 1, 2, \ldots, I$ – Equations (9) and (10);
4. find I nearest pairs of clusters, each pair is defined by $\mathbf{c}^{(k)(1)}$, $\mathbf{s}^{(k)(1)}$ and $\mathbf{c}^{(n)(2)}$, $\mathbf{s}^{(n)(2)}$ for $k, n = 1, 2, \ldots, I$;
5. compute $F^{(i)}(\mathbf{x}_k)$ (11) and $\mathcal{G}\left(F^{(i)}(\mathbf{x}_k), w^{(i)}\right)$ [11] for $k = 1, 2, \ldots, N$ and $i = 1, 2, \ldots, I$;
6. fix regularization constant τ;
7. train classifiers $g_i(\mathbf{x}) = \mathbf{w}^{(i)T}\mathbf{x}'$, $i = 1, \ldots, I$ in accordance with absolute error minimization procedure for the given τ and $\mathbf{D}^{(i)}$.

Speaker is represented in a speaker recognition system by the parameters $\mathbf{w}^{(i)}$ and $\mathbf{c}^{(i)(1)}$, $\mathbf{s}^{(i)(1)}$, $\mathbf{c}^{(i)(2)}$, $\mathbf{s}^{(i)(2)}$ for $i = 1, 2, \ldots, I$. Achieved results of speaker recognition for this classifier were shown in [10].

Introducing some expressions, Equation (13) may be reformulated to the more suitable in pattern recognition form. Let us define formulas

$$S^{(i)}(\mathbf{x}) = \frac{\mathcal{G}\left(F^{(i)}(\mathbf{x}), w^{(i)}\right)}{\sum\limits_{k=1}^{I} \mathcal{G}\left(F^{(k)}(\mathbf{x}), w^{(k)}\right)} , \tag{15}$$

$$\mathbf{d}(\mathbf{x}) = \left[S^{(1)}(\mathbf{x})\mathbf{x}'^{T}, S^{(2)}(\mathbf{x})\mathbf{x}'^{T}, \ldots, S^{(I)}(\mathbf{x})\mathbf{x}'^{T}\right]^{T} , \tag{16}$$

$$\mathbf{W} = \left[\mathbf{w}^{(1)T}, \mathbf{w}^{(2)T}, \ldots, \mathbf{w}^{(I)T}\right]^{T} \tag{17}$$

then (13) may be expressed as

$$g(\mathbf{x}) = \mathbf{d}(\mathbf{x})^{T}\mathbf{W} . \tag{18}$$

3   Fuzzy Kernel Ho-Kashyap Classifier FKHK

In pattern recognition techniques based on structural risk minimization and kernel functions in almost most cases the most typical kernel functions like gaussian and polynomial are applied. Function of the kernel in classification process and its influence on generalization abilities is one of the least recognized problems of classification [12]. It may be expected that proper selection of kernel function and kernel matrix, should allow to achieve lower classification error rates in comparison with techniques based on "standard" kernel functions.

This section presents new approach to kernel matrix construction based on its linguistic interpretation. As a result new type of classifier denoted as Fuzzy Kernel Ho-Kashyap FKHK is obtained. Idea of this approach was inspired by paper [13], where it was applied to fuzzy modeling. It can be shown that learning of a fuzzy system based on logical interpretation of if-then rules and with parametric conclusions may be presented as learning of kernel Ho-Kashyap KHK classifier [8, 2] with a special type of the kernel matrix.

The decision function of the classifier for the feature vector $\mathbf{x}_n$ can be defined as

$$g(\mathbf{x}_n) = \mathbf{d}(\mathbf{x}_n)^T \mathbf{W} \begin{cases} \geq 0, & \mathbf{x}_n \in \omega_1, \\ < 0, & \mathbf{x}_n \in \omega_2, \end{cases} \tag{19}$$

where $\mathbf{d}(\mathbf{x}_n)$ and $\mathbf{W}$ are given by (16) and (17) respectively. After multiplying by -1 all patterns from ω_2 class the Equation (19) may be rewritten in the form

$$y_n \mathbf{d}(\mathbf{x}_n)^T \mathbf{W} \geq 0, \qquad n = 1, 2, \ldots, N \ . \tag{20}$$

After introducing margin classifier, inequality (20) in a canonical form is given by

$$y_n \mathbf{d}(\mathbf{x}_n)^T \mathbf{W} \geq 1, \qquad n = 1, 2, \ldots, N \ , \tag{21}$$

which is equivalent to expression

$$y_n \left[\widetilde{\mathbf{d}}(\mathbf{x}_n)^T \widetilde{\mathbf{W}} + w_0 \right] \geq 1, \qquad n = 1, 2, \ldots, N \ , \tag{22}$$

where $\widetilde{\mathbf{d}}(\mathbf{x}_n)$ is a ($It \times 1$) column vector specified by

$$\widetilde{\mathbf{d}}(\mathbf{x}_n) = \left[S^{(1)}(\mathbf{x}_n)\mathbf{x}_n^T, S^{(2)}(\mathbf{x}_n)\mathbf{x}_n^T, \ldots, S^{(I)}(\mathbf{x}_n)\mathbf{x}_n^T \right]^T \ , \tag{23}$$

and $\widetilde{\mathbf{W}}$ is a column vector of dimension ($It \times 1$)

$$\widetilde{\mathbf{W}} = \left[\widetilde{\mathbf{w}}^{(1)T}, \widetilde{\mathbf{w}}^{(2)T}, \ldots, \widetilde{\mathbf{w}}^{(I)T} \right]^T \ . \tag{24}$$

The last factor w_0 equals to

$$w_0 = S^{(1)}(\mathbf{x}_n)w_0^{(1)} + S^{(2)}(\mathbf{x}_n)w_0^{(2)} + \cdots + S^{(I)}(\mathbf{x}_n)w_0^{(I)} \ . \tag{25}$$

Assuming for the simplicity of notation that patterns $\mathbf{x}_n$ are ordered according to their class membership and $\widetilde{\mathbf{D}}_1$ is a ($N_1 \times It$)-dimensional matrix

$$\widetilde{\mathbf{D}}_1 = \begin{bmatrix} \widetilde{\mathbf{d}}(\mathbf{x}_1)^T \\ \widetilde{\mathbf{d}}(\mathbf{x}_2)^T \\ \vdots \\ \widetilde{\mathbf{d}}(\mathbf{x}_{N_1})^T \end{bmatrix}, \tag{26}$$

where $\mathbf{x}_1, \ldots, \mathbf{x}_{N_1}$ belong to class ω_1 and $\widetilde{\mathbf{D}}_2$ is a $(N_2 \times It)$-dimensional matrix

$$\widetilde{\mathbf{D}}_2 = \begin{bmatrix} \widetilde{\mathbf{d}}(\mathbf{x}_{N_1+1})^T \\ \widetilde{\mathbf{d}}(\mathbf{x}_{N_1+2})^T \\ \vdots \\ \widetilde{\mathbf{d}}(\mathbf{x}_{N_1+N_2})^T \end{bmatrix} , \tag{27}$$

where $\mathbf{x}_{N_1}, \ldots, \mathbf{x}_{N_1+N_2}$ belong to class ω_2, inequality (22) may be rewritten in a form

$$\begin{cases} \widetilde{\mathbf{D}}_1 \widetilde{\mathbf{W}} + w_0 \mathbf{1}_{N_1 \times 1} \geq \mathbf{1}_{N_1 \times 1} \\ -\widetilde{\mathbf{D}}_2 \widetilde{\mathbf{W}} - w_0 \mathbf{1}_{N_2 \times 1} \geq \mathbf{1}_{N_2 \times 1} \end{cases} \tag{28}$$

where $\mathbf{1}_{N_1 \times 1}$ is a N_1-dimensional column vector with all entries equal to 1.

Construction of classifier based on kernel function requires that unknown vector $\widetilde{\mathbf{W}}$ is a linear combination of all patterns

$$\widetilde{\mathbf{W}} = \sum_{n=1}^{N} y_n \gamma_n \widetilde{\mathbf{d}}(\mathbf{x}_n) = \left[\widetilde{\mathbf{D}}_1^T, -\widetilde{\mathbf{D}}_2^T \right] \boldsymbol{\Gamma} , \tag{29}$$

where

$$\boldsymbol{\Gamma} = [\gamma_1, \gamma_2, \ldots, \gamma_N]^T , \tag{30}$$

is a N-dimensional vector obtained during learning.

Defining $(N \times N)$-dimensional kernel matrix as

$$\mathbf{K} = \begin{bmatrix} \widetilde{\mathbf{D}}_1 \widetilde{\mathbf{D}}_1^T & -\widetilde{\mathbf{D}}_1 \widetilde{\mathbf{D}}_2^T \\ -\widetilde{\mathbf{D}}_2 \widetilde{\mathbf{D}}_1^T & \widetilde{\mathbf{D}}_2 \widetilde{\mathbf{D}}_2^T \end{bmatrix} = \left[y_n y_j \widetilde{\mathbf{d}}(\mathbf{x}_n)^T \widetilde{\mathbf{d}}(\mathbf{x}_j) \right]_{n,j=1}^{N} , \tag{31}$$

and

$$\boldsymbol{\Theta} = [\mathbf{1}_{N_1 \times 1}^T, -\mathbf{1}_{N_2 \times 1}^T]^T , \tag{32}$$

then (28) may be given by

$$\mathbf{K}\boldsymbol{\Gamma} + w_0 \boldsymbol{\Theta} - \mathbf{1} \geq \mathbf{0} , \tag{33}$$

where $\mathbf{1}$ and $\mathbf{0}$ are column vectors with N rows and all entries equal to 1 or 0 respectively. As a result the same expression was obtained as in [8, 2] but with different, data dependent, kernel matrix $\mathbf{K}$. Since product $y_n y_j \widetilde{\mathbf{d}}(\mathbf{x}_n)^T \widetilde{\mathbf{d}}(\mathbf{x}_j)$ may be expressed as

$$y_n y_j \widetilde{\mathbf{d}}(\mathbf{x}_n)^T \widetilde{\mathbf{d}}(\mathbf{x}_j) = y_n y_j \sum_{i=1}^{I} S^{(i)}(\mathbf{x}_n) S^{(i)}(\mathbf{x}_j) \mathbf{x}_n^T \mathbf{x}_j , \tag{34}$$

then introducing

$$\mathbf{K}^{(i)} = \left[S^{(i)}(\mathbf{x}_n) S^{(i)}(\mathbf{x}_j) \mathbf{x}_n^T \mathbf{x}_j \right]_{n,j=1}^{N} , \tag{35}$$

kernel matrix $\mathbf{K}$ may be written in a form

$$\mathbf{K} = \Theta\Theta^T \otimes \sum_{i=1}^{I} \mathbf{K}^{(i)} , \tag{36}$$

where $\otimes$ is an element by element multiplication. Further classifier construction reduces to known KHK procedure described in [8]. Obtained, after learning of KHK classifier, parameters Γ and w_0 are used to calculate vector $\widetilde{\mathbf{W}}$ in accordance with (29) and final discriminative function is given by

$$g(\mathbf{x}) = \widetilde{\mathbf{d}}(\mathbf{x})^T \widetilde{\mathbf{W}} + w_0 . \tag{37}$$

Summarizing, the training procedure of this classifier denoted as FKHK consists of the following steps:

1. choose interpretation of if-then rules – function $\mathcal{G}$ [11];
2. fix number of if-then rules I and $w^{(i)} = 1$ for $i = 1, 2, \ldots, I$;
3. fuzzy c-means clustering of data belonging to ω_1 and ω_2; compute parameters $\mathbf{c}^{(i)(1)}$, $\mathbf{s}^{(i)(1)}$, $\mathbf{c}^{(i)(2)}$, $\mathbf{s}^{(i)(2)}$ for $i = 1, 2, \ldots, I$ – Equations (9) and (10);
4. find I nearest pairs of clusters, each pair is defined by $\mathbf{c}^{(k)(1)}$, $\mathbf{s}^{(k)(1)}$ and $\mathbf{c}^{(n)(2)}$, $\mathbf{s}^{(n)(2)}$ for $k, n = 1, 2, \ldots, I$;
5. compute $F^{(i)}(\mathbf{x}_k)$ (11), $\mathcal{G}\left(F^{(i)}(\mathbf{x}_k), w^{(i)}\right)$ and $S^{(i)}(\mathbf{x}_k)$ (15) for $k = 1, 2, \ldots, N$ and $i = 1, 2, \ldots, I$;
6. calculate kernel matrix $\mathbf{K}$ (36);
7. fix regularization constant τ;
8. train KHK classifier in accordance with procedure described in [2] for the given $\mathbf{K}$ and τ – Γ and w_0 are returned;
9. calculate $\widetilde{\mathbf{W}}$ (29).

Speaker is represented in a recognition system by the parameters $\widetilde{\mathbf{W}}$, w_0 and obtained from fuzzy c-means procedure $\mathbf{s}^{(i)(1)}$, $\mathbf{c}^{(i)(2)}$, $\mathbf{s}^{(i)(2)}$ for $i = 1, 2, \ldots, I$. Proper choose of I, τ and interpretation of if-then rules $\mathcal{G}$ requires application of rotation method.

4 Speaker Identification in Matlab

All research was done on Polish database ROBOT [14]. This database consists of 2 CD with 1 GB of speech data. The speech utterances were collected from 30 speakers of both sex in a several time-separated sessions to catch intraspeaker variability. There were 10 speaker models. Each model was trained by 5 utterances of the speaker (approximately 5 s of speech after silence removing). Text dependent speaker identification was implemented. The test utterances came from 20 speakers of which 10 were unseen for the system during its training. This procedure enables to obtain more realistic results when most of the speakers are not previously registered in the system and represent impostors. Each speaker provided 11 test sequences of approximately 5 s each. There were 110 (10 · 11) genuine attempts (possible false rejection FR error), 990 ($9 \cdot 11 \cdot 10$) impostor attempts (possible false acceptance FA error, speaker has its own model in a system – seen impostor) and 1100 ($10 \cdot 11 \cdot 10$) impostor tests (possible false acceptance FA error, speaker does not have model in a system – unseen impostor).

In order to obtain FKHK models, LBG algorithm was used to reduce cardinality of the training set. As a result feature parameters extracted from training utterances were grouped into 8, 16, 32 and 64 codevectors. Approach "one from many" was applied, e.g. FKHK 8 means that the training set consisted of 8 codevectors belonging to verified speaker (class ω_1) and 72 ($9 \cdot 8$) codevectors of all other speakers (class ω_2).

Each speaker model was obtained for the following fuzzy implications: Fodor, Lukasiewicz, Reichenbach, Kleene-Dienes, Zadeh, Goguen, Gödel and Rescher. Obtained values of identification rate for the FKHK classifier are shown in Table 1.

Table 1. Identification rate in percentages for FKHK classifier

Implication	Order of the model			
	8	16	32	64
Lukasiewicz	61.82	96.36	92.73	98.18
Fodor	73.64	95.45	91.82	97.27
Reichenbach	54.55	94.55	90.91	96.36
Kleene-Dienes	56.36	96.36	90.91	98.18
Zadeh	52.73	90	86.36	97.27
Goguen	65.45	81.82	90.91	94.55
Gödel	58.18	87.27	83.64	97.27
Rescher	56.36	92.73	90	94.55

Table 2. Identification rate in percentages for classical models

Model	Id. rate	Model	Id. rate
VQ 4	87.27	GMM 4 diag	90
VQ 8	87.27	GMM 8 diag	90.91
VQ 16	91.82	GMM 16 diag	90
VQ 32	90	GMM 32 diag	90
VQ 64	91.82	GMM 2 full	90.91
NN	91.82	GMM 4 full	90

5 Conclusion

In speaker identification task decision of the identity of the speaker is based on the whole utterance which is different from the typical classification task, where the decision about membership of the single vector to one of many classes needs to be done. However, achieved results indicate that proposed kernel Ho-Kashyap classifier with "fuzzy" kernel matrix has better generalization properties than classical approaches shown in Table 2. The highest identification rate equal to 98.18 % was achieved for FKHK 64 classifier with Lukasiewicz and Kleene-Dienes implications. The more codevectors in learning sequences, the higher identification rate.

Presented speaker identification system is based only on acoustic features. It seems that obtained results could be significantly improved if some higher level information was used, e.g. pronunciation, vocabulary, accent. However, this requires completely different approach to speaker modeling and will be the aim of further research.

References

1. Kłosowski, P.: Speech processing application based on phonetics and phonology of the polish language. In: Kwiecień, A., Gaj, P., Stera, P. (eds.) CN 2010. CCIS, vol. 79, pp. 236–244. Springer, Heidelberg (2010)
2. Dustor, A.: Voice verification based on nonlinear Ho-Kashyap classifier. In: International Conference on Computational Technologies in Electrical and Electronics Engineering, SIBIRCON 2008, Novosibirsk, pp. 296–300 (2008)
3. Dustor, A.: Speaker verification based on fuzzy classifier. In: Cyran, K.A., Kozielski, S., Peters, J.F., Stańczyk, U., Wakulicz-Deja, A. (eds.) Man-Machine Interactions. AISC, vol. 59, pp. 389–397. Springer, Heidelberg (2009)
4. Beigi, H.: Fundamentals of speaker recognition. Springer, New York (2011)
5. Togneri, R., Pullella, D.: An overview of speaker identification: Accuracy and robustness issues. IEEE Circuits and Systems Magazine 11(2), 23–61 (2011)
6. Fazel, A., Chakrabartty, S.: An overview of statistical pattern recognition techniques for speaker verification. IEEE Circuits and Systems Magazine 11(2), 62–81 (2011)
7. Vapnik, V.N.: The nature of statistical learning theory. Springer, New York (1995)
8. Łęski, J.: Kernel Ho-Kashyap classier with generalization control. Int. J. Appl. Math. Comput. Sci. 14(1), 101–109 (2004)
9. Łęski, J.: A fuzzy if-then rule based nonlinear classifier. Int. J. Appl. Math. Comput. Sci. 13(2), 215–223 (2003)
10. Dustor, A.: Speaker identification and verification based on cepstral features and fuzzy nonlinear classifier. In: International Conference Mixed Design of Integrated Circuits and System, MIXDES 2006, Gdynia, pp. 692–697 (2006)
11. Czogała, E., Łęski, J.: Fuzzy and neuro-fuzzy intelligent systems. Physica-Verlag (2000)
12. Schölkopf, B.: Statistical learning and kernel methods. Technical Report MSR-TR-2000-23, Microsoft Research Limited (2000)
13. Łęski, J.: On support vector regression machines with linguistic interpretation of the kernel matrix. Fuzzy Sets and Systems 157(3), 1092–1113 (2006)
14. Adamczyk, B., Adamczyk, K., Trawiński, K.: Zasób mowy ROBOT. Biuletyn Instytutu Automatyki i Robotyki WAT 12, 179–192 (2000)

Automatic Speech Segmentation for Automatic Speech Translation

Piotr Kłosowski and Adam Dustor

Silesian University of Technology, Akademicka Str. 16,
44-100 Gliwice, Poland
{Piotr.Klosowski,Adam.Dustor}@polsl.pl

Abstract. The article presents selected, effective speech signal processing algorithms and their use in order to improve the automatic speech translation. Automatic speech translation uses natural language processing techniques implemented using algorithms of automatic speech recognition, speaker recognition, automatic text translation and text-to-speech synthesis. It is very possible to improve the process of automatic speech translation by using effective algorithms for automatic segmentation of speech signals based on speaker recognition and language recognition.

Keywords: speech recognition, speech translation, speech synthesis.

1 Introduction

Division of Telecommunication, a part of the Institute of Electronics and Faculty o Automatic Control, Electronics and Computer Science Silesian University of Technology, for many years has been specializing in advanced fields of telecommunication engineering [1–5]. One of them is speech signal processing [6–8]. The one of many research areas aims to gain new knowledge in the field of the basic phenomena of perception and processing of human speech such as understanding and translation of speech made by a person. The main scientific objective of this research area is development of selected, effective speech signal processing algorithms and their use in order to improve the automatic speech translation. Automatic speech translation system uses natural language processing techniques implemented using algorithms of automatic speech recognition, speaker recognition, automatic translation of text and text-to-speech synthesis [9–11]. Research hypothesis can be formulated as follows: It is possible to improve the process of automatic speech translation by using efficient algorithms for automatic segmentation of speech signals coming from different speakers.

2 Automatic Speech-to-Speech Machine Translation

Field of automatic speech translation (called SSMT – Speech-to-Speech Machine Translation) is part of a long-established area of research on speech processing

A. Kwiecień, P. Gaj, and P. Stera (Eds.): CN 2013, CCIS 370, pp. 466–475, 2013.
© Springer-Verlag Berlin Heidelberg 2013

and natural language [12]. This is an area of great importance, which is associated with high hopes, because it relates to the basic problems and needs of the modern information society, such as communication between people and access to information in different languages, which is essential in today's globalized world [9]. The catalog of languages lists over 7000 living languages [13, 14]. Although currently available technology provides many ways of global communication, it is the variety of different language speakers that can be a serious barrier to communication. Basic research in the field of digital signal processing of speech can make a significant contribution to solving basic technical problems in the field of automatic speech translation, which is one of the main priorities for the development of the information society. The importance of the research field of automatic speech-to-speech translation is mirrored by a multitude of recent or ongoing large-scale research projects [15–17].

Automatic speech-to-speech translation systems can play a critical role in empowering people to communicate with speakers of a different language and to access or present information in a cross-lingual way. Speech translation is the process by which conversational spoken phrases are instantly translated and spoken aloud in a second language. A speech translation system would typically integrate the following three software technologies: automatic speech recognition (ASR), automatic machine translation (AMT) and voice synthesis (TTS). Tasks of typical automatic speech-to-speech translation is presented in Fig. 1. The tasks are as follows:

- **Automatic Speech Recognition** – translation of spoken words into text,
- **Automatic Machine Translation** – translation of text from source language to destination language,
- **Automatic Speech Synthesis** – artificial production of human speech based on text-to-speech (TTS) conversion of normal language text into speech.

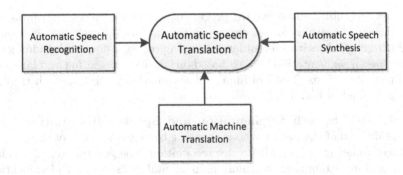

Fig. 1. Tasks of typical automatic speech-to-speech translation system

Figure 2 presents block diagram of typical automatic speech-to-speech translation from speech-to-speech in language A to speech in language B.

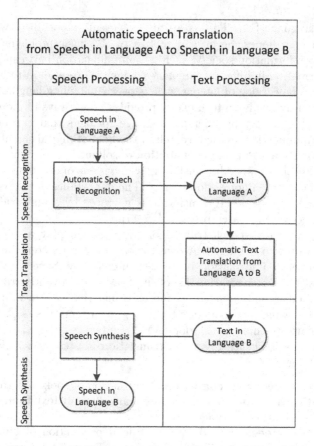

Fig. 2. Block diagram of typical automatic speech-to-speech translation from speech in language A to speech in language B

It is very possible to improve the process of automatic speech translation by using efficient algorithms for automatic segmentation of speech signals coming from different speakers in different languages. Improving is possible by adding automatic speech segmentation process based on speaker recognition and language recognition algorithms. Tasks of improved automatic speech-to-speech translation is presented in Fig. 3. The tasks are as follows:

– **Automatic Speech Segmentation and Speaker Recognition** – the identification of the person who is speaking by characteristics of their voices (voice biometrics), also called voice recognition. The general area of speaker recognition encompasses two more fundamental tasks: speaker identification and speaker verification. Speaker identification is the task of determining who is talking from a set of known voices or speakers.
– **Automatic Language Recognition** – process of determining which natural language given content is in speech.
– **Automatic Speech Recognition** – translation of spoken words into text.

- **Automatic Machine Translation** – translation of text from source language to destination language.
- **Automatic Speech Synthesis** – artificial production of human speech based on text-to-speech (TTS) conversion normal language text into speech.

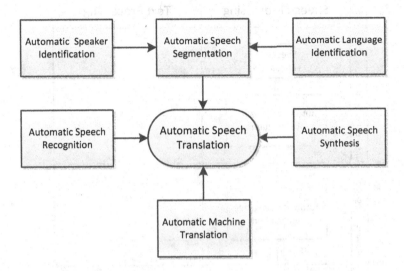

Fig. 3. Tasks of improved automatic speech translation system

Figure 4 presents block diagram of improved automatic speech-to-speech translation from speech in language A to speech in language B.

3 Automatic Speech Segmentation and Speaker Recognition

The purpose of the automatic segmentation is to separate the speaker from the audio signal containing speech fragments and to allocate them to the speakers. It is also desirable before the speech translation process to remove untranslatable fragments as music and other sounds, ambient noise and noise. This necessity stems from the fact that most of the speech recognition systems only work well when the recognized speech is free of noise [18]. In addition to segmentation in many applications it is also necessary to identify the speakers, which is separated from the signal assignment parts of speech to individuals whose identity is unknown. This occurs in applications including teleconferencing, where in addition to the transcription of speech it is important to the identity of the speaker.

Automatic speaker recognition is field of knowledge akin to speech recognition. An important element in distinguishing these two issues is the fact that speech recognition is important to extract the contents of the analyzed linguistic expression, while recognizing the speaker characteristics of the speech signal

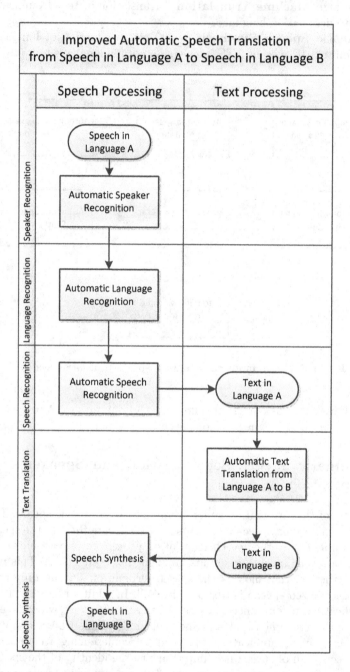

Fig. 4. Block diagram of improved automatic speech translation from speech in language A to speech in language B

output specific to the person who will recognize it in the future. Automatic speaker recognition includes automatic speaker verification, automatic speaker identification and speaker authentication. During the verification process, a user must initially declare their identity by entering their personal identification number and is then obligated to provide one or more statements. The result of the verification process is confirmation or rejection of the identity declared by the user. Such a decision is based on comparison of similarity between the model voice (already registered in the system) and the recognized utterance of a fixed threshold.

In the process of speaker identification, identity is not predeclared and speaker, whose voice is subject to examination, may have been previously registered in the system (has its own model of voice), or is someone completely unknown to the recognition system. During the identification of a set of closed (called closed set identification), it is assumed that access to the system is granted to those whose voice models were developed. The recognition system makes a choice type 1 from N, where N is the number of registered users. When this assumption is not true, there is identification in the open set (called open-set identification). It may happen that similarities of the unknown speaker's speech to the characteristics of one of the models of speakers registered in the system is large enough that you can decide to identify a person or to regard it as not belonging to any of the speakers in the system. In the second of these situations the system may decide to reject the speaker or his registration.

The last of the speaker recognition procedures is speaker authentication, which consists of determining whether the statement is one of the speakers already registered in the system or not. Speaker recognition systems are also divided as text dependent or text independent. The relationship of the text means that when trying to identify, the system requires that a person diagnosed uttered a word or words that were in sequence learning, which is used to create a model of the speaker. The system without requirements on pre-entered dictionary is called "independent of the text". Due to the same vocabulary learning and test sequences the effectiveness of recognition process from the text-dependent systems is greater. These systems typically rely on a fixed password assigned to each user. In the case of an incorrect recognition, the system often requires keyword repeated several times. Automatic speech segmentation used in automatic speech translation systems must be based on text independent speaker recognition algorithms.

4 Automatic Language Recognition

One of the important components of an automatic speech translation module is automatic language recognition. The task of spoken language recognition is to determine the language of an utterance. Since multilingual applications become increasingly popular due to globalization, spoken language recognition has become an essential technology in areas such as multilingual conversational systems, spoken language translation and multilingual speech recognition. Most of the systems rely on two types of features: the acoustic features and the phonotactic features. The acoustic features reflect low-level spectral characteristics, while the phonotactic features represent the phonological constraints that govern a spoken language. Both features have been shown to be effective in spoken language recognition. The most important factors affecting the level of language recognition system errors are: the duration of the test expression and a limited amount of learning material, the problem of the diversity of speakers, the low quality of the speech signal and the large variation in the time to vote. The impact of these factors in an ideal language recognition system should be kept to a minimum and the recognition system should work with the smallest possible error rate.

Automatic language recognition process can be divided into: the language identification and language verification. Language identification is to determine speakers language on an open set of languages. Language verification to determine whether the speaker actually speaks in a language that speaker declares. The most important characteristics of the language are: a set of phonemes, vocabulary and grammar rules specifying the connection between the words and sentence structure. In addition, each language has a significant acoustic patterns such pronounciation and melody of words and sentences (prosody). The fact that the vocabulary and grammar are almost too extensive source of information, however, leads to the fact that most of the developed systems use features such as phonotactics and simpler properties as acoustic and prosody.

Process of automatic language recognition consists of several stages. Statement in an unknown language is divided into short, on the order of 20 ms, segments called frames. Each frame is subjected to parameters extraction process. The most frequently used parameters are Linear Prediction Cepstral Coefficients (LPCC) or Mel Frequency Cepstral Coefficients (MFCC). The sequence of multidimensional parameters is used later in the process of calculating the similarity between the model and the language of this sequence. Maximum similarity is the criterion for recognizing the decision-making system. Constructing models of language during learning phase is very important and determines the effectiveness of recognition. Two families of models are used in language recognition. The first group of models is based on vector quantization. Each language is represented by a collection of multidimensional vectors defined using clustering techniques (k-means algorithm) based on training speech sentence. This collection creates a codebook. During recognition for each of the test vectors a distance to the nearest neighbor from codebook is calculated. Total distance, normalized to the length of speech is the basis for the decision of the recognition system. The second approach to the modeling language is through utilization of parametric models using statistical properties of the voice. In this case, the language is represented by sum of Gaussian Mixture Models (GMM) [19].

It seems that the obtained results could be significantly improved if some higher level information was used, e.g. pronunciation, vocabulary or accent. However, this requires a broader approach to language modeling and will be the aim of further research.

5   Research Methodology

The specific objectives of the research project can be defined as follows:

1. Development of the structure, construction and functional modules of automatic speech translation, using automatic segmentation of the speech signal derived from a variety of speakers who speak different languages.
2. Development of efficient algorithms for the identification / verification of the speakers allowing for segmentation of speech coming from different speakers.
3. Development of efficient algorithms for automatic language identification of speech fragments.

Objectives will be achieved by the following research tasks:

- Record multimedia content in the form of recordings and speech samples from different speakers in different languages. No studio condition recordings are required.
- Development of algorithms for the identification / verification of the speakers allowing for segmentation of speech coming from different speakers.
- Implementation of developed speaker identification / verification algorithms allow segmentation of speech coming from the various speakers in the MAT-LAB. Evaluation of algorithms in action. The MATLAB is also used for speech feature extraction.
- Developing algorithms for automatic language identification of speech fragments.
- Implementation of algorithms for automatic language identification of parts of speech in the MATLAB. Evaluation of algorithms in action. The MATLAB is also used for speech feature extraction.
- Development of design and simulation environment that allows to assess the effectiveness of the developed algorithms.
- Experimental research. Evaluation of results and their statistical analysis. Preparing articles for publication in scientific conferences and journals.

The research methodology consists of three steps. The first stage is the formulation made in a thesis of the theoretical development of the algorithm. The second stage is the experiment, as the practical implementation of the proposed algorithms in the selected environment. The third step is thesis verification by evaluation of the algorithm effectiveness, leading to the confirmation or rejection of the thesis formulated in the first step. It is expected that the developed algorithms will improve the automatic speech translation. Their implementation in the MATLAB computing environment and analyze the effectiveness of their actions will evaluate the usefulness in automatic speech translation. Research project will also require collection of source audio files as multimedia recordings of speech from different speakers in different languages. The culmination of the project research will be a series experiments testing the effectiveness evaluation of the developed solutions and their potential use in automatic speech translation.

6 Summary

An expected result of the research project is the development of efficient algorithms which allow to improve the automatic speech translation. The use of automatic speech translation systems can be versatile. The research project is very innovative, because the problem of automatic speech translation has not been effectively resolved. Each solution to improve the performance of automatic speech translation seems to be very important. Number of research projects carried out in the field of automatic speech translation shows that this research area is very actively developing and research projects in this area are supported by the governments of many countries around the world. Many of these projects have an international character. The combination of these efforts in many countries has a chance to create an interesting prospect for future research projects aimed at solving the fundamental problems of global communication multilingual societies in a globalized world.

References

1. Dziwoki, G.: An analysis of the unsupervised phase correction method in quadrature amplitude modulation systems. Przeglad Elektrotechniczny 88(7a), 245–249 (2012)
2. Izydorczyk, J., Izydorczyk, M.: Limits to microprocessor scaling. Computer 43(8), 20–26 (2010)
3. Sułek, W.: Pipeline processing in low-density parity-check codes hardware decoder. Bulletin of the Polish Academy of Sciences Technical Sciences 59(2), 149–155 (2011)
4. Zawadzki, P.: Security of ping-pong protocol based on pairs of completely entangled qudits. Quantum Information Processing 11(6), 1419–1430 (2012)
5. Kucharczyk, M.: Blind signatures in electronic voting systems. In: Kwiecień, A., Gaj, P., Stera, P. (eds.) CN 2010. CCIS, vol. 79, pp. 349–358. Springer, Heidelberg (2010)
6. Dustor, A.: Speaker verification based on fuzzy classifier. In: Cyran, K.A., Kozielski, S., Peters, J.F., Stańczyk, U., Wakulicz-Deja, A. (eds.) Man-Machine Interactions. AISC, vol. 59, pp. 389–397. Springer, Heidelberg (2009)
7. Kłosowski, P.: Speech processing application based on phonetics and phonology of the polish language. In: Kwiecień, A., Gaj, P., Stera, P. (eds.) CN 2010. CCIS, vol. 79, pp. 236–244. Springer, Heidelberg (2010)
8. Kłosowski, P., Pułka, A.: Polish Semantic Speech Recognition Expert System Supporting Electronic Design System. In: Prooccedings of The International Conference on Human Systems Interactions, HSI 2008, Kraków, Poland. IEEE Eurographics Technical Report Series, pp. 479–484 (2008)
9. Stuker, S., Herrmann, T., Kolss, M., Niehues, J., Wolfel, M.: Research Opportunities In Automatic Speech-To-Speech Translation. IEEE Potentials 31(3), 26–33 (2012)
10. Koehn, P.: Statistical Machine Translation. Cambridge Univ. Press, Cambridge (2009)
11. Waibel, A., Fügen, C.: Spoken language translation-enabling crosslingual human-human communication. IEEE Signal Processing Mag. 25(3), 70–79 (2008)

12. Huang, X., Acero, A., Hon, H.W.: Spoken Language Processing. Prentice-Hall, Englewood Cliffs (2001)
13. Gordon Jr., R.G.: Ethnologue, Languages of the World, 15th edn. SIL International, Dallas (2005)
14. Janson, T.: Speak-A Short History of Languages. Oxford Univ. Press, London (2002)
15. Hutchins, J.: International Association for Machine Translation compendium of translation software (2010), http://www.hutchinsweb.me.uk/Compendium.htm
16. A new framework strategy for multilingualism, Communication from the Commission to the Council, the European Parliament, the European Economic and Social Committee, and the Committee of the Regions. Commission of the European Communities (November 2005)
17. Steinbiss, V.: Human language technologies for Europe. Work Comissioned by ITC-irst, Trento, Italy to Accipio Consulting, Aachen, Germany (April 2006)
18. Rabiner, L.R., Juang, B.H.: Fundamentals of speech recognition. Prentice-Hall (1993)
19. Reynolds, D.A., Rose, R.C.: Robust text-independent speaker identification using Gaussian mixture speaker models. IEEE Transactions on Speech and Audio Processing 3(1), 72–82 (1995)

WSN Power Conservation Using Mobile Sink for Road Traffic Monitoring

Marcin Bernaś

Institute of Computer Science, University of Silesia,
Bedzinska 39, 41-200 Sosnowiec, Poland
marcin.bernas@gmail.com

Abstract. The paper discusses the possibility of application and modification of a sink movement strategy for multiple object tracking, with use of a wireless sensor network (WSN). Based on an extended Nagel-Schreckenberg 2D road traffic model various traffic conditions was obtained. Next, current methods of mobile sink control were used to research the energy consumption, its balance and efficiency. Finally, two modification was proposed to improve the WSN network utilisation. Simulation was performed on over 160 thousand nodes and compared against state-of-the-art approaches. The results show that the proposed modification to control strategies decreases energy consumption and improves uniform utilisation of nodes in WSN.

Keywords: wireless sensor networks, multiple object tracking, energy consumption.

1 Introduction

The fundamental purpose of wireless sensor networks (WSNs) is data gathering task. In areas where no sensor array is present, usage of mobile sensors is preferred. Deployed WSN nodes collect sensor readings from selected area and are sending them to the sink, which is usually a gate for other network technologies. Research proved that sensors near data sinks are depleting their energy resources faster than those far away from them [1]. A faster energy consumption in selected areas (usually network centre) degrade network ability to collect or send data. In consequence, life-span of the WSN is reduced. To overcome this issue two solutions were proposed: usage of multiple sinks or a mobile sink technology. Although sink mobility has been explored theoretically and practically to reduce and balance energy expenditure among sensors [2], no unified solution for road traffic monitoring in various traffic conditions was proposed. The sensors deployed in an area of low or high traffic inten-sity requires different sink behaviour, which will be described in this paper.

The remainder of this paper is organized as follows. Related works concerning road traffic and WSN mobile sink issue are thoroughly described in Sect. 2. Section 3 presents a detailed description of simulated road traffic network and WSN

A. Kwiecień, P. Gaj, and P. Stera (Eds.): CN 2013, CCIS 370, pp. 476–484, 2013.
© Springer-Verlag Berlin Heidelberg 2013

used for traffic monitoring. Finally, Section 4 discusses a modification proposition for the phenomenon of uneven energy depletion around a sink. Section 5 concludes the paper with ideas for further research directions.

2   Related Works

The traffic flow can be modelled in mesoscopic, macroscopic and microscopic scale. The paper focuses on monitoring of individual vehicles, therefore a microscopic model will be considered. There are many traffic models that evaluate the positions and velocity of vehicles. The paper will use the basic kinetic traffic equa-tions and cellular automata to provide data for verification purposes.

Cellular automata have become a useful tool for microscopic modelling of road traffic processes, due to their low computational complexity and high performance in computer simulations [3]. Cellular automata models are limited to discrete time, space and state representation. However, despite several limitations, a traffic process can be simulated with sufficient precision. The detailed implementations was thoughtfully described in [4]. The model has many applications and extensions for:

- urban road networks [5],
- signalised urban networks [6],
- traffic modelling [6],
- the fuzzy cellular model [3].

For further analysis, the Nagel-Schreckenberg (NaSch) cellular model was chosen.

The WSN network model was implemented for traffic monitoring purposes. So far, many routing procedures and sink movement strategies have been presented. This paper focuses on control of mobile sink movement, while receiving data from multiple sensors at once (WSNs collecting data in real time). The goal is to find a compromise between movement of a sink toward data sources to shorten path length and a balance in energy consumption between network nodes. To fulfil this task, [7] suggested that the sink should be as close to the monitored object as possible, but it should perform temporary change in moving direction, if it enters the zone of the energy-depleted sensors. The event-based approach was described in [8], where single-sink relocation is executed, if sensors detect an object in their radius. They become data sources and trigger an event. Data sources, on event, wake up sensors on the shortest path to the sink and transmit data. In [9] position of a single-sink is determined by a mixed integer linear programming (MILP) problem. According to [10] research, when shortest path routing is used, optimal sink mobility strategy is to move along the periphery of the network. This research was followed by [2] where relocation scheme was further investigated to deal with sink partition problem caused by fast energy depletion of nodes around sinks.

Extensive research was also performed to use the randomized re-routing (RRR) al-gorithm [11,12] to alternate the paths toward a fixed sink node in a presence of dynamic traffic. These issue will be considered in aspect of merging proposed and RRR algorithm.

In this research we assume that moving pattern of vehicles is unknown (random) and therefore a model based on tracking cannot be applied. The paper research will focus on two moving strategies of the sink: a strategy that moves a sink directly toward mean average of tracked object positions avoiding the depleted sensors, and one that moves along periphery of the zone.

3   Model Definition

The entry point is a NaSch cellular automata model, which will be used [13] to construct a traffic matrix $P(t)$. The matrix contains cells $\{p_{ij}, i, j \in N^2\}$ with binary value. Value 1 represents vehicle presence at given cell at t step. To evaluate vehicle presence the model parameters like vehicle position u_i, velocity $v_i = (v_x, v_y)$ and acceleration value must be tracked. The data are collected as a fusion of several NaSch one-dimensional models. Each one-dimensional model is cast into P matrix at a randomly selected starting point chosen from the border of tracking area. The calculations of vehicle parameters follows the cellular model rules:

$$v_i(t) = \min\{v_i(t-1) + 1,\ g_i(t),\ v_{\max}\}\ , \tag{1}$$

$$g_i(t) = u_j(t) - u_i(t) - 1, \text{ if } \xi < z \text{ then } v_i(t) = \max\{0,\ v_i(t) - 1\}\ , \tag{2}$$

$$u_i(t) = u_i(t-1) + v_i(t)\ , \tag{3}$$

where:

i – no of vehicle in t time step,
z – probability threshold (usually set to 0.19),
j – no of vehicle in front of vehicle at i-th position,
j is selected basing on evaluated direction from previous frames and velocity,
$u_i(t) = (x_i^c, y_i^c)$ – no of cell occupied by vehicle i at time step t,
ξ – random variable with uniform distribution (values within range $[0, 1]$).

Vector operations are performed for X and Y axis separately. The simulation process in P matrix is presented in Fig. 1.

A traffic is monitored by a WSN sensors deployed as a regular mesh. Each sensor is at the centre of traffic cells $\{p_{ij}\}$ defined by P matrix. To simulate the energy consumption the energy matrix $E(t)$ is used. Discrete Cartesian coordinates $(i, j) \in N^2$ are used to identify the sensors.

Each matrix cell e_{ij} represents a sensor, and its value – the energy reserves of the sensor. The network is equally distributed and transmits data to a fixed range and therefore energy depletion level is represented by a relation of number

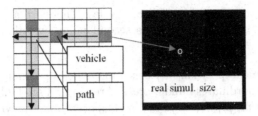

Fig. 1. Two-dimensional road traffic simulation as P matrix

of data transmissions that node still can perform to its initial value. The energy reserves take value from 0 to 1, where 0 represents depleted node and 1 fully charged node. To simulate vast area applications network for 160 thousand nodes ($4\,km^2$ area) were designed, at the cost of simplification of WSN simulation procedures. Communication range of each sensor node covers the four nearest neighbouring segments $(x+1,y)$, $(x-1,y)$, $(x,y+1)$, and $(x,y-1)$. At a given time, only those sensor nodes which monitor vehicle in its radius are in active state. The rest are put into sleep state to save energy. The steering procedure is executed in discrete time steps. At each time step the vehicles can move 2 cells according to the traffic model calibrated for $50\,km/h$ and sink speed is defined as s_{vel} in cells per simulation step. The sink movement is restrained to four directions only: north, west, south or east. The sink moving strategy for multiple targets was defined for two groups. First group represents moving strategy aiming to shorten the distance to tracking objects – its gravity centre (Alg. 1) with author's modification balancing sensor utilisation (Alg. 4). Second group is a implementation of periphery moving strategy (Alg. 2) with author's modification (Alg. 3).

3.1 Shorten the Distance to Tracking Objects

According to this strategy a distance to the tracking objects needs to be reduced. The algorithm are defined based on sink position s_{ij}, velocity svel and positions of separate vehicles $\{p_{ij}\}$ The strategy for distance minimization is presented in Alg. 1:

1. Calculate the mean position m_{ij} based on vehicles position $\{p_{ij}\}$.
2. Move sink s_{ij} toward m_{ij} to s'_{kl} position with velocity equal s_{vel} according to the following equation:

$$s_{ij} \rightarrow s'_{kl} \quad i,k \in X, \quad j,l \in Y \quad \text{for} \quad |i-k,j-l| \leq s_{vel} \text{ and } e_{kl} > 0 \ . \quad (4)$$

3. Gather data from sensors and store vehicles current position $\{p_{ij}\}$.
4. While not end of simulation return to 1.

The algorithm minimises the path between the sink and the sensors which detected a target. The main drawback of the presented algorithm, however, is unbalanced energy consumption of WSN nodes. Therefore a modification was proposed to Alg. 1. Algorithm extends second point by following pseudocode (Alg. 4):

1. Calculate the mean position m_{ij} based on vehicles position $\{p_{ij}\}$.
2. Based on m_{ij} coordinates build 5 areas. Areas $a \in A$ are build over following points $M[m_{i,j}, m_{i,j+par1}, m_{i,j-par1}, m_{i+par1,j}, m_{i-par1,j}]$ with radius equal to $par2$.
3. Calculate the energy of the areas using following equation:

$$f(m'_{kl} \in M) = \sum_{x=k-par2}^{k+par2} \sum_{y=l-par2}^{l+par2} \begin{cases} e_{xy} & \text{for } e_{xy} > par3 \\ -penalty & \text{for } e_{xy} < par3 \end{cases} . \quad (5)$$

4. Increase the energy value of central (most favoured) area: $f'(m_{i,j}) = par4 * f(m_{i,j})$.
5. Select area a' with highest energy factor: $\max(f)$.
6. Select randomly one cell within a' area as new m_{ij} and follow 2'nd point of Alg. 1.

The proposed modification reduces the effect of depletion of sensor energy. Algorithm defines the alternative areas. Once, main area total energy (close to gravity centre of monitored vehicles) is dropping to low the sink is moved to one of alternative areas. $Par1$ allows to set distance of alternative areas, while $par2$ define its size. $Par3$ is used to give additional penalty to the areas that are nearly depleted. The penalty value increases area shift probability and more unified energy distribution. Finally, $par4$ allows to favour the central area – the most optimal in terms of path length.

3.2   Moving Around Periphery Strategy

The major advantage over previously presented algorithms is a fixed path. The sensor knows the position of the sink in every n-th steps of the process and can easily calculate shortest path. However, this strategy does not reduce the distance to a detected vehicle, but is focused on balancing the energy usage of WSN. The simplest algorithm implementation is presented as Alg. 2:

1. Define starting position of a sink to m_{00}.
2. Move sink from position s_{ij} to s_{kl} along the border area clockwise:

$$s_{ij} \rightarrow s_{kl} = \begin{cases} k = i, & l = j + s_{\text{vel}} & \text{for } x = \text{sizeof}(P) \\ k = i, & l = j - s_{\text{vel}} & \text{for } x = 0 \\ k = i + s_{\text{vel}}, & l = j & \text{for } y = \text{sizeof}(P) \\ k = i - s_{\text{vel}}, & l = j & \text{for } y = 0 \end{cases} . \quad (6)$$

3. Gather data from sensors.
4. While not end of simulation return to 2.

The algorithm tends to equally distribute sensor energy usage at a the cost of the number of transmissions. Therefore, a modification was proposed, where the sink decreases the distance (rad) to the centre of detection area with every full rotation around it by $par5$ from $\frac{sizeof(P)}{2}$ up to $\frac{sizeof(P)}{4}$ value. In Alg. 3. The Equation 6 must be modified to the following one:

$$s_{ij} \rightarrow s_{kl} = \begin{cases} k = i, & l = j + s_{vel} & \text{for} & x = \frac{sizeof(P)}{2} + rad \\ k = i, & l = j - s_{vel} & \text{for} & x = \frac{sizeof(P)}{2} - rad \\ k = i + s_{vel}, l = j & \text{for} & y = \frac{sizeof(P)}{2} + rad \\ k = i - s_{vel}, l = j & \text{for} & y = \frac{sizeof(P)}{2} - rad \end{cases} \quad (7)$$

4 Simulation Results

A simulation was performed to examine the two strategies in various traffic condi-tions and to discover its pros and cons. The simulation size was based on NaSch model, which allow to simulate vast traffic networks thanks to low computation complexity. Therefore, the $N^2 = 400$ matrix simulated $4\,km^2$ area could be constructed. On this basis an E matrix for sensor network was created. Unfortunately 160 thousand nodes were difficult to process in commonly used simulation packages like ns2, therefore the author created a simplified WSN simulation environment. The algorithm was calibrated based on 100, 800-steps simulation in various traffic conditions. Each sensor's energy is sufficient to transmit data 1000 times. Simulation was set to 800 steps. The algorithm parameters after calibration are defined as: Alg. 1: $s_{vel} = 4$, Alg. 2: $s_{vel} = 4$, Alg. 3: $s_{vel} = 4$, $par5 = 10$, Alg. 4: $s_{vel} = 4$, $par1 = 20$, $par2 = 10$, $par3 = 0.1$, $pen = -1$, $par4 = 1.25$.

To make the results clearer, they are limited to two representative groups, which were determined during a simulation process: low traffic density and high traffic density. The numbers of generated paths in NaSch model for these two groups are 30 and 120. The performance results are shown in Fig. 2 and Fig. 3.

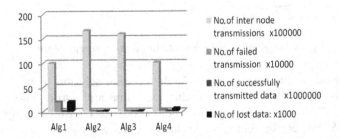

Fig. 2. Algorithm performance (low traffic)

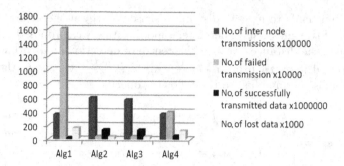

Fig. 3. Algorithm performance (dense traffic)

The results show in both cases that Alg. 1 reduces number of transmissions to the minimum. On the other hand, Alg. 2 and 3 reduce the probability of depleted nodes, after relatively short period of time. The proposed algorithm Alg. 4 presents a balanced solution which contains features of both Alg. 1 and Alg. 2. The histogram of depleted tracks and its representation on the maps was shown in Fig. 4 for high traffic.

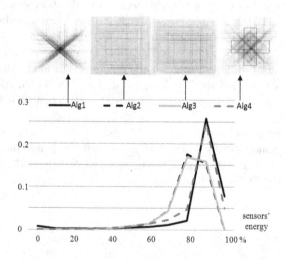

Fig. 4. Maps and histogram of the sensors' energy state

The obtained results shows that it is the optimal solution to use proposed algorithm modification for changing traffic conditions and border algorithms for dense traffic. The usage of Alg. 1 over Alg. 2 decreases transmission operations by nearly 50 % but deploy central nodes faster. In case of equal density network, this causes to decrease overall WSN energy. Using moving radius modification in Alg. 3 achieved decrease in transmission load by 5 % on average.

Last analysis consider the life-span of WSN network. Figure 5 presents the relation of simulation steps and sum of remaining energy in nodes.

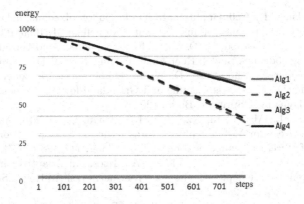

Fig. 5. WSN energy loss in simulation steps

The modification allows to save up to 5 % of a energy in all sensor network. Alg. 2 and Alg. 3 is more energy-consuming, however sensor energy is used in a more balanced way.

5   Conclusion

The paper analyses the usefulness of two strategies for traffic monitoring by a single sink and multiple vehicles. Results of the simulations are encouraging for further investigation. The proposed modifications implemented in Alg. 3 and 4 decrease the transmission load in case of Alg. 3 and strong unbalance in depletion of central nodes in case of Alg. 4. Moreover, the proposed Alg. 4 can be adopted to work as Alg. 1 and Alg. 2 by changing its parameters.

In further research, interesting results could be obtained by merging mobile sink technology with dynamic routing. In this case, the disadvantage of the slow-moving sink could be reduced by changing paths selection using an algorithm like RRR. Vital information could be sent via optimal paths and less vital ones by longer alternative paths that avoid nearly depleted nodes.

Further research will also consider the multi-agent approach to manage more than one mobile sink. Finally, field tests will be performed.

References

1. Banerjee, T., Xie, B., Jun, J.H., Agrawal, D.P.: Increasing Lifetime of Wireless Sensor Networks Using Controllable Mobile. Cluster Heads, A manuscript (2008)
2. Hashish, S., Karmouch, A.: Deployment-based Solution for Prolonging Lifetime in Sensor Networks with Multiple Mobile Sinks. Ad Hoc & Sensor Wireless Networks (2008)
3. Płaczek, B.: Fuzzy cellular model for on-line traffic simulation. In: Wyrzykowski, R., Dongarra, J., Karczewski, K., Wasniewski, J. (eds.) PPAM 2009, Part II. LNCS, vol. 6068, pp. 553–560. Springer, Heidelberg (2010)

4. Lo, S.C., Hsu, C.H.: Cellular automata simulation for mixed manual and automated control traffic. Mathematical and Computer Modelling 51, 1000–1007 (2010)
5. Kyungnam, K., Harwood, D., Davis, L.: Real-time foreground-background segmentation using codebook model. Real-Time Imaging Journal 11(3) (June 2005)
6. Schadschneider, A., Chowdhury, D.: A new cellular automata model for city traffic. In: Helbing, D., et al. (eds.) Traffic and Granular Flow 1999: Social, Traffic, and Granular Dynamics. Springer, Berlin (2000)
7. Bi, Y., Sun, L., Ma, J., Li, N., Khan, I.A., Chen, C.: HUMS: An Autonomous Moving Strategy for Mobile Sinks in Data-Gathering Sensor Networks. EURASIP Journal of Wireless Communication & Networking, 117–128 (2007)
8. Vincze, Z., Vass, D., Vida, R., Vidacs, A., Telcs, A.: Adaptive sink mobility in event-driven densely deployed wireless sensor networks. Ad Hoc & Sensor Wireless Networks 3(2-3), 255–284 (2007)
9. Basagni, S., Carosi, A., Melachrinoudis, E., Petrioli, C., Wang, Z.M.: Controlled sink mobility for prolonging wireless sensor networks lifetime. ACM Wireless Networks 14(6), 831–858 (2008)
10. Luo, J., Hubaux, J.P.: Joint mobility and routing for lifetime elongation in wireless sensor networks. In: Proceedings of the 24th Annual Joint Conference of the IEEE Computer and Communications Societies (INFOCOM), vol. 3, pp. 1735–1746 (2005)
11. Gelenbe, E., Ngai, E.: Adaptive QoS routing for significant events in wireless sensor networks. In: 5th IEEE International Conference on MASS 2008, pp. 410–415 (2008)
12. Gelenbe, E., Ngai, E.: Adaptive Random Re-Routing for Differentiated QoS in Sensor Networks. Comput. J. 53(7), 1052–1061 (2010)
13. Bernaś, M.: Objects detection and tracking in highly congested traffic using compressed video sequences. In: Bolc, L., Tadeusiewicz, R., Chmielewski, L.J., Wojciechowski, K. (eds.) ICCVG 2012. LNCS, vol. 7594, pp. 296–303. Springer, Heidelberg (2012)

Optimizing Data Collection for Object Tracking in Wireless Sensor Networks

Bartłomiej Płaczek[1] and Marcin Bernaś[2]

[1] Faculty of Transport, Silesian University of Technology,
Krasińskiego 8, 40-019 Katowice, Poland
placzek.bartlomiej@gmail.com
[2] Institute of Computer Science, University of Silesia, Będzińska 39, 41-200
Sosnowiec, Poland
marcin.bernas@gmail.com

Abstract. In this paper some modifications are proposed to optimize an algorithm of object tracking in wireless sensor network (WSN). The task under consideration is to control movement of a mobile sink, which has to reach a target in the shortest possible time. Utilization of the WSN resources is optimized by transferring only selected data readings (target locations) to the mobile sink. Simulations were performed to evaluate the proposed modifications against state-of-the-art methods. The obtained results show that the presented tracking algorithm allows for substantial reduction of data collection costs with no significant increase in the amount of time that it takes to catch the target.

Keywords: wireless sensor networks, data collection, object tracking.

1 Introduction

The task of object tracking in wireless sensor network (WSN) is to detect a moving object (target), localize it and report its location to the sink. Usually, it is assumed that the actual location of the object has to be determined continuously with a predetermined precision. The tracking capabilities of WSNs have been used in many applications, such as battlefield monitoring, wildlife habitat monitoring, intruder detection, and traffic control [1,2]. In this paper we consider an application of the object tracking WSN for target chasing. It means that the location information is delivered to a mobile sink, which has to follow and catch the target. Thus, the objective of target chasing is to control the movement of a mobile sink which has to reach the target in the shortest possible time.

The design of object tracking applications over WSNs raise great challenges due to the bandwidth-limited communication medium, energy constraints, data congestion and transmission delays. These issues are particularly important when dealing with the target chasing task, which require reliable real-time data delivery [3]. In order to ensure efficient execution of this task, the data collection procedures for WSN should be optimized, taking into account the above-mentioned limitations. Continuous data collection scheme is not suitable for developing

A. Kwiecień, P. Gaj, and P. Stera (Eds.): CN 2013, CCIS 370, pp. 485–494, 2013.
© Springer-Verlag Berlin Heidelberg 2013

the target chasing applications, because periodical transmissions of the target location to the sink would drain sensors energy rapidly. Therefore, the target chasing task requires dedicated data collection methods to ensure the amount of transmitted data is as low as possible.

In this paper the existing WSN-based target chasing methods are discussed and some modifications are proposed to optimize the use of sensor nodes by transferring only selected data readings (target locations) to the mobile sink. Simulation experiments were performed to evaluate the proposed modifications against state-of-the-art methods. The experimental results show that the presented tracking algorithm allows for substantial reduction in the amount of transmitted data with no significant negative effect on performance of the target chasing application.

The remainder of this paper is organized as follows. Related works are reviewed in Sect. 2. Section 3 presents a detailed description of the considered target chasing problem. WSN-based target tracking algorithms and data collection strategies are presented in Sect. 4. Section 5 contains results of the experiments on the target chasing in WSN. Finally, in Sect. 6, conclusion is given.

2   Related Works

In the literature a number of methods have been introduced for wireless sensor networks that enable the tracking of moving objects. A comprehensive and detailed review of these approaches can be found in [1,4]. The majority of the available methods aim at delivering the real-time information about trajectory of a tracked object to a single sink. However, in most approaches the sink is assumed to be stationary and only few publications deal with the problem of chasing the target by a mobile sink.

A basic formulation of the target chasing problem postulates that the target performs a simple random walk in a two dimensional grid, moving to one of the four adjacent grid points with equal probability every time step [5]. The chasing strategy presented in [5] was designed for the case of static sensors able to detect the target, with no communication between them. A static sensor can deliver the information about when the target was detected to the mobile sink only if the sink is located on the same grid point as the sensor.

More realistic model of the WSN was used by Tsai et al. [6] to develop the dynamical object tracking protocol (DOT). This protocol allows the sensor network to assist in detecting the target and collecting the target's trajectory information. The trajectory information is stored by an intermediate (beacon) node, which guides the sink to chase the target. A similar method is the target tracking with monitor and backup sensors [7], which additionally take into account the effect of a target's variable velocity and direction.

Complex scenarios with multiple targets and multiple pursuers are also analysed in the literature [3,8]. For such scenarios a centralized coordination of the pursuers has to be performed by a control module, i.e. by a base station or one of the pursuers. This task requires both communication among pursuers and high computational resources. The tracking algorithms discussed in this paper can contribute to the complex scenarios by optimizing the data transmission between the control module and particular pursuers.

The proposed approach extends the DOT protocol by providing heuristic rules to reduce the amount of data transmitted in WSN during target chasing. Moreover, in order to decrease the number of activated sensor nodes the introduced data collection strategies adopt the prediction-based tracking method [9]. According to this method a prediction model is used, which anticipates the future location of the target so only the sensor nodes expected to detect the target are periodically activated.

3   Problem Formulation

The considered target chasing problem deals with controlling movement of a mobile sink, which has to catch the single target in the shortest possible time. The target moves in a closed area, which is divided into square segments of equal dimensions. Discrete Cartesian coordinates $(x, y) \in \mathbb{N}^2$ are used to identify the segments. For each segment there is a static sensor node deployed that can detect presence of the target in this particular segment. Communication range of each sensor node covers the four nearest neighbouring segments $(x+1, y), (x-1, y), (x, y+1)$, and $(x, y-1)$. At a given time, only the selected sensor nodes are in active state to track the target and other nodes are put into sleep state to save energy consumption.

The chasing procedure is executed in discrete time steps. At each time step both the target and the sink move in one of the four directions: north, west, south or east. Their velocities (in segments per time step) are determined as parameters of the simulation. Target changes its movement direction randomly. The probability that the target moves to an adjacent segment depends on the direction. Moving direction of the sink is decided on the basis of information delivered from WSN.

Main objective in target chasing is to minimize time-to-catch, i.e., the number of time steps in which the sink reaches the moving target. However, due to the limited energy resources also the minimization of data collection costs (data transmission and sensing cost) has to be taken into consideration. The data collection costs are measured by: number of data transfers to the sink, hop counts, and active times of sensor nodes. An obvious trade-off exists between the time-to-catch minimization and the data collection costs minimization. In this study heuristic rules are proposed that enable considerable reduction of the data collection costs with no significant increase of time-to-catch.

4   Tracking Algorithms

In this study, three object tracking algorithms are compared in application to the target chasing problem. The first two algorithms presented below were developed on the basis of the methods available in literature, i.e. the prediction-based tracking and the dynamical object tracking. The last part of this section describes the proposed algorithm, which utilizes heuristic rules to select data readings that have to be transmitted.

4.1 Prediction-Based Tracking

According to Algorithm 1 (Fig. 1), which uses the prediction-based tracking method, the target location is discovered and reported to the sink at each time step. The sink moves toward segment (x_T, y_T), where the target is detected. It means that movement direction is selected which minimizes distance between the sink location (x_S, y_S) and the target location (x_T, y_T). Because the sink can move in one of the four directions (N, W, S, E), the city-block metric was used to determine the distance D between segments:

$$D\left[(x_1, y_1), (x_2, y_2)\right] = |x_1 - x_2| + |y_1 - y_2| \ . \tag{1}$$

```
1      set node(xC,yC) to be target node
2      repeat
3        at target node do
4          determine P
5          collect data from each node (x,y):(x,y) ∈ P
6          determine (xT,yT)
7          if (xT,yT) changed then
8            communicate (xT,yT) to the sink
9            set node(xT,yT) to be target node
10         at sink do
11           move toward (xT,yT)
12       until (xS,yS)=(xT,yT)
```

Fig. 1. Pseudocode of Algorithm 1

Thus, the selected segment (x_S^*, y_S^*) into which the sink will move, has to fulfil the following condition:

$$(x_S^*, y_S^*) \in M \wedge D\left[(x_S^*, y_S^*), (x_T, y_T)\right] = \min_{(x,y) \in M} \{D\left[(x, y), (x_T, y_T)\right]\} \ , \tag{2}$$

where $M = \{(x_S + v_S \cdot \Delta T, y_S), (x_S - v_S \cdot \Delta T, y_S), (x_S, y_S + v_S \cdot \Delta T), (x_S, y_S - v_S \cdot \Delta T)\}$ is the set of segments that can be selected, v_S denotes value of the sink velocity, and $\Delta T = 1$ time step.

Prediction of the possible target locations is based on a simple model of the target movement, which is consistent with the above assumptions on available directions and predetermined maximum velocity. Let us denote maximum value of the target velocity by v_T. If for previous time step $(t - 1)$ the target was detected in segment (x_T, y_T), then at time t there is a set P of possible target locations:

$$P = \{(x,y) : D[(x,y),(x_T,y_T)] \leq v_T \cdot \Delta T\} \,, \tag{3}$$

where $\Delta T = 1$ time step.

Sensor nodes for all possible target locations $(x, y) \in P$ are activated, and the discovered target location is transmitted to the sink. The transmission is suppressed if the target location is the same as at the previous time step. At the beginning of the tracking procedure a central segment (x_C, y_C) of the monitored area is assumed to be a hypothetical target location.

An important feature of the above algorithm is that the collected information about target trajectory has the maximum available precision. Moreover, the information is delivered to the sink with the highest attainable frequency (at each time step of the tracking procedure).

4.2 Dynamical Object Tracking

Pseudocode of Algorithm 2 is shown in Fig. 2. This algorithm is based on the tracking method which was proposed for the DOT protocol [6]. In this algorithm the location of target is discovered at each time step using the same approach as in Algorithm 1.

```
1   in case of Algorithm 2
2      condition:= Δt=τ or (xS,yS)=(xI,yI)
3   in case of Algorithm 3
4      condition:= dIT/dSI>α or Δt/(dST/vS)>β or (xS,yS)=(xI,yI)
5   set node(xC,yC) to be intermediate node
6   Δt:=0
7   repeat
8      at intermediate node do
9         determine P
10        collect data from each node(x,y):(x,y)∈P
11        determine (xT,yT)
12        if (xT,yT)<>(xI,yI) then
13           Δt:=Δt+1
14           if condition then
15              Δt:=0
16              set node(xT,yT) to be intermediate node
17              communicate new (xI,yI) to the sink
18     at sink do
19        move toward (xI,yI)
20   until (xS,yS)=(xT,yT)
```

Fig. 2. Pseudocode of Algorithms 2 and 3

The target location (x_T, y_T) is determined and stored at the intermediate node. Sink moves toward location of the intermediate node (x_I, y_I). A new intermediate node is set if the sink enters segment (x_I, y_I) or a predetermined time τ passes since the last update. The update means that the sensor node, which currently detects the target in segment (x_T, y_T), becomes new intermediate node and its location is communicated to the sink.

By using the intermediate node, the cost of data transmission in WSN is reduced because the data transfers to sink are executed less frequently. Data readings from the activated sensor nodes are collected by the intermediate node, which is closer to the segments $(x, y) \in P$ than the sink. Therefore, a lower number of hops is required to complete the data transmission.

4.3   Application of Heuristic Rules

Algorithm 3 implements heuristic rules that are proposed to improve performance of the object tracking task in terms of data collection costs. The major difference between Algorithm 2 and Algorithm 3 lies in the condition, which determines when the intermediate node has to be updated. The following symbols are used to formulate this condition (Fig. 2, line 4 of the pseudocode): d_{ST} – distance between sink and target, d_{IT} – distance between intermediate node and target, d_{SI} – distance between sink and intermediate node.

The heuristic rules were motivated by an observation that the sink does not need the precise information about target location to chase the target effectively when the distance to target is large. The closer to target, the higher precision of the localization has to be obtained. This fact is illustrated by examples in Fig. 3. Locations of the target and the sink are indicated by 'T' and 'S'. The numbers describe distance to target from segments into which the sink can move during the analysed time step. Optimal moves are shown by arrows. It was assumed that $v_S = 2$ and $v_T = 0$ for these examples. Required precision of the target localization corresponds to the shaded regions. It should be noted here that the sink moves toward intermediate node. The shaded regions indicate segments where the intermediate node can be located to ensure that the sink selects the optimal movement direction. It can be seen from this illustration that for a greater distance between sink and target there is a larger area, which includes allowable locations of the intermediate node.

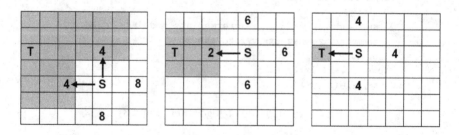

Fig. 3. Required precision of the target location information

On the basis of the above insights two heuristic rules were introduced. The first rule says that the update of intermediate node is necessary if the distance between intermediate node and target is relatively high in comparison to the distance between sink and intermediate node. This rule was translated into the elementary condition $d_{IT}/d_{SI} > \alpha$. According to the second rule, the update of intermediate node has to be executed if the time elapsed since the last update is relatively long in comparison to the time in which the sink could reach the current target location. The second rule can be written as a formula: $\Delta t/(d_{ST}/v_S) > \beta$. Values of the parameters α and β were chosen experimentally.

5   Experimental Results

Simulation experiments were performed to compare effectiveness of the three tracking algorithms presented in Sect. 4. The comparison was made with respect to data collection costs and tracking performance. Hop counts, active times of sensor nodes and numbers of data transfers to sink were analyzed to evaluate the cost of data collection in WSN. The tracking performance was measured as time-to-catch, i.e., the time in which the sink reaches the moving target. Both the active time and the time-to-catch are measured in time steps of the control procedure. Hop counts were determined assuming that the shortest path is used for each data transfer.

In the experiments, it was assumed that target velocity v_T equals 3 and sink velocity v_S equals 4. It should be noted that the velocities are expressed in segments per time step. The monitored area is a square of 200×200 segments. Thus, the number of sensor nodes equals 40 000. The results presented below were averaged for 100 simulation runs. Each simulation run starts with the same locations of sink and target. During simulation, random trajectory of the target is generated. The simulation stops when target is caught by the sink. Experiments were performed using simulation software that was developed for this research.

Initial experiments were carried out in order to examine influence of the parameters α and β on effectiveness of Algorithm 3. The results in Fig. 4 illustrate the effect of these parameters on hop count as well as on time-to-catch. Black colour in these charts corresponds with low level of the analyzed quantities. A similar analysis was conducted for total active time of sensor nodes and number of data transfers to sink. Based on the results, the optimal values of parameters were determined: $\alpha = 0.20$ and $\beta = 0.25$. In case of Algorithm 2 the preliminary experiments have shown that the best results can be obtained for $\tau = 6$. The above settings were used for all the simulations reported in this section.

In Figure 5 simulation results are compared for the three examined algorithms. From these results it is apparent that the proposed algorithm (Algorithm 3) provides short time-to-catch values with low data collection cost. In comparison, Algorithm 1 involves a much higher number of data transfers and hops. Algorithm 2 needs significantly longer time to reach the moving target than the other considered algorithms. As it could be expected, the shortest time-to-catch

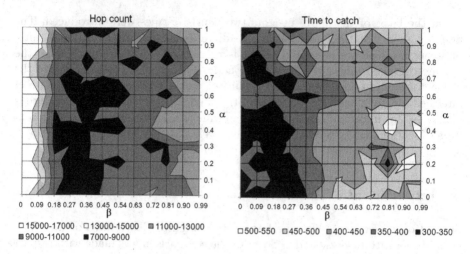

Fig. 4. Impact of parameters α and β on hop count and time-to-catch for Algorithm 3

was obtained for Algorithm 1, in which the extracted information about target locations has the highest available precision. However, for Algorithms 1 and 3 the difference of time-to-catch values is negligible.

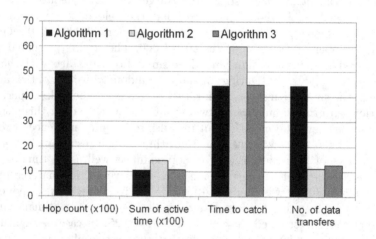

Fig. 5. Comparison of data collection costs and tracking performance

An example of a simulated target trajectory and resulting sink trajectories are presented in Fig. 6. The different trajectories of a sink were obtained by using the three examined algorithms. At the beginning of the simulation, target is located in segment with coordinates $(66, 66)$ and sink is in segment $(160, 160)$. For Algorithm 1 as well as for Algorithm 3 the target is caught after 62 time steps in segment $(75, 12)$. When using Algorithm 2 the sink reaches target in 85-th time step at segment $(105, 51)$. In this example, the hop counts for Algorithms 1, 2, and 3 are 7442, 1975, and 1440 respectively.

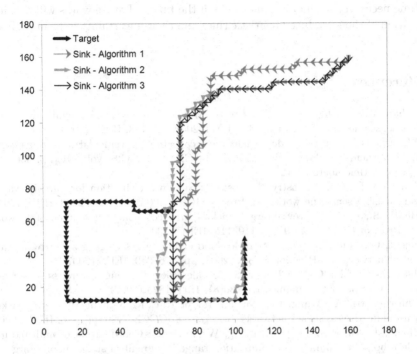

Fig. 6. Target trajectory and sink trajectories for compared algorithms

According to the presented results, it could be concluded that Algorithm 3, which is based on the proposed heuristic rules, enables a significant reduction of the data collection costs and ensures good performance of the target chasing application. The comparison of simulation results for Algorithms 1 and 3 shows that Algorithm 3 reduces hop count by about 75 % and increases time-to-catch by 1 % on average.

6   Conclusion

The WSN-based target chasing task requires dedicated methods for data collection. In order to optimize the utilization of WSN, the scope of the collected data has to be dynamically adjusted to the variable circumstances. In this paper, a tracking algorithm was proposed, which uses heuristic rules to decide when data transfers are necessary for achieving the chasing objectives. The heuristic rules are based on an observation that in some circumstances the sink does not need the precise information about target location to chase the target effectively.

Effectiveness of the proposed algorithm was compared against state-of-the-art methods. The experimental results show that the introduced heuristic rules enable substantial reduction in the data collection costs (hop count, sensor active time, and number of data transfers) with no significant increase in the amount of time necessary for mobile sink to catch the target. Future works will consider definition of the introduced heuristic rules and their uncertainty in terms of fuzzy sets.

References

1. Tahan, M.N., Dehghan, M., Pedram, H.: Mobile object tracking techniques in wireless sensor networks. In: Proc. of ICUMT 2009, pp. 1–8. IEEE (2009)
2. Płaczek, B.: Uncertainty-dependent data collection in vehicular sensor networks. In: Kwiecień, A., Gaj, P., Stera, P. (eds.) CN 2012. CCIS, vol. 291, pp. 430–439. Springer, Heidelberg (2012)
3. Schenato, L., Oh, S., Sastry, S., Bose, P.: Swarm coordination for pursuit evasion games using sensor networks. In: Proc. of ICRA 2005, pp. 2493–2498. IEEE (2005)
4. Bhatti, S., Jie, X.: Survey of target tracking protocols using wireless sensor network. In: Proc. of ICWMC 2009, pp. 110–115. IEEE (2009)
5. Kosut, O., Lang, T.: The nose of a bloodhound: target chasing aided by a static sensor network. In: Proc. of ACSSC 2007, pp. 376–380. IEEE (2007)
6. Tsai, H.-W., Chu, C.-P., Chen, T.-S.: Mobile object tracking in wireless sensor networks. Computer Communications 30(8), 1811–1825 (2007)
7. Bhuiyan, M.Z.A., Wang, G., Wu, J.: Target tracking with monitor and backup sensors in wireless sensor networks. In: Proc. of ICCCN 2009, pp. 1–6. IEEE (2009)
8. Zheng, J., Yu, H., Zheng, M., Liang, W., Zeng, P.: Coordination of multiple mobile robots with limited communication range in pursuit of single mobile target in cluttered environment. Journal of Control Theory and Applications 8(4), 441–446 (2010)
9. Samarah, S., Al-Hajri, M., Boukerche, A.: An energy efficient prediction-based technique for tracking moving objects in WSNs. In: Int. Conf. on Communications, ICC 2011, pp. 1–5. IEEE (2011)

Data Security in Microprocessor Units

Andrzej Kwiecień[1], Michał Maćkowski[1], and Marcin Sidzina[2]

[1] Silesian University of Technology, Institute of Computer Science
[2] University of Bielsko-Biała, Department of Mechanical Engineering Fundamentals
{akwiecien,michal.mackowski}@polsl.pl, msidzina@ath.bielsko.pl

Abstract. Protection of computer systems from an unauthorized access to the classified information is a very essential issue. Security of IT systems can concern various aspects, such as software security, connected with ensuring its confidentiality, as well as preventing its modification. Thanks to the developed methods, it is possible to analyze the program code based on the disturbances of voltage supply, which occur during the execution of the program. Moreover, it also enables to recognize the numbers of bits changes on microcontroller data bus, as a result of realized instruction. In such context, it can constitute a potential threat for data processed by microcontroller program or embedded systems. The presented method is very similar to simple power analysis method, which is very effective in relation to cryptographic algorithms, whose execution in many cases depends on processed data. The results indicate that presented method is an effective and low-cost attack, due to its simplicity in many real applications. Moreover, the research results inspire to study carefully the ways and methodology for developing software and hardware, which should reduce the possibility of software reverse engineering.

Keywords: reverse engineering, data security, program code, microcontroller, conducted emission, electromagnetic disturbances, electromagnetic interference, simple power analysis.

1 Introduction

The emission of electromagnetic field is a phenomenon that is contributed to electric current flow, which in turns is the base for working the whole electronic and electric devices. In many cases, based on the electromagnetic field changes, it is possible to deduce about the work of devices being its source. What is more, the properties of electromagnetic field allow for its remote registration and further analysis. The phenomenon of electromagnetic emission, which can provide the information about the work of electric and electronic devices, is called compromising emanation or compromising radiation [1, 2]. As the authors notice, the security of processing and storing information [3–5], and also safety of communication protocols [6, 7] is one of the most important problems of information technology.

Protection against remote, non-invasive reading of information based on detection of electromagnetic waves, constitute an important area, which first of

A. Kwiecień, P. Gaj, and P. Stera (Eds.): CN 2013, CCIS 370, pp. 495–506, 2013.
© Springer-Verlag Berlin Heidelberg 2013

all is connected with creating of software and designing computer devices. The use of electronic devices for processing the information, very often confidential, caused that compromising emanation has become of a special meaning.

For decades, issues connected with the reduction of electromagnetic emission of disturbances that can adversely affect the work of other devices, as well as ensuring the suitable immunity of devices to electromagnetic interference are the subject of intense study, described extensively in [8, 9]. It is obvious, that the level of signals which enable to reconstruct the information processing by a device, can be considerably lower than the level of disturbances causing an incorrect work of other devices. Contrary to the issues related to the electromagnetic compatibility, the information about compromising emanation based on the electromagnetic emission is not very often published or is confidential, and the access to it in many cases is very difficult.

Microprocessor units, currently used in network devices as network controllers may also be considered as advanced chips responsible for data processing and reconstruction of transmitting frames. Such units can also be source of electromagnetic disturbances and can, for example provide information about work of the network controller.

The research deals with some aspect of this problem resulting from the fact, that for instance, an author of a software for embedded system is not aware that it is possible to identify partly a program code (the following executed instruction), or to provide some information about the data processed by the system, without any direct interference into the program memory of microcontroller. The research results presented in the previous authors studies [10, 11] concern the analysis of the microprocessor program code based on the nature of disturbances in the voltage course. However, the research did not focus on details of data processing, but mostly on the possibility of reconstructing the program code.

This study presents the analysis of data processed by microprocessor during execution of program code, based on the voltage supply changes. The expected results of the research may lead to a broader look at the ways and methodologists of software development for microprocessor systems and more widely used of embedded systems.

2   Security of Microprocessor Systems

In recent years the issue connected with the security of microcontroller systems has become of special meaning. It is due to the fact that the number of electronic devices, which use microprocessor units of various manufacturers increase continuously. The systems responsible for processing the strictly confidential data, are particular noteworthy.

Obtaining information about the operation of a device through the influence on its work or monitoring the parameters of its activity, is called side-channel attack. The existence of the side-channel through which such information is obtained, is usually unintended and results from the construction of a device or technology, in which it was built.

The research on side-channel attacks were introduced in the second half of 90's and were related mainly to the attacks on cryptographic systems. The assumptions of those days as to the security of such systems, were based mainly on the correct execution of an algorithm implemented in the cryptographic system, and the lack of data distortion during its work. As it was proved, those assumptions were wrong [12, 13], and the possibility to impact on the device work and forcing mistakes in its work may simplify the cryptanalysis in a significant way. Side channel attacks can be divided according to: methods of influence, the way of interference into a device, and also the way of analysis the received results. Considering the way of interference into device the attacks can be divided into 2 groups:

1. Active attacks involving the attempt to take control over the system and observation of its response by:
   - voltage supply changes (voltage dips, short interruptions and voltage variations),
   - generating fault states, for example, due to the disturbances of clock signal timing microcontroller,
   - interacting with electromagnetic field of determined frequency and intensity.
2. Passive attacks involving the observation and measurement the microcontroller parameters during its normal work. The most frequent parameters to be monitored are as follows:
   - power consumption by a device,
   - signal levels on interface lines,
   - time for executing a particular part of algorithm,
   - electromagnetic disturbances spreading via conducted and radiated emission.

The presented method of analysis of data processed by microprocessor, which is based on voltage supply changes can be included into group of non-invasive attacks, which means with no direct physical access to the internal components included in the microcontroller. Although there is an access to a device (system), but any attempts to reconstruct the information processed by the central unit cannot break the structure of integrated circuit and leave the traces of activity. This type of attack is reduced to measure working parameters and observing inputs and outputs of the unit.

Moreover, the presented method is very similar to simple power analysis method (SPA) [14–17]. Such analysis requires very often the exact knowledge of the algorithm implemented into the device – which is regarded, at the same time, as its main disadvantages. As mentioned above, the method is very effective in relation to cryptographic algorithms, whose execution in many cases depends on processed data (e.g. in smartcards). A good example can be an algorithm, in which some instructions are executed only when a particular condition is performed. In this case, if the following instructions are characterized by various power consumption, then SPA analysis can enable to determine for example the value of the encryption key.

In papers [16, 17] it was noticed that the level of disturbances (noise) has a great impact on the effectiveness of extracting sensitive information, for example the encryption key. In current work, this problem was solved by placing the whole test bench in shielded area which helped to minimize the influence of electromagnetic disturbances on parameters to be measured. The test bench is presented in Sect. 3. Such solution increases the effectiveness of data analysis processed by the microcontroller.

What is more, some authors [15, 17] often use the Hamming weight parameter, which provides the information about the number of bits in the string being different from zero. This paper, in contrast, uses Hamming distance parameter. In information theory, the Hamming distance between two strings of equal length is the number of positions at which the corresponding symbols are different. Put another way, it measures the minimum number of substitutions required to change one string into the other. The advantages of using it are described in the further part of the paper.

This paper presents the analysis of data processed by the microcontroller, based on the voltage supply changes, which is free of these defects. In this case, change of the character of power consumption by the system is not a result of executing the other instructions (e.g. during program branching in IF conditional instruction), but is due to the processing the other data by the system. At the same time, the number of attempts required to restore the data processed by the processor is less than in the method of differential power analysis (DPA) [18, 19].

3   Test Bench and Research Procedure

Test bench consisted of Microchip microcontroller with PIC16F84A signature. This processor represents very numerous group of Mid-Range processors. There are about 85 models in this serie of processors. The main differences among microcontrollers included in Mid-Range class concern: the size of data memory (RAM/EEPROM) and Flash program memory, the maximum work frequency of circuits, and available numbers of communication modules. The access to the program memory, which can store 1024 instructions is via 14 bits program bus. To supply the microprocessor Agilent stabilized power supply was used. Oscilloscope probe was connected to microcontroller supply lines to monitor voltage drop during realization of following instructions. The test bench, for the period of research was placed in shielded cell – GTEM (*Gigahertz Transverse ElectroMagnetic*), which provided total separation of measuring area from external electromagnetic influences. The exact description of the test bench, methods used to measure voltage disturbances and ways to analyze the obtained results in the time and frequency domain, were presented in previous paper of the authors [20].

In paper [11] the authors presented and described the method, which enable to analyse the program code based on the voltage supply changes. Depending on a program code and instruction argument, the different disturbances on power supply lines in a particular machine cycles were noticed.

The first step to do this is to measure the microprocessor voltage supply waveform while running the entire program. The next step is to cut the part of time waveform referring to the instruction being tested. Then the minimum and maximum value of the voltage for the first three machine cycles is saved – a total of six values are saved. In this way the sample database was created, in which each microprocessor instruction is characterized by 6 points – three maximum and three minimum values of voltage, measured in particular machine cycles $Q1$, $Q2$, and $Q3$. Moreover, the fourth machine cycle was not considered due to the fact that in the last cycle ($Q4$) another (next) instruction is also fetched from the program memory, thus not only currently realized instruction but also next instruction has the influence on the current flow (shape).

Presented method based on measuring the voltage changes during the realization of each microprocessor instruction, and then on creating a database of samples. The database of voltage disturbances samples was next used to recognize the instructions for any program executed by the microprocessor (reverse engineering). The detailed description of a manner for instruction recognition was presented in papers [10, 11]. The authors achieved the effectiveness of instruction recognition on any arguments at the level of 72 %.

During the research, it became clear that voltage waveform during instruction execution depends not only on the instruction code, but also on instruction argument, especially when the first (fetching the argument) and the third (executing instruction by arithmetic-logic unit) machine cycle is realized.

Suppose that the result of some operation is zero (such state is maintained on data bus till receiving another argument). Then, while monitoring the voltage change during executing another instruction, i.e. MOVLW (move constant value to W register), regardless whether it is instruction MOVLW 1, MOVLW 2, MOVLW 4, MOVLW 8...MOVLW 128, it can be seen that the voltage waveform is the same – in each of these cases only one bit is changed on data bus.

In order to explain the changes that occur in the voltage waveform during instruction realizations, it is helpful to use the parameter called Hamming distance (HD). Figures 1a–c presents the voltage waveforms on the microprocessor supply lines in the middle of executing MOVLW 0, MOVLW 15, MOVLW 255 instructions (the instructions arguments were selected to present the values of HD, which equal 0, 4 and 8). Comparing the voltage waveforms in the figures it can be seen that mainly the differences during the first machine cycle (first 4th μs) occur. The voltage shapes for 2, 3 and 4 cycle are the same for all considered instructions. In this case MOVLW instruction does not involve ALU, hence the third machine cycle is the same in all cases. The similar analysis can be conducted for other instructions with various arguments. These issues were presented in more details in paper [20].

In this case, in order to create a database of samples a schema based on Hamming Distance was used. This knowledge enabled to simplify significantly the creation of samples database, which was used afterwards in the process of instruction recognition. Hence, it is not necessary to create a sample for an instruction with each possible argument, but only include Hamming Distance.

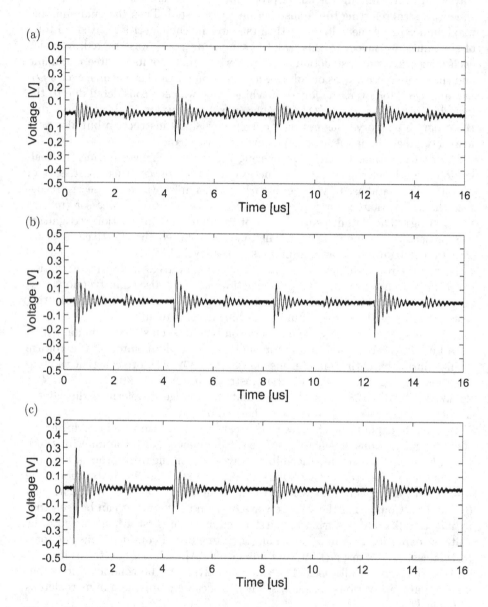

Fig. 1. The voltage waveforms on the microprocessor power supply lines during the execution: of (a) `MOVLW 0` – {HD = 0}, (b) `MOVLW 15` – {HD = 4}, (c) `MOVLW 255` – {HD = 8} instructions

All programs used for creating a database (excising a part of time waveform of a tested instruction, searching for a minimum and maximum value of voltage, estimating the voltage average value, etc.) were implemented in Matlab [20]. A serious problem was the acquisition of individual waveforms for each instructions, including all Hamming distance. At the current stage of the research this process was performed manually. However, in the future the authors intend to automate this process, which may reduce significantly time of creating the database of samples.

Figure 2 presents the schema of compiling database of samples and the process of instructions recognition executing the arguments with the value of any kind. The database consists of samples describing 10 instructions, where samples from 1 to N describe instruction 1, then M samples describe instruction 2, etc. According to the above schema, it was possible to create a database for the previously mentioned instructions, consisting of 1935 samples and used next in the process of microprocessor program code recognition. Then, each of the instructions of a test program (in the form of 6 minimum and maximum voltage values) was compared to all samples in database using the method of the least squares and correlation [10]. It enabled to reveal that a sample from database, which was the most similar to the tested instruction, was then typed as a recognized instruction.

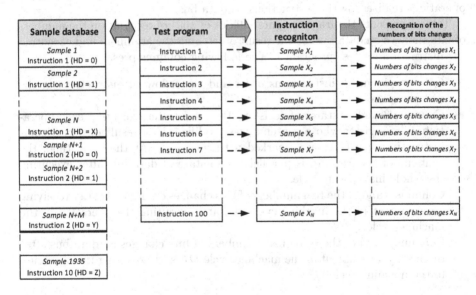

Fig. 2. The process of instruction recognition and the numbers of bits changes on data bus as a result of fetching and executing instructions

Next, as the result of fetching the argument and executing operation, it was possible to recognize the numbers of bits changes on data bus for each instruction. In order to correlate the database trace with the voltage trace in real program, the procedure in Matlab environment was written. This procedure returns such

502   A. Kwiecień, M. Maćkowski, and M. Sidzina

samples and the Hamming distance values from database, which are the most similar to the instruction in the test program.

The usefulness of this information is presented in the further part of the paper.

4   The Research Results

Having compiled the database three test programs consisting of 100 instructions were generated, and were used for determining the effectiveness of the method of recognition the numbers of bits changes on data bus, as the result of fetching and executing the instruction. Both, the order of instructions in the test programs and instruction argument were selected at random.

The database of samples besides using it for recognising following instructions of a program written into the microcontroller memory, can be also used for receiving some information on data processed by the microprocessor. As mentioned before, in the process of compiling the database one of the useful information was that Hamming Distance had the influence on the power supply course during execution of instruction cycle, between the state of data bus and instruction argument (machine cycle *Q1*) and between operation result and instruction argument (machine cycle *Q3*). In such case, the correct instruction recognition in test program, except the instruction name, provides also information about operations realized by the instructions on data bus.

Table 1 presents the fragment (first 10 out of 100) of recognition the numbers of bits changes on data bus, as the result of fetching the argument and executing the instruction for test program 1. The following columns presents:

- Columns 1 and 2: instructions order and the following instructions in the test program.
- Columns 3 and 4: states of registers W and F, after executing each operation. The register whose contents was modified, as a result of a particular instruction execution, was marked with bold font. At the same time, the contents of this register is parallel to the state of data bus after executing the whole instruction cycle.
- Columns 5 and 6: the real numbers of bits changes on data bus after receiving an argument in the machine cycle $Q1$, and executing the operation in the machine cycle $Q3$.
- Columns 7 and 8: the recognized numbers of bits changes on data bus after receiving an argument in the machine cycle $Q1$, and executing the operation in the machine cycle $Q3$.

The numbers of bits changes on data bus during receiving an argument and executing operation is analysed on the example of the three first instructions of test program 1 (Table 1):

- Before the instruction MOVLW .196 is executed, the state on data bus has the value 0. In the first machine cycle occurs fetching of argument 196 (binary 11000100) and setting it on data bus, where 3 bits change their state on

Table 1. Results of recognition the numbers of bits changes on data bus, as the result of fetching the argument and executing instructions for test program 1

Following instructions of test program 1	State of register		The real numbers of bits changes		The recognized numbers of bits changes	
	W	F	The numbers of bits changed on data bus after fetching the instruction argument	The numbers of bits changed in data bus after executing the instruction	The numbers of bits changed on data bus after fetching the instruction argument	The numbers of bits changed in data bus after executing the instruction
1 MOVLW .196	196	0	3	0	3	0
2 BSF 15,1	196	2	3	1	3	1
3 ADDLW .224	164	2	4	2	4	3
4 BSF 15,2	164	6	4	1	3	1
5 ADDLW .107	15	6	5	3	5	3
6 COMF 15,1	15	249	2	8	2	8
7 NOP	15	249	6	4	6	5
8 MOVLW .247	247	249	5	0	5	0
9 MOVF 15,1	247	249	3	0	4	0
10 MOVF 15,1	247	249	0	0	0	0
11 ...	...	...	...	...	...	...

the value of 1 (3 bits changes), in the third machine cycle ALU sets the operation result (value 196) on data bus, which maintains the same state (0 bits changes).

- In the BSF 15,1 instruction, occurs overwriting of the state on data bus 196 (binary 11000100), through the contents of F register, with the address 15 equals 0 – resetting the state of data bus (3 bits changes). Next, BSF instruction set a chosen bit (1) of register F in the machine cycle $Q3$ and set the result (binary 00000010) on the data bus (1 bit change).

- Before the instruction ADDLW .224 is executed, the state on data bus has the value of 2 (binary 00000010), then the argument 224 (binary 11100000) is received and the new value is overwritten on the data bus (4 bits changes). Next, argument 224 is added into the contents of W register with the value of 196, the result – 164 (binary 10100100) is set on data bus, and the argument fetched in the first machine cycle is overwritten (2 bits changes).

The analysis of all instructions in test programs 1, 2 and 3 was conducted in the similar way, and then the real numbers of bits changes, for each instruction on data bus, were compared to recognised numbers of changes (correctly recognition is marked in gray color in Table 1). The results of this comparison are presented in Table 2. For the following 100 instructions of test programs 1, 2 and 3 the effectiveness of proper recognition of bits changes on data bus after fetching instruction argument is as follows: 81 %, 85 % and 80 %. For the remaining instructions, the maximum mistake in the numbers of correctly recognised bits changes is one. The similar remarks refer to the numbers of bits changes achieved on data bus in the third machine cycle. For three test programs the numbers of correctly recognised bits changes on data bus being the result of fetching the instruction argument and executing the operation are approximately 82 % and 71.67 %.

Table 2. Numbers of correctly recognized bits changes on data bus after fetching the instruction argument and executing the instructions for test program 1, 2 and 3

	Numbers of correctly recognized bits changes on data bus after fetching the instruction argument (out of 100 instructions)	Numbers of correctly recognized bits changes on data bus after executing the instruction (out of 100 instructions)
Program 1	81	69
Program 2	85	72
Program 3	80	74
Average value	82	71.67

5 Conclusion

Presented results show the high efficiency of the numbers of correctly recognised bits changes on data bus resulting from fetching the argument and executing the instruction. During the research, it was noticed that the numbers of bits changes on data bus after fetching the argument in the machine cycle 1, and after executing the instruction and setting the result on data bus in the machine cycle 3, caused the significant changes of voltage amplitude measured on microprocessor power supply, during the instruction cycle. Based on these observations, an attempt was made in order to render the information about the amount of bits changes on data bus, focusing only on the voltage waveforms measured for the following instructions in test programs. For three test programs the numbers of correctly recognised bits changes on data bus during the machine cycle *Q1* and *Q3* are approximately 82 % and 71.67 %. In this case, it is of course impossible to answer directly the question concerning the argument value, for which a particular instruction was executed. However, this method can be used if the implementation of algorithm is known (presented method recognises the algorithm in the first step), and if there is a possibility to select the input data, i.e. in cryptographic algorithm. Based on the several iteration for different values of input vector (plain text) and the observation of the result, there is a possibility to recognise the instruction argument, which in case of cryptanalysis may be an additional source of information about the encryption key. For example XOR operation, being the part of many developed cryptographic algorithms, especially many stream and block ciphers, is undoubtedly prone to such kind of attacks. The essential properties of this function is its commutativity. This feature causes that XOR instruction is very often used in cryptography.

The paper presents the problem of security of microprocessor systems, and it discusses also the possibility of threats for programs written into the system memory and the information processed. The developed method for analysing the microprocessor program code based on the voltage supply changes, enables also to recognise the numbers of bits changes on microcontroller data bus, as a result of realized instruction. In such context, it can constitute a potential threat for data processed by microcontroller program or embedded systems. What is more, such threat can be the issue for cryptology research, where the security of cryptographic algorithms is analysed not only from the mathematical point of view, but also as the real hardware and software implementations.

Acknowledgments. This work was supported by the European Union from the European Social Fund (grant agreement number: UDA-POKL.04.01.01-00-106/09).

References

1. Kuhn, M.G.: Security limits for compromising emanations. In: Rao, J.R., Sunar, B. (eds.) CHES 2005. LNCS, vol. 3659, pp. 265–279. Springer, Heidelberg (2005)
2. Tanaka, H.: Information leakage via electromagnetic emanations and evaluation of tempest countermeasures. In: McDaniel, P., Gupta, S.K. (eds.) ICISS 2007. LNCS, vol. 4812, pp. 167–179. Springer, Heidelberg (2007)
3. Kwiecień, A., Stój, J.: The cost of redundancy in distributed real-time systems in steady state. In: Kwiecień, A., Gaj, P., Stera, P. (eds.) CN 2010. CCIS, vol. 79, pp. 106–120. Springer, Heidelberg (2010)
4. Pieprzyk, J., Hardjono, T., Seberry, J.: Fundamentals of computer security. Springer, Heidelberg (2003) ISBN 978-3-540-43101-5
5. Stera, P.: Company's data security – case study. In: Kwiecień, A., Gaj, P., Stera, P. (eds.) CN 2010. CCIS, vol. 79, pp. 290–296. Springer, Heidelberg (2010)
6. Gaj, P., Jasperneite, J., Felser, M.: Computer communication within industrial distributed environment – a survey. IEEE Transactions on Industrial Informatics 9(1), 182–189 (2013)
7. Sidzina, M., Kwiecień, B.z.: The algorithms of transmission failure detection in master-slave networks. In: Kwiecień, A., Gaj, P., Stera, P. (eds.) CN 2012. CCIS, vol. 291, pp. 289–298. Springer, Heidelberg (2012)
8. Clayton, P.: Introduction to electromagnetic compatibility, 2nd edn. John Wiley and Sons, New Jersey (2006) ISBN: 978-0-471-75500-5
9. Montrose, M., Nakauchi, E.: Testing for EMC compliance: approaches and techniques. Wiley-IEEE Press, Canada (2004) ISBN: 978-0-471-43308-8
10. Kwiecień, A., Maćkowski, M., Skoroniak, K.: Instruction prediction in microprocessor unit. In: Kwiecień, A., Gaj, P., Stera, P. (eds.) CN 2011. CCIS, vol. 160, pp. 427–433. Springer, Heidelberg (2011)
11. Kwiecień, A., Maćkowski, M., Skoroniak, K.: Reverse engineering of microprocessor program code. In: Kwiecień, A., Gaj, P., Stera, P. (eds.) CN 2012. CCIS, vol. 291, pp. 191–197. Springer, Heidelberg (2012)
12. Bao, F., Deng, H., et al.: Breaking public key cryptosystems on tamper resistant devices in the presence of transient faults. In: Christianson, B., Crispo, B., Lomas, M., Roe, M. (eds.) IW 1997. LNCS, vol. 1361, pp. 115–124. Springer, Heidelberg (1998)

13. Biham, E., Shamir, A.: Differential fault analysis of secret key cryptosystems. In: Kaliski Jr., B.S. (ed.) CRYPTO 1997. LNCS, vol. 1294, pp. 513–525. Springer, Heidelberg (1997)
14. Ahn, M., Lee, H.-J.: Experiments and hardware countermeasures on power analysis attacks. In: Gavrilova, M.L., Gervasi, O., Kumar, V., Tan, C.J.K., Taniar, D., Laganá, A., Mun, Y., Choo, H. (eds.) ICCSA 2006. LNCS, vol. 3982, pp. 48–53. Springer, Heidelberg (2006)
15. Dabbish, E.A., Messerges, T.S., Sloan, R.H.: Investigations of power analysis attacks on smartcards. In: WOST 1999, Proceedings of the USENIX Workshop on Smartcard Technology, Chicago (1999)
16. Vermoen, D., Witteman, M., Gaydadjiev, G.N.: Reverse engineering java card applets using power analysis. In: Sauveron, D., Markantonakis, K., Bilas, A., Quisquater, J.-J. (eds.) WISTP 2007. LNCS, vol. 4462, pp. 138–149. Springer, Heidelberg (2007)
17. Mayer-Sommer, R.: Smartly analyzing the simplicity and the power of simple power analysis on smartcards. In: Paar, C., Koç, Ç.K. (eds.) CHES 2000. LNCS, vol. 1965, pp. 78–92. Springer, Heidelberg (2000)
18. Broujerdian, M., Doostari, M., Golabpour, A., et al.: Differential power analysis in the smart card by data simulation. In: Proceedings of the 2008 International Conference on MultiMedia and Information Technology, MMIT 2008, USA, pp. 817–821 (2008)
19. Kocher, P.C., Jaffe, J., Jun, B.: Differential power analysis. In: Wiener, M. (ed.) CRYPTO 1999. LNCS, vol. 1666, pp. 388–397. Springer, Heidelberg (1999)
20. Kwiecień, A., Maćkowski, M., Skoroniak, K.: The analysis of microprocessor instruction cycle. In: Kwiecień, A., Gaj, P., Stera, P. (eds.) CN 2011. CCIS, vol. 160, pp. 417–426. Springer, Heidelberg (2011)

The Concept of Software-Based Techniques of Increasing Immunity of Microprocessor Unit to Electromagnetic Disturbances

Andrzej Kwiecień, Michał Maćkowski, and Krzysztof Skoroniak

Silesian University of Technology, Institute of Computer Science,
Akademicka 16, 44-100 Gliwice, Poland
{akwiecien,michal.mackowski,krzysztof.skoroniak}@polsl.pl
http://www.polsl.pl/

Abstract. The essential feature of modern IT systems, related to a device reliability, is their immunity to electromagnetic disturbances. The paper deals with the software-based techniques used for detecting disturbances, which can appear during the microcontroller work. The use of presented techniques allows the system to response appropriately to the threat.

Keywords: software prevention, embedded software, program code, microcontroller, EMC, conducted emission, electromagnetic disturbances, electromagnetic interference, measurements, immunity.

1 Introduction

Nowadays, the disturbances which are constantly introduced into the electromagnetic environment has become greater because of the commonly used wireless transmission and electronic devices. Therefore, the very important feature of modern IT systems, which ensures operation reliability, is their immunity to electromagnetic disturbances. It can be defined as the property of an appliance, device or a system, referring to the ability of the equipment to function without lowering its quality, when the electromagnetic disturbances occur [1, 2]. The immunity is the fundamental concept in electromagnetic compatibility, which defines the relationship between test system and electromagnetic disturbance.

The absence of appropriate resources used for protecting distributed IT systems, in particular network infrastructure, may lead to disturbances of data transmission [3, 4]. These disturbances, appearing in the form of transmission errors, contribute directly to the decrease of system performance, and in the worst case may prevent its operation.

The aim of electromagnetic compatibility is to increase the reliability of devices and whole IT systems that work in an environment, where lots of electromagnetic disturbances occur [5–7]. Thus, the very essential aspect of this research is the use of software-based techniques designed for increasing the security of devices against the influence of electromagnetic disturbances.

A. Kwiecień, P. Gaj, and P. Stera (Eds.): CN 2013, CCIS 370, pp. 507–516, 2013.
© Springer-Verlag Berlin Heidelberg 2013

Microprocessor units currently used in network devices as network controllers, can be also considered as advanced chips being responsible for data processing and reconstruction of transmitting frames. Such units can also be affected by the various kind of electromagnetic interferences.

There are many software methods used for increasing the immunity of microprocessor units [8–10]:

- The most important is the use of watchdogs, allowing for restart of the processor in case of system hang, without user interference. Its intention is to bring the system back from an unresponsive state into normal operation.
- Redundancy of systems allows processing the information in few modules, and next comparing the result. However, it makes the system design more sophisticated, and at the same time may increase the cost of the system.
- Triggering via signal level instead of the edge, enables to avoid the system reaction on temporary change of signal state, being the result of electromagnetic disturbances.
- "Resetting" the unused program memory by inserting NOP instruction in all unused memory cell or jump instruction to the beginning of a program. Thanks to such solution, it is possible to predict the processor work in case of appearing an unauthorized address.
- Saving the sensitive data in several places in non-volatile memory and implementing the method for data selection when an error occurs.
- Using digital transmission, which together with the mechanism of detecting and correcting errors, is a much higher resistant than analogue one.
- The unused interruptions should indicate the safe procedures (service routine), which for instance inform users about the system malfunction.
- There are also methods used for increasing the immunity, which consist of analysis of currently receiving results and rejection of the incorrect ones, or those being beyond the tested area.

The main goal of all presented method is one – to decrease the sensitivity of microprocessor systems to electromagnetic disturbances. However, there may be a situation where, despite of using the above methods, a device work may be disturbed, and neither the system itself nor the user can detect it. In consequence may lead to unexpected errors.

The issue connected with the use of software to increase the system immunity to electromagnetic disturbances has been discussed in many papers [11–15]. Such technique is often called defensive programming. The biggest attitude of using defensive programming is that it is inexpensive to implement, and if done correctly can save hardware costs in PCB layout [12].

Papers [11, 15] describe and discuss the method of increasing the immunity of microcontroller to conducted continuous-wave interference. This method uses the analog digital converter (ADC), which measures the reference voltage 2.5 V. If the ADC conversion result is different from a fixed known value, then the microcontroller will know that there is interference at the power supply or input port of ADC. The current paper (contrary to paper [15]) presents the method of detecting pulse disturbances, which means with the short duration time. In many cases detection of pulse disturbances is much more difficult than a continuous disturbances.

Many studies [13, 14] contains only theoretical considerations and guidelines for the use of hardware and software to increase the immunity of the microcontroller to electromagnetic disturbances. A very interesting solution is a technique described in [13], which focuses on passing token between procedures in a program. In this case, the called method may check the authority of calling routine.

The study presents the software-based techniques for disturbance detection, which under certain circumstances can detect incorrect system work. It gives the opportunity for an appropriate reaction of a system or stuff, supervising a given system. This method, contrary to others papers, has also been tested with the use of pulse electromagnetic disturbances of ESD (*Electrostatic Discharge*) kind. The research results are presented in the paper.

2   Software-Based Techniques for Disturbances Detection

The developed method allows determining the incorrect work of processor, caused by the influence of electromagnetic disturbances of suitably high intensity. This method focuses on repeating the sequence of instructions that operate on particular processor registers, when the result of operation is predicted in advance. If the result, after executing a particular sequence of instructions, is not compatible with the expected one, then it is a sign that in this moment the microcontroller is being exposed to electromagnetic disturbances, which cause its fail work. Sequence of instructions can be calling periodically, e.g. inside the interrupt service routine from the timer.

An exemplary sequence of instructions realized periodically may consist of instructions for moving constant value to the register, its negation and subtraction from it the constant value (the process is described more precisely in research procedure). After executing each sequence, the final register should indicates the value 0. If the value is different than 0, then it probably means that in one or couple of steps some kind of distortion occurred.

Research presented in the paper were realized with the use of a special equipment for generating pulses similar to the static electricity discharges, on the test bench prepared according to PN-EN 61000-4-2 standard *Testing and measurement techniques – Electrostatic discharge immunity test*. This standard includes the requirements referring to the immunity of electronic or IT equipment, which is exposed to static electricity discharges. The discharge may occurs either via direct contact of the system with the charged object – contact discharge, or via the electric arc in the air – air discharge. The test levels are between 500 V and 15 kV. The use of test levels and research procedure are presented below in details.

3   Test Bench and Research Procedure

The test bench was prepared according to requirements of PN-EN 61000-4-2 norm, and the schema of test bench is presented on Fig. 1. Equipment under test (EUT) was 8-bit PIC16F84A microcontroller manufactured by Microchip. This processor represents very numerous group of Mid-Range processors. There are about 85 models in this series of processors. The main differences among microcontrollers included in Mid-Range class concern: the size of data memory (RAM/EEPROM) and Flash program, the maximum work frequency of circuits, and available numbers of communication modules. The processor used in test, is free of advanced peripheral blocks. PIC16F84A processor is built on the basis of CMOS technology (*Complementary Metal-Oxide-Semiconductor*) and Harvard architecture, which means that its data memory and program memory are separated. The access to the memory of a program, which is produced in Flash-ROM technology and can store 1024 instructions is via 14 bits program bus. 68 bytes of RAM and 64 bytes of EEPROM memory can be used for data storing. The access to them is via 8 bits data bus. Microcontroller was powered by the battery with the voltage of 5.5 V and clocked with frequency of 1 MHz.

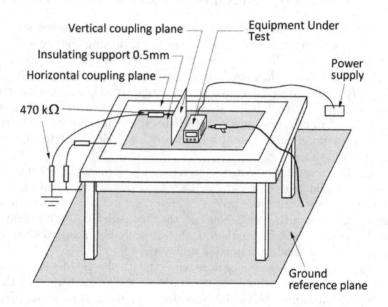

Fig. 1. Test setup for electrostatic discharge immunity test

Standard PN-EN 61000-4-2 determines a number of test levels (Table 1), which requires the device immunity test for electrostatic discharges. A pulse with a determined current shape and amplitude is given with the hand-held generator (*ESD Gun*). The ESD simulator consists of a capacitor in series with a resistor. The capacitor is charged to a specified high voltage from an external source, and then suddenly discharged through the resistor into an electrical terminal of the device under test.

Table 1. Test levels for electrostatic discharge immunity test

Contact discharge		Air discharge	
Level	Test voltage [kV]	Level	Test voltage [kV]
1	2	1	2
2	4	2	4
3	6	3	8
4	8	4	15
x[1]	Special	x[1]	Special
[1] "x" can be any level, above, below or in between the others. The level shall be specified in the dedicated equipment specification. If higher voltages than those shown are specified, special test equipment may be needed.			

Following the standard, the most preferable are contact discharges, which are about putting the end of ESD generator to the conducted part of a device and releasing the pulse. Such test is the best for reproducing the real conditions and makes majorly, that the results do not depend on such parameters as pressure, temperature and humidity. Therefore during the test, contact discharges and voltage at the level of 4 kV were used. The level of amplitude is established by the standard PN-EN 55024 *Information technology equipment – Immunity characteristics – Limits and methods of measurement* to which microcontroller under test can be qualified.

At the beginning of the research, disturbing signals were generated in all metal parts of the microcontroller. It was noticed then, that errors in microcontroller work occurred mainly when electrostatic discharges were generated directly into the microcontroller clock lines. Hence, the further research were conducted using the above idea.

In order to check the immunity of microcontroller to ESD pulse disturbances at the level of 6 and 8 kV (levels 3 and 4 in Table 1) additional tests were also conducted. However, it turned out that after several ESD discharges the processor stopped working (was broken). In this case probably the internal silicon structure was damaged, and inner protection circuit against electromagnetic disturbances were insufficient. As mentioned before microcontroller belongs to the group of IT equipment, which according to the standard PN-EN 55024 should be immune to contact ESD disturbances at the level 2 (4 kV). Therefore, the further research were conducted according to the above requirement.

The research goal was to develop a method which allow determining an incorrect work of microcontroller resulting from electromagnetic disturbances, and consequently a proper system reaction. Figure 2 presents a flow chart of a program implemented in processor under test. The program is saved into internal program memory of microcontroller and is divided into three basic parts:

1. Initialization of processor registers (setting inputs, outputs and registers states).
2. The main part of a program is responsible for microcontroller error detection, which results from interference of electromagnetic disturbances. In this part a set of instructions, such as: moving the value to register, complementary and subtract operation is realized.
3. If an error occurs during the program execution, it may then indicates that processor was exposed to electromagnetic disturbances. In that case, the states of registers used in program in point 2 are written into EEPROM memory.

Writing the registers contents into EEPROM memory enables to reproduce their states on PC and make an appropriate analysis in further stage. It is possible then, to indicate a particular place in the program or even a particular instruction, where bits errors occurred.

A LED diode, which signalises the program end is turned on after writing data into EEPROM internal memory. In some cases, an error occurs after second or even third electrostatic discharge into microcontroller clock lines. Hence, the turned diode signalises a situation, where electromagnetic disturbances caused microcontroller malfunction.

4   The Research Results

In order to test the effectiveness of the developed method a short test program, equipped with the procedure of disturbances detection (Fig. 2) was written. Next, this program was executed 10 times, generating each times ESD disturbances and checking the correctness of its work.

In each program realization the processor was exposed to the electrostatic discharges of the value 4 kV. ESD discharge was put as many times as the disturbances in microprocessor work appeared. Table 2 presents the contents of registers with the addresses 30, 31 and 32, where the results of particular instructions were saved. An incorrect contents of registers, which means the one that is different from the expected one after executing a particular processor instruction, is marked in Table 2 with gray color. For instance, during the first realization of test program in case of negation of register contents with address 30 (01010101), the expected value is 10101010. However, as a consequence of exposing the processor on ESD disturbances the received values was 01010100, which may indicate to 7 bits errors while the COMF instruction is executed (instruction of negation).

If the error occurs while the first instruction is being executed, then the fail result is also propagated for the following instructions, such as during tests 2 and 4 of the program. In test 7 on the other hand, the errors occur during the execution of two following instructions. Although the nature of electrostatic

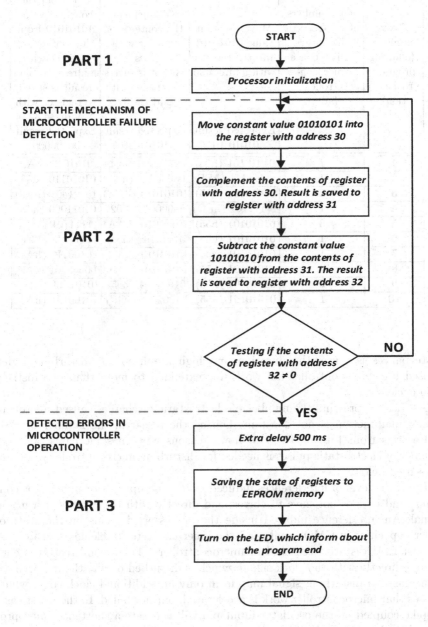

Fig. 2. Flow chart of developed program, responsible for microcontroller failure detection

Table 2. Electrostatic discharge test results

Test number (following program realization)	The numbers of electrostatic discharges which cause microcontroller failure	Move constant value 01010101 into the register with address 30	Complement the contents of register with address 30. Result is saved to register with address 31	Substract the constant value 10101010 from the contents of register with address 31. The result is saved to register with address 32
		Expected result 01010101	Expected result 10101010	Expected result 00000000
1	1	01010101 (h55)	01010100 (h54)	10101010 (hAA)
2	1	11111111 (hFF)	00000000 (h00)	01010110 (h56)
3	2	01010101 (h55)	10101010 (hAA)	10101001 (hA9)
4	1	11111101 (hFD)	00000010 (h02)	01011000 (h58)
5	3	01010101 (h55)	10101010 (hAA)	11111101 (hFD)
6	1	01010101 (h55)	01010100 (h54)	11111111 (hFF)
7	1	01010101 (h55)	10101010 (hAA)	10100110 (hA6)
8	3	01010101 (h55)	10101010 (hAA)	00000001 (h01)
9	2	01010101 (h55)	01010101 (h55)	10101011 (hAB)
10	1	01010101 (h55)	01001111 (h4F)	10100101 (hA5)

disturbances is pulse, the duration and high amplitude of disturbing signal caused the microcontroller fail work was extended to more than one instruction cycle.

As Table 2 presents the number of electrostatic discharges caused microcontroller malfunction is not constant. During the research, the number of ESD pulses was from 1 to 3. The two main reasons why, in some cases, a larger number of electrostatic pulses is needed for disturbing microcontroller work, are presented below.

First, the level of interferences induced inside microprocessor unit is not constant, and it does not have to correspond directly with the level of disturbing signal. An interference induced inside the processor, depends on the state of inner capacities in CMOS unit, and the inner character of elements protecting against EMI disturbances used in microcontrollers. The second reason is connected directly with the clock signal, which is disturbed during the research. In some cases, a disturbing signal may form only an additional clock edge, which causes that microcontroller work is accelerated, but not failed. In the worst case, a spike coupled on the oscillator input pin will generate a shortened microprocessor clock period, which is liable to cause a spurious instruction/data access, and corrupting the program counter. The unexpected clock transition will cause a microprocessor machine cycle to be shorter than normal, which may give inadequate time for the program memory to respond. Then, the next instruction fetch would be incorrect, causing the system error.

5   Conclusion

One of the main feature of systems based on various microcontrollers units (e.g. measuring systems) are strict requirements as to their work reliability. Usually, the system reaction on incorrect work, both of hardware and software, is very short, and the delayed response can have serious consequences in a technological process. Therefore, the issue related to the device ability to work in an independent and correct way is very essential.

The presented research describes one of the aspect for developing software, which enables to increase the reliability of the whole system. The developed method allows determining an incorrect work of microprocessor, which is the effect of electromagnetic disturbances. It gives the system itself or the stuff supervising a particular system an opportunity to response properly. In case of the lack of presented method, the temporary system malfunction is possible not to be noticed. Such situation may carry serious consequences in system operation.

The sequence of instructions that verifies the appropriate work of a microcontroller can be called periodically, e.g. inside the interrupt service routine that comes from timer. Thus, the higher frequency of calling the procedure for error detection, the chance to detect an error is greater. Obviously, the frequency of calling the presented method is some kind of compromise between the rate of a proper program work executed by microprocessor, and the probability of error detection, such as the effect of electromagnetic disturbances.

It is worth to mention the fact that the disturbances in microprocessor work, for test levels, occurred only when the electrostatic pulses were discharge to the clock lines. It means that those lines are especially prone to such pulse influences, which have very short signal rise time – about few nanoseconds.

The conducted research and the research results are a kind of introduction to the further research connected with the use of software for increasing reliability of microprocessor systems and improving their immunity to various electromagnetic disturbances.

Acknowledgments. This work was supported by the European Union from the European Social Fund (grant agreement number: UDA-POKL.04.01.01-00-106/09).

References

1. Bernardon, D.: Improved electromagnetic immunity circuit design. In: Roermund, A., Casier, H., Steyaert, M. (eds.) Analog Circuit Design, pp. 203–217. Springer, Netherlands (2006)
2. Deutschmann, B.: Improvement of system robustness through EMC optimization. In: Roermund, A., Steyaert, M., Huijsing, J.H. (eds.) Analog Circuit Design, pp. 227–242. Springer, US (2004)
3. Maćkowski, M.: The influence of electromagnetic disturbances on data transmission in USB standard. In: Kwiecień, A., Gaj, P., Stera, P. (eds.) CN 2009. CCIS, vol. 39, pp. 95–102. Springer, Heidelberg (2009)

4. Maćkowski, M., Skoroniak, K.: Electromagnetic emission measurement of microprocessor units. In: Kwiecień, A., Gaj, P., Stera, P. (eds.) CN 2009. CCIS, vol. 39, pp. 103–110. Springer, Heidelberg (2009)
5. Clayton, P.: Introduction to Electromagnetic Compatibility, 2nd edn. John Wiley and Sons, New Jersey (2006) ISBN: 978-0-471-75500-5
6. Montrose, M., Nakauchi, E.: Testing for EMC compliance: approaches and techniques. Wiley-IEEE Press, Canada (2004) ISBN: 978-0-471-43308-8
7. Carlton, R., Racino, G., Suchyta, J.: Improving the Transient Immunity Performance of Microcontroller-Based Applications Freescale Semiconductor. Application Note, AN2764 (2005)
8. Ong, H.L.R., Pont, M.J., Peasgood, W.: Do software-based techniques increase the reliability of embedded applications in the presence of EMI? Microprocessors and Microsystems 24(10), 481–491 (2001)
9. Ganssle, J.: Watching the watchdog. Embedded Systems Programming 16 (2003)
10. Vick, R., Habiger, E.: The dependence of the immunity of digital equipment on the hardware and software structure. In: Electromagnetic Compatibility, Beijing, pp. 383–386 (1997)
11. Baffreau, S., BenDhia, S., Ramdani, M., et al.: Characterisation of microcontroller susceptibility to radio frequency interference. In: Proceedings of the Fourth IEEE International Caracas Conference on Devices, Circuits and Systems, pp. I031-1–I031-5 (2002)
12. Campbell, D.: Defensive software programming with embedded microcontrollers. In: Electromagnetic Compatibility Of Software, IEE Colloquium, London, pp. 3/1–3/5 (1998)
13. Coulson, D.R.: EMC techniques for microprocessor software. In: Electromagnetic Compatibility Of Software, IEE Colloquium, London, pp. 2/1–2/5 (1998)
14. McKeever, K.M.: Design for effective EMC in microprocessor based systems. In: Interference and Design for EMC in Microprocessor Based Systems, IEE Colloquium, London, pp. 1/1–1/3 (1990)
15. Wan, F., Duval, F., Cao, H., et al.: Increase of immunity of microcontroller to conducted continuous-wave interference by detection method. Electronics Letters 46(16), 1113–1114 (2010)

Hardware Aspects of Data Transmission in Coal Mines with Explosion Hazard

Marek Kryca

Institute of Innovative Technologies EMAG,
Leopolda 31, 40-189 Katowice, Polska
marek.kryca@emag.pl
http://www.emag.pl

Abstract. The diversity and commonness of the sensors used for monitoring industrial processes and the widespread use of the digital data processing methods for the measurements are crucial issues related to the communication and data collection. The popular technical solutions for fast and reliable transmission of large packets of information cannot be directly applied in an environment where the risk of explosion appears. The paper shows the physical layer limitations for the transmission links with the potential danger of explosion of methane and coal dust that are characteristic for coal mines. Areas of mine-specific applications for data transmission were summarized and an example of safety parameters calculation for the RS485 communication standard was presented.

Keywords: explosion hazard, coal mines, wireless transmission, intrinsically safe.

1 Introduction

High diversity of communication technologies available on the market today reflects the customers' demands. The industrial applications need reliability and robustness but some of them need also safety. A coal mine is a good example of such a customer – high performance requirements coexist with the crucial role of transmitted data and extremely difficult environmental conditions. These environmental issues are very important not only for long and trouble-free devices exploitation but also for the safety of the working miners. Methane, coal dust and air make explosive mixtures, therefore the usage of any electrical equipment must be strictly controlled. An outline of electrical requirements defined in Polish standards referred to the ATEX Directive can explain parameter limitations. A wide range of areas of mine-specific applications for data transmission shows that technologies used in standard devices will be transferred to the coal mines.

2 European Legislative Backgrounds

Each kind of device before implementation must be checked for correctness of its design and it is carried out according to the corresponding standard. Measuring

A. Kwiecień, P. Gaj, and P. Stera (Eds.): CN 2013, CCIS 370, pp. 517–530, 2013.
© Springer-Verlag Berlin Heidelberg 2013

and communication equipment intended to be used in underground parts of coal mines must be designed and manufactured according to the Low Voltage Directive (LVD) 2006/95/EC, Electromagnetic Compatibility (EMC) Directive 2004/108/EC and the ATEX 95 Equipment Directive 94/9/EC. There are sets of standards connected with each directive but only few are discussed in this paper.

2.1   Low Voltage Directive (LVD)

The Low Voltage Directive is a guide for designers to produce safe electrical equipment with supply voltage between 50 and $1000\,V_{AC}$ or between 75 and $1500\,V_{DC}$. Measuring and communication devices are often supplied by external power supply with low voltage output so in that case standards connected with LVD are unnecessary.

2.2   Electromagnetic Compatibility (EMC) Directive

The Directive applies to all electronic or electrical products liable to cause or be disturbed by electromagnetic interference (EMI). The range of this directive covers devices used on the surface of the ground, however for underground parts of mines the same standards must be followed based on a governmental act. The discussion continues on disabling the underground mining workings areas from complying with regulations related to the use of the radio spectrum and thereby obtaining radio licenses. The lack of licenses can lead to the disruption of important functions of underground systems but certainly will not influence devices in use on the surface.

2.3   ATEX 95 Equipment Directive 94/9/EC

The Directive covers equipment to be used in potentially explosive areas. The equipment is treated as a source of ignition – harmonized standards help constructors to estimate potential risk of explosion during normal work and faults as well. ATEX marking describes five features of the device:

- area of application: group I (mining) or group II (industrial). Group II is subdivided into IIA, IIB and IIC according to the gases ignition energy,
- equipment category: for group I level of protection M1 (Very High) or M2 (High); for group II level of protection 1 (Very High), 2 (High) or 3 (Standard),
- atmosphere: G (gas) or D (combustible dust),
- type of protection applied: d-flameproof enclosure, ia, ib- intrinsic safety, etc.,
- temperature: maximum surface temperature reached in the worst condition that can provoke the explosion. Group II has 6 temperature classes ($T1 > 450\,°C$ up to $T6 < 100\,°C$) Hard coal ignition temperature for deposit is $245\,°C$ and for cloud $590\,°C$.

The comparison of methane and other flammable gases shows that methane is relatively "safe" – the ignition temperature is about 540 °C, nevertheless devices dedicated for use in chemical industry (group II) could not be applied in coal mines. ATEX marking for group II is interpreted as "not suitable for mining application marked as group I". Standards which are useful during the communication equipment certification process belong to the group EN-60079 which consists of 35 standards.

3   Areas of Application

The communication issues are connected with almost all devices used in the underground parts of the coal mines. Modern solutions developed for typical applications become popular in a short time, yet the underground coal industry is always behind the main stream of changes. The environmental restrictions have strong influence on implementing advances – there are many solutions developed, verified and applied for commercial applications desired by miners, which should be modified in such a way as to be installed in the underground areas where the explosion hazard occurs.

3.1   Process Control

The natural application of a communication link is the remote process control. A suitable example is a longwall shearer which works in an extremely risky area. The shearer is a large mining machine which cuts coal along the wall between two parallel service roads. Extracted coal is transported by a chain conveyor. Along the wall hydraulically powered supports are installed which prevent the roof from collapsing on the shearer. After each pass the shearer and its rail system are moved towards an uncut coal seam. The same move is made by the roof supporting system and unsupported area behind collapses. The excavation works are dangerous for the personnel due to falling rocks, proximity of the shearer moving parts, rock and methane bursts or possibility of methane and coal dust explosion. The objective to evacuate miners from the area of excavation is reflected in remote control systems developed by several companies. An example of the shearer remote control developed by EMAG is shown in Fig. 1.

3.2   Video Transmission

A vital element of remote control is a video surveillance system. Requirements for this kind of transmission are not very high – usually sending 20–30 pictures in one minute is enough to control the correctness of the process. This kind of data is useful for the shearer operators and can be the source of information for image analysis tools. The shearer is a crucial element of the coal excavation system so the abnormalities in its work should be detected as soon as possible, therefore the video signals and data received from other sensors are transmitted to the dispatch room. Cameras are installed in heavy flameproof enclosures (Ex d protection)

Fig. 1. Remote control for the shearer

or pressurized ones (EX p protection) filled with nitrogen. Power supply and transmission lines are intrinsically safe (ia or ib protection).

3.3   Voice Transmission

Voice communication has high requirements – the amount of data sent with time delay constraints makes this kind of communication challenging. Traditional telephones are still the backbone of communication systems in the underground mines. The development process in that area made them more universal – telephones have become a part of an emergency dispatcher's systems. A programmable telephone signaling device makes possible except for the standard telephone communication, announcing alarm signals, as well as previously recorded and stored in the inside memory verbal and acoustic messages. They also can be used to call the dispatcher by pushing buttons ALARM and DYSP, in the alarm and regular modes. The switching of broadcasting and listening functions enables to carry out a simplex conversation with persons who are in close vicinity of the signaling device. After picking up the microtelephone it performs telephone functions. Ruggedized construction is dedicated for dusty environments with extremely high level of noise. Voice communication between shearer operators increase the safety and accuracy of their work. This application is usually based on the Bluetooth standard.

3.4   Measurements

Technological progress in coal excavation methods, measurements of coal quality and the development of sensors measuring environmental parameters have generated a strong demand for efficient communication channels. Information about methane concentration can generate a warning signal for workers and in the dispatch room. Sensors monitoring others parameters of atmosphere like humidity, temperature, CO, O_2 concentration, ventilation efficiency, dust in air concentration, etc, are connected in the dedicated networks. Information about excavated coal quality is useful for economical analyses so data must be transferred to offices located on the surface. A methane concentration monitoring devices usually have the transmission link and outputs switching off electricity.

3.5 Tagging Systems

Identification systems using radio frequency – RFID (Radio Frequency Identification) are applied also in hazardous areas like coal mines. The system consists of intrinsically safe tags, readers and antennas. The applied anti-collision protocols allow recognition up to 50 transponders in the antenna field at the same time. Apart from standard applications like time keeping or access control for secure areas, the system can be used for vehicle tracking, asset tracking and evacuation systems. The reader ability of determining the movement direction enables automatic identification, counting and storing information about any number of people in the given area. This information is vital during the rescue actions. RFID readers and antennas are installed in strategic locations and integrated with other transmission systems sending data to the surface. Personal tags are usually embedded into lamps due to two reasons – this is a source of energy for active tags and, more importantly, the lamps as a source of indispensable light are always carried by miners and are always attached to their belts. An example of an RFID gate developed in cooperation between Davis-Derby Company and Sevitel is shown in Fig. 2.

Fig. 2. RFID control point which consists of two antennas and a reader (source www.sevitel.pl)

One of the most challenging applications is tagging explosives and detonators. The use of mobile and stationary readers enables to trace explosives during a transportation process and to find misfired ones. The RFID tags will not be destroyed in the case of a misfire thus before drilling new boreholes in rocks a worker can check the RFID signals and this way minimize the risk of a fatal accident. The energy transmitted by the reader antenna should activate a microcontroller in the tag but the wires of blasting circuits or the wires connected to detonators can aggregate enough energy to initiate an explosion. The most dangerous situation exists when wiring is as long as one-half wavelength of the transmitted radio wave [1]. The approvals which describe the rules of application for that kind of equipment define the minimum safe distance between a single reader and explosives [2]. The user is responsible for fulfilment of this requirement.

4   Transmission Methods

The communication systems for underground mines can be classified as follows [3]:

- TTW (through-the-wire) systems for which the medium of transmission are copper wires or fiber optic cables,
- TTA (through-the-air) systems based on the propagation of radio waves through the air in the excavated areas,
- TTE (through-the-earth) systems transmitting data through the surrounding strata.

There are some solutions which combine different methods of transmission – for example a leaky feeder applied as a long antenna for short range wireless phones. Each of these methods is dedicated to strictly defined applications, although energy limits are common. The approving authorities check safety features of the equipment, the safety resulting from electrical and mechanical construction and this way the hazard in the case of faults is estimated.

4.1   The Wireless Transmission

This kind of transmission covers TTA and TTE systems. Frequencies are in the range from a single kHz up to a few GHz. The propagation of radio waves in coal mines is very unique because the rule of straight line does not usually work [4]. The electrical properties of coal, wide range of frequencies and complicated pattern of service roads make an estimation of the transmission range very difficult. The attenuation of radio waves depends on the wave frequency and amount of steel elements in the area of communication and that fact determines the selection of a proper solution. Transmission channels exploiting low frequencies have lower throughput but better propagation parameters.

Electromagnetic waves as a possible ignition source together with explosive atmosphere and metallic object which functions as an antenna can provoke the explosion. The special kind of antenna can be created by a mechanical failure of coil or buzzer having interruption at a soldering point. To eliminate this risk the limit values of the radiated power were defined. For the explosion group I (mining industry) the power limit value for continuous radiation is 6 W averaged by 200 μs and for pulsed radiation 1500 μJ. These limits are pretty high but are concerned with the fault analysis and refer to EIRP (equivalent isotropically radiated power). Some applications require concentrating emitted energy in one direction while others are used to cover a certain area. The ability to concentrate energy is called antenna gain. The EIRP of the complete device is calculated as a sum of ERP (effective radiated power) of a transmitter, the gain of the antenna and is diminished the attenuation of the antenna cable and connectors.

The new edition of the IEC 60079-0 standard says that radio signals which in normal operation are clearly below the above-mentioned limit values do not have to be subjected to failure analysis of a fault that has caused excessive power of the radio signal [5]. The application of a device using ISM band (Industrial, Scientific and Medical unlicensed radio bad) like Bluetooth, Zigbee, RFID or WiFi could be easily performed by encapsulating a standard device in a metal and flameproof enclosure and connecting it with an external and certified antenna. The radiated power by these devices is usually in the range between 10 and 100 mW, so it is clearly below the limit value.

Antennas and RFID tags could not be encapsulated in metal enclosures, however using plastics is also problematic – there arises a hazard of electrostatic discharge, and surface resistance must be checked. The resistance index of the circulation on the surface current is defined by the CTI parameter (comparative tracking index). Isolating materials inside an apparatus, like a solder mask, should have high value of CTI on the contrary to the enclosure material.

The special group of communication solutions consists of systems intended to be used after collapses and explosions. Tagging systems help to determine how many workers were in a dangerous area but it is essential to establish communication between trapped miners and the rescue team. Communication infrastructure is usually destroyed after the collapse or explosion. The construction of emergency communication channel utilizes the TTE technology and is performed by personal transceivers carried by miners and a loop antenna deployed near the trapped miners or even on the surface of a coal mine. This kind of communication uses very low frequency in the range between 300 Hz and 3 kHz and requires up to 1.2 kW of power to transmit data from the surface by a large antenna which can have up to 12 km circumference [6]. Smaller systems have limited communication distance but work in the semi-duplex mode. The TTE transmission is sensitive to interference with any electrical equipment.

4.2 Leaky Feeder Based Systems

This method can be classified as a semi wireless method. The leaky feeder cable is a coaxial cable which contains holes in its outer copper shielding. These holes allow leaking out electromagnetic waves along its length. Wireless devices working in short distances from the cable can hold duplex communication with high bandwidth.

Covering underground mines with this kind of transmission is possible with the use of amplifiers after each 400 meters of the cable. The bandwidth depends on the frequency of radio waves – for UHF frequencies (450 mHz up to 1 GHz) it is possible to transmit 2 Mbps in a range up to 200 meters from the cable. The complete system requires the base station, leaky feeder cable, amplifiers, walkie-talkies, and radio modems.

4.3   Transmission through the Optical Fiber Cable

The fiber optics cable system increases quality of communication because it is unaffected by electromagnetic interferences. This method of data transmission has a wide bandwidth and can be applied for long distances. Optical radiation can be a source of ignition when the light beam is focused on a small point, e.g. a piece of coal dust, because at a focal point the energy is concentrated and can ignite an explosion. The IEC 60079-28 standard defines 4 possible mechanisms of ignition [7]:

- heating particles and reaching a surface temperature which can start an explosion,
- thermal ignition of the flammable gas due to matching optical wavelength with absorption band of the gas (resonance effect),
- photochemical ignition (dissociation of oxygen molecules),
- direct laser-inducted electric discharge.

The most popular method of protection is inherently safe optical radiation marked as "op is" which is very similar to intrinsic safety. It is based on limiting the optical energy during normal operation and in fault conditions. Energy requirements are defined for continuous and pulsed radiation but radiation source with pulse intervals fewer than 5 s are regarded as continuous wave sources. The output parameters for laser shall not exceed $20 \, mW/mm^2$ or $150 \, mW$.

4.4   The Twisted Pair Transmission

The most popular method of sending data is transmission based on twisted pair and it is used by different transmission standards, like RS422, RS485, CAN, LIN Ethernet, and others. Underground applications must provide the data transfer for long distances, often more than 2 000 meters. Repeaters are not easy to apply due to power supply constrains. The problem is resolved with intrinsically safe analog modems or media converters working with data concentrators. Another type of hardwired communications system is the trolley wire used in mines with extensive rail system [8].

A very important issue is the separation between an underground, intrinsically safe communication system and the surface infrastructure for data processing. Isolation between devices in flameproof enclosures and communication lines is done with a standard solution – the use of optocouplers. ATEX requirements are fulfilled by providing a distance between input and output greater than 3 mm.

5 Calculation of Safety Parameters for the RS485 Transmission System

The basic requirement for an electronic device installed in such environment is to prove the impossibility to generate the spark with energy which may ignite the explosion. The second solution is to protect the device in a flameproof enclosure or use encapsulation, oil immersion or powder filling, in such a way that an explosion inside device should not ignite an explosion outside. An example of calculations of an intrinsically safe ash meter communication module using external power supply is shown below. For the purposes of calculations based on Ohm's law, the following symbols apply:

U_i – input voltage,
I_b – fuse nominal current,
K – safety coefficient dfined in
    EN-60079-11 standard,
T, T_1 – tolerance coefficient defined by the manufacturer,
U_z – reverse Zener voltage,
U_f – forward voltage,
I_z – reverse current,
I_{zmax} – maximal reverse current,
I_f – forward current,
I_o – output current,
I_{fmax} – maximal forward current,
$R_{\theta jA}$ – thermal resistance from junction-to-lead,
T_{max} – maximal temperature of the element,
T_a – ambient temperature,
P_1, P_2, P_3, P_4 – dissipated power.

5.1 Determining the Maximal Power Provided to the Device

The calculations concern only input and output circuits regardless of the elements used inside apparatus. The external power supply is a source of 12 V applied to dedicated connectors. In the certificated parameters of the power supply unit it is shown that in case of its failure maximal output voltage will not exceed 13.6 V, therefore this is the value which should be used in the calculations. The elements which influence safety parameters are shown in Figs. 3 and 4.

Diodes D1, D2 and D3 protect against the effects of bad polarity. Standards define that safety parameters must provide the possibility of double fault in the device – in that case this is short-circuit of two diodes. The D5 Zener diode limits voltage so the over voltage will lead to additional current flow. The F1 fuse limits the current, eliminating dangerous situations.

The maximal power dissipated at power supply input can be calculated according to Formula (1):

$$P_1 = U_i \cdot I_b \cdot K = 13.165 \cdot 0.063 \cdot 1.7 = 1.46 \, \text{W} \ . \tag{1}$$

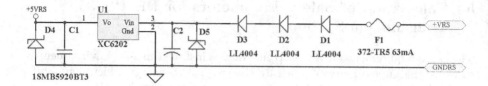

Fig. 3. The simplified scheme of the power supply circuit

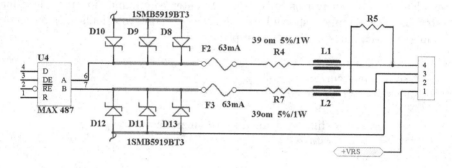

Fig. 4. The electrical scheme of the RS485 transmission circuit

The communication channel in the electrical standard RS485 consists of L1 and L2 inductors for EMC noise reduction, R4 and R7 resistors for current limiting, F2 and F3 fuses, and two groups of Zener diodes.

Maximal power supplied to input GNDRS-485A and GNDRS-485B can be calculated as below:

$$P_2 = (U_z \cdot T + U_F) \cdot I_b \cdot K = (5.6 \cdot 1.05 + 1.5) \cdot 0.063 \cdot 1.7 = 0.79\,\text{W} \ . \qquad (2)$$

Total power supplied to intrinsically safe device is equal to

$$P = P_1 + P_2 = 1.46 + 0.79 = 2.25\,\text{W} \qquad (3)$$

and is less than the threshold value 3.3 W, therefore small elements which are installed in the device are not able to achieve temperature which can provoke the ignition.

5.2   The Output Current Limits and Thermal Analysis

The overcurrent protection at 485A and 485B inputs is performed by:

- Applying a fault-free protection unit which consists of three D8, D9, D10 Zener diodes for 485A input and D11, D12, D13 diodes for 485B connected by copper tracks 2.16 mm wide and 35 μm thick.
- Applying a current limiter which consists of fault-free R4 (R5) resistors and F2 (F3) fuses.

Short circuit current for D8, D9, D10 and D11, D12, D13 Zener diodes is equal to

$$I_z = I_b \cdot K = 0.063 \cdot 1.7 = 107.1\,\text{mA} \ . \tag{4}$$

The calculated maximal reverse currents for Zener diodes should be greater than

$$I_{max} \geq I_z \cdot 1.5 = 161\,\text{mA} \ . \tag{5}$$

The catalogue value of I_{zmax} for 1SMB5920 (ONSEMI) is equal to 241 mA and for SMBJ5920 (MICROSEMI) is equal to 482 mA. This example shows that it is very important to apply elements from the manufacturer specified in documents. The value of I_{Fmax} was not clearly specified in the catalogue card. It was written that U_f is less than 1.5 V while $I_f = 200$ mA. The calculated power dissipation for each diode must be greater than

$$P_3 = U_f \cdot T \cdot I_z \cdot K = 5.6 \cdot 1.05 \cdot 107.1 \cdot 1.5 = 944\,\text{mW} \ . \tag{6}$$

The thermal resistance from junction-to-lead for SMB case installed on PCB made of FR4 with 35 μm copper thick is equal to $R_{\theta jA} = 226\,°\text{C/W}$ so maximal power dissipated on the diode at 40 °C ambient temperature is

$$P = (T_{max} - T_a)/R_{\theta jA} = (150 - 40)/226 = 0.48\,\text{W} \ . \tag{7}$$

This calculation shows that a radiator must be installed and during the power dissipation calculated in (6) the temperature growth must be checked.

The maximal output current at 485A-GNDRS, 485B-GNDRS connectors is equal to

$$I_o = U_f \cdot T/R4 \cdot T_1 = 5.6 \cdot 1.05/39 \cdot 0.95 = 158\,\text{mA} \ . \tag{8}$$

For short circuit of 485A and 485B outputs and active Max 487 transmitter

$$I_o = U_f \cdot T/(R7 \cdot T_1 + R4 \cdot T_1 + 2 \cdot R_T) \tag{9}$$

$$I_o = 6.2 \cdot 1.05/(2 \cdot (39 \cdot 0.95 + 3.5)) = 80.3\,\text{mA} \ . \tag{10}$$

The necessary wattage for the current limiting R4 (R7) resistor can be calculated as below

$$P_4 = (I_b \cdot K)^2 \cdot R4 \cdot T = (0.063 \cdot 1.7)^2 \cdot (39 \cdot 1.05) = 0.47\,\text{W} \ . \tag{11}$$

R4 and R7 resistors have nominal wattage 1 W and fulfill the requirements.

5.3   The Output Parameters

The IEC 60079-11 standard limits parameters of the devices connected to the analyzed circuit. Maximal output currents calculated in (9) and (10) can be connected to inductance $L_o < 20$ mH with voltage equal to 8 V. The inductance can be read out of the nomogram. The similar calculations must be done to determine output capacitance Co, maximal output voltage Uo and output power Po.

Connecting the devices with the RS485 link must be preceded by careful comparison of their output and input parameters. The values of R4 and R7 resistors are very important – high values make the connection safe but cut back the range of communication. The parasitic capacitances of protective elements like Zener diodes reduce transmission speed.

6   The Application

The calculations shown above relate to a single RS485 transmission channel and can be used for safety analysis of the system which consists of at least two devices. This analysis does not cover functional parameters like maximal transmission speed or transmission range connected with it. Getting a large bandwidth, large ranges and interference immunity requires additional energy supplied to the transmission system, which is in contradiction with the requirements of explosion hazard management. The certification process is usually carried out before application requirements can be defined in detail. The role of the system designer is the appropriate selection of devices, so as to get proper functional parameters and maintain explosion safety.

The example above is a part of the coal quality analyzer consisting of 5 programmable units connected as shown in Fig. 5.

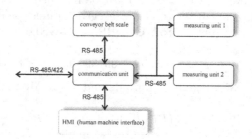

Fig. 5. The block scheme of the coal quality analyzer

Each of the devices in the system has their own parameters (maximum output voltage, output current, capacity and inductance output) and the maximal permissible output parameters of the connected device. The connection can be carried out without loose of the explosion safety if the limit values are not exceed. The RS485 transmission links are independent of each other and their lengths do not exceed a dozen or so meters. A small amount of data makes it possible to lower transmission speed and achieve low error rate. The examples of observed transmission signals at the beginning and at the end of the transmission line in 80 m and 400 m range are shown in Fig. 6. The impact of time constant defined by protective resistors R4 and R7 and the parasitic capacity of the cable can be observed. The YnTKGX $3 \times 2 \times 0.8$ cable applied for measurements is dedicated for use in coal mines and has 55 nF/km capacitance and 73.6 Ω/km resistance.

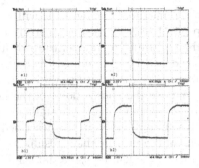

Fig. 6. The signal distortion for the distances 80 m and 400 m, with transmission speed 115 200 Bd. The line is not terminated at an impedance-matched load resistor. (a1), (a2) Signal at the beginning and at the end of a transmission line (80 m). (b1), (b2) Signal at the beginning and at the end of a transmission line (400 m).

The used resistors correspond to the transmission line with a length of more than 500 meters. The system designer should note that on the other end of the transmission line there is a similar system installed, which has its own limiting resistances.

Signal degradation depends on the time constant resulting from the current limiting resistances and the used cable parameters. Knowledge of these parameters allows a designer to estimate the maximum transmission speed, but taking measurements in situ are difficult because of explosion hazard. The signal-to-noise ratio degrades toward the end of a trace but impedance mismatch has much more stronger influence on the error rate (Fig. 7).

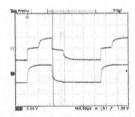

Fig. 7. Signal at the beginning and at the end of a transmission line (400 m) with and without an impedance-matched load resistor

During communication the transmitter and receiver know the signaling speed. The receiver cannot determine when a packet will be received with respect to its clock (protocol is called as asynchronous). The transmitter simply has to send "1" and "0" with the defined length but the receiver has to recognize the start of the frame (synchronization) and therefore determine the best data sampling point for the bit stream. The bits are typically sampled at the middle of the bit time and it is defined by hardware. The timing is synchronized at the falling edge of the start bit so the maximal timing error will be at the last sampling point, which is the stop bit. The clock mismatch error should be less than 2 % and usually it is less than 1 % due to temperature drift and tolerances. The maximum achievable speed can be determined by the estimation of stop bit degradation. The transmission link will not be stable if the rising or falling edge of received bit is shifted outside the middle point of the expected stop bit.

Wherever it is necessary to obtain greater transmission speed or greater transmission range, additional solutions apply. In the described case it was necessary to ensure the transmission of the measurement results to the surface of the coal mine. It was possible to use a modem or use a data hub nearby. The hub was installed at a distance of 150 meters and contained a built-in RS485 converter for fiber optic transmission. The intrinsically safe parameters analysis showed the ability to connect both devices. Transmission speed was reduced to 9600 bits per second maintaining the functionality and proper immunity to interference.

7 Conclusion

In Polish coal mines fiber optic communication is successively installed in shafts and key localizations. A significant number of working systems were installed with traditional copper-wired interfaces so a big job for engineers is to integrate different kinds of transmission and build efficient and robust systems which will be open for new solutions. In the near future the increasing number of WLAN application are expected. This method of data transmission is suitable for video surveillance, voice transmission, workers and equipment tagging, and sensors connection.

References

1. Mishra, P., Bolic, M., Yagoub, M., Stewart, R.: RFID technology for tracking and tracing explosives and detonators in mining services applications. J. Appl. Geoph. 76, 33–43 (2012)
2. MINER Act Compliant Communication and Tracking Systems and Peripherals (2012), http://www.msha.gov/techsupp/commoandtracking.asp
3. Schiffbauer, W., Brune, J.: Coal mine communications, http://stacks.cdc.gov/view/cdc/8606/
4. Bandyopadhyay, L., Mishra, P., Kumar, S., Narayan, A.: Radio Frequency Communication Systems in Underground Mines, http://www.ursi.org/Proceedings/ProcGA05/pdf/CP4.5(0771).pdf
5. Hauke, R., Schultz, S.: Wireless Applications in Hazardous Areas. Ex-Magazine, 56–60 (2012)
6. Laliberte, P.: Summary Study of Underground Communications Technologies. Final Project Report 603478-00-0 (2009), http://www.wvminesafety.org/PDFs/UndergroundCommunicationsReport.pdf
7. Fritsch, A.: Industrial Ethernet in process automation. Ex-Magazine, 10–19 (2008)
8. NIOSH National Institute for Occupational Safety and Health: Tutorial on Wireless Communication and Electronic Tracking. Part1: Technology Overview, http://www.msha.gov/techsupp/PEDLocating/WirelessCommandTrack2009.pdf

Planning-Based Method for Communication Protocol Negotiation in a Composition of Data Stream Processing Services

Paweł Stelmach, Paweł Świątek, Łukasz Falas, Patryk Schauer, Adam Kokot, and Maciej Demkiewicz

Institute of Informatics, Wrocław University of Technology
Wyb. Wyspiańskiego 27, 50-370 Wrocław, Poland
{pawel.stelmach,pawel.swiatek,lukasz.falas,patryk.schauer,
adam.kokot,maciej.demkiewicz}@pwr.wroc.pl

Abstract. Data streaming is often used for video and sensor data delivery, nowadays gaining in popularity along with the development of mobile devices. In this paper we briefly describe a platform for automated composition of distributed data stream processing services. It decreases the complexity of composite service building process from the users point of view by introducing automation mainly in appropriate service selection and their communication protocols negotiation. This paper is focused on automated negotiation of communication protocols, which will be further used to transfer data among services. Data stream processing services are designed independently, often with many possible communication methods and not pointing directly to other services. With the use of the platform, they can be loosely coupled, forming a composite service od-demand. We present several approaches to communication protocol negotiation in a composition of data stream processing services and introduce planning-based approach. Finally, we discuss consequences of various approaches on an example composite service from image processing domain.

Keywords: Service Oriented Architecture, data stream processing services, service management.

1 Introduction

With the development of Internet more and more applications are available via Web. Also, they no longer have to be monoliths but developers outsource many functions to web services. This service orientation became more popular over the recent decade with the SOA paradigm (Service Oriented Architecture), which introduced a way to build large distributed systems. It uses standards, like WS-* for SOAP-based web services and WSDL for their description. However, the Web Service standard is based on request and result behaviour. After that the service is expected to stop working, waiting for the next request. This seems natural, however, the need for continuous access to ever-changing data, like video

A. Kwiecień, P. Gaj, and P. Stera (Eds.): CN 2013, CCIS 370, pp. 531–540, 2013.
© Springer-Verlag Berlin Heidelberg 2013

feeds or sensor data, requires a different kind of service, one that puts more strain on the Internet transporting capabilities. Such streaming services can deliver audio/video surveillance, stock price tracing, sensor data [1], etc. With time we have introduced specialized middleware for data stream processing [2,3], changing video format or colour, calculating on-line whether a patient did not have a heart attack and more.

Distributed stream processing, adding more intermediary services forwarding the stream from one peer to another, is growing in popularity [2,4,5], especially in eHealth, rehabilitation and recreation fields where distributed measurement data acquisition is natural [6,12] or in computational science and meteorological applications, where there are multiple data sources and multiple recipients interested in the processed data stream [7].

In this paper we describe the ComSS Platform (COMposition of Streaming Services) that is a result of ongoing work in the area of Future Internet [8]. It offers management over compositions of any data stream processing services, provided that they are compatible (that is each two services communicating have to be able to communicate via a common protocol). The communication negotiation is of extreme importance and is reflected in negotiation process that takes place during creation of a distributed data stream processing service. Many papers, which take on the subject of composition of services, often refer to services in the WS-* standard [9] or consider stream processing services [10,11], but still treat them similarly to WS-* service, omitting their unique characteristics (namely the flexibility in various formats and communication protocols employment).

2   Platform Description

2.1   Platform Overview

The main goal of the platform is to manage data stream processing composite services (also called streaming services). Provided the designer of such services would like to utilize the platform automated management capabilities, he can delegate the all tasks of assembling, disassembling or monitoring of such services to the platform simply by using its basic services.

The ComSS Platform consists of several autonomous software components but also requires atomic streaming services to implement specific communication libraries for control purposes (see Fig. 1 for reference).

The basic scenario for the ComSS Platform is to create a new composite streaming service given a graph of atomic services (composite service plan). It is assumed that those services have been implemented using the provided framework (Fig. 1.1) and are registered in the service registry (Fig. 1.2). The platform searches for appropriate atomic streaming services to fulfil the user composite service request and forwards them information on neighbour services (Fig. 1.4), with which they will have to negotiate the communication protocol. Streaming services start negotiating (Fig. 1.5) and create new service instances to handle the new composite service request.

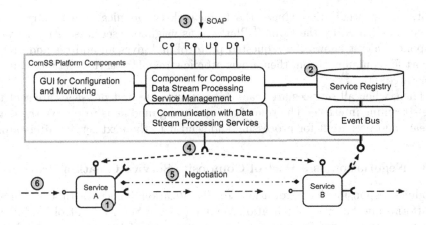

Fig. 1. ComSS Platform overview

2.2 Framework for Data Stream Processing Service

Part of the effort to make the streaming service composable lies with the service designer himself. He has to follow conventions for the service design and implement necessary libraries for control, negotiation and communication.

Using the framework provided with the ComSS Platform allows him to focus on implementation of the data stream processing algorithms alone (Fig. 2).

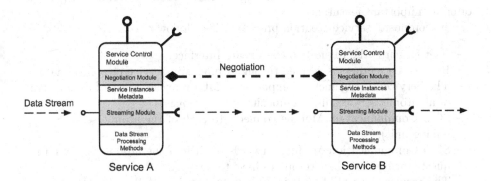

Fig. 2. Overview of the Data Stream Processing Service Framework

It should be noted that in contrast to the Web Services using the SOAP protocol, data stream processing services are not limited to one single protocol or format – most popular uses, and completely different protocols, are video stream processing and sensor data stream processing.

In the basic scenario of creating a new composite streaming service a request for a new service instance is sent via the Web Service interface (Fig. 2). Before that, in the negotiation phase, the framework communicates with streaming services indicated by the ComSS Platform as neighbour services. If the service responds that it knows the requested protocol and gives an address and opens a port for communication, then a new atomic streaming service instance can be generated.

Then, when all services are ready to communicate first messages are sent to the streaming interface. The role of the streaming module is to receive the data stream and prepare it for processing and send it to the next service afterwards.

2.3   Negotiation as a Part of Composite Service Creation Process

Modules responsible for negotiation are: Signalization Supervisor (in the ComSS Platform) and Service Signalization Module (part of Service Control Module of the streaming service framework). Former is a part of Platform Core and the latter is part of atomic service application. There are two types of communication inside the ComSS Platform: signalization (between core component and atomic services) and negotiation (between pairs of services). The composite service request is a fundament for these two types of communication. The service composition plan is used to establish which what services are neighbours in the sense of the composition and what negotiation requests to send to each of them. Secondly, some negotiation approaches may use the information on the order of services in the composition to maximize the chance for success. In our model negotiation information is distributed directly to services and then among pairs of neighbours. Services don't need to use core component to mediate between them. In distributed systems minimizing traffic through the central point it is often an important feature.

The composite service creation procedure is as follows:

- User forwards the request to the Create Interface.
- Request correctness is verified through Composite Service Graph validation.
- The Service Management Component (SMC) requests a list of descriptions of all atomic services in a composite service request.
- SCM Communication Module connects to each of the atomic services (signalling unit).
- SCM Communication Module forwards to each of the atomic services a request stating that their resources have to be reserved.
- The Service Control Module in each atomic service (AS) verifies the type of request and its correctness.
- After all Atomic Services confirm the reservation then SCM sends each of them a list of services they should start negotiating with.
- AS Service Control Module (sender) starts negotiations with recipient services, providing them a list of types and formats in which it is able to transmit (this list can vary depending on the received request, the list is sorted by priority).
- AS Service Control Module (receiver) selects the data type and format it can receive and sends back this information to the inquiring AS.

- AS Service Control Module (sender) confirms the determination of the type and format of the recipient and communicates the outcome of negotiations to the SCM communication module.
- SCM Communication Module collects the answers from the AS.
- SCM sends to the User a confirmation along with information on the newly established composite service.

3 Motivation

Author of the atomic streaming services is unaware of what composite services will be built using them. With the ComSS Platform he can register his services and their capabilities: location, type of processing if offers, what kind of input data it can process – what format, codec etc. Provided with this information, the ComSS Platform can automatically determine how to connect services in a composite service requested by the user.

In most papers on streaming services composition authors neglect the unique nature of streaming services, namely that they can accept a specific data stream in various forms: with different coding and format. Some services are capable of converting those formats (for a cost) while other cannot, and offer simple processing or monitoring capabilities on a range of formats, incapable of their modification. Using services with conversion capabilities offers more flexibility but also can increase processing costs over more simple versions of services. However, using only processing services can lead to bottlenecks in service compositions where two services cannot communicate due to incompatibility in the format of the data stream.

In this work it assumed that a service composition has been defined either automatically or by hand and now this service has to be physically initiated. In order to do this, appropriate resources have to be reserved, service instances created and communication among atomic services directed to other services ports, suitable of processing a given format.

A simple, greedy approach used to date can lead to rejection of service composition request due to incompatibility of formats among some atomic services. In the next section a planning-based approach is presented to guarantee that if there exists a feasible solution than a communication between atomic services can be established. What is more, this solution will be optimal due to both processing and communication cost.

4 Communication Negotiation

4.1 AI Planning in Negotiation

Typically AI Planning is used in service composition. In this situation all services and a sequence in which they are connected are already defined, but – as stated in the previous section – an incorrect choice of streamed data formats can lead to incompatibility of services in the sequence.

The proposed approach is in opposition to currently implemented greedy approach, where each atomic service can determine its own communication with another service. We suggest that formats and communication protocols were defined centrally in the ComSS Platform and then propagated to all services.

The approach we describe is based on backward search. Below you can see both the main body of the algorithm as well the recursive plan search procedure in a form of pseudocode. In this approach a general goal is defined based on user format requirements and while iterating through all services (starting from the last), a temporary goal format is defined, based on the input format of currently selected service. The algorithm utilizes information on the user input and output format requirements (F_{in},F_{out}) – which are based on the format of the external source data stream that will be transferred to the composite web service and the format of the goal of the data stream that the composite service will deliver the stream to. Finally, the algorithm is not a service composition algorithm and thus it is limited to the services and their order defined prior to the negotiation.

The algorithm gives the guarantee that appropriate formats, ensuring communication at each step of the data stream processing will be selected if possible and that the selection will minimize both the processing and transport cost.

- **Input:**
   F_{in} – format of the external input data stream,
   F_{out} – format of the data stream to be delivered on the output,
   AS – a set of atomic services, which communication formats have to be
      negotiated; those services comprise the composite.
- **Output:**
   Optimal plan determining communication formats for each of the services
   in a composite service.

1. $P = \oslash$
2. $AS \leftarrow sortServicesInIncreasingOrder$
3. $Plans = findPlans(P, F_{out}, F_{in}, AS)$
4.   **if** $Plans = \oslash$ **then**   "Communication cannot be established"
5.   **else**
6.   **foreach** $Plan$ in $Plans$
7. $Q = calculatePlanQuality(Plan)$
8.   **if** $Q > Q_{max}$ **then** $Q_{max} = Q$ **and** $P_{max} = Plan$
9.   **end foreach**
10.   "Optimal plan is P_{max} with quality Q_{max}"
11.   **end if**

The above algorithm comprises of two parts. Firstly, it initializes the search for valid plans that, with the services in AS, transform the input data stream with format F_{in} to the output data stream of format F_{out}. Secondly, it determines the quality of each of the plans, selecting the plan with the best quality. The quality is calculated based on the processing cost of each service and cost of transporting data stream between them. The former takes in to consideration that converting the format usually leads to increased processing cost and the latter takes into account that some formats can decrease the volume of data, thus decreasing the cost of transport of such data.

Algorithm: Plans Backwards Search

- **Input:**

  P – current plan to achieve the goal format F_{out},

  $F_{current}$ – current goal format, for selecting appropriate input formats of a service,

  F_{in} – format of the input data stream,

  AS – the remaining web services, which formats have to be established.

- **Output:** Valid plans determining communication formats for each of the services in the current AS

  $Plans = findPlans(P, F_{current}, F_{in}, AS)$:

```
1.    as = AS.last; AS = AS \ as; Plans = ⊘
2.    V = generateVersionsOf(as)    such that each v ∈ V : v_out = F_current
3.    if V ≠ ⊘ then foreach v ∈ V
4.        if AS == ⊘ and v_in == F_in then Plans := Plans ∪ (P ∪ v)
5.        else if AS ≠ ⊘
6.            Plans* = findPlans(P ∪ v, F_current = v_in, F_in, AS)
7.            if Plans* ≠ ⊘ then Plans := Plans ∪ Plans*
8.        end if
9.    end foreach
10.   end if
11. return Plans
```

The algorithm above takes the last atomic service in AS and will establish possible format transformations. A service as can be treated as a container for multiple abstract services that each can take one input format and generate one output format. Each of whose abstract versions of service as has its own processing cost. In (step 2) only those abstract versions of the service are generated which output format is equal to the current goal format $F_{current}$.

If there are versions that meet the requirements and there are still services left in AS (step 6) then each version is considered separately in an alternative plan searched recursively.

The procedure stops if no services are left in AS and returns a valid plan only if the last version can take as an input the format defined in F_{in}.

All valid plans are returned and compared in the main body of the algorithm.

4.2   Example and Comparison to Greedy Approach

The differences of various approaches will be briefly described using examples of negotiation procedure in those approaches and possible outcomes.

The current implementation does not enforce any order of negotiation requests thus it can provide unexpected results. Below we show two most popular approaches: forward and backward negotiation requests propagation and their outcomes, compared to the planning-based approach.

In Fig. 3 and Fig. 4 we observe that each service is provided with information about services it has to negotiate the communication format with. In contrast, in

Fig. 5 we see that requests are not typical negotiation requests but the planner enforces the format on each service (in any order). Additionally, both forward and backward negotiation are extended with initial pre-configuration phase in which appropriate services (first and last in the composition) are configured with information about requested input and output stream, thus preventing the situations when the service would choose a format incompatible with the request.

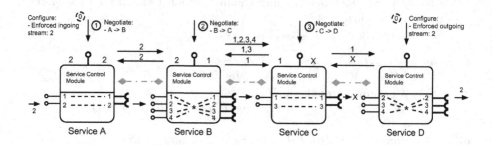

Fig. 3. Example of forward negotiation without planning

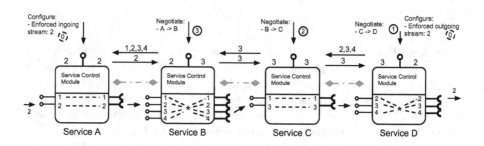

Fig. 4. Example of backward negotiation without planning

In Figure 3 we observe forward propagation of negotiation requests. Service A negotiates with service B. It sends formats in which it can communicate with service B. Notice that service A cannot transform the input format and following the requirement from external data source it limits its formats list. Service B responds with a format from its list that is on service A's list. Here, the negotiation is trivial but typically service B would select a format it "prefers" – usually higher in a static rank. Now service B has its input format set to "2", but considering that it is capable of converting it, it is not limiting its output format. Following the next negotiation request service B sends all its output formats to service C and service C responds with the same list limited to the formats in can process. Service B selects format and relays its decision to service C. Next,

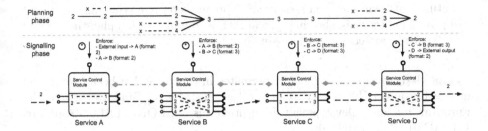

Fig. 5. Example of negotiation with planning

service C sends to service D a limited list of formats, which has no common format with service D and service D has to report an error. Please note that if a different format were negotiated between services B and C then the negotiation would end in success. However, neither service B nor C had any knowledge that could point the negotiation in that direction.

In Figure 4 we observe the inverse of the previous process. Following the initial configuration of the first and last service in the composition, the negotiation starts with service D. It relays its formats to service C and then service C to service B etc. The simple change in direction of negotiation resulted in negotiation success. However, were it not for the initial configuration of service A, services B and A would negotiate a different format ("1" had the priority) and the final verification with the external source would end in failure.

The planning-based approach in Fig. 5 allows for analysis of possible connections and enforces the communication for each service pair. Additionally, it allows for selection of optimal formats from the both processing and transport cost point of view. In this example there was only one valid solution, but if service D could process format "1" on the input then there would be two possible solutions: one transporting the data through services B, C and D in format "1" and another in format "3".

5   Conclusions and Further Work

Research presented in this paper describes the communication negotiation phase during construction of a composite data stream processing service. A ComSS Platform has been briefly described as an example of a tool that delivers such capability, relaying negotiation requests in accordance with the composite service structure.

In this work a planning-based approach to negotiation was presented. This approach extends basic capabilities of the ComSS Platform and limits negotiations in the atomic services. As a result, more calculations have to be performed centrally, but – as presented in examples – the proposed negotiation approach guarantees not only finding a valid result if one exists but also the solution is optimal with regard to data stream processing and transport time.

Future work will focus on testing all approaches on real streaming data and real processing services, showing typical cost of communication plans negotiated – both calculation cost and processing and transfers costs.

Acknowledgments. The research presented in this paper has been co-financed by the European Union as part of the European Social Fund and within the European Regional Development Fund programs no. POIG.01.01.02-00-045/09 & POIG.01.03.01-00-008/08.

References

1. Gu, X., Yu, P.S., Nahrstedt, K.: Optimal component composition for scalable stream processing. In: Proceedings of 25th IEEE International Conference on Distributed Computing Systems, ICDCS 2005, pp. 773–782 (June 2005)
2. Chen, L., Reddy, K., Agrawal, G.: Gates: a grid-based middleware for processing distributed data streams. In: Proceedings of 13th IEEE International Symposium on High performance Distributed Computing, 2004, pp. 192–201 (June 2004)
3. Schmidt, S., Legler, T., Schaller, D., Lehner, W.: Real-time scheduling for data stream management systems. In: Proceedings of 17th Euromicro Conference on Real-Time Systems (ECRTS 2005), pp. 167–176 (July 2005)
4. Gu, X., Nahrstedt, K.: On composing stream applications in peer-to-peer environments. IEEE Trans. Parallel Distrib. Syst. 17(8), 824–837 (2006)
5. Rueda, C., Gertz, M., Ludascher, B., Hamann, B.: An extensible infrastructure for processing distributed geospatial data streams. In: 18th International Conference on Scientific and Statistical Database Management, 2006, pp. 285–290 (2006)
6. Świątek, P., Klukowski, P., Brzostowski, K., Drapała, J.: Application of wearable smart system to support physical activity. In: Advances in Knowledge-based and Intelligent Information and Engineering Systems, pp. 1418–1427. IOS Press (2012)
7. Liu, Y., Vijayakumar, N., Plale, B.: Stream processing in data-driven computational science. In: 7th IEEE/ACM International Conference on Grid Computing, pp. 160–167 (September 2006)
8. Grzech, A., Juszczyszyn, K., Świątek, P., Mazurek, C., Sochan, A.: Applications of the future internet engineering project. In: 2012 13th ACIS International Conference on Software Engineering, Artificial Intelligence, Networking and Parallel Distributed Computing (SNPD), pp. 635–642 (August 2012)
9. Świątek, P., Stelmach, P., Prusiewicz, A., Juszczyszyn, K.: Service composition in knowledge-based soa systems. New Generation Computing 30, 165–188 (2012)
10. Riabov, A., Liu, Z.: Planning for Stream Processing Systems. In: Proceedings of the National Conference on Artificial Intelligence (2005)
11. Riabov, A., Liu, Z.: Scalable Planning for Distributed Stream Processing Systems. In: Proceedings of ICAPS (2006)
12. Brzostowski, K., Drapała, J., Grzech, A., Świątek, P.: Adaptive Decision Support System For Automatic Physical Effort Plan Generation–Data-Driven Approach. Cybernetics and Systems: An International Journal 44(2-3), 204–221 (2013)

Users in IT Product Development Process

Malgorzata Pankowska

University of Economics, Katowice, Poland
malgorzata.pankowska@ue.katowice.pl

Abstract. The rise of the Internet has increased the business transformation. Companies want to communicate with customers and partners in real-time. On the other side, Internet users demand speedy, well fitted access to actual information as well as to business applications. The paper focuses on user participation in the development of business information system (BIS). The paper includes the analysis of the empirical research results on user support software preferences.

Keywords: business information system, system development, IT users, IT prosumption, user support application acceptance.

1 Introduction

Encouraging users to be value creators in IT development processes is an important step in the competitive effectiveness increase activity. Shapiro argues that in e-economy the focus on users instead of customers is necessary [1]. Users are employees working with business applications, but also customers, Internet Websites reviewers, job candidates, business prospects and partners, brand fans, media members and other stakeholders, who interact with a company through digital media and Internet technology. According to Claycomb et al. [2] users can actively participate in creating solutions, when IT service failures occur by applying specialized skills and knowledge. For example, a user can diagnose his laptop problems based on the product's user manual. Shapiro is adding that today's customers have a strong self-service mentality – they want to be able to do everything themselves and solve their problems without professional intervention [1]. On the other side, they have a strong full-service mentality – when self-service fails, they demand immediate, personal support with a real person.

Research on user participation so far has focused on how to employ users to increase productivity in the service delivery context. The purpose of this paper is to bridge the gap in the literature by investigating user participation in IT development processes and to present some suggestions in users' future co-creative behaviours. The paper is to show that nowadays users are no longer passive audience, but under certain circumstances they are active co-producers. The computer literate users are observed as able to develop simple software applications to increase their personal productivity. User applications development has evolved to include complex application development by groups of users and shared across departmental boundaries. However, there still prevails a common

A. Kwiecień, P. Gaj, and P. Stera (Eds.): CN 2013, CCIS 370, pp. 541–551, 2013.
© Springer-Verlag Berlin Heidelberg 2013

opinion that users applications are not significant, but rather transient and disposable as not production oriented.

The main goal of the paper is to explain the role of users in the business information system exploitation and development processes. The paper consists of two main parts. The first part covers analysis of the works from other researchers on user innovativeness and their participation in system development. The second part includes considerations connected with empirical research on the user willingness to utilize software tools to support their business system development and exploitation.

2   User Involvement in BIS Development Process

Historically, there are several kinds of adaptive methodologies of information system development that build a model of users' knowledge and their involvement in the development process. Prahalad and Ramaswany presented a co-creation model based on dialogue, access, risk reduction and transparency of information exchange between users and software company [3]. Von Hippel developed the lead user approach and he says that some users are more appropriate to co-develop new products and services than others [4]. Therefore, the IT suppliers and end users have created opportunities to exchange their knowledge and competencies needed for joint information system development. Tidd et al. [5] emphasize the unique characteristics of lead users, who are able to recognize requirements early, be ahead of the market in identifying and planning for new requirements, expect special benefits due to their company market position and complementary assets, and develop their own innovations and applications. Similarly, Cartlige introduced the informed customer (IC) concept [6]. Typical areas of involvement of ICs are:

– the alignment of business and IT plans and strategies,
– the development of business unit objectives and requirements for information
   systems,
– the establishment and co-ordination of user groups,
– the development and negotiation of SLRs and SLAs,
– managing the provision of the IT services,
– sharing the IT risk and gains.

In PRINCE2 project management methodology, the idea of Senior User represents the interests of all users, who will use the final products of the project, those for whom the product will achieve an objective, or those who will use the product to deliver benefits [7].

Barki and Hartwick proposed a distinction of user involvement from user participation. They define user participation as the assignments, activities and behaviours that users perform during the system development process. User involvement refers to the subjective psychological state reflecting the importance and personal relevance that a user attaches to a given system [8]. System designers have promoted techniques requiring user participation, such as prototyping, collaborative engineering, rapid application development and joint application design.

Collaborative engineering is an emerging approach to designing collaborative work practices for high value recurring tasks [9]. To implement a collaborative work practice, groups need to be trained or require facilitation support. A key requirement is the users' willingness to change. In the information system life cycle, the designers analyse the system and basing on design patterns derived from their expertise, they propose changes that are evaluated by the user community and then implemented. Involving end users in the development process requires that end users and developers can communicate in a common language to identify and specify requirements as well as solutions. However, in collaborative engineering the end users are still dominated by the producer. They are still considered less experienced. Although they do not always behave in the same way. They are characterized by the following criteria:

- involvement in the information system development process: strongly involved or just watching and not very useful,
- environment: personal (home) users, worker (corporate, organizational, enterprise) users,
- frequency of use: occasional, frequent, extensive,
- software used: word editors, email, graphics, accounting,
- educational level: basic, intermediate, advanced,
- relationship: internal users (co-workers), external users (clients) [10].

Taking into account the presented above methods, it should be noted that they are developed by IT professionals and in all of them the active role belongs to IT developers instead of users, who are passive. So, for quality and proficiency of IT products, it is an advantage, but stimulation of user to their creativity is limited and disadvantageous. Differentiation of expertise and experience of users creates the need to support them in the information system development process. Although help desk systems and computer user support provide IT services, the organizational strategy of user support often depends on the organization size, type, functions, location, financial condition, people skills and competencies as well as of the goals of computer support services. A challenge that continually confronts user support staff is the need to develop an IT platform to enable software applications creation by users and for them.

3   Empirical Research on User Involvement in System Development

The literature studies created the need to empirically verify that very optimistic attitude of academic publications' authors towards user involvement. Therefore, the empirical research was done in 2011. The research covered the interviews with Chief Information Officers (CIOs) from 270 firms in Poland. The similar research results in other countries were published in [11,12], and [13]. Characteristics of the surveyed in Poland firms are presented in Table 1.

User involvement in this paper is considered as participation in the business
information system development process measured as a set of activities that
users have performed. In this research, CIOs answered the questions concerning
the activities of users at their companies. Historically, there are several kinds
of adaptive methods of information system development that build a model of
users' knowledge and their involvement in that process. Active participation
of a person in a community is a powerful indicator of the person's interests,
preferences, beliefs and social and demographic context. Community members
are a part of users' model and can contribute to tasks like personalized services,
assistance and recommendations.

Table 1. Surveyed companies features

Feature	N=270
Number of employees	
Micro Enterprises (1–9 employees)	44.4%
Small Enterprises (10–49 employees)	29.3%
Medium Enterprises (50–250 employees)	15.2%
Big Companies (more than 250 employees)	11.1%

Involvement of the end users in IT projects covering IS development is presented in Table 2.

Table 2. Participation of users in IT projects

	User					
	Passive	Evaluator	Co-creator	Partner	Producer	Prosumer
GS PC	15%	17%	33%	24%	10%	1%
BLA BPM	32%	22%	19%	20%	5%	1%
RE	37%	19%	17%	16%	10%	1%
ISD	34%	20%	19%	15%	10%	1%
ISI	39%	16%	19%	15%	9%	1%
IST	18%	20%	24%	21%	15%	2%
ISE	22%	27%	24%	14%	10%	2%
ISM	20%	23%	24%	19%	12%	3%
SIS	33%	18%	20%	14%	13%	1%
ISU	7%	23%	20%	30%	15%	4%

In Table 2, the following activities of users have been specified: goal specification and project concepts (GSPC), business logic analysis and business process
modelling (BLA BPM), requirements engineering (RE), information system design (ISD), information system implementation (ISI), information system testing
(IST), information system installation and migration to a new IT environment
(ISE), information system maintenance (ISM), security of information system

(SIS), information system usage (ISU). In Table 2, six different profiles of users has been included. Passive users and users-evaluators are oriented towards the observation and acceptance of other people efforts for IT projects and business information system (BIS) development. Co-creator supports IT staff in the BIS design and implementation works. User as the partner plays equally important role as IT professionals in the system development process. User as the producer is self-dependent and self-organized and has got sufficient competencies to utilize IT independently of the IT staff help. Self-organization is the property of well functioning complex systems that allows the relationships among individuals to shape the nature of an evolving group knowledge. The last i.e., prosumers are able to develop the BIS by themselves and utilize them for their work purposes and individual goals. Taking into account the results included in Table 2 you can notice that users are rather inactive. In this survey, CIOs evaluate users as inactive at business analysis and business process modelling stages as well as at requirements engineering, system design and implementation stages. IT people do not demand the technical expertise from users, but they should be helpful at the initial stages of business information system development process. Users were evaluated as co-creator in project concepts specification stage, information system testing and maintenance. Security of BIS is the domain of IT professionals. The strong activity of users is revealed in the business information system exploitation process.

Further analyses were realized separately for each of the groups included in Table1. In micro and small companies users are assumed to have direct, informal and face-to-face (F2F) contact with IT staff, therefore they know more about requirements of each individual. In medium and big companies, the contact between the user and IT personnel is indirect, online and occasional, therefore the formal procedures of registration of user needs are implemented and the users have no chance to be personally involved in the BIS development process.

In the realized survey, the first question concerns the expected benefits and potential impediments. At big companies, over 70 % of respondents admit that the most important benefits of user involvement in the BIS development process cover better understanding of user requirements (83.3 %), reduction of costs of research and development works (73.3 %), opportunities for market offer differentiation (73.3 %), and improvement of company image (over 70 %). Similarly at medium enterprises, over 70 % of respondents argue that the most important benefits of end-user involvement in BIS development process comprise better understanding of user requirements (80.5 %), development of strong relationship with user (80.5 %), reduction of the cost of knowledge acquisition (75.6 %) and improvement of company image (75.6 %).

At small companies, the most important benefits of user involvement in the BIS development include: supporting user education (72.2 %), better understanding of user requirements (72.2 %), taking better market position (67.1 %). At micro companies, the most important benefits of user involvement in the BIS development process comprise: better understanding of the user requirements (over 83 %), moving to the better market position (over 71 %), and improvement of company image (70 %). It should be noticed, that CIOs do not perceive that user involvement is helpful to reduce IT product time to market.

The most important impediments of user involvement in the BIS development process comprise:

- at micro companies, user lack of knowledge and computer skills (73 % of respondents mention that), inevitability to learn new technologies (66 %), lack of incentives and encouragement from the BIS producer (65 %),
- at small companies, necessity to learn new technologies (72 % of respondents within this group state it), user lack of skills and knowledge (71 %) and user lack of incentives provided by BIS producers (62 %),
- at medium companies, user lack of knowledge and skills (according to 80 % of respondents), necessity to learn new technologies (68 %), and the threat of theft of user ideas (61 %),
- at big companies, necessity to know new technologies (80 %), user lack of knowledge and skills (73 %), and lack of incentives provided by BIS producers (57 %).

At companies, mostly oriented towards corporate customers (business-to-business, B2B sector), the most important benefits of users' involvement into the BIS development process are as follows:

- better understanding of customer needs (83 %),
- improvement of corporate image (80 %),
- personal satisfaction of users (83 %),
- opportunity to learn new IT products (80 %).

At companies, mostly oriented towards individual customers (business-to-customer, B2C sector) the important benefits of users' involvement are following:

- better understanding of user needs (73 %),
- opportunity to get business knowledge in a low cost way (67 %),
- improvement of corporate image (66 %).

The most important impediments of user involvement in the BIS development process comprise:

- user lack of knowledge and skills (at B2B companies 76 % of all asked respondents mentioned it, and respectively at B2C companies – 67 %),
- obligation to learn new IT products and technologies (at B2B companies 65 % of respondents argue that, and at B2C companies – 67 %),
- lack of incentives and encouragement from the BIS producers to support user activation (52 % of responses at B2B companies, and 67 % – at B2C firms).

The second question in this survey concerns methods of user activation to encourage them to the cooperation for BIS development. So, in the survey the following methods have been identified:

- at big companies, participation of users in trainings, courses and workshops (90 % of respondents emphasize that), constant discussions of IT personnel with users (86.7 %) and participation of users in reviewing processes covering interfaces reviews and use case analyses (77 %),
- at medium companies, participation of end-users in BIS testing (indicated by 90 % of respondents), constant discussions of IT staff with users (85 %), and user involvement into the quality management team work (76 %),
- at small companies, participation of users in trainings, courses and workshops (75 % of respondents answered that), constant discussions of IT personnel with users (72 %) and occasional interviews and meetings with users (72 %),
- at micro companies, interviews and meetings of IT staff with users (75 %), participation of users in trainings, courses and workshops (73 %), user involvement into the quality management team works (71 % of respondents), and distribution of free and open source software (70 %).

None of the respondent groups emphasize agile methods application for software development or for project management. IT personnel and users are observed as conservatively minded persons. Similarly, the corporate architecture model discussions as well as IT product customisation opportunities have not been perceived as valuable for user encouragement. Beyond that, users are not interested in a control and evaluation of BIS administrator works.

At the B2C companies the following user activation methods have been considered as the most important:

- interviews and discussions with users (72 %),
- participation of users in trainings, courses and workshops (75 %),
- distribution of freeware and demo version of software tools and business applications (69 %),
- user participation in software testing during prototyping or pilot version implementation (69 %).

At the B2B companies the CIOs as survey respondents emphasized the following users' activation methods:

- interviews and constant discussion of IT staff with users (82 %),
- participation of users in software testing (81 %),
- presentation of UML use cases to enable and support users in the process of the BIS construction understanding (81 %),
- presentation of screen layouts of business application to discuss with users the software usability (78 %),
- participation of users in trainings, courses and workshops (81 %).

The last question asked for this survey concerned attitudes of users towards traditional solutions implemented for their support i.e., customer relationship management (CRM) systems, insourced and outsourced help desk, IT service anticipation systems and providing the consultancy by CIOs. The user support mechanisms are considered as required. The results are visible in Tables 3–6.

Table 3. Acceptance of user support applications at micro companies – IT personnel attitude [%]

Media	For high quality of BIS				
	C	R	N	U	I
Analytical CRM	13	**42**	40	5	0
Transactional CRM	20	**43**	34	2	1
Operational CRM	13	**48**	29	8	3
Online insourced help desk	14	**49**	29	8	0
Online outsourced help desk	10	**44**	40	6	0
User Requirement Forecasting system	13	38	**41**	8	1
CIO department for users	13	38	**39**	10	0

Table 4. Acceptance of user support applications at small companies – IT personnel attitude [%]

Media	For high quality of BIS				
	C	R	N	U	I
Analytical CRM	32	**34**	29	4	1
Transactional CRM	20	**61**	15	3	1
Operational CRM	22	**56**	20	1	1
Online insourced help desk	23	**53**	22	1	1
Online outsourced help desk	20	**48**	27	4	1
User Requirement Forecasting system	39	**34**	20	4	3
CIO department for users	17	**52**	20	10	1

Table 5. Acceptance of user support applications at medium companies – IT personnel attitude [%]

Media	For high quality of BIS				
	C	R	N	U	I
Analytical CRM	**39**	34	27	0	0
Transactional CRM	24	**39**	37	0	0
Operational CRM	32	29	**39**	0	0
Online insourced help desk	**42**	34	24	0	0
Online outsourced help desk	22	**29**	**29**	15	5
User Requirement Forecasting system	22	**39**	37	2	0
CIO department for users	22	32	**41**	5	0

Table 6. Acceptance of user support applications at big companies – IT personnel attitude [%]

Media	For high quality of BIS				
	C	R	N	U	I
Analytical CRM	27	**47**	23	3	0
Transactional CRM	27	**37**	36	0	0
Operational CRM	23	**54**	23	0	0
Online insourced help desk	**37**	36	27	0	0
Online outsourced help desk	20	**43**	27	10	0
User Requirement Forecasting system	30	**40**	30	0	0
CIO department for users	30	**37**	30	0	3

For the surveyed companies, the presented in Tables 3–6 user support systems are considered as required, but they are not compulsory. In Tables 3–6 the following categories of user support software applications were distinguished:

- Analytical CRM system, covering user knowledge databases and data warehouses, enabling data mining, concluding on the user experiences and requirements, and recommending them the necessary trainings and workshops,
- Transactional CRM system, enabling IT services registration and management, service level management and service level agreement (SLA) management,
- Operational CRM systems, implemented to register all possible contacts with users by phones, emails and mobile devices, as well as F2F meetings,
- Online insourced help desk, developed and maintained internally by IT staff for corporate internal (e.g., employees) and external (i.e., Internet customers) users,
- online outsourced help desk to ensure IT services for internal and external users,
- information system for forecasting the users' needs basing on their actual usage of the business applications,
- establishing the CIO office for development of the formal procedures of user support.

For the companies, the presented in Tables 3–6 user support systems are considered by the respondents as required, but not compulsory solutions. Generally, CIOs argue that proposed user support applications are positively accepted, there are no opinions that presented systems are impediments. The CIOs perceive that CRM systems, particularly analytical CRM systems are helpful to recognize the unique user requirements and encourage them into the BIS development activities. The perception of usefulness of CRM systems is similar in different size companies. The very positive attitude towards user support systems suggest that user knowledge management approach could be further developed basing just on the analytical CRM systems. The sector analysis of user involvement in information system development process is presented in [14].

4   Conclusion

Literature studies lead to the conclusion that business organizations are beginning to realize the potential benefits that can be captured when users and IT people co-create values. So far, different methods have been developed to strongly involve users in the BIS development. Methods i.e., collaborative engineering, participatory design, UCSD (user-centred system design), customer knowledge management, user experience design are based on the assumption that users must be pulled into action and they must be encouraged to reveal their requests. On the other side, approaches i.e., user centric management, actor-network theory, user as innovator emphasize the central and active role of users. In many cases, companies are not conscious of the opportunities and potential benefits that could be realized when the users will be more active and involved in the BIS development and exploitation processes. Companies perceive some benefits from a large cooperation with users. They get benefits such as marketing insights, cost savings, brand awareness and idea generation. User benefits from the fulfillment of personal needs, opportunities to get new knowledge and satisfaction, because of doing something for themselves. Nowadays, cloud computing creates opportunities for stronger user involvement in information system development [15].

In the survey done in 270 firms, the CIOs or IT professionals have been observed as very skeptical about the usabilities of new media and social networking for the business application development. The research revealed important problems of lack of skills and knowledge of users as well as lack of incentives necessary for their involvement in the BIS development. Therefore, a huge social capital is unused and development of user knowledge codification and management would be needed. The future research will cover the analysis of prerequisites for decentralized management and architecture modelling of end user applications. Particularly, the huge opportunities are hidden in cloud computing, which is a new way of IT services' distribution and a new way of business application user activity.

Acknowledgement. The work is funded by the National Science Centre in Poland. The grant number is 4100/B/H03/2011/40.

References

1. Shapiro, A.: Users not Customers. Penquin Books Ltd., London (2011)
2. Claycomb, C., Lengnick-Hall, C.A., Inks, L.W.: The customer as a productive resource: A pilot study and strategic implications. Journal of Business Strategies 18(1), 47–69 (2001)
3. Prahalad, C.K., Ramaswamy, V.: Co-opting customer competence. Harvard Business Review 78(1), 79–87 (2000)
4. Von Hippel, E.: Lead users: A Source of Novel Product Concepts. Management Science 32(7), 791–805 (1986)

5. Tidd, J., Bessant, J., Pavitt, K.: Managing innovation, Integrating technological, market and organizational change, 3rd edn. John Wiley and Sons, Chichester (2005)
6. Cartlidge, A.: Best Practice for Business Perspective: The IS View on Delivering Services to the Business, ITIL, the key to Managing IT services. TSO, Norwich (2004)
7. Pieper, M., van Bon, J.: Project Management Based on PRINCE2. Van Haren Publishing, Zaltbommel (2005)
8. Barki, H., Hartwick, J.: Rethinking the concept of user involvement, and user attitude. MIS Quarterly 18(1), 59–79 (1994)
9. Kolfschoten, G.L., Briggs, R.O., de Vreede, G.J.: A technology for Pattern-Based Process Design and its Application to Collaboration Engineering. In: Rummler, S., Bor, N.G.K. (eds.) Collaborative Technologies and Applications for Interactive Information Design, Emerging Trends in user Experiences, pp. 1–19. Information Science Reference, Hershey (2010)
10. Beisse, F.: A Guide to Computer user support for help desk and support specialist. Course Technology, Cengage Learning, Boston (2010)
11. Ottmann, G., Laragy, C., Damonze, G.: Consumer Participation in Designing Community Based Consumer-Directed Disability Care: Lessons from a Participatory Action Research-Inspired Project. Syst. Pract. Action Res. 22, 31–44 (2009)
12. Terry, J., Standing, C.: The Value of User Participation in e-Commerce Systems Development. Informing Science Journal 7 (2004), http://2003.insite.nu
13. Xie, C., Bagozzi, R.P., Troye, S.V.: Trying to prosume: toward a theory of consumers as co-creators of value. Journal of the Acad. Mark. Sci. 36, 109–122 (2008)
14. Pankowska, M.: Information Technology Prosumption Acceptance by Business Information System Consultants. In: Frameworks of IT Prosumption for Business Development. IGI Global, NY (in press)
15. Pankowska, M.: Cloud Computing as Information Technology Prosumption Environment. In: User-Driven Information System Development. Research Papers. University of Economics in Katowice (in press)

Automatic Customer Segmentation
for Social CRM Systems

Adam Czyszczoń and Aleksander Zgrzywa

Politechnika Wrocławska, Faculty of Computer Science and Management,
Institute of Informatics,
Wybrzeże Wyspiańskiego 27, 50370 Wrocław, Poland
{adam.czyszczon,aleksander.zgrzywa}@pwr.wroc.pl
http://www.zsi.ii.pwr.wroc.pl

Abstract. This paper attempts to address the problem of the automatic customer segmentation by processing data collected in Social Customer Relationship Management (Social CRM) systems using Kohonen networks. Presented segmentation approach comprises classic loyalty-profitability link model that is explicit for CRM, and new social media components direct to Social CRM. The result of presented approach is an analysis tool with data visualization for managers which significantly improves the process of customer segmentation. Presented research is supported by implementation of proposed approach by which experiments were conducted. Additionally, the experimental results showed that proposed method performed very close to k-means algorithm which indicate the correctness of the proposed approach.

Keywords: customer segmentation, CRM, Social CRM, clusterization, SOM, unsupervised learning, ANN, data mining.

1 Introduction

To acquire competitive advantage many companies use the strategy of Customer Relationship Management (CRM) what can be observed in growing interest in this domain. However, in recent years new element of strategic importance appeared called social media. In order to meet the changing expectations of customers needs, Social CRM (SCRM) systems represent new branch of CRM systems which is oriented on the use of social media.

With the emergence of a new family of CRM systems, there has arisen the need for developing tools supporting these systems. Although both CRM and SCRM systems have many analytical tools, still a lot of them impose the necessity of extensive data management and using external software packages [1]. This in turn causes that many analyses are carried out semi-automatically or even manually. Such a situation results in not only a loss of valuable time, but also a lack of focus on the most important components of customer relationship management systems and their benefits. This comes down to the fact that the management is unable to keep up with the rapidly changing customers trends, particularly in the area of social networks.

A. Kwiecień, P. Gaj, and P. Stera (Eds.): CN 2013, CCIS 370, pp. 552–561, 2013.
© Springer-Verlag Berlin Heidelberg 2013

The aim of this work is to propose an approach to solve the problem of automatic segmentation of customers in the SCRM systems. The purpose of the method is to support the CRM strategy by providing applicable tools of data analysis for managerial staff.

Presented segmentation approach is based on well-known model, linking customer profitability and loyalty, which are also the two most important components of CRM strategy. Moreover, the presented approach has been extended to include elements related to social media, which are crucial to SCRM systems. It was also assumed that each of the main segmentation components can be composed of many features. In addition, an adequate representation of analyzed data that provides management with clear results in form of diagrams is required. Therefore, for customer segmentation the Self Organizing Maps (SOM) algorithm is proposed which is commonly used for clusterization and visualization of high-dimensional data.

2 Related Work

This paper is a continuation to our research on intelligent tools supporting CRM systems using their information potential. The research presented in [2] includes definitions of some indicators, which were also used in this paper. This includes RFM (Recency Frequency Money), LTV (customer LifeTime Value), and NPP (Next Purchase Probability) used for customer loyalty estimation.

The theoretical foundation on clusterization using SOM is [3] whereas the basic concepts for customer segmentation using data mining techniques in CRM are presented in [1]. There are also numerous studies devoted to the area of customer segmentation in CRM. Some of them use genetic algorithms (for example [4]), however this technique is not suitable for automatic segmentation since it requires lot of data-specific parametrization and complex fitness function for high-dimensional data. On the other hand, most of the studies on customer segmentation for CRM focus on solving classification or clusterization task. They can be divided into three main classes according to the data mining model they apply: (i) supervised learning models, (ii) unsupervised learning models, (iii) and hybrid models.

First group which addresses the classification problem (e.g. [5]) could be applied for segmentation of high-dimensional data or even take into account imbalanced data [6]. However, classification requires establishing a training set that will represent particular segments. For this reason the classification methods are not applicable for automatic segmentation.

Second group that use clusterization techniques seem to be appropriate for automatic segmentation since there is no target variable to classify or estimate so the resulting segments are not known in advance. In [7] authors presented intelligent method which uses SOM. However, the whole process of segmentation is preceded by linear programming method, of which output is later used for SOM training. In such a case SOM is used as a classifier. In result the method requires setting a few parameters for segmentation (for linear programming and

SOM classifier) and it cannot be applied for automatic segmentation. Moreover, the method puts focus on profitability only, which seems not to treat the problem of segmentation in CRM as a whole. Another study on segmentation using clusterization techniques was presented in [8] and [9], however both methods require that some of the parameters should be determined empirically, which limits their automation capabilities.

The last group represents methods that combine supervised learning models with unsupervised ones. In [10] authors presented segmentation method utilizing both SOM and back-propagation ANN (Artificial Neural Networks). The approach itself proposes a classifier where part of the input data comes from SOM. Although this method is not suitable for automatic segmentation, it allows expanding the number of features. However, the feature expansion is possibly at the cost of re-parameterization of the network.

All of the above studies use different segmentation methods and different data sets suitable for different purposes. Moreover, vast majority of mentioned studies were application-specific whereas different areas of application affect selection of different segmentation components. Despite the fact that most of mentioned studies use similar features where RFM and LTV are the most common ones, some of them focus on profitability only while other take it into account only indirectly. In result, it is difficult to compare presented approaches of customer segmentation to each other. Additionally, none of above research considers the Social CRM segmentation problem and none of above works propose data visualization tools as the outcome of their methods.

3   Segmentation Problem

The aim of this work is to propose an approach to solve the problem of automatic segmentation of customers in the SCRM systems and visualize it in such a manner that it provide a complete tool of data analysis for management. In order to create automated tools for customer segmentation, presented approach requires intelligent techniques of data mining.

One of such a techniques are Self Organizing Maps (SOM) which is commonly used for clusterization and visualization of high-dimensional data. Self Organizing Maps, also referred to as Kohonen network, is an Artificial Neural Network approach which uses unsupervised learning algorithm. It is composed of a map or grid of neural cells where each of them is associated with a n-dimensional vector. Cells are adjusted through a learning process using n-dimensional training data. No supervision means that there is no human expert who must assign input data to particular class and provide it as training data. In clustering, it is the distribution of the data that will determine cluster membership. The usage of SOM also gives the possibility to make low-dimensional representation of possibly high-dimensional data set which is crucial for proposed approach of customer segmentation. This is possible because SOM neighborhood function preserves the topological properties of the input data space. Based on the above-mentioned reasons, this method is selected for the approach presented in this paper.

Proposed in this section segmentation methods is based on classic model of customer loyalty-profitability segments. This model became very popular after article published in 2002 by W. Reinartz and V. Kumar [11]. Authors indicated that there is evident link between loyalty and profitability that can by illustrated by Fig. 1 [12].

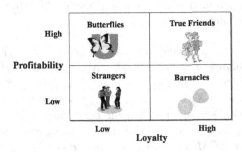

Fig. 1. Customer segments in the context of loyalty-profitability link

This model assumes that customers are sorted according to their profitability and longevity [11] where longevity translates to loyalty. By dividing those two indicators by high and low groups, the resulting four categories represent customer segments to which different marketing strategies should be applied.

Since loyalty and profitability are the most important components of CRM strategy, in CRM systems there are measures that allow the calculation of both of those factors. Presented approach uses classic methods to measure the level of customer loyalty which was presented in [2]. This includes calculating loyalty using three features: the RFM (Recency Frequency Money), LTV (customer LifeTime Value), and NPP (Next Purchase Probability).

Moreover, the presented in this paper approach considers modifying and extending the classic segmentation model in order to include elements related Social CRM systems. The modification is based on proposing new method of measuring customer's profitability in the sphere of social media and it is called Social Media Profitability (SMP). The extension is based on measuring the level of client's Social Media Engagement (SME), which in result adds a new dimension to standard segmentation model presented on Fig. 1.

For the purpose of this paper the calculation of Social Media Profitability is done using only one feature called Social Media Return of Investment (SM-ROI). Measuring the return of investment is the most common profitability ratio. Therefore, translating it into social media seem to be an accurate measure of profitability for SCRM. The SMROI is calculated per customer as profit from social networking events (for example on Facebook) divided by money invested on those events. The SMP can be later extended with more features to better reflect profitability, especially in the long-term view.

The calculation of Social Media Engagement so far also includes only one feature called Weighted Event Participation (WEP) calculated as customer's average event participation multiplied by $1 +$ Twitter Follow Frequency (TFF). The TFF is used to measure how often a person attracts new followers on Twitter and is calculated as average number of new followers per day. Applied as a multiplier allows to emphasize customer's influence on others in social networking services.

Considering the above, presented segmentation problem can be reduced to hard clusterization problem which is solved using SOM function. This can be formally defined as follows:

Definition 1. *Given a customer data set $C = \{c_1, c_2, \ldots, c_n\}$, where each $c_i \in C$ is represented by following tuple $c_i = \langle X_1, X_2, X_3 \rangle$. Elements X_j ($j = 1 \ldots 3$) represent SCRM segmentation components such as loyalty, SMP, SME, and are defined as feature groups $X_j = \{x_1, x_2, \ldots, x_m\}$ composed of real-valued $x_l \in X_j$ feature vector in Euclidean space. Then the clustering of C is the partitioning it into k sets $K = \{K_1, K_2, \ldots, K_k\}$ called clusters using an assignment $s : C \to K$ where s is surjective SOM function (none of the clusters in K is empty). It is also required that function s is not bijective (there is no one to one mapping) and $\cap_{k=1}^{|K|} K_k = \emptyset$.*

Problem definition formulated in such a manner allows to extend feature groups X_j to any size, and thereby increasing the number of dimensions of the input vectors. Therefore, the loyalty, SMP and SME measures can be easily expanded by adding more factors and better reflecting models they represent. In addition, the definition of customer data c_i as three-tuples gives the possibility of representation multidimensional data by the means of colors, using the RGB color model.

The second advantage is the ease of handling this type of data on two- and three-dimensional graphs for data visualization used by managers. To do that it is enough to calculate the mean of feature vector components in particular feature group X_j, and scale it into range between 0 and 255 using min-max normalization. Clustered data is later plotted on graphs using cluster color to which it was assigned.

Three types of graphs are proposed – first one representing data in loyalty-SMP relation, second one in loyalty-SME relation, and the last one in the loyalty-SMP-SME relation. Furthermore, data on first two graphs is divided into halves using vertical and horizontal lines. In result, clustered data is additionally divided into four areas A, B, C, D. This procedure allows to reflect the classical segmentation methodology presented in [11] and link together clustered data with proposed by authors marketing strategies. Moreover, this allows to compare clusters in terms relation to particular areas. Despite the fact that data visualization using graphs is based on two or three dimensions only, the SOM clusterization operates on multi-dimensional data. In this paper the number of dimensions is limited to five, according to presented loyalty, SMP and SME schemes.

In order to make the segmentation process on the basis of the above model ran automatically, SOM parameter values must be fixed or depend on the parameters of the model. The bigger the map the better the clustering results, however the number of resulting clusters grows and the algorithm execution time significantly increases. Smaller maps are more generalized but number of classes is smaller and execution time is much shorter. For the purpose of this paper it is assumed that the SOM parameters are fixed and in order to get appropriate outcome for presented segmentation method, that is to keep number of resulting classes reasonably low with good clusterization quality and low algorithm execution time, the SOM map was proposed to be of size 14x14, number of interactions equal to 1000 and learning rate set to 0.05.

However, it should be noted that the presented model assumes that the number of features can be arbitrarily increased, which can affect the quality of clustering for SOM with fixed parameters. In addition, the number of input data may also have impact on the results. Initial empirical studies have shown that for the assumed constant parameter values of SOM the size and number of dimensions of input data had negligible impact on the number of resulting clusters. Such an assumption, however, requires verification and thus assumed SOM constant values are preliminary. Their optimization, finding limit values and determining dependencies to the model are the subject of further research.

4 Experimental Results

Based on implementation of presented approach an experiment was conducted. The aim of the experiment was to apply presented approach of automatic customer segmentation on a case study example. The input data for the experiment consisted of 50 customers collected from SCRM system of a textile industry company. Clustering results are also evaluated by the means of comparison to k-means algorithm. The data set did not contain any personal information about customers. All collected data concerned one year period and all feature values are calculated according to this timespan.

Given input data was divided into 7 clusters and presented on Fig. 2. First cluster (with centroid position in [0, 7] on Fig. 2) contained 22 customers, second (pos. [8, 13]) 3 customers, third (pos. [9, 3]) 2 customers, fourth (pos. [3, 3]) 14 customers, fifth (pos. [0, 13]) 6 customers, sixth (pos. [13, 9]) 2 customers, and seventh (pos. [13, 3]) contained 1 customer.

It is important to notice that some clusters that are located near one another and seem to share similar color (as for example clusters with centroid pos. [0, 13] and [8, 13]) are significantly different because their Euclidean distance is big. Such a situation may happen because SOM map visualization is done using only three parameters (RGB colors), whereas nodes are composed of five features. The true border between clusters is visible on the U-matrix (Unified Distance Matrix) illustrated on Fig. 3 which gave a two-dimensional representation of high-dimensional data (in presented case five-dimensional). The u-matrix presents the distance between the adjacent neurons and depicts it in gray-scale.

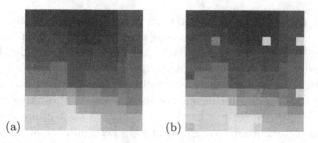

Fig. 2. Two-dimensional SOM node layer of size 14x14. Left side of the figure (a) presents map without plotted data, whereas map on the right (b) includes clusterized data. The more red the more objects particular class contains. Yellow field indicates that there is only one object in that cluster.

A dark color between the neurons corresponds to a large distance and thus is considered as cluster separator. A light color between the neurons indicates that they are close to each other and thus light areas represent clusters. The value of a particular node is the average distance between the node and its closest neighborhood. There are two types of u-matrix graph presented on Fig. 3. The square grid u-matrix allows to present up to 8 neighbors of a node, whereas hexagonal grid allows to consider up to 6 neighbors.

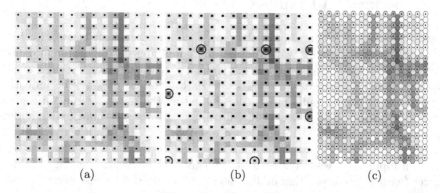

Fig. 3. U-matrix SOM representation illustrating distances between neurons (marked with black dots). Left side of the figure (a) presents square grid u-matrix, the middle (b) presents square grid u-matrix with highlighted colors of corresponding clusters, and the right side (c) presents hexagonal grid u-matrix.

Secondly, the clustered data is plotted on subsequent graphs representing subsequent aspects of customers segmentation – the loyalty-SMP link (Fig. 4a), loyalty-SME link (Fig.4b), and three-dimensional graph of loyalty-SMP-SME (Fig. 5). Each segment is automatically assigned a color that corresponds to the color of its cluster. As it was mentioned this data is expressed in RGB color model, where the first component represents loyalty, the second customer profitability in the area of social media and the third of his/her engagement. As a result, clusters can be represented in the full range of colors. Figure 4 present the customer segmentation results of 50 customers.

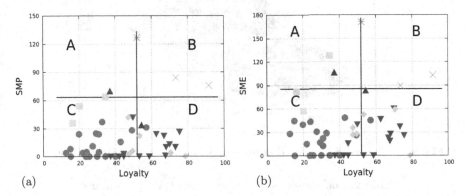

Fig. 4. Clusterized data plotted on loyalty-SMP graph (a) and loyalty-SME graph (b). Vertical and horizontal lines divide data by halves.

Vertical and horizontal lines divide the data in halves to reflect the classical idea of the segmentation shown in Fig. 1 and allow to combine it with the identified clusters. As a result, customers are divided into seven segments in eight areas (four areas A, B, C, D for each dimension). A complete picture of resulting segmentation is presented on the graph as shown in Fig. 5. Based on this kind of data visualization, managers can immediately plan separate marketing activities for each segment. For example, customers of the segment [0, 7] in the area C are characterized by relatively low loyalty, low engagement to social media events, and low profitability. This group can be certainly classified as strangers. For this group, a good strategy might be to encourage more frequent purchases by offering a loyalty program. Much better group of customers belong to segment [8, 13] in the same area. However, their relatively high SME is not reflected in profitability. Perhaps these are the customers who prefer on-line shopping. Similar element of classical approach to segmentation presented in [11] can be applied to other identified segments.

As can be seen automatic segmentation using SOM clustering technique gives much more detailed client groups than in the case of manual segmentation by determining ranges for each segment, especially in the case of multi-dimensional data. Proposed approach allows to save time and better meet customer needs so the company can achieve greater benefits.

In the last stage of the experiment the SOM clustering was compared to clustering generated using k-means algorithm with the same number of iterations (1000) and number of clusters ($k = 7$). The results are presented in Table 1.

The Silhouette Score allows to measure of how tightly grouped is data in all clusters and thus informs how appropriately the data has been clustered [13]. It

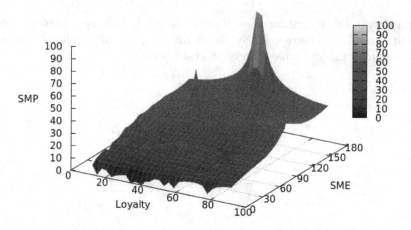

Fig. 5. Clusterized data plotted in three-dimensional loyalty-SMP-SME space

Table 1. Comparison of clustering using SOM and k-means

SOM Silhouette Score:	0.222
K-means Silhouette Score:	0.238
Normalized Mutual Information:	0.667
Rand Index Score:	0.458
Adjusted Mutual Information:	0.538

ranges from -1 to 1 therefore result close to 0.2 is satisfying. Remaining metrics allow to measure a similarity between two clusterings [13]. Comparison results show that SOM clusterization performed very close to k-means and resulting clusterings are relatively similar. This indicates that proposed approach of customer segmentation for CRM is valid.

5   Conclusions and Future Work

The aim of this work was to propose an approach to solve the problem of automatic segmentation of customers in the Social CRM systems using Self Organizing Maps. The purpose of the method was to support the CRM strategy by providing tools of data analysis for managerial staff. Presented segmentation approach was based on well-known model, linking customer profitability and loyalty, which represent the two most important components of CRM strategy. Moreover, the presented approach included new elements related to social media, which are crucial to SCRM systems. Presented segmentation model also assumed that each of the main segmentation components can be composed of many features. Additionally, the model allows further feature expansion by adding more factors to loyalty, social media profitability (SMP) and social media engagement (SME) components to better reflect information they carry. Based on presented approach of automatic customer segmentation an experiment was conducted in a form of a case study example. The input data for the experiment considered 50 customers collected from

SCRM system of a textile industry company. Additionally, the experimental results showed that proposed method performed very close to k-means algorithm which indicate the correctness of the proposed approach. However, the advantage of SOM algorithm is that it does not require defining number of clusters which in case of automatic approach is not known a priori.

On the other hand, presented approach requires SOM parameters optimization, finding limit values and determining dependencies to presented clusterization model. Additionally, in order to improve clusterization results number of neurons should be increased. Because this would generate too many classes, presented method requires Hierarchical Clustering or other hybrid approach that would successfully decrease number of classes while keeping total number of clusters low. Both mentioned problems are the subjects of further research.

References

1. Tsiptsis, K., Chorianopoulos, A.: Data Mining Techniques in CRM: Inside Customer Segmentation. Wiley Publishing (2010)
2. Czyszczoń, A., Zgrzywa, A.: Consensus as a tool supporting customer behaviour prediction in social CRM systems. Computer Science 13(4) (2012)
3. Larose, D.T.: Discovering Knowledge in Data: An Introduction to Data Mining. Wiley InterScience (2005)
4. Chan, C.C.H.: Intelligent value-based customer segmentation method for campaign management: A case study of automobile retailer. Expert Syst. Appl. 34(4), 2754–2762 (2008)
5. Kim, Y., Street, W.N.: An intelligent system for customer targeting: a data mining approach. Decis. Support Syst. 37(2), 215–228 (2004)
6. Napierala, K., Stefanowski, J.: BRACID: a comprehensive approach to learning rules from imbalanced data. J. Intell. Inf. Syst. 39(2), 335–373 (2012)
7. Lee, J.H., Park, S.C.: Intelligent profitable customers segmentation system based on business intelligence tools. Expert Syst. Appl. 29(1), 145–152 (2005)
8. Kuo, R.J., Ho, L.M., Hu, C.M.: Integration of self-organizing feature map and k-means algorithm for market segmentation. Comput. Oper. Res. 29(11), 1475–1493 (2002)
9. Wu, R.S., Chou, P.H.: Customer segmentation of multiple category data in e-commerce using a soft-clustering approach. Electron. Commer. Rec. Appl. 10(3), 331–341 (2011)
10. Yan, X.B., Li, Y.J.: Customer segmentation based on neural network with clustering technique. In: Proc. of the 5th WSEAS Int. Conf. on AI, Knowledge Engineering and Data Bases. AIKED 2006, pp. 265–268. World Scientific and Engineering Academy and Society, WSEAS (2006)
11. Reinartz, W., Kumar, V.: Mismanagement of customer loyalty. Harvard Business Review (2002)
12. Tyler, T.: Loyalty programs: Making sure you invest in only the right customers (January 2013), http://www.genroe.com/blog/loyalty-programs-making-sure-you-invest-in-only-the-right-customers/306
13. Wagner, S., Wagner, D.: Comparing Clusterings – An Overview. Technical Report 2006-04, Universität Karlsruhe (TH) (2007)

Practical Aspects of Log File Analysis
for E-Commerce

Grażyna Suchacka[1] and Grzegorz Chodak[2]

[1] Institute of Mathematics and Informatics, Opole University,
Opole, Poland
[2] Institute of Organisation and Management, Wroclaw University of Technology,
Wroclaw, Poland
gsuchacka@uni.opole.pl, grzegorz.chodak@pwr.wroc.pl

Abstract. The paper concerns Web server log file analysis to discover knowledge useful for online retailers. Data for one month of the online bookstore operation was analyzed with respect to the probability of making a purchase by e-customers. Key states and characteristics of user sessions were distinguished and their relations to the session state connected with purchase confirmation were analyzed. Results allow identification of factors increasing the probability of making a purchase in a given Web store and thus, determination of user sessions which are more valuable in terms of e-business profitability. Such results may be then applied in practice, e.g. in a method for personalized or prioritized service in the Web server system.

Keywords: Web server, log file analysis, statistical inference, e-commerce, B2C, Business-to-Consumer, Web store.

1 Introduction

The analysis of historical data recorded in Web server log files is the basic way of capturing knowledge of the Web server workload and the behavior of Web users. In e-commerce environment such analyses have been performed at multiple levels, including the lowest, protocol level (corresponding to HTTP requests), the application level (corresponding to Web page requests or business-related Web interactions), and the user level (corresponding to user sessions) [1–3]. From the online retailers' point of view the application and user level analyses are of the highest practical value because understanding the way in which customers use the site and navigate through the store, especially in the context of successful purchase transactions, may lead to better organization of e-commerce service and more efficient business decisions.

This was our motivation for exploring dependencies between different characteristics of user visits to the site of a Web store and their probability of being ended with a purchase. Three groups of visits differing in user behavior were considered: visits of anonymous users, visits of users who logged on but did not buy anything, and visits of users who decided to make a purchase.

A. Kwiecień, P. Gaj, and P. Stera (Eds.): CN 2013, CCIS 370, pp. 562–572, 2013.
© Springer-Verlag Berlin Heidelberg 2013

The rest of the paper is organized as follows. Section 2 overviews related work on analyses of historical data for Web stores. Section 3 discusses some practical aspects and possibilities of data analysis for different user groups in the Web store. Section 4 presents our research methodology and Sect. 5 discusses results of the research. Key findings are summarized in Sect. 6.

2   Related Work

An important aspect of e-commerce workload characterization is a click-stream analysis, which concerns discovering user navigational patterns at a Web site [1]. The click-stream analysis is often combined with segmentation or clustering methods to determine different customer profiles [4, 2, 5, 3]. Discovery of meaningful usage patterns characterizing the browsing behavior of Internet users realized by applying data mining techniques is called Web usage mining.

Segmentation methods have been especially intensively explored for Web store historical data. Some techniques used in CRM (*Customer Relationship Management*) in traditional commerce have been also applied in e-commerce. For example, RFM (*Recency, Frequency, Monetary*) analysis was combined with such data mining techniques as Apriori algorithm [6], rough sets and K-means algorithm [7], and approximate reasoning from the set of fuzzy rules [8]. The vector quantization-based clustering and the "Apriori"-based association rule mining algorithm was applied to classify e-customers based on their RFM values and to find out relationships among their purchases [9]. A design of the Web mining algorithm based on variable precision rough sets to classification tasks was proposed in [10].

As a consequence, some models of a user session in a Web store have been proposed as probabilistic finite state machines for different user profiles. Examples include Customer Behavior Model Graph [11], TPC-W session model [12], and Extended Finite State Machine [13].

Association rules have also been successfully applied to e-commerce, combined with collaborative filtering methods [14, 15]. Association rules for products viewed or purchased by different customers are widely used in contemporary Web stores in product recommendation. In this paper, we focus on applying statistical analysis and association rules to identify these sessions attributes and characteristics which increase the probability of making a purchase.

3   Key Questions in the Context of Different Web Store User Groups

Although most online retailers today use analytical applications (e.g. Google Analytics), which provide much valuable information, they are not able to answer many questions concerning a detailed behavior of three user groups: users, who do not log on, users who log on but do not buy anything, and users who log on and finalize purchase transactions.

1. **Anonymous users.**
   Questions concerning the behavior of unlogged users aim at finding reasons
   for which the users – potential customers – do not log on and do not place
   an order. This group is the most numerous. It contains many users who visit
   only one page in the session. Some information on these users is available in
   Google Analytics, e.g. their share in overall traffic (called the bounce rate)
   and sources of their visits. Since these users do not stay in the Web store,
   no information on their behavior can be read from log files. Therefore, such
   sessions were not under consideration in this paper.
   However, some anonymous users navigate through the site for some time and
   it gives the chance to analyze their usage patterns: sequences of visited pages,
   usage of specific tools, visited featured places of the Web store (new products,
   bestsellers, promotions). The analysis should aim at finding reasons for which
   the user do not decide to log on and make a purchase.

2. **Users who log on but do not buy anything.**
   The analysis of these users' behavior should be more detailed because they
   put some effort into the interaction, which shows they are interested in the
   offered products, but they did not decide to purchase for unknown reasons.
   First of all, one should analyze the last pages statistics in their sessions,
   including the following page types: a product page, a product category page,
   a product review page, the shopping cart page, the page informing about
   shipping charges and conditions, the page containing the store regulations,
   the contact page, the home page, or a page containing search results. One
   aspect of possible practical analysis concerns the percentage of logged users
   who added products to the carts but did not buy them. Other aspect is
   connected with the usage of the search engine at the store site and the
   analysis of search results: maybe users did not find products satisfying their
   searching criteria, or maybe they got some results which were not satisfactory
   for them.

3. **Users who log on and finalize purchase transactions.**
   These analyses are the most valuable for the online retailer. Customers may
   be classified in terms of their navigational patterns, viewed and/or purchased
   products, value of the shopping cart, source of the customer visit, attributes
   captured in the registration process (age, sex, geographical location, etc.).
   Not all that information may be read from logs – some of them require
   installing additional software able to intercept data from registration forms
   (like Google Analytics does).

Some questions which can be answered based on log file analysis are the following.
Is there a relationship between the number or value of purchased products and
the source of the visit? Is there a relationship between the time users spend
in the Web store and the percentage of customers making a purchase? Is there
a relationship between the number of viewed products and the percentage of
customers making a purchase? What is the mean time users spend on individual
pages of the checkout process? What percentage of customers, who had checked
information about shipping costs before starting the checkout process, decide to
confirm the purchase?

The comparison of the aforementioned user groups may provide information on significant differences between them. It is especially important to identify features distinguishing customers in the third group and to determine factors increasing the probability of making a purchase.

To illustrate the practical aspect of such research we performed analyses in order to answer the following questions:

1. How many logged users confirm a purchase and how many users give up after reading shipping charges in the checkout process? How many users from these both groups checked shipping charges before the checkout process?
2. How many users, who had checked information on shipping charges before the checkout process, decided to make a purchase?
3. How many (logged and unlogged) users, who added products to the shopping carts, give up purchasing them?

4 Research Methodology

The analysis was performed for one month data in Web bookstore log files, recorded in December 2011. The Web server was an Apache HTTP server on Linux with PHP and MySQL support. Using a C++ computer program, data was read from log files, preprocessed, and cleaned; user sessions were reconstructed, key session states were distinguished, and data was analyzed by means of statistical analysis and association rules.

4.1 Data Preprocessing

First, raw data was read from Web server log files and for each HTTP request the following data was distinguished: the IP address of the Web client, the identifier of the Web client, the user identifier, the timestamp, the HTTP method, the URI of the resource requested, the version of HTTP protocol, the HTTP status code, the size of the object sent to the client, the address of the referring page, and the client browser information. Some data was then transformed, e.g. the timestamps had to be converted to integer values so that the comparison of request interarrival times could be performed.

4.2 Data Cleaning

First stage of data cleaning was connected with elimination of useless data. Since our analysis concerns the behavior of users and involves a click-stream analysis, the following requests have been excluded from analysis: hits for embedded objects (e.g. images), automatically generated by Web client browsers, requests generated by Web bots (e.g. Web crawlers), and requests connected with administrative tasks.

The original data set contained 1 600 964 HTTP requests. 1 540 812 requests were excluded from analysis and only 60 152 requests (3.76 %) were left, as directly resulting from users' clicks.

4.3   Reconstruction of User Sessions

To analyze user behavior, Web users were identified based on IP addresses combined with the client browser information, included in each request. Then user sessions were reconstructed taking into consideration request arrival times. A *user session* means a sequence of Web page requests issued by the user during their single visit to the site. We assumed that the intervals between Web page requests in a session does not exceed 30 minutes.

16 081 user sessions were identified. For each session three basic characteristics were determined:

- the *session length*, i.e. the number of pages requested in the session,
- the *session duration*, i.e. the time interval between arrivals of the first and the last Web page requests in the session (with one second accuracy),
- the *mean time per page*, i.e. the average time of browsing a page by user in the session, calculated as the session duration divided by the session length.

The determined session duration is shorter than the actual time of the user-Web site interaction, because it does not include the time of browsing the last page in session, which cannot be determined at the server side. Therefore, the *session duration* and the *mean time per page* could not be determined for sessions containing only one page.

4.4   Elimination of Outlier Sessions

Session lengths and durations were used to find and exclude outlier sessions, diverging from a general trend for the sessions. In our case, such sessions may represent incorrect entries in log files or may suggest robot-generated interactions, which have not been identified. Using a graphical method [16], four sessions were eliminated. Thus, 16 077 user sessions were analyzed.

4.5   Identification of Key Session States

A user visits multiple Web pages in a session and performs multiple Web interactions, such as searching products according to given keywords, browsing detailed information on a selected product, etc. Pages may be grouped by functionality into different session states. The most desirable state is connected with making a purchase, i.e. with purchase confirmation:

- *Checkout_ success* state occurs when a user successfully realizes the checkout process, accepting total transaction cost, giving the shipping address, and finally confirming the purchase.

Taking into account goals of our analysis, we identified four other key session states, which may affect the probability of making a purchase by a user:

- *Login* means user's registration or logging into the site. Users may browse the site content without being logged on; logging into the site is necessary if they want to finalize a purchase transaction. However, not every user who is logged on will decide to buy something finally.

- *Shopping_ cart* means adding a product to the virtual shopping cart. Adding at least one item to the cart is a prerequisite for purchase transaction; however, products in cart do not guarantee their purchase. In fact, shopping cart abandonment is a huge problem for many online retailers [17].
- *Shipping_ info* means the situation when a user checks the information on shipping charges and conditions before starting the checkout process. Distinguishing this state is important because it may affect a purchase transaction: a user who is aware of shipping charges, may take them into consideration while adding products to cart and thus they will not be surprised by the total transaction cost at checkout.
- *Shipping_ info_ at_ checkout* is an integral stage of the purchase transaction and it occurs when a user is informed about possible means and charges of shipping during the checkout process. The shipping cost is automatically added to product costs.

Our analysis revealed that the prevailing majority of Web interactions in the Web store is connected with browsing and searching operations, whereas very few users perform Web interactions connected with key session states. For all 16 077 user sessions, visits to key session states have been rare (Fig. 1).

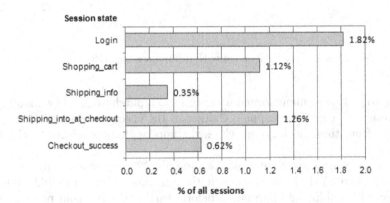

Fig. 1. Percentages of sessions with visits to key session states

As it can be seen, only 1.12 % of users added some products to the shopping carts and 0.62 % of users decided to buy them. The checkout process was started by at least 1.26 % of users (it can be inferred from the percentage of visits to the session state *Shipping_ info_ at_ checkout*, which is a part of this process), however, more than half of them gave up confirming the purchase.

It seems surprising that although only 1.12 % of users added products to the carts, as much as 1.26 % started the checkout process (because having products in carts is essential for starting this process). The explanation is the fact that products added to the user's cart during one visit are remembered till the next user's visit (unless a cookie mechanism is blocked in the user's Internet browser). Thus, some users had to add products to the carts during their former visits.

5   Analysis of Dependencies between Session States

The next step of analysis concerned discovery of relationships between the key session states in order to answer three questions formulated in Sect. 3. Main results are presented in Fig. 2.

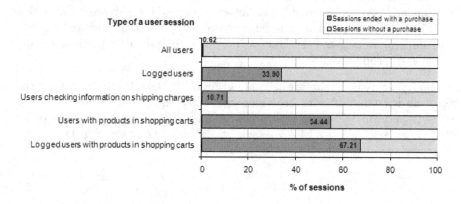

Fig. 2. Percentages of sessions ended with a purchase depending on a session type

- Question 1: How many logged users confirm a purchase and how many users give up after reading shipping charges in the checkout process? How many users from these both groups checked shipping charges before the checkout process?

  From among all users entering the site only 292 logged on. Only 99 logged users (33.9 %) decided to confirm the purchase: 6 of them (6.06 %) had checked the shipping information before starting the checkout process; 93 of them (93.94 %) had not checked the shipping information before.

  These results suggest that when a user logs on, the probability of making a purchase by them increases: in the set of all users only 0.62 % confirmed the purchase, whereas in the set of logged users as much as 33.9 % confirmed the purchase, which is more than one third.

  Moreover, 103 logged users (i.e. 35.52 % of all logged users) started the checkout process (they visited state *Shipping_ info_ at_ checkout*) but did not finish it: 26 of them (25.24 %) had checked the shipping information before starting the checkout process; 77 of them (74.76 %) had not checked the shipping information before, so maybe they were surprised by the additional shipping cost at checkout.

  However, only three from among these 103 users finished their sessions just after checking shipping information at checkout.

– Question 2: How many users, who had checked information on shipping charges before the checkout process, decided to make a purchase?
56 users read the shipping information before starting the checkout process. From among them only six (10.71 %) decided on a purchase. Thus, checking the shipping information during the navigation through the site before starting the checkout process increase the probability of making a purchase from 0.62 % to 10.71 %.

– Question 3: How many (logged and unlogged) users, who added products to the shopping carts, give up purchasing them?
Products were added to carts by 180 users; 98 of them (54.44 %) did not decide on the purchase: 122 logged users added products to the carts and 40 of them (32.79 %) did not buy the products; 58 unlogged users added products to the carts and 58 of them (100%) did not buy the products.
It can be noticed that products have been added to carts by much more logged users than the unlogged ones. Moreover, much more logged users decided to buy the products. One can infer that for sessions with not empty carts the fact of user's logging into the site significantly increases the probability of purchase.

To formally describe relationships between the key session states, we applied association rules. We implemented the a priori algorithm using the methodology and notation used in [18].

We determined eight characteristics of a user session which may matter in the context of the purchase probability. The first group of characteristics include visits to four key session states, defined above. The second group of characteristics was determined based on the *session length*, the *session duration*, and the *mean time per page*. Based on results of analyses comparing characteristics of all sessions and the sessions ended with a purchase (which were not discussed in this paper due to limited place), we replaced numeric values of the session characteristics with qualitative values in the following way:

$$
session\ length = \begin{cases} short, & \text{if the number of requests } = 1 \\ medium, & \text{if the number of requests } \in [2,13] \\ long, & \text{if the number of requests } \geq 14 \end{cases} \quad (1)
$$

$$
session\ duration = \begin{cases} short, & \text{if the session duration } < 60\,s \\ medium, & \text{if the session duration } \in [60,960]\,s \\ long, & \text{if the session duration } > 960\,s \end{cases} \quad (2)
$$

$$
mean\ time\ per\ page = \begin{cases} short, & \text{if mean time per page } < 60\,s \\ medium, & \text{if mean time per page } \in [60,180]\,s \\ long, & \text{if mean time per page } > 180\,s \end{cases} \quad (3)
$$

We define a set I containing key session characteristics: $I = \{$"*Login* = true", "*Shopping_ cart* = true", "*Shipping_ info* = true", "*Checkout_ success* = true", "*session length* = medium", "*session length* = long", "*session duration* = medium", "*session duration* = long", "*mean time per page* = short", "*mean time per page*

= medium", "*mean time per page* = long"}. Characteristics "*session length* = short" and "*session duration* = short" were excluded from the set because it is impossible to make a purchase in a session containing only one page request or lasts less than one minute.

Let D denote a set of all user sessions. Each session in D is represented as a set of session characteristics from I. For example, if a user logged on, searched for a few books, added two books to the shopping cart, and then finished the session; the session consisted of 12 page requests and was lasting 10 minutes (which gives the mean time per page equal to 54.5 s), the session will be represented as the set {"*Login* = true", "*Shopping_ cart* = true", "*session length* = medium", "*session duration* = medium", "*mean time per page* = short"}.

An association rule is the implication $A \Rightarrow B$ (*if A then B*), where the antecedent $A \subseteq I$, the consequent $B \subseteq I$, and $A \cap B = \varnothing$. Moreover, each association rule is described with two measures: *support* and *confidence* [18].

The *support* for a given association rule $A \Rightarrow B$ is the percentage of sessions in D which contain both A and B:

$$support = P(A \cap B) = \frac{number\ of\ sessions\ containing\ both\ A\ and\ B}{number\ of\ all\ sessions} . \quad (4)$$

The *confidence* for a given association rule $A \Rightarrow B$ is the percentage of sessions containing A which also contain B:

$$confidence = P(B \mid A) = \frac{number\ of\ sessions\ containing\ both\ A\ and\ B}{number\ of\ sessions\ containing\ A} . \quad (5)$$

A rule is *strong* if it meets or surpasses certain minimum support and confidence criteria. Since we consider associations between session states in the context of making a purchase and very small percentage of all sessions (0.62 %) ends with a purchase, we assumed the minimum support of 0.5 % and the minimum confidence of 70 %. Rules resulting from the application of a priori algorithm for such conditions are presented in Table 1.

The support of the first rule, R_1, is 0.5 %, meaning that the rule applies to 0.5 % of sessions in the analyzed dataset. It is relatively low level of support, however we have to keep in mind that only 0.62 % of all sessions ended with a purchase. 100 sessions ended with a purchase and of these sessions, 80 fulfilled the antecedent condition, i.e. the user logged on, added some products to cart (what is obvious as these events occur for each purchase transaction), the user opened at least 14 pages, and the mean time per page did not exceeded one minute.

The confidence of the rule R_1 equal to 73 % means that of all 16 077 sessions some group of sessions (0.68 %, i.e. 109 sessions) fulfilled the antecedent condition; 73 % of these 109 sessions (i.e. 80 sessions) ended with a purchase. Taking into account that for all sessions in data set the probability of purchase is only 0.62 %, one can say that the rule has very high level of confidence.

Table 1. Strong association rules determined for the analyzed data set

Rule	Antecedent	Consequent	Support [%]	Confidence [%]
R_1	{"*Login* = true", "*Shopping_cart* = true", "*session length* = long", "*mean time per page* = short"}	"*Checkout _success* = true"	0.5	73
R_2	{"*Login* = true", "*Shopping_cart* = true", "*session length* = long"}	"*Checkout _success* = true"	0.5	72
R_3	{"*Login* = true", "*Shopping_cart* = true", "*mean time per page* = short"}	"*Checkout _success* = true"	0.5	70

6   Conclusions

Motivated by the need of precise characterization of e-customer behavior, we analyzed data for a Web bookstore applying statistical analysis and association rules. The main goal of the analysis was identification of these sessions characteristics which increase the probability of making a purchase and thus, determination of user sessions which are more valuable in terms of e-business profitability.

Results show that only 0.6 % of user sessions ends with a purchase. When a user checks the shipping information, the probability of a purchase increases to almost 11 %, whereas the fact of the user's logging on increases this probability to 34 %. It can be observed that logged users are much more likely to add products to carts than unlogged users. Unlogged users usually do not decide to buy the products they have in carts, whereas from among logged users with not empty shopping carts almost two thirds finalize purchase transactions.

Application of association rules confirmed that factors increasing the probability of making a purchase include user's logging on, adding a product to the shopping cart, the mean time per page not exceeding one minute, and at least 14 page requests in the session. One can formulate some relationships, e.g. a logged user with not empty shopping cart, spending on average less than one minute per page, who visited at least 14 pages in the session, will decide to confirm the purchase with probability of 73 %.

Though our results are promising, the analysed data set was rather small so the correctness of our approach should be verified for a bigger data set. The findings may be used in practice, e.g. in a method for personalized user service in the Web store, encouraging "likely buyers" to finalize purchase transactions, or in a method for prioritized service in the Web server system, aiming to offer the best quality of service to probable buyers.

References

1. Kurz, C., Haring, G.: E-Business Benchmarking Based on Hierarchical Customer Behavior Characterization. In: 5th ICECR (2002)
2. Menascé, D.A., Almeida, V.A.F., Riedi, R., Ribeiro, F., Fonseca, R., Meira Jr., W.: A Hierarchical and Multiscale Approach to Analyze E-Business Workloads. Perform. Eval. 54(1), 33–57 (2003)
3. Wang, Q., Makaroff, D.J., Edwards, H.K.: Characterizing Customer Groups for an E-commerce Website. In: 5th ACM EC, pp. 218–227. ACM Press, New York (2004)
4. Joshi, A., Joshi, K., Krishnapuram, R.: On Mining Web Access Logs. In: ACM SIGMOD Workshop DMKD, pp. 63–69 (2000)
5. Song, Q., Shepperd, M.: Mining Web Browsing Patterns for E-commerce. Comput. Ind. 57(7), 622–630 (2006)
6. Chen, Y.-L., Kuo, M.-H., Wu, S.-Y., Tang, K.: Discovering recency, frequency, and monetary (RFM) sequential patterns from customers' purchasing data. Electron. Commer. R. A. 8(5), 241–251 (2009)
7. Cheng, C.H., Chen, Y.S.: Classifying the segmentation of customer value via RFM model and RS theory. Expert Syst. Appl. 36(3), 4176–4184 (2009)
8. Chan, C.-C.H., Cheng, C.-B., Hsu, C.-H.: Bargaining strategy formulation with CRM for an e-commerce agent. Electron. Commer. R. A. 6(4), 490–498 (2007)
9. Tanna, P., Ghodasara, Y.: Exploring the Pattern of Customer Purchase with Web Usage Mining. Advances in Intelligent Systems and Computing 174, 935–941 (2012)
10. Zhang, Z., Zhang, S.: The Research of Web Mining Algorithm Based on Variable Precision Rough Set Model. In: Jin, D., Lin, S. (eds.) Advances in FCCS, Vol. 1. AISC, vol. 159, pp. 573–578. Springer, Heidelberg (2012)
11. Menascé, D.A., Almeida, V.A.F., Fonseca, R., Mendes, M.A.: A Methodology for Workload Characterization of E-Commerce Sites. In: 1st ACM EC, Denver, CO, USA, pp. 119–128 (1999)
12. García, D.F., García, J.: TPC-W E-Commerce Benchmark Evaluation. IEEE Computer 36(2), 42–48 (2003)
13. Krishnamurthy, D., Shams, M., Far, B.H.: A Model-Based Performance Testing Toolset for Web Applications. Engineering Letters 18(2), 92–106 (2010)
14. Adomavicius, G., Tuzhilin, A.: Toward the Next Generation of Recommender Systems: A Survey of the State-of-the-Art and Possible Extensions. IEEE Trans. Knowl. Data Eng. 17(6), 734–749 (2005)
15. Kim, J.-K., Cho, Y.H.: Using Web Usage Mining and SVD to Improve E-commerce Recommendation Quality. In: Lee, J.-H., Barley, M.W. (eds.) PRIMA 2003. LNCS (LNAI), vol. 2891, pp. 86–97. Springer, Heidelberg (2003)
16. Larose, D.T.: Discovering Knowledge in Data: An Introduction to Data Mining. Wiley-Interscience (2004)
17. Maravilla, N.: 8 Tips to reduce shopping cart abandonment (2012), http://www.powerhomebiz.com/online-business/shop/shopping-cart-abandonment.htm (access date: January 20, 2013)
18. Markov, Z., Larose, D.T.: Data Mining the Web: Uncovering Patterns in Web Content, Structure, and Usage. Wiley-Interscience (2007)

Multi-criteria Index Selection
for Grouped SQL Queries

Radoslaw Boroński and Grzegorz Bocewicz

University of Technology of Koszalin,
ul. Sniadeckich 2, 75-453 Koszalin, Poland
radoslaw.boronski@tu.koszalin.pl, bocewicz@ie.tu.koszalin.pl
http://tu.koszalin.pl

Abstract. Indexing is a key element of optimization of relational database systems (RDBMS). Commercial tools supporting index selection (e.g. SQL Access Advisor, Toad, SQL Server Database Tuning Advisor, DB2 Advisor) are based on the methods dedicated to individual queries. This paper presents a new approach to tables indexing for the SQL queries group that take into account the size of the indexes and their creation time. Examples illustrate that the use of the group concept reduces the query execution time by 44 % compared to classical methods.

Keywords: index, indexing, database, query, optimization, SQL, RDBMS.

1 Introduction

The index selection problem has been discussed in the literature [1–5]. Several standard approaches have been formulated for the optimal single-query and multi-query index selection [2, 3, 6]. Some past studies have developed rudimentary on-line tools for index selection in relational databases, but the idea has received little attention until recently [7, 8]. In the past year, on-line tuning came into the spotlight and more refined solutions were proposed [9, 10] (e.g. query execution plan modifications). Although these techniques provide interesting insights into the problem of selecting indexes on-line, they are not robust enough to be deployed in real production systems [5]. The reason for this may be a consideration of the problem of index selection without the size of the index or the index creation time aspects. Another obstacle may be selecting an index for a single query only. Omitting other queries (often relating to the same table) in an index selection process may lead to creation of too many similar indexes.

Above limitations are significant for production database systems for which response time as well as disk space usage are very important [9, 4].

The aim of this work is to illustrate an approach of multi-queried SQL group with the index size and the index creation time criteria. In some cases such approach may give better results than some classic methods in which every query in a group is treated individually.

A. Kwiecień, P. Gaj, and P. Stera (Eds.): CN 2013, CCIS 370, pp. 573–581, 2013.
© Springer-Verlag Berlin Heidelberg 2013

The rest of the paper is organized as follows: in Sect. 2, we describe a problem statement. In Section 3, we briefly present a classic index selection approach together with simple examples that illustrate the subject. In Section 4, we demonstrate a new method of grouped queries index selection and compare examples results with classic approach. Section 5 presents our conclusions and further works.

2   Problem Statement

The problem is known in a literature as Index Selection Problem (ISP). In the classical sense, the problem is the search for indexes set that minimize database response time for a given SQL query. According to [11] problem is NP-hard. It is worth noting that in practice the space limit in the ISP is soft (P-hard), because databases usually grow in linear way, thus the disk space limit is specified in such way that a significant amount of storage space remains free for the ISP use [9].

The time criterion is not the only index selection problem gauge. In practice, often it is the case where indexes are evaluated in terms of occupied space and time needed for index creation and its further update/rebuild (index maintenance). Additionally, index selection mechanisms are mainly dedicated to individual queries. The indexes size, together with its maintenance and ability to analyse grouped queries are particularly important for production support systems. Databases operating in such structures are characterized by the necessity to comply with thousands of SQL queries, referencing hundreds of tables which consist dozens of columns. The indexes search space is huge and grows exponentially with the size of the input workload.

It is noted in the introduction that index size is very important in practice. In such context, the problem discussed in this paper extends the ISP problem of index size, its creation/maintenance time requirements and the ability to process grouped queries.

Considered case of Index Selection Problem can be defined in following way. Given is a set of tables:

$$T = (T_1, \ldots, T_i, \ldots, T_n) \,, \tag{1}$$

described by a set of columns included in the tables:

$$K = \{k_{1,1}, \ldots, k_{1,l(1)}, \ldots, k_{i,j}, \ldots, k_{n,1}, \ldots, k_{n,l(n)}\} \tag{2}$$

where: $k_{i,j}$ is a j-th column of table T_i. Each column $k_{i,j}$ corresponds to set of values $V(k_{i,j})$ (tuples set) included in this column.

For the set of tables T various queries Q_i can be formulated (in SQL these are SELECT queries). These queries are put against the specified set of columns $K_i^* \subseteq K$. The result of query Q_i is set as:

$$A_i \subseteq \prod_{k_{i,j} \in K^*} V(k_{i,j}) \tag{3}$$

where: $\prod_{i=1}^{n} Y_i = Y_1 \times Y_2 \times \ldots \times Y_n$ is a cartesian product of sets $Y_1, \ldots, Y_n$.

For a given database DB it is taken into account that A_i is a result of following function:

$$A_i = Q_i\left(K_i^*, Op(DB)\right) \tag{4}$$

where: K_i^* is a subset of columns used in query Q_i, $Op(DB)$ is a set of operators available in database DB of which relation describing query Q_i is built.

The time associated with the determination of the set A_i is depended on the DB database used (search algorithms, indexes structures) and set of indexes adopted $J \subseteq P(K_i^*)$ (where $P(K_i^*)$ – is a power set of K_i^*). It is therefore assumed that the query execution time Q_i in given database DB is determined by the function: $t(Q_i, J, DB)$.

Due to the fact that above considerations apply to the J indexes only (database work state is constant), query Q_i execution time is defined as: $t_i(J)$.

Furthermore, indexes set $J \subseteq P(K_i^*)$ together with tables referenced in query Q_i are determined by space occupied on disk respectively: $S(J)$ – index size, $S(Q_i)$ – indexed tables size (tables used in Q_i queries). The ratio of these volumes is determined by the J indexes size against tables size in query Q_i: $m(J, Q_i) = \frac{S(J)}{S(Q_i)}$.

In practice, apart from the database response time $t_i(J)$ and the $m(J, Q_i)$ ratio, it is important to consider creation time $C_R(J, DB)$ and maintenance time $C_U(J, DB)$ (time needed for the index update/rebuild/coalesce) of an index J in a DB database. These figures are of particular importance in databases with changing data, where index is dropped before the data load and rebuilt after, or where index is updated together with the table. The rest of the paper ignores the index maintenance time because only the databases with rarely changing tables are considered (i.e. warehouses with a specific amount of data). For such databases, the data extraction process occurs more often and the process of the index update is very rare $\left(C_U(J, DB) \approx 0\right)$.

In the context of the so-defined parameters, a typical problem associated with the ISP responds to the question: *What set of indexes $J \subseteq P(K_i^*)$ minimizes the query Q_i execution time: $t_i(J) \to min$?*

When a multi-component set of queries $Q = Q_1, \ldots, Q_m$ is considered together with indexes size and index creation time criteria, question takes the form: *What set of indexes $J \subseteq P(K_i^*)$, that satisfies the condition:*

1) $m(J, Q_i) < m_h$ *(rate of indexes/tables size is less that given value m_h)*,
2) $C_R(J) < C_h$ *(index creation time is less that given value C_h)*,

minimizes the query Q_i execution time: $\sum_{Q_i \in Q} t_i(J) \to min$?

3   Classic Index Selection

Classic index selection approach applies to an individual query and tries to find a good index or indexes set for tables in a single query in a given block.

Such approach does not take into consideration queries in a block as a whole. Each query from a queries group is considered individually. The disadvantage of this method is the possibility of creating excessive number of indexes. Additionally, similar indexes may be created or those that satisfy one query from the revised queries group only.

This may lead to the situation in which disk resources and time needed for indexes creation is misused.

Let us consider three examples where given is a group of four database queries $Q = Q_1, Q_2, Q_3, Q_4$:

Q_1: SELECT * FROM T_1, T_2, T_3 WHERE $k_{1,1}$ = $k_{2,2}$ AND $k_{3,1}$= $k_{1,2}$,
Q_2: SELECT * FROM T_2, T_3 WHERE $k_{2,2}$ = $k_{3,2}$,
Q_3: SELECT * FROM T_2 WHERE $k_{2,1}$ = [const],
Q_4: SELECT * FROM T_3 WHERE $k_{3,1}$ > [const].

Interpretation of this type of queries (according to (4)) is as following:

Q_1: searching for a set of triples: $A_i = \{(a,b,c,d): a \in V(k_{1,1}), b \in V(k_{2,2}), c \in V(k_{3,1}), d \in V(k_{1,2}); a = b, c = d\}$, set $K_1^* = \{k_{1,1}, k_{2,2}, k_{3,1}\}$ – set of columns used in a query.

Q_2: searching for a set of pairs: $A_i = \{(a,b): a \in V(k_{2,2}), b \in V(k_{3,2}); a = b\}$, set $K_2^* = \{k_{2,2}, k_{3,2}\}$ – set of columns used in a query.

Q_3: searching for a set: $A_i = \{a: a \in V(k_{2,1}); a = [const]\}$, set $K_3^* = \{k_{2,1}\}$ – set of columns used in a query.

Q_4: searching for a set: $A_i = \{a: a \in V(k_{3,1}); a > [const]\}$, set $K_4^* = \{k_{3,1}\}$ – set of columns used in a query.

Tables T_1, T_2, T_3 contain 10^7 records each. Total tables size is 960 MB. For such defined tables set and queries Q group, J indexes set is wanted, that minimizes queries block Q execution time: $\sum_{Q_i \in Q} t_i(J) \to min$. Additionally, it is assumed that the index/tables size rate in queries Q is not higher than 0.4: $m(J, Q) < 0.4$ (i.e. total indexes size must not exceed 50 % of total T_1, T_2, T_3 tables size), and the index creation time is not higher than 480 s: $C_R(J) < 480$). In practice, database administrators usually don't create indexes which total size exceeds half of the size of the indexed table.

With the first test run, no indexes are created (index set is empty: $J = \emptyset$). In such context, database response time for queries Q_1, Q_2, Q_3, Q_4 is as following: $t_1(J) = 2040$ s, $t_2(J) = 3611$ s, $t_3(J) = 345$ s, $t_4(J) = 615$ s. Lack of indexes require full table scan for each table [12]. Presented experiments run on Oracle database 11.2.0.3 version, installed on a server with Redhat Enterprise Linux 6 operating systems, with 64 GB memory and ASM file system governing SAN disk storage.

For the next test run new database response times $t_1(J), t_2(J), t_3(J), t_4(J)$ are evaluated for J indexes with the classic approach use. The classic approach requires treating every database query individually. Hence indexes are built: $k_{1,1}, k_{1,2}$ on table T_1; $k_{2,1}, k_{2,2}$ on table T_2; $k_{3,1}, k_{3,2}$ on table T_3. Such indexes are represented by the set: $J = \{\{k_{1,1}\}, \{k_{1,2}\}, \{k_{2,2}\}, \{k_{3,1}\}, \{k_{3,2}\}, \{k_{2,1}\}\}$ containing six subsets. Each element (set) of J contains columns which are used to build the indexes. For example, the set $\{k_{2,2}\}$ means that one index for column $k_{2,2}$ is built. It is possible to create a composite indexes on T_1, T_2, T_3 tables that contain relevant columns from a table (i.e. $J = \{\{k_{1,1}, k_{1,2}\}, \{k_{2,1}, k_{2,2}\}, \{k_{3,1}, k_{3,2}\}\}$, however for the illustrative purposes no such operation is undertaken.

New J indexes set is built for four tables and occupies 610 MB of additional disk space $(m(J, Q) = 0.63)$. It takes 893 s to create the index set $(C_R(J) = 893)$. With the second test run, database returns following response times: $t_1(J) = 2612$ s, $t_2(J) = 2580$ s, $t_3(J) = 5$ s, $t_4(J) = 23$ s respectively. The database response time is better of approximately 11 %, the index/table size rate condition $m(J, Q) > 0.5$ is not met. Also, the indexes creation time condition $C_R(J) < 480$ is not met.

By creating the J indexes, the response time for Q_1 query increases. This is due to the need of reading the entire index for $k_{1,1}$ column. Because of the absence of the $k_{1,2}$ column values, full T_1 table scan operation is performed. Database system engines are not able to use two different indexes for one table in one query (with condition that no sub-query or self-join on the same table is present).

Presented examples show that in some cases classical index selection method (usually implemented in production environments) may result in database response time decrease because of the extensive index size or number. Furthermore, indexes creation time must also be considered.

It is worth noting that sometimes revalued index may create more delay in database response than in no-indexes $(J = \emptyset)$ situations. This is caused by a two-stage index and table read operations.

4   Grouped Queries Index Selection

Grouped queries index selection bases on mutual relationship of grouped SQL queries. It is assumed that SQL queries are related if they reference the same table columns in the same queries group. Indexing of Q queries group, in which queries are related through references to mutual tables, may lead to the indexes which cover several queries from the group. Unlike the classical methods, proposed approach takes into account search for an indexes set for all the queries. The ability to determine the index set for a queries group not only prevents from index redundancy but also reduces the database response time (the significance of skilful indexing was discussed in [1]). The proposed approach may be an alternative to the classic index selection method, where one common index set may be found. The benefits of grouped queries index selection are presented in following examples.

Analysed in the previous example, a group of Q contains queries connected together by a common set of tables T_1, T_2, T_3. Let KW is a pair sequence $(k_{i,j}, c_{i,j})$ that determines number of occurrences $c_{i,j}$ of each column $k_{i,j}$ in Q queries. In considered example, sequence KW takes form:

$$KW = \left((k_{1,1}, 1), (k_{1,2}, 1), (k_{2,1}, 1), \boxed{(k_{2,2}, 2)}, \boxed{(k_{3,1}, 2)}, (k_{3,2}, 1)\right) . \tag{5}$$

For example, for $k_{2,2}$ column (marked box in (5)), value $c_{2,2} = 2$, which means, $k_{2,2}$ column appears twice in Q queries (queries Q_1 i Q_2). Number of a column occurrence in a queries group may be used for J index selection. Assuming, that indexes are built based on $k_{i,j}$ column occurrence number, J index set for a given group Q takes form: $J = \{ \{k_{2,2}\}, \{k_{3,1}\} \}$. J index set contains two single-column indexes built on columns $k_{2,2}$ of table T_2 and $k_{3,1}$ of table T_3.

New J indexes set is built for four tables and occupies 172 MB of additional disk space ($m(J, Q) = 0.17$). It takes 296 s to create the index set ($C_R(J) = 296$). In such context, two predetermined criteria are met.

Such solution not only decreases database response time for a given Q queries block, but also greatly reduces the volume of disk space and time needed for the index creation.

With the third test run, database returns following response times: $t_1(J) = 1235$ s, $t_2(J) = 2430$ s, $t_3(J) = 5$ s, $t_3(J) = 21$ s, not only decreasing the database response time of 44 % but also saving disk space and time needed for the indexes creation. Indexes created for this run meet the index/table size ratio condition: $m(J, Q) > 0.5$ (17 %) which is better of 72 % than in previous run (classical approach). This is due to the fact that only indexes are used for tables from Q_1, Q_2, Q_4 queries or a full table scan for non-indexed table (from Q_3 query) resulting in smaller response times for Q_1, Q_2, Q_4 queries. There is no need for a two-stage index and full table scan operation for the same table within the same query. This proves that indexes should be selected with care.

It is worth noting that for proposed solution, index creation time is shorter of 67 % than the result obtained in the previous test run.

Presented example illustrates situation where database response time is decreased for a given Q queries group, where whole queries group is considered for the index selection process.

Grouped queries index selection method may be an advantage over classical approach, only if queries in the group satisfy the condition of a mutual relationship. Queries Q_1, Q_2, Q_4, from considered example are mutually related, thus satisfy the condition.

Mutual relationship Q of queries group may be illustrated by $G(Q)$ hypergraph, of which hyperedge set is determined by a Q form.

Example of a hypergraph for considered queries Q is presented on Fig. 1.

In this type of graph vertices represent the columns used in queries Q, edges connect those vertices which together make T_a table (dashed line hyper edge) or related Q_i query (solid line hyper edge). For example, hyper edge connecting vertices $k_{1,1}, k_{2,2}, k_{1,3}$ represents relation with query Q_1.

It is assumed that the query set Q is related if corresponding hypergraph $G(Q)$ is consistent.

In this context, the group queries indexes set creation can benefit compared to classic index selection only for related sets.

As a counterexample, given is a group of three database queries $Q^* = \{Q_1^*, Q_2^*, Q_3^*, Q_4^*\}$:

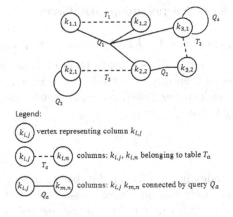

Fig. 1. Hypergraph for considered Q queries group

Q_1^*: SELECT * FROM T_1 , T_2 WHERE $k_{1,1}$ = $k_{1,2}$,
Q_2^*: SELECT * FROM T_2 , T_3 WHERE $k_{2,1}$!= $k_{3,2}$,
Q_3^*: SELECT * FROM T_4 WHERE $k_{4,1}$ > [const] ,
Q_4^*: SELECT * FROM T_3 WHERE $k_{3,3}$ < [const] .

Example of a hypergraph for considered queries Q^* is presented on Fig. 2. Presented hypergraph is inconsistent. For this reason, queries Q^* are treated as unrelated queries.

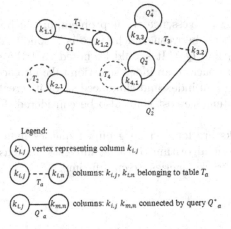

Fig. 2. Hypergraph for considered Q^* queries group

Unrelated queries term means that queries shall not be treated as a group in an index selection process. In such case, best index set is a set determined for each query individually:

$$J^* = \big\{ \{k_{1,1}\}, \{k_{1,2}\}, \{k_{2,1}\}, \{k_{3,2}\}, \{k_{4,1}\}, \{k_{3,3}\} \big\} \ . \tag{6}$$

Pair sequence KW^* for Q^* queries group takes form:

$$KW^* = \big((k_{1,1}, 1), (k_{1,2}, 1), (k_{2,1}, 1), (k_{3,2}, 1), (k_{4,1}, 1), (k_{3,3}, 1)\big) . \qquad (7)$$

One should notice that for KW^* elements (candidate columns) there are no mutual relationships between queries (number of occurrences $c_{i,j}$ of each $k_{i,j}$ column equals 1). In such context, each table T_i must be indexed separately for each individual query in the Q^* group.

5   Summary

Searching for a good index in a relational database is a response to the need of a query execution time minimization, the cost associated with the index creation and the disk space usage. Index may reduce queries execution time, but wrongly selected – may also increase it.

Finding an index that decreases a query execution time, occupy little space and its creation time is little, is often a difficult task for complex database systems.

Presented examples show that there is a need for finding an automatic index selection mechanism for queries group rather than a single query only. This is particularly important for the regularly repeating queries block in production databases. Practice shows that for repetitive large queries group which consist of a large number of tables, grouped queries index selection approach gives better results and enables user to save time needed for the index creation. This may not only result in a better response time from the database, but also in a disk space saving.

The examples shows the essence of the problem, in which the group approach makes sense for queries, between which there is a condition of a mutual relationship (queries Q_1 and Q_2). It should be noted that the presented example, and the proposed approach to an index selection take into account queries block execution time, size of an index and time needed for its creation. In the general case the index maintenance cost shall also be considered. Future research will take it into account.

Our current works are focused on grouped queries index selection method with the use of genetic algorithm [13] that analyses database queries, suggests index structure and tracks queries execution time.

References

1. Boronski, R.: Automatyzacja i optymalizacja procesu doboru indeksow dla dowolnego wycinka czasowego w relacyjnej bazie danych (na przykladzie Oracle 11g). Studia Informatica 33(2A)(105), 229 (2012)
2. Chaudhuri, S., Narasayya, V.: An efficient Cost-Driven Index Selection Tool for MS SQL Server. Very Large Data Bases Endowment Inc. (1997)
3. Frank, M., Omiecinski, M.: Adaptive and Automated Index Selection in RDBMS. In: Pirotte, A., Delobel, C., Gottlob, G. (eds.) EDBT 1992. LNCS, vol. 580, pp. 277–292. Springer, Heidelberg (1992)

4. Maggie, Y., Ip, L., Saxton, L.V., Raghavan, V.: On the Selection of an Optimal Set of Indexes. IEEE Transactions on Software Engineering 9(2), 135–143 (1983)
5. Schnaitter, K.: On-line Index Selection for Physical Database Tuning. ProQuest, UMI Dis-sertation Publishing (2011)
6. Gupta, H., Harinarayan, Y., Rajaraman, A., Ullman, J.D.: Index Selection for OLAP. In: Proceedings of the Internatoinal Conference on Data Engineering, Birmingham, pp. 208–219 (1997)
7. Schkolnick, M.: The Optimal Selection of Indices for Files. Information Systems V.1 (1975)
8. Wedekind, H.: On the selection of access paths in a data base system. In: Klimbie, J.W., Koffeman, K.L. (eds.) Data Base Management, pp. 385–397. North-Holland, Amsterdam (1974)
9. Kołaczkowski, P., Rybiński, H.: Automatic index selection in RDBMS by exploring query execution plan space. In: Ras, Z.W., Dardzinska, A. (eds.) Advances in Data Management. SCI, vol. 223, pp. 3–24. Springer, Heidelberg (2009)
10. Kratica, J., Ljubić, I., Tosic, D.: A genetic algorithm for the index selection problem. In: Raidl, G.R., et al. (eds.) EvoWorkshops 2003. LNCS, vol. 2611, pp. 280–290. Springer, Heidelberg (2003)
11. Finkelstein, S., Schkolnick, M., Tiberio, P.: Physical database design for relational databases. ACM Trans. Database Syst. 13(1), 91–128 (1988)
12. Boronski, R.: Wplyw ustawien parametru wieloblokowego sekwencyjnego czytania danych na czas wykonania zapytania SQL w bazie danych Oracle. In: Materialy VI Krajowej Konferencji Naukowej Infobazy 2011, Gdansk, p. 135 (2011)
13. Back, T.: Evolutionary algorithms in theory and practice: evolution strategies, evolutionary programming, genetic algorithms. Oxford University Press, Oxford (1996)

Applying the Bidding Mechanism
in Web Services with Quality of Service

Jolanta Wrzuszczak-Noga and Leszek Borzemski

Institute of Informatics, Wroclaw University of Technology
Wroclaw, Poland
{jolanta.wrzuszczak,leszek.borzemski}@pwr.wroc.pl

Abstract. In this paper an auction based scheduling algorithm with the options of three different pricing policies is presented and investigated. The Quality of Service as assuring bandwidth on a proper level during transmission of resources is analyzed. Presented experiments were performed in a real-life web service and three performance indexes were evaluated (service income, mean value of RFM and number of offers with lower account price than offered price).

Keywords: web service control, admission control, auctions algorithms, scheduling policies in web service, quality of service.

1 Introduction

Nowadays web services are used for many kinds of business, especially they are used in B2B (Busines to Busines) and B2C (Busines to Customer) systems. Modern web services are designed for delivering more and more data and the web client requirements rise. It is required they will be served in a proper time or with some Quality of Service profile (QoS) i.e. defined transmission parameters. The aim of web resource control is to deliver them to some clients according to some given business strategy and taking into account limits in web resources causing that there is not enough power to serve all user requests [1–9].

In many cases web users pay charges for delivering web resources. Economical investigations showed that efficiency of web services based on serving regular customers, so that analyzing and providing client loyalty, plays very important role. An example of evaluating client loyalty called RFM index, was proposed in [10] and it depends on three parameters: recency of buying, frequency of buying and monetary value (amount of money, which was spent).

In this paper a new scheduling algorithm based on auction mechanism (called SBO) will be developed and presented combining with three different pricing policies for charges. The pricing idea complies a bidding mechanism.

The next section describes the state of the art, whereas then the problem formulation and the solution implementation (new scheduling algorithm and pricing algorithms) are presented. The fourth section contains the real-life experiments and the results. The last section includes conclusions and discussion of idea of further extending the scope of study.

A. Kwiecień, P. Gaj, and P. Stera (Eds.): CN 2013, CCIS 370, pp. 582–591, 2013.
© Springer-Verlag Berlin Heidelberg 2013

2   The State of the Art

In the literature many works related to scheduling of access to web resources were proposed. Most of them were based on the FIFO scheme, so that requests are allocated according to the timestamp of arriving to the web system. Rapidly growing of using web resources effects overload of the system, and as a consequence it causes that some requests have to be rejected [11, 12, 7–9].

In last years works treating with dynamic aspects in scheduling were proposed and implemented. In domain of energy, telecommunication or system science the static schemas are very seldom used, because they do not guarantee of competition and increasing of the income of servicing system. This leaves of static algorithms due to market mechanisms (auctions policies) [13, 14, 9].

There are given systems for providing exchange transaction of energy which use agent software and they gain products among bidding mechanism. In [3] a system for delivering bandwidth for reservation based network was presented which is based also on bidding mechanisms and agent software.

The current trend of managing of web services is to analyze the client behavior to predict of buying and future behavior of them.

Our proposal is to manage of web service resources (bandwidth pieces) in context of delivering data, for which charges are expected with assuring QoS [15] and client loyalty based on bidding mechanisms. The main contribution of the work is to develop and prove the novel scheduling algorithm and pricing algorithms comparing to the FIFO and common pricing schemas.

3   Problem Formulation: Scheduling and Pricing Algorithms

The aim of the web service is to deliver data files, such like video data or text file or software with volume of several MBytes, eg. 10, 20, ..., 200 MBytes. The duty of the web service will be serving requests on a proper level, with guarantee of bandwidth pieces (multiply of 256 kbps). Web service users will be obliged to send their requests in form of offers. The offer $o_{ijn} = \langle u_i, r_j, p_n, b_n \rangle$ (in abb. o_n) is described by four parameters: user id u_i, requested resource id r_j, optionally requested bandwidth p_n and offer price p_n.

Offers without requested bandwidth will be treated as standard offers, for which transmission parameters are not important, so this will distinguish two kinds of offers: standard and extended, featured with requested bandwidth.

Web clients can send many offers for a single resource, and for increasing requested bandwidth volume the offered price must also increase.

Process of collecting offers takes places periodically in scheduling intervals (T). The web service may serve only some offers, as web resources are limited. Offers will be chosen according to the index of offers φ_{ijn} (in abb. φ_n), which determines the income of the service obtained in time.

The problem will be formulated as follows, for given:

- Set of offers $O = \{o_1, o_2, \ldots, o_n, \ldots, o_N\}$.

- Time interval T, scheduling interval with defined slots Z (unit of the time interval).
- Bandwidth B for serving standard (B_1) and additional offers (B_2), for example B is devided as $B_1 = 0.1\,B$, $B_2 = 0.9\,B$.
- Criterions – performance indexes, defined as income of the web service, defined as (1),

$$\varphi_p = \sum_n^{N'} p_n^*$$ (1)

client loyalty (2),

$$\varphi_{RFM} = \sum_n^{N'} \frac{1}{N'}(RFM_n)$$ (2)

number of offers with price reduction (3),

$$\varphi_{cn} = \sum_n^{N'} (x_n)$$ (3)

when p_n^* different from p_n, where: p_n^* is the account price for n-th offer, RFM_n – the value of RFM index for users submitted n-th offer, N – number of submitted offers, N' – number of serviced offers.

The scheduling algorithm should be found for scheduling time interval, which increases the income of the web service and assures client loyalty by assumptions:

- requested bandwidth in n-th offer don't exceed the available bandwidth $(b_n \leq B)$,
- that every scheduled request is transmitted continuously (for following slots) $t_{n''} \leq t_{n'} + l_n Z$ and every scheduled request is will be serviced in scheduling time $t_{n''} \leq T$,

where $t_{n''}$ – determines the end time of transmitting resources, $t_{n'}$ – determines the start time of transmitting resources (start time of slot), l_n – determines the number of slots for transmitting n-th resources.

Scheduling process determines the start time $t_{n'}$ for requests (offers) and number of slots, which are needed to transmission of n-th offer for given bandwidth b_n. An example of scheduling offers is presented in Fig. 1.

The described problem is NP-complete and it is multidimensional knapsack problem. In solution of knapsack problem a Greedy approximation algorithm was proposed, where an index was designed for solving the knapsack problem in a proper time. The index depends on price and weight. It is defined as a price/weight ratio. The products are selected among the highest value of the index until free space is given.

In our problem we distinguish three parameters: price, weight (size of requested resources) and bandwidth. We propose an profit index for offers which determine the price in a time and is defined as (4),

$$\varphi_{ijn} = \frac{p_n}{t_{ijn}}$$ (4)

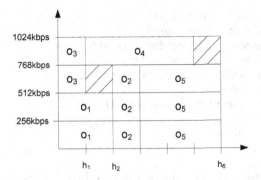

Fig. 1. Scheduled offers

by assumption (5)

$$t_{ijn} = \frac{\text{sizeof}(r_j)}{b_n} \tag{5}$$

where t_{ijn} in abb. (t_n) – transfer time for serving n-th offer, it depends on requested resource and requested bandwidth, for standard offers the bandwidth is assigned as default 0.5 of bandwidth piece, it means 128 kbps.

To solve the problem a new scheduling algorithm was proposed named SBO (Scheduling Algorithm Based on Offers). The pseudokod of SBO algorithm is depicted below:

```
n=1
while (n <=N){
calculate the profit index for every offer
n = n +1}

sort offers according to profit index desc
n=1
while(available slots and !end of queue of offers){
get requested bandwidth for n-th offer
if (pair user-resource not serve){
        calculate number of slots for n-th offer l_n
 check if l_n slots continuously given
             if(available continuously slots >= l_n)
{   Schedule n-th offer
          mark l_n slots as occupated}
  }//if
n = n +1//get next offer
}//while
```

This approach was compared with FIFO scheme, which is depicted as follows

```
Sort offers among arriving time desc
n =1
while(available slots and !end of queue of offers){
```

```
get requeted bandwidth for n-th offer
   if(pair user-resource not serve){
      calculate number of slots for n-th offer l_n
   check if l_n slots continuously given
            if(available continuously slots >=l_n)
{   Schedule n-th offer
            mark l_n slots as occupated}
   }//if
n = n +1//get next offer
}//while
```

Two scheduling algorithms were distinguish, the first one selects offers according to the profit index and the second one selects offers according to the arriving time of request to the web service.

Three different pricing algorithms are developed:

- first price policy (6),

$$p^*_{n(I)} = p_n \ , \tag{6}$$

- second price policy (7),

$$p^*_{n(II)} = \frac{p_n \Phi_{n+1}}{\Phi_n} \ , \tag{7}$$

- the price policy builds according to the regular customer (RC) presented in Table 1 where $p_r b$ describes the mean price of the resource r and bandwidth b known from history. RC is customer, who bought products for at least 1000 \$.

Table 1. Conditions for pricing algorithm based on RFM index

Conditions	Pricing policy
$RC = 1$ and $RFM \geq 70\%$	$\max\left(p_{rb}, p^*_{n(II)}\right)$ when calculated price lower than declared. In other case $p^*_{n(II)}$
$RC = 1$ and $RFM < 70\%$	$\min\left(p_{rb}, p^*_{n(II)}\right)$
$RC = 0$ and $RFM \geq 70\%$	$p^*_{n(I)}$
$RC = 0$ and $RFM < 70\%$	$\max\left(p_{rb}, p^*_{n(II)}, 0.9p^*_{n(I)}\right)$ when calculated price lower than declared. In other case $\max\left(p^*_{n(II)}, 0.9p^*_{n(I)}\right)$

Summarising, the SBO algorithm selects offers according to the profit index, where the FIFO does not, so generally the income for SBO algorithm will be higher comparing to the FIFO manner. The pricing algorithm – the first price determines that every user pays offered price, the second one determines, that every user will pay always price lower than offered. The last algorithm – RFM based – gives discount only to group of clients, so only them will pay lower price than offered.

4 Experiments and Results

Experiments have been performed in a web service installed in the laboratory of Institute of Informatics, Wroclaw University of Technology. At first, a set of offers for lognormal and Poisson population was generated during Matlab simulations. Then the generated set was imported into MySql Database.

Two scheduling algorithms and three pricing algorithms were implemented in the php language and applied into web service Apache. Experiments were performed for different assumptions, including different lengths of the scheduling periods, different lengths of time slots, different number of offers and for different available bandwidth.

A suite of "experiment runs" was developed. An experiment run E is described by five attributes: population (lognormal or Poisson), available bandwidth in Gbps, scheduling time interval measured in seconds, length of slot (unit of scheduling time interval) measured in seconds, number of submitted offers, for example, entity E contains information about experiment run characterized by lognormal population (LN), 5 Gbps available bandwidth (B), scheduling interval 300 s (T), slot 60 s (Z) and 1000 submitted offers (N): The first experiment was executed for entity $E = \langle LN, B = 5, T = 300, Z = 60, N = 1000 \rangle$ and for different available bandwidth $B = \{1, 2, 3, 4, 5, 6, 7, 8, 9, 10\}$ Gbps.

The next experiment was performed for entity $E = \langle LN, B = 5, T = 900, Z = 10, N = 1000 \rangle$ and for different available slots $Z \in \{10, 20, \ldots, 100, 110, 120, 150, 180, 210, 240, 270, 300\}$ s.

5 Conclusions and Future Work

Experiments showed, that the highest value of performance index – web service income was measured for SBO scheduling algorithm with the first price policy (Fig. 2, Fig. 3). For available bandwidth 10 Gbps the income reaches abut 1000 $ for SBO scheduling with first price and for FIFO only 600 $. Figure 3 shows that for increasing length of slots, the service income decreases, because most of slots are not used, the scheduling process assign offers only to several number of slots. For relatively long slots the service income for both scheduling policies is similar.

The second performance index – mean value of RFM index is presented in Fig. 4 and Fig. 5. In both cases the value of RFM index was higher for SBO scheduling algorithm comparing to the FIFO one. Value above 70 % means, that client is really satisfied.

The third performance index (number of offers with account price lower than offer price) for different available bandwidth and length of slots – measured for SBO and FIFO policy showed, that SBO scheduled more offers than FIFO, and give discount for lower number of offers comparing to the FIFO one (Fig. 6, Fig. 7). This confirms that the SBO policy is very effective.

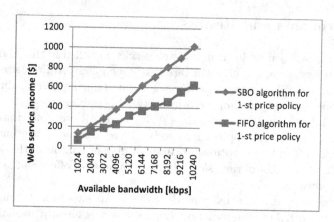

Fig. 2. Income of the web service for different available bandwidth for $E = \langle LN, B = 5, T = 900, Z = 60, N = 1000 \rangle$

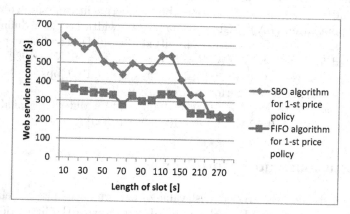

Fig. 3. Income of the web service for different length of slot $E = \langle LN, B = 5, T = 900, Z = 10, N = 1000 \rangle$

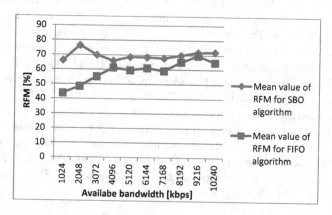

Fig. 4. Mean RFM value for different available bandwidth for $E = \langle LN, B = 5, T = 900, Z = 60, N = 1000 \rangle$

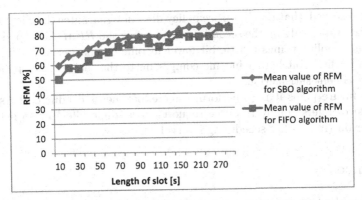

Fig. 5. Mean RFM value for different slot length for $E = \langle LN, B = 5, T = 900, Z = 10, N = 1000 \rangle$

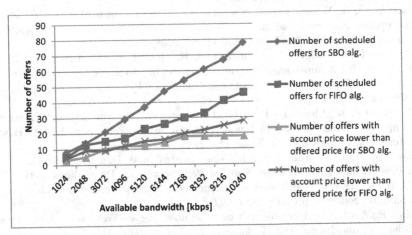

Fig. 6. Number of scheduled offers for different available bandwidth for $E = \langle LN, B = 5, T = 900, Z = 60, N = 1000 \rangle$

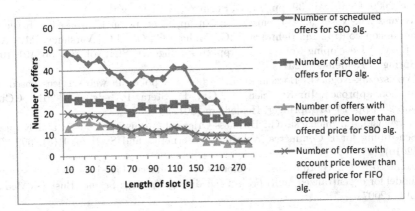

Fig. 7. Number of scheduled offers for different slot length for $E = \langle LN, B = 5, T = 900, Z = 10, N = 1000 \rangle$

It was observed, that the first price policy does not guarantee the client loyalty. This is guaranted only by the second price policy or the RFM based policy. The RFM based policy brings a little bit lower income comparing to the first price policy and a little bit higher income comparing to the second price policy, but guarantees client satisfaction.

In the future plans it is to perform some experiments conducted by sending a boundle of offers (many resource requests in a single offer) or to delivering data later on (in another scheduling interval).

References

1. Anthony, P., Law, E.: Reserve price strategy for seller agent in multiple simultaneous auctions. Internationel Journal of Knowledge Based and Intelligent Engineering Systems (KES) 16, 163–175 (2012)
2. Borzemski, L., Wrzuszczak, J., Kotowski, G.: Management of Web service delivering multimedia files based on the bid strategy. In: Information Systems Architecture and Technology ISAT 2008, pp. 13–23 (2008)
3. Brazier, F., Cornelissen, F., Gustavsson, R., Jonker, C., Lindeberg, O., Polak, B., Treur, B.: A multi-agent system performing one-to-many negotiation for load balancing of electricity use. Electronic Commerce Research and Applications 1, 208–222 (2002)
4. Cherkasova, L., Phaal, P.: Peak Load Management for commercial Web servers using adaptive session-based admission control. In: Proceedings of the 34th Hawaii International Conference on System Sciences (2001)
5. Lee, M., Lui, J., Yau, D.: Admission Control and Dynamic Adaptation for a Proportional-Delay DiffServ-Enabled Web Server. In: Proceedings of the 2002 ACM SIGMETRICS International Conference on Measurement and Modeling of Computer Systems, pp. 172–182 (2002)
6. Nah, F.: A study on tolerable waiting time: how long are Web userswilling to wait. Behaviour and Information Technology 23(3), 153–163 (2004)
7. Wrzuszczak, J., Borzemski, L.: Management of web services based on the bid strategy using the user valuation function. In: Kwiecień, A., Gaj, P., Stera, P. (eds.) CN 2009. CCIS, vol. 39, pp. 19–25. Springer, Heidelberg (2009)
8. Wrzuszczak-Noga, J., Borzemski, L.: An approach to auction-based web server admission control. In: Mehrotra, K.G., Mohan, C., Oh, J.C., Varshney, P.K., Ali, M. (eds.) Developing Concepts in Applied Intelligence. SCI, vol. 363, pp. 101–106. Springer, Heidelberg (2011)
9. Wrzuszczak-Noga, J., Borzemski, L.: Admission policy in web services based on auction approach. In: Kwiecień, A., Gaj, P., Stera, P. (eds.) CN 2012. CCIS, vol. 291, pp. 24–31. Springer, Heidelberg (2012)
10. Borzemski, L., Suchacka, G.: Business-Oriented Admission Control and Request Scheduling for E-Commerce Websites. Cybernetics and Systems 41(8), 592–609 (2010)
11. Confessore, G., Giordani, S., Rismondo, S.: A market-based multi-agent system model for decentralized multi-project scheduling. Springer Science Business Media, LLC (2007)

12. Pérez-Bellido, Á.M., Salcedo-Sanz, S., Portilla-Figueras, J.A., Ortíz-García, E.G., García-Díaz, P.: An Agent System for Bandwidth Allocation in Reservation-Based Networks Using Evolutionary Computing and Vickrey Auctions. In: Nguyen, N.T., Grzech, A., Howlett, R.J., Jain, L.C. (eds.) KES-AMSTA 2007. LNCS (LNAI), vol. 4496, pp. 476–485. Springer, Heidelberg (2007)
13. Lubacz, J.: Mechanizmy aukcyjne i gieldowe w handlu zasobami telekomunikacyjnymi. Wyd. WKL, Warszawa (2011)
14. Toczylowski, E.: Optymalizacja procesów rynkowych przy ograniczeniach. Akademicka Oficyna Wydawnicza EXIT, wyd. 2 rozszerzone (2003)
15. Fiedler, M., Hossfeld, T., Tran-Gia, P.: A generic quantitative relationship between quality of experience and quality of service. Network 24(2), 36–41 (2010)

Author Index

Adamczyk, Błażej   86

Bernaś, Marcin   476, 485
Bilski, Tomasz   124
Bocewicz, Grzegorz   573
Boroński, Radosław   573
Borsali, Riad Ahmed   252
Borzemski, Leszek   45, 55, 582
Brachman, Agnieszka   105

Chodak, Grzegorz   562
Chrost, Lukasz   268
Chydzinski, Andrzej   446
Cupek, Rafał   189
Czachórski, Tadeusz   363, 426
Czubak, Adam   64
Czyszczoń, Adam   552

Danielak, Michał   55
Demkiewicz, Maciej   531
Domańska, Joanna   363
Domański, Adam   363
Droniuk, Ivanna   38
Dudin, Alexander   406, 416
Dudin, Sergey   406
Dudina, Olga   406
Dustor, Adam   456, 466
Dziwoki, Grzegorz   222, 290

Falas, Łukasz   531
Fedevych, Olga   38
Fojcik, Marcin   210
Folkert, Kamil   189
Fortuna, Tomasz   446

Garraux, Gaëtan   114
Ghadikolayi, Ali Azami   277
Gorawska, Anna   300
Gorawski, Marcin   300, 312, 372
Gorawski, Michal   312, 372
Grochla, Krzysztof   258, 268

Hanczewski, Sławomir   436

Jamro, Marcin   200

Kamińska-Chuchmała, Anna   45, 55
Kawecka, Magdalena   394
Keshavarz-Haddad, Alireza   114
Kim, Chesoong   406
Klimenok, Valentina   416
Kłosowski, Piotr   456, 466
Kokot, Adam   531
Kozielski, Stanislaw   11
Kryca, Marek   517
Kryshchuk, Andrii   146
Kucharczyk, Marcin   222, 290
Kwiecień, Andrzej   177, 495, 507
Kwiecień, Błażej   166

Leontyeva, Olga   95
Lis, Damian   312
Lysenko, Sergii   146

Maćkowski, Michał   177, 495, 507
Małysiak-Mrozek, Bożena   323, 334
Marks, Pawel   372
Martyna, Jerzy   232, 240
Mazoochi, Mojtaba   277
Medykovsky, Mykola   38
Mohajer, Amin   277
Mrozek, Dariusz   323, 334

Nabipour, Mohammad   277
Nazarkevich, Maria   38
Niasar, Freshteh Atri   277
Nikodem, Maciej   76
Nycz, Monika   426
Nycz, Tomasz   426

Obelovska, Kvitoslava   95
Olejnik, Remigiusz   31

Pankowska, Małgorzata   541
Pasterak, Krzysztof   300
Pekergin, Ferhan   426
Piórkowski, Adam   21
Płaczek, Bartłomiej   485
Pomorova, Oksana   146

Rouissat, Mehdi   252
Rząsa, Wojciech   1
Rzońca, Dariusz   200

Sakulin, Hannes   11
Sande, Joar   210
Savenko, Oleg   146
Sawerwain, Marek   344
Schauer, Patryk   531
Sharei-Amarghan, Hoda   114
Sidzina, Marcin   166, 177, 495
Simon, Michal   11
Skoroniak, Krzysztof   507
Skrzewski, Mirosław   157
Słabicki, Mariusz   76
Sochor, Tomas   136
Stanek, Dawid   323
Starzyk, Mateusz   189
Stasiak, Maciej   436
Stelmach, Paweł   531
Stolarz, Piotr   258
Suchacka, Grażyna   562

Sulek, Wojciech   222
Surmacz, Tomasz   76
Świątek, Paweł   531
Szemla, Przemysław   21

Tikhonenko, Oleg   394
Trybus, Bartosz   200

Vishnevsky, Vladimir   416

Weissenberg, Joanna   436
Wiśniewska, Joanna   344
Wojciechowski, Bartosz   76
Wrzuszczak-Noga, Jolanta   582

Zatwarnicka, Anna   384
Zatwarnicki, Krzysztof   384
Zawadzki, Piotr   354
Zgrzywa, Aleksander   552